Simultaneous Mass Transfer and Chemical Reactions in Engineering Science

Simultaneous Mass Transfer and Chemical Reactions in Engineering Science

Bertram K. C. Chan

WILEY-VCH

Author

Bertram K. C. Chan
1534 Orillia Court
CA
United States

Cover Image: © Pattadis Walarput/Getty Images

Library of Congress Card No.: applied for

British Library Cataloguing-in-Publication Data
A catalogue record for this book is available from the British Library.

Bibliographic information published by the Deutsche Nationalbibliothek
The Deutsche Nationalbibliothek lists this publication in the Deutsche Nationalbibliografie; detailed bibliographic data are available on the Internet at <http://dnb.d-nb.de>.

Print ISBN: 978-3-527-34665-3
ePDF ISBN: 978-3-527-82350-5
ePub ISBN: 978-3-527-82351-2
oBook ISBN: 978-3-527-82352-9

Typesetting Straive, Chennai, India
Printing and Binding CPI Group (UK) Ltd, Croydon, CR0 4YY

C096423_180123

Dedicated to

- *the glory of God,*
- *my better half Marie Nashed Yacoub Chan, and*
- *the fond memories of my high school physical science teacher, the Rev. Brother Vincent Cotter, BSc, at the De La Salle Catholic College, Cronulla, Sydney, New South Wales, Australia, as well as my former professors in Chemical Engineering in Australia, including:*
  - *at the University of New South Wales:*
    Professor Geoffrey Harold Roper and Visiting Professor Thomas Hamilton Chilton, from the University of Delaware, USA
  - *and at the University of Sydney:*
    Professor Thomas Girvan Hunter and Professor Rudolf George Herman Prince.

Contents

Preface *xiii*
Author Biography *xv*

1 Introduction to Simultaneous Mass Transfer and Chemical Reactions in Engineering Science *1*
1.1 Gas–Liquid Reactions *1*
1.1.1 Simultaneous Biomolecular Reactions and Mass Transfer *2*
1.1.1.1 The Biomedical Environment *2*
1.1.1.2 The Industrial Chemistry and Chemical Engineering Environment *3*
1.1.2 Conclusions *6*
1.1.3 Summary *6*
1.2 The Modeling of Gas–Liquid Reactions *7*
1.2.1 Film Theory of Mass Transfer *7*
1.2.2 Surface Renewal Theory of Mass Transfer *9*
1.2.3 Absorption into a Quiescent Liquid *11*
1.2.3.1 Absorption Accompanied by Chemical Reactions *13*
1.2.3.2 Irreversible Reactions *13*
1.2.4 Absorption into Agitated Liquids *17*
1.2.4.1 An Example of a First-Order Reaction *19*
1.2.4.2 The Film Model *20*
1.3 The Mathematical Theory of Simultaneous Mass Transfer and Chemical Reactions *20*
1.3.1 Physical Absorption *21*
1.3.2 Chemical Absorption *21*
1.3.2.1 Preliminary Remarks on Simultaneous Mass Transfer (Absorption) with Chemical Reactions *21*
1.3.2.2 Some Solutions to the Mathematical Models of the Theory of Simultaneous Mass Transfer and Chemical Reactions *22*
1.3.2.3 Approximate Closed Form Solutions *23*
1.3.3 Numerical Solutions *29*
1.4 Diffusive Models of Environmental Transport *30*
Further Reading *30*

2 Data Analysis Using R Programming *31*
2.1 Data and Data Processing *32*
2.1.1 Introduction *32*
2.1.2 Data Coding *33*
2.1.2.1 Automated Coding Systems *34*
2.1.3 Data Capture *34*
2.1.4 Data Editing *35*
2.1.5 Imputations *35*
2.1.6 Data Quality *36*
2.1.7 Quality Assurance *36*
2.1.8 Quality Control *36*
2.1.9 Quality Management in Statistical Agencies *36*
2.1.10 Producing Results *37*
2.2 Beginning R *38*
2.2.1 R and Statistics *38*
2.2.2 A First Session Using R *40*
2.2.3 The R Environment (This is Important!) *52*
2.3 R as a Calculator *54*
2.3.1 Mathematical Operations Using R *54*
2.3.2 Assignment of Values in R, and Computations Using Vectors and Matrices *56*
2.3.3 Computations in Vectors and Simple Graphics *57*
2.3.4 Use of Factors in R Programming *57*
2.3.4.1 Body Mass Index *59*
2.3.5 Simple Graphics *59*
2.3.6 x as Vectors and Matrices in Statistics *62*
2.3.7 Some Special Functions that Create Vectors *64*
2.3.8 Arrays and Matrices *65*
2.3.9 Use of the Dimension Function dim() in R *65*
2.3.10 Use of the Matrix Function matrix() in R *66*
2.3.11 Some Useful Functions Operating on Matrices in R: colnames, rownames, and t (for transpose) *66*
2.3.12 NA "Not Available" for Missing Values in Datasets *67*
2.3.13 Special Functions that Create Vectors *68*
2.4 Using R in Data Analysis in Human Genetic Epidemiology *73*
2.4.1 Entering Data at the R Command Prompt *73*
2.4.1.1 Creating a Data-Frame for R Computation Using the EXCEL Spreadsheet (on a Windows Platform) *73*
2.4.1.2 Obtaining a Data Frame from a Text File *75*
2.4.1.3 Data Entry and Analysis Using the Function data.entry() *77*
2.4.1.4 Data Entry Using Several Available R Functions *77*
2.4.1.5 Data Entry and Analysis Using the Function scan() *79*
2.4.1.6 Data Entry and Analysis Using the Function Source() *81*
2.4.1.7 Data Entry and Analysis Using the Spreadsheet Interface in R *82*

2.4.1.8 Human Genetic Epidemiology Using R: The CRAN Package Genetics 83
2.4.2 The Function list() and the Construction of data.frame() in R 84
2.4.3 Stock Market Risk Analysis 87
2.4.3.1 Univariate, Bivariate, and Multivariate Data Analysis 87
2.A Appendix. Documentation for the Plot Function 109
2.A.1 Description 109
2.A.2 Usage 109
2.A.3 Arguments 109
2.A.4 Details 109
2.A.5 See Also 110
Further Reading 110

3 A Theory of Simultaneous Mass Transfer and Chemical Reactions with Numerical Solutions 111
3.1 Introduction 111
3.1.1 A Classical Experimental Study of Simultaneous Absorption of Carbon Dioxide and Ammonia in Water 111
3.1.2 Physical Absorption 112
3.1.2.1 Results 113
3.2 Biomolecular Reactions 114
3.2.1 Occurrences of Simultaneous Biomolecular Reactions and Mass Transfer Are Common in Many Biomedical Environments 114
3.3 Some Examples in Chemical Engineering Sciences 115
3.3.1 Simultaneous Chemical Reactions and Mass Transfer 115
3.4 Some Models in the Diffusional Operations of Environmental Transport *Unaccompanied* by Chemical Reactions 116
3.4.1 Diffusion Models of Environmental Transport 116
3.4.2 Advection–Diffusion Models of Environmental Transport 116
3.5 The Concept of Diffusion 116
3.5.1 Publishers' Remarks 116
3.5.2 Fick's Laws of Diffusion 117
3.5.2.1 Fick's First Law of Diffusion (Steady-State Law) 117
3.5.2.2 Fick's Second Law of Diffusion 119
3.5.3 Derivation of Fick's Laws of Diffusion 120
3.5.3.1 Remarks: Additional Remarks on Fick's Laws of Diffusion 120
3.5.3.2 Example Solution in One Dimension: Diffusion Length 122
3.6 The Concept of the Mass Transfer Coefficient 122
3.7 Theoretical Models of Mass Transfer 123
3.7.1 Nernst *One-Film* Theory Model and the Lewis–Whitman *Two-Film* Model 123
3.7.1.1 Gas Transfer Rates 123
3.7.1.2 The Nernst One-Film Model 123
3.7.1.3 Mass Transfer Coefficients 123
3.7.1.4 The Lewis–Whitman Two-Film Model 124

3.7.1.5 The Two-Film Model 124
3.7.1.6 Single-Film Control 126
3.7.1.7 Applications 126
3.7.2 Higbie's Penetration Theory Model 127
3.7.3 Danckwerts' Surface Renewal Theory Model 129
3.7.4 Boundary Layer Theory Model 131
3.7.4.1 Fluid–Fluid Interfaces 131
3.7.4.2 Fluid–Solid Interfaces 131
3.7.4.3 Example: Prandtl's Experimental Mass Transfer from a Flat Plate 131
3.7.5 Mass Transfer Under Laminar Flow Conditions 132
3.7.6 Mass Transfer Past Solids Under Turbulent Flow 132
3.7.7 Some Interesting Special Conditions of Mass Transfer 132
3.7.7.1 Equimolar Counter-Diffusion of A and B ($N_A = -N_B$) 132
3.7.7.2 For Liquid-Phase Diffusion 133
3.7.7.3 Conversions Formulas for Mass Transfer Coefficients in Different Forms 134
3.7.8 Applications to Chemical Engineering Design 134
3.7.8.1 Designing a Packed Column for the Absorption of Gaseous CO_2 by a Liquid Solution of NaOH, Using the Mathematical Model of Simultaneous Gas Absorption with Chemical Reactions 134
3.7.8.2 Calculation of Packed Height Requirement for Reducing the Chlorine Concentration in a Chlorine–Air Mixture 141
3.8 Theory of Simultaneous Bimolecular Reactions and Mass Transfer in Two Dimensions 144
3.8.1 Numerical Solutions of a Model in Terms of Simultaneous Semi-linear Parabolic Differential Equations 144
3.8.1.1 Theory of Simultaneous Bimolecular Reactions and Mass Transfer in Two Dimensions 144
3.8.2 Existence and Uniqueness Theorems of First-Order Linear Ordinary Differential Equations 174
3.8.2.1 Differential Equations 174
3.8.2.2 Contraction Mappings on a Banach Space 174
3.8.2.3 Application to Differential Equations 177
3.8.3 An Existence Theorem of the Governing Simultaneous Semi-linear Parabolic Partial Differential Equations 183
3.8.4 A Uniqueness Theorem of the Governing Simultaneous Semi-linear Parabolic Partial Differential Equations 188
3.9 Theory of Simultaneous Bimolecular Reactions and Mass Transfer in Two Dimensions: Further Cases of Practical Interests 192
3.9.1 Case of Stagnant Film of Finite Thickness – Second-Order Irreversible Reactions 192
3.9.2 Case of Unsteady-State Absorption in the Stagnant Liquid – Slow First-Order Reaction (S&P 325, 328) 196
3.9.3 Simultaneous Absorption of Two Gases in a Liquid in Which Each Then Reacts With a *Third* Component in the Liquid 198

3.9.3.1 Mathematical Modeling *199*
3.9.3.2 Analysis of the Model: A + B → *201*
3.9.3.3 Discussions *201*
3.9.3.4 Further Theoretical Analysis *202*
3.9.4 Simultaneous Absorption of Two Gases in a Liquid in Which Each Then Reacts with a *Third* Component in the Liquid *210*
3.9.4.1 The Mathematical Model *210*
3.9.4.2 Analysis of the Model *210*
3.9.4.3 Boundary Conditions *212*
3.9.4.4 Mass Transfer Coefficients *212*
3.9.5 Cases of Slow First-Order Reactions *213*
3.9.5.1 Case of Unsteady-State Absorption in the Stagnant Liquid *213*
3.9.5.2 Case of Unsteady-State Absorption in the Stagnant Liquid – Slow First-Order Reactions *216*
3.10 Further Theoretical Analysis *218*
Further Reading *219*

4 Numerical Worked Examples Using R for Simultaneous Mass Transfer and Chemical Reactions *221*
4.1 Advection and Convection *221*
4.1.1 Advection *221*
4.1.2 Advection vs. Convection *222*
4.1.2.1 Meteorology *222*
4.1.2.2 The Mathematics of Advection *222*
4.1.2.3 The Advection Equation *223*
4.1.2.4 The Advection Operator in the Incompressible Navier–Stokes Equations *224*
4.2 Worked Examples *224*
4.3 Partial Differential Equations *386*
4.4 A Parabolic PDE *387*
4.4.1 Steady-State Solution *388*
4.4.2 The Method of Lines *389*
Further Reading *390*

5 More Numerical Worked Examples Using R for Simultaneous Mass Transfer and Chemical Reactions *391*
5.1 Introduction *391*
5.2 Advection *391*
5.2.1 Advection vs. Convection *392*
5.2.1.1 Meteorology *392*
5.2.1.2 The Mathematics of Advection *392*
5.2.1.3 The Advection Equation *393*
5.2.1.4 Solving the Advection Equation *393*
5.2.1.5 The Advection Operator in the Incompressible Navier–Stokes Equations *394*

5.3 Solving Partial Differential Equations Using the R Package ReacTran *395*
5.3.1 Worked Examples *395*
5.4 Some Final Remarks on Solving Partial Differential Equations Using the R Package ReacTran *555*
5.4.1 Partial Differential Equations *555*
5.4.2 A Parabolic PDE *557*
5.4.2.1 Steady-State Solution *558*
5.4.2.2 The Method of Lines *559*
Further Reading *560*

6 Solving Partial Differential Equations, Generally Applicable to Modeling Simultaneous Mass Transfer and Chemical Reactions, Using the R Package ReacTran *561*
6.1 Partial Differential Equations (PDE) *561*
6.2 A Parabolic PDE *562*
6.3 Steady-State Solution *563*
6.3.1 The Method of Lines *565*
6.3.2 A Hyperbolic PDE *566*
6.4 The General 3D Advective–Diffusive Transport PDE *568*
6.4.1 An Elliptic PDE *568*
6.5 The General 3D Advective–Diffusive Transport PDE *577*
6.5.1 The Advection Equation *577*
6.5.2 Solving the Advection Equation *578*
6.5.3 The Advection Operator in the Incompressible Navier–Stokes Equations *579*
Further Reading *641*

References *643*
Further Reading *647*
Index *649*

Preface

This book aims to provide a comprehensive theoretical reference for students, professors, design and practicing engineers in the chemical, biomolecular, and process engineering industries a thorough and modern scientific approach to the design of major equipment for processes involving simultaneous mass transfer and chemical reactions.

Key Features

- Presents the basic scientific and computational models of diffusional processes involving mass transfer with simultaneous chemical reactions.
- Provides a vigorous theoretical and computational approach to processes involving simultaneous mass transfer and chemical reactions.
- Involves the use of the open-sourced computer programming language R, for quantitative assessment in the analysis of models for simultaneous mass transfer and chemical reactions.

What Problems Does this Book Solve?

This book is a complete resource for

- A fundamental description of the scientific basis for diffusional processes and mass transfer operations in the presence of simultaneous chemical reactions. Several models are presented, assessed, and showcased for engineering design applications.
- Based on a vigorous assessment of several theoretical models for mass transfer, a selected preferred methodology is demonstrated and recommended as a firm basis for engineering design.

Author Biography

Bertram K.C. Chan, PhD, PE (California, USA), Life Member–IEEE, Registered Professional Chemical Engineer in the State of California, completed his secondary education in the De La Salle College, Cronulla, Sydney, New South Wales, Australia, having passed the New South Wales State Leaving Certificate Examination (viz the state-wide university matriculation public examination) with excellent results, particularly in pure and applied mathematics, and in Honors Physics and Honors Chemistry. He then completed both a Bachelor of Science degree in Chemical Engineering with First Class Honors, and a Master of Engineering Science degree in Nuclear Engineering at the University of New South Wales, followed by a PhD degree in Chemical and Biomolecular Engineering at the University of Sydney, both universities are in Sydney, New South Wales, Australia.

This was followed by two years of working as a Research Engineering Scientist (in Nuclear Engineering) at the Australian Atomic Energy Commission Research Establishment, Lucas Heights, New South Wales, and two years of a Canadian Atomic Energy Commission Post-doctoral Fellowship (in Chemical and Nuclear Engineering) at the University of Waterloo, Waterloo, Ontario, Canada.

He had undertaken additional graduate studies at the University of New South Wales, at the American University of Beirut, and at Stanford University, in mathematical statistics, computer science, and pure and applied mathematics (abstract algebra, automata theory, numerical analysis, etc.), and in electronics, and electromagnetic engineering.

His professional career includes over 10 years of full-time, and 10 years of part-time, university-level teaching and research experiences in several academic and industrial institutions, including a Research Associateship in Biomedical and Statistical Analysis, Perinatal Biology Section, ObGyn Department, University of Southern California Medical School, teaching at Loma Linda University, Middle East College (now University), and San Jose State University, and had held full-time industrial research staff positions, in the Silicon Valley, California, for 27 years – at Lockheed Missile & Space Company (10 years), Apple Computer (7 years), Hewlett-Packard (3 years), and as a research and design electromagnetic compatibility engineer at a start-up company: Foundry Networks (7 years).

In recent years:

- He supported the biostatistical work of the Adventist Health Studies II research program at the Loma Linda University Health (LLUH) School of Medicine, California, and consulted as a forum lecturer for several years in the LLUH School of Public Health (Biostatistics, Epidemiology, and Population Medicine). In these lectures, Dr. Chan introduced the use of the open-sourced programming language R and designed these lectures for the biostatistical elements for courses in the MPH, MsPH, DrPH, and PhD programs, with special reference to epidemiology and biostatistics in particular, and public health and population medicine in general.
- Dr. Chan had been granted three US patents in electromagnetic engineering, had published over 30 engineering research papers, and authored a 16-book set in educational mathematics [1], as well as 5 monographs entitled: "Biostatistics for Epidemiology and Public Health Using R" [2], "Applied Probabilistic Calculus for Financial Engineering: An Introduction Using R" [3], "Biostatistics for Human Genetic Epidemiology" [4], "Simultaneous Mass Transfer and Chemical Reactions in Engineering Science – Solution Methods and Chemical Engineering Applications" [5], and "Fundamental System Design Principles for Simultaneous Mass Transfer and Chemical Reactions in Chemical Engineering Science – including a Computational Approach with R" [6].
- He is a registered Professional Chemical Engineer (**PE**) in the State of California, USA, as well as a Life Member of the Institute of Electrical and Electronic Engineers (**IEEE**).

References

1 Chan, B. (1978). *A New School Mathematics for Hong Kong*. Hong Kong: Ling Kee Publishing Co. 10 Volumes of Texts: 1A, 1B, 2A, 2B, 3A, 3B, 4A, 4B, 5A, 5B. 6 Volumes of Workbooks: 1A, 1B, 2A, 2B, 3A, 3B.

2 Chan, B.K.C. (2016). *Biostatistics for Epidemiology and Public Health Using* R. New York: Springer Publishing Company (with additional materials on the Publisher's website).

3 Chan, B.K.C. (2017). *Applied Probability Calculus for Financial Engineering: An Introduction Using* R. Hoboken, NJ: Wiley.

4 Chan, B.K.C. (2018). *Biostatistics for Human Genetic Epidemiology*. New York/Cham, Switzerland: Springer International Publishing AG.

5 Chan, B.K.C. (2020). *Simultaneous Mass Transfer and Chemical Reactions in Engineering Science – Solution Methods and Chemical Engineering Applications*. Cambridge, MA/Amsterdam, Holland, The Netherlands: Elsevier.

6 Chan, B.K.C. (2021). *Fundamental System Design Principles for Simultaneous Mass Transfer and Chemical Reactions in Chemical Engineering Science – Including a Computational Approach with* R. Cambridge, MA/Amsterdam, Holland, The Netherlands: Elsevier.

1

Introduction to Simultaneous Mass Transfer and Chemical Reactions in Engineering Science

In many biochemical, biomedical, and chemical processes, in both the chemical industry and in physiological systems, including environmental sciences, mass transfer, accompanied by reversible, complex biochemical, or in chemical reactions in gas–liquid systems, is frequently found. From the viewpoint of biochemical and/or chemical design purposes, it is very important that the absorption rates of the transferred reactants may be estimated accurately.

Moreover, the mass transfer phenomena can also affect substantially important process variables like selectivity and yield. Considerable research effort has been expended in describing these processes and in the development of mathematical models that may be used for the computation of the mass transfer rates and other parameters.

For example, the description of the absorption of a gas followed by a single first-order reversible reaction is simple and straightforward. For all mass transfer models, e.g. film, surface renewal, and penetration, this process may be analytically solved. For other processes, however, only for a limited number of special cases analytical solutions are possible, and therefore numerical techniques must be used for the description of these phenomena. Besides numerically solved absorption models, the mass transfer rates often may be calculated, with sufficient accuracy by simplifying the actual process by means of approximations and/or linearizations. In this book, an overview will be given of the absorption models that are available for the calculation of the mass transfer rates in gas–liquid systems with (complex) reversible reactions. Both numerically solved and approximated models will be treated and conclusions on the applicability and restrictions will be presented.

1.1 Gas–Liquid Reactions

It is well known that many biochemical and chemical processes involve mass transfer of one or more species from the gas phase into the liquid phase. In the liquid phase, the species from the gas phase are converted by one or more (possibly irreversible) biochemical or chemical reactions with certain species present in the liquid phase.

Typical of such examples are provided in Sections 1.1.1.1 and 1.1.1.2.

Simultaneous Mass Transfer and Chemical Reactions in Engineering Science, First Edition. Bertram K. C. Chan.

1.1.1 Simultaneous Biomolecular Reactions and Mass Transfer

1.1.1.1 The Biomedical Environment

In epidemiologic investigations, occurrences of simultaneous biomolecular reactions and mass transfer are common in many biomedical environments. Some typical examples are:

(1) **Intestinal Drug Absorption Involving Bio-transporters and Metabolic Reactions with Enzymes** [1]: The absorption of drugs via the oral route is a subject of on-going and serious investigations in the pharmaceutical industry since good bio-availability implies that the drug is able to reach the systemic circulation via the oral path. Oral absorption depends on both the *drug properties* and the physiology of the gastrointestinal tract, or *patient properties*, including drug dissolution, drug interaction with the aqueous environment and membrane, permeation across membrane, and irreversible removal by organs such as the liver, intestines, and the lung.

(2) **Oxygen Transport via Metal Complexes** [1]: On average, an adult at rest consumes 250 ml of pure oxygen per minute to provide energy for all the tissues and organs of the body, even when the body is at rest. During strenuous activities, such as exercising, the oxygen needs increase dramatically. The oxygen is transported in the blood from the lungs to the tissues where it is consumed. However, only about 1.5% of the oxygen transported in the blood is dissolved directly in the blood plasma. Transporting the large amount of oxygen required by the body, and allowing it to leave the blood when it reaches the tissues that demand the most oxygen, require a more sophisticated mechanism than simply dissolving the gas in the blood. To meet this challenge, the body is equipped with a finely tuned transport system that centers on the metal complex ***heme***. The metal ions bind and then release ligands in some processes, and to oxidize and reduce in other processes, making them ideal for use in biological systems. The most common metal used in the body is iron, and it plays a central role in almost all living cells. For example, iron complexes are used in the transport of oxygen in the blood and tissues. Metal–ion complexes consist of a metal ion that is bonded via "coordinate-covalent bonds" to a small number of anions or neutral molecules called ligands. For example, the ammonia (NH_3) ligand is a mono-dentate ligand; i.e. each mono-dentate ligand in a metal–ion complex possesses a single electron-pair-donor atom and occupies only one site in the coordination sphere of a metal ion. Some ligands have two or more electron-pair-donor atoms that can simultaneously coordinate to a metal ion and occupy two or more coordination sites; these ligands are called polydentate ligands. They are also known as chelating (Greek word for "claw") agents because they appear to grasp the metal ion between two or more electron-pair-donor atoms. The coordination number of a metal refers to the total number of occupied coordination sites around the central metal ion (i.e. the total number of metal–ligand bonds in the complex). This process is another important example of biomolecular reaction and transport.

(3) **Carotenoid Transport in the Lipid Transporters SR-BI, NPC1L1, and ABCA1**: The intestinal absorption of carotenoids in vivo involves several crucial steps:
   (1) release from the food matrix in the lumen
   (2) solubilization into mixed micelles
   (3) uptake by intestinal mucosal cells
   (4) incorporation into chylomicrons
   (5) secretion into the lymph.

   Research has shown that:
   (A) EZ is an inhibitor of the intestinal absorption of carotenoids, an effect that decreased with increasing polarity of the carotenoid molecule
   (B) SR-BI is involved in intestinal carotenoid transport
   (C) EZ acts not only by interacting physically with cholesterol transporters as previously suggested but also by downregulating the gene expression of three proteins involved in cholesterol transport in the enterocyte, the transporters SR-BI, NPC1L1, and ABCA1.

   The intestinal transport of carotenoid is thus a facilitated process resembling that of cholesterol; therefore, carotenoid transport in intestinal cells may also involve more than one transporter.

   Hence, the study of biomolecular reaction and transport is an area of importance in biomedical processes and their occurrences in epidemiologic investigations.

   In this section, one applies the facilities available in the R environment to solve problems arisen from these processes. This study is being approached from two directions:
   - Using the R environment as a ***support*** to numerical analytical schemes that may be developed to solve this class of problems.
   - Applying the R functions in the CRAN package ReacTran [2].

1.1.1.2 The Industrial Chemistry and Chemical Engineering Environment

Typical examples of industrial chemical and chemical engineering processes in which this phenomenon occurs include chlorination, gas purification, hydrogenation, and oxidation processes. To undertake the process and equipment design of new reactors and the optimization of existing reactors, applicable theoretical models for reactors are helpful and most likely needed. In general, models of liquid–gas contactors consist of two main parts: the ***micro model*** and the ***macro model***:

- the **micro** model describes the interphase mass transfer between the gas phase and the liquid phase,
- the **macro** model describes the mixing behavior in both phases.

Both parts of the overall model may be solved sequentially, but solving micro and macro models simultaneously is preferred because of optimization of computational time.

Gas–liquid mass transfer modeling has been well studied. The Whitman stagnant film model was first described in 1923 by W.G. Whitman, and it was concluded that some phenomena of gas–liquid mass transfer may be regarded as nearly incompletely explained. Moreover, the Higbie penetration model has been used as a basis for the development of some new reactor models. The influence of the bulk liquid on the mass transfer process has been studied in some detail. More attention has been paid to the dynamical behavior and stability of gas–liquid reactors and the influence of mass transfer limitations on the dynamics. Also, some important differences between the results of the Higbie penetration model and the Whitman stagnant film model are found.

Analytical solution of micro models for mass transfer (accompanied by chemical reactions) is restricted to asymptotic cases in which many simplifying assumptions had to be made (e.g. reaction kinetics are simple and the rate of the reaction is either very fast or very slow compared to the mass transfer). For all other situations, numerical–computational techniques are required for solving the coupled mass balances of the micro model.

In general, it seems that mostly numerical solution techniques have been applied. Wherever possible, analytical solutions of asymptotic cases have been used to check the validity of the numerical solution method.

For example, by modifying one of the boundary conditions of the Higbie penetration model, it had been found that the mass transfer may be affected by the presence of the bulk liquid. For example, in a packed column, the liquid flows down the column as a thin layer over the packings. It has been examined whether or not the penetration model may be applied for these configurations. Both physical absorption and absorption accompanied by first- and second-order chemical reactions have been investigated.

From model calculations, it is concluded that the original penetration theory, by assuming the presence of a well-mixed liquid bulk, may be applied also to systems where no liquid bulk is present, provided that the liquid layer is sufficiently thick!

- For packed columns this means, in terms of the Sherwood number, $N_{Sh} = 4$, for both physical absorption and absorption accompanied by a first-order reaction.
- In case of a second-order 1,1-reaction, a second criterion: $N_{Sh} \geq 4\sqrt{(D_b/D_a)}$ has to be fulfilled.
- For very thin liquid layers ($N_{Sh} < 4$, or $N_{Sh} < 4\sqrt{(D_b/D_a)}$), the original penetration model may give erroneous results, depending on the exact physical and chemical parameters, and a modified model is required.

Analytical solution of models for gas–liquid reactors is restricted to a few asymptotic cases, while most numerical models make use of the physically less realistic stagnant film model – this is relatively simplistic and easy to apply using the "**hinterland model**." *The hinterland model assumes the reaction phase to consist of ONLY a stagnant film and a well-mixed bulk. Inflow and outflow of species to and from the reactor proceeds via the non-reaction phase or via the bulk of the reaction phase, but never via the stagnant film. ("Hinterland" is a German word meaning "the land behind" [a port, a city, ...] in geographic usages!)*

By modifying one of the boundary conditions of the Higbie penetration model, it illustrated how the mass transfer may be affected by the presence of the liquid bulk. Thus, for example, in a packed column, the liquid flows as a thin layer over the structured or dumped packing. It has been examined whether or not the penetration model can be applied for these situations. Both physical absorption and absorption accompanied by first- and second-order chemical reactions have been investigated.

From model calculations, it is concluded that the original penetration theory, which assumes the presence of a well-mixed liquid bulk, may be applied also to systems where no liquid bulk is present, provided that the liquid layer has sufficient thickness.

For packed columns, this means, in terms of Sherwood number, $Sh > 4$ for both physical absorption and absorption accompanied by a first-order reaction. In case of a second-order 1,1-reaction a second criterion $Sh \geq 4\sqrt{(D_b/D_a)}$ has to be fulfilled. For very thin liquid layers, $Sh < 4$ or $Sh < \sqrt{(D_b/D_a)}$, the original penetration model may give erroneous results, depending on the exact physical and chemical parameters, and the modified model is required.

Most numerical models of gas–liquid reactors make use of the physically less realistic stagnant film model because implementation of the stagnant film model is relatively easy using the hinterland concept. The combination of stagnant film model and Hinterland concept may successfully predict many phenomena of gas–liquid reactors.

The Higbie penetration model is however preferred as a micro model because it is physically more realistic. Direct implementation of the hinterland concept is not possible with the Higbie penetration model. Nevertheless, numerical techniques have been applied to develop a new model that implements the Higbie penetration model for the phenomenon mass transfer accompanied by chemical reaction in well-mixed two-phase reactors: assuming the stagnant film.

A model was developed that simulates the dynamic behavior of gas–liquid tank reactors by simultaneously solving the Higbie penetration model for the phenomenon of mass transfer accompanied by chemical reaction and the dynamic gas and liquid phase component balances. The model makes it possible to implement an alternative for the well-known hinterland concept, which is usually used together with the stagnant film model. In contrast to many other numerical and analytical models, the present model can be used for a wide range of conditions, the entire range of Hatta numbers, (semi-)batch reactors, multiple complex reactions, and equilibrium reactions, components with different diffusion coefficients, and also for systems with more than one gas phase component. By comparing the model results with analytical asymptotic solutions, it was concluded that the model predicts the dynamic behavior of the reactor satisfactorily. It had been shown that under some circumstances, substantial differences exist between the exact numerical and existing approximate results. It is also known that for some special cases, differences can exist between the results obtained using the stagnant film model with hinterland concept and the implementation of the Higbie penetration model.

Analytical solution of models for gas–liquid reactors is restricted to a few asymptotic cases, while most numerical models make use of the physically less realistic stagnant film model.

1.1.2 Conclusions

1. The penetration model is preferred for the phenomenon mass transfer accompanied by chemical reaction in well-mixed two-phase reactors.
2. By comparing the model results with analytical asymptotic solutions, it is concluded that the model predicts the reactor satisfactorily. It is shown that for many asymptotic cases, the results of this new model coincide with the results of the stagnant film model with hinterland concept.
3. For some special conditions, differences may exist between the results obtained using the stagnant film model with hinterland concept and the implementation of the Higbie penetration model.
4. An important result is that for 1,1-reactions, the saturation of the liquid phase with gas phase species does not approach zero with increasing reaction rate (increasing Hatta number), contrary to what is predicted by the film model with Hinterland concept. Another important deviation may be found at the specific conditions of a so-called instantaneous reaction in combination with the absence of chemical enhancement of mass transfer.
5. Application of the penetration model does not provide any numerical difficulties, while application of the stagnant film model would lead to a discontinuity in the concentration gradient.
6. Another disadvantage of the hinterland concept is that it can strictly only be applied to isothermal systems, whereas in the systems investigated in this thesis the reaction enthalpy is an important parameter that may significantly influence the phenomena of gas–liquid mass transfer.

A rigorous model may be developed that simulates the dynamic behavior of stirred nonisothermal gas–liquid reactors by simultaneously solving the Higbie penetration model for the phenomenon mass transfer accompanied by chemical reaction and the dynamic gas and liquid phase component and heat balances. This is achieved by coupling the ordinary differential equations of the macro model mass and heat balances to the partial differential equations of the penetration model. This model is not yet published!

Using the newly developed rigorous reactor model, it is shown that dynamic instability (limit cycles) can occur in gas–liquid reactors. The influence of mass transfer limitations on these limit cycles has been studied, and it has been found that mass transfer limitations make the process more stable.

1.1.3 Summary

Although the rigorous model is believed to be a very accurate model, it has the disadvantage that owing to the complex numerical methods applied it is a rather time-consuming model. On behalf of a more efficient prediction of the possible occurrence of limit cycles, the reactor model was simplified. The simplified model is suited for the prediction of limit cycles using a stability analysis. A stability analysis is a very efficient method to predict the dynamic behavior and stability of a system of ordinary differential equations by linearization of the governing nonlinear ODEs in the neighborhood of the steady state and analyzing the Eigenvalues. This method is very powerful for attaining design rules for stable operation of stirred gas–liquid reactors. The influence of mass transfer limitations on the limit cycles is predicted very well using the simplified model, though small discrepancies are found with the more accurate rigorous model.

The developed reactor models have been used to model the dynamics of a new, to-be-developed, industrial hydro-formylation reactor. At a certain design of the reactor, the model predicts serious and undesired limit cycles. These conditions have to be avoided by an appropriate reactor design. Hydro-formylation reactions are often characterized by a negative reaction order in carbon monoxide. Model calculations showed that this may lead to interesting phenomena: at certain process conditions, an improvement of the mass transfer (higher $k_L a$, for example, owing to improved mixing) may give rise to a less stable reactor, without increasing the conversion. This unusual phenomenon is explained by the negative reaction order of carbon monoxide. Apparently, the increasing hydrogen and carbon monoxide concentrations cancel each other out and the overall reaction rate remains unchanged. The increasing hydrogen and carbon monoxide concentrations do however make the process more sensitive for the occurrence of limit cycles.

Finally, a start has been made with studying the influence of macro-mixing on the dynamical behavior of gas–liquid reactors. For this purpose, a cascade of two reactors in series is compared to a single reactor. Initial results indicate that a cascade of reactors in series provides a dynamically more stable design. The total required cooling surface to prevent the occurrence of temperature–concentration limit cycles decreases significantly with increasing number of reactors in series. The first reactor in the cascade is the one with the highest risk of dynamic instability.

1.2 The Modeling of Gas–Liquid Reactions

This process has evolved through a number of theoretical processes, including:

1.2.1 Film Theory of Mass Transfer

In typical industrial absorption processes, one should consider the absorption of gases into liquids which are agitated such that the dissolved gas is transported from the surface to the interior by convective motions. The agitation may occur in various ways, including:

(i) The gas, or vapor, may be blown through the liquid as a stream of bubbles – as, for example, on a perforated plate or in a sparged vessel.

(ii) The liquid may be running in a layer over an inclines or vertical surface, and the flow may be turbulent (as, for example, in a wetted-wall cylindrical column operating at a sufficiently high Reynolds number), or ripples may develop and enhance the absorption rate by convective motion. Discontinuities on the surface may cause periodic mixing of the liquid in the course of its flow, or strings of discs or of spheres.

(iii) The liquid may be advantageously agitated by a mechanical stirrer, which may also entrain bubbles of gases into the liquid.

(iv) The liquid may be sprayed through the gas as jets or drops. First consider a steady-state situation in which the composition of the liquid and gas, averaged over a specified region and also with respect to any temporal fluctuations, are statistically constant. For example, one may consider an agitated vessel through which liquid and gas flow steadily, both being so thoroughly mixed that their time-average compositions are the same at all points; or one may consider a short vertical section of a packed column (or sphere or disc or wetted-wall column) operating at steady state, such that the average compositions of the liquid and gas in the element remain *constant* with time.

Clearly, the situation is a complicated one: the concentrations of the various species are not uniform or constant when measured over short length and time scales. Diffusion, convection, *and* reaction proceed simultaneously. The nature of the convective movements of liquid and gas is difficult to define: any attempt to describe them completely would encounter considerable complications. Thus, to obtain useful predictions about the behavior of such systems for practical purposes, it is necessary to use simplified models which simulate the situation sufficiently well, without introducing a large number of unknown parameters. This approach may take a number of simplifying steps, as follows:

(A) Physical Absorption [2]

Consider first physical absorption, in which the gas dissolves in the liquid *without* any reaction; it is found experimentally that the rate of absorption of the gas is given by

$$\underline{R}_a = k_L a(A^* - A^0) \tag{1.1}$$

in which A^* is the concentration of dissolved gas at the interface between gas and liquid, assuming this partial pressure to be uniform throughout the element of space under discussion. The area of interface between the gas and liquid, per unit volume of the system, is a and k_L is the "physical mass-transfer coefficient." $\underline{R}$ is the rate of transfer which may vary from point to point and from time to time. $\underline{R}$ is the average rate of transfer of gas per unit area; the actual rate of transfer may vary from point to point, and from time to time. A^0 is the average concentration of dissolved gas in the bulk of the liquid.

It is usually not possible to determine k_L and a separately, by measurements of physical absorption. For example, in a packed column, the fraction of the surface of packings which is effectively wetted is unknown, and in a system containing bubbles, the interfacial area is ***not*** generally known! Thus, the quantity directly measurable by physical absorption measurements is the combined quantity $k_L a$.

Hence, the validity of Eq. (1.1) has been established in numerous experimental studies, and an expression of this form would be predicted from first principles, provided that certain conditions are met. The chief of these are that the temperature and diffusivity at the surface (where the concentration is A^*) should be very different from those in the bulk of the liquid; and that no chemical reaction occurs, so that all molecules of dissolved gas are in the same condition. It is sometimes difficult to decide whether a solute reacts chemically with a liquid or merely interacts with it physically. From the present purpose, "physical" solution means that it molecules are indistinguishable.

1.2.2 Surface Renewal Theory of Mass Transfer

Models evolved under this theory take as their basis the replacement at time intervals of liquid at the surface by liquid from the interior which has the local mean bulk composition. While the liquid element is at the surface and is exposed to the gas, it absorbs gas as though it were infinitely deep and quiescent: the rate of absorption, R, is then a function of the exposure time of the liquid element and will be described by a suitable expression such as those to be described by the reaction kinetics of the system. In general, the rate of absorption is fast or infinite initially, decreasing with time. The replacement of liquid at the surface by fresh liquid of the bulk composition may be due to the turbulent motion of the body of the liquid. Moreover, when liquid runs over the surface of a packing, it may be in a state of undisturbed laminar flow at the top of each piece of packing, except at the discontinuities between pieces of packing, where it may mix thoroughly: at the top of each piece of packing, a fresh surface would then be developed and moved discontinuity, when it would then be replaced again by fresh liquid.

With this scenario, the surface-renewal models propose that the surface of an agitated liquid, or a liquid flowing over a packing, is a collection of elements which have been exposed to the gas for different durations of time, and which may well be, in general, absorbing at different specific rates. Thus, different versions of the model will lead to different specific rates. Moreover, different versions of the model will lead to different distributions of surface ages about the mean value. The form of the surface-renewal model first proposed by Higbie, in 1935, assumed that every element of surface is exposed to the gas for the ***same*** duration of time, θ, before being replaced by liquid of the bulk composition. During this time, the element of liquid absorbs the same amount Q of gas per unit area as though it were infinitely deep and stagnant. The average rate of absorption is therefore Q/θ, and this is also the rate of absorption $\underline{R}$ per unit area averaged over the interface in a representative region of a steady-state absorption system in which the bulk composition is statistically uniform – e.g. in a small, but representative, volume element of a packed column.

The exposure-time θ may be determined by the hydrodynamic properties of the system, and is the only parameter required to account for their effect on the transfer coefficient k_L. The relation between θ and k_L is derived herebelow – in physical absorption.

Under such circumstances, the variation in time and space of the concentration a of dissolved gas in the liquid in the absence of reaction is governed by the diffusion

$$D_A\, \partial^2 a/\partial x^2 = \partial a/\partial t \tag{1.2}$$

And the rate of transfer of dissolved gas initial concentration passage of gas across the interface, then the concentration of the surface might vary with time. For the present, it is assumed that the diffusion of dissolved gas into the latter. This assumption generally holds when the solubility of the gas is not very great, so that A^* represents a mole fraction much less than unity.

It would not be true, for example, if ammonia at atmospheric pressure were diffusing into pure water (in which there will be a substantial temperature rise). Under these conditions, the variation in time and space of the concentration, a, of dissolved gas in the liquid in the absence of reaction as governed by the diffusion from bubbles or absorption by wetted-wall columns, the mass transfer surface is formed instantaneously and transient diffusion of the material takes place. Assuming that a bubble is rising in a pool of liquid (where the liquid elements are swept on its surface) and remains in contact with it during their motion and finally detached at the bottom.

The basic assumptions of the penetration theory are:

- Unsteady-state mass transfer occurring to a liquid element as long as it is in contact with the gas bubbles
- Equilibrium existing at gas–liquid interface
- Each liquid element staying in contact with the gas for same period of time. (The liquid elements are moving at the same rate, and there is not a velocity gradient within the liquid.)

Under these assumptions, the convective terms in the diffusion may be neglected and the unsteady-state mass transfer of gas (penetration) into the liquid element may be written from Fick's second law for unsteady-state diffusion as (Figure 1.1)

$$\partial c/\partial t = D_{AB}(\partial^2 c/\partial z^2) \tag{1.3}$$

and the boundary conditions are

$$t = 0, z > 0: \quad c = c_{Ab}$$

and

$$t > 0, z = 0: \quad c = c_{Ai}$$

where

c_{Ab} = The concentration of solute A at an infinite distance from the surface (viz. the bulk concentration)

c_{Ai} = The interfacial concentration of solute A at the surface

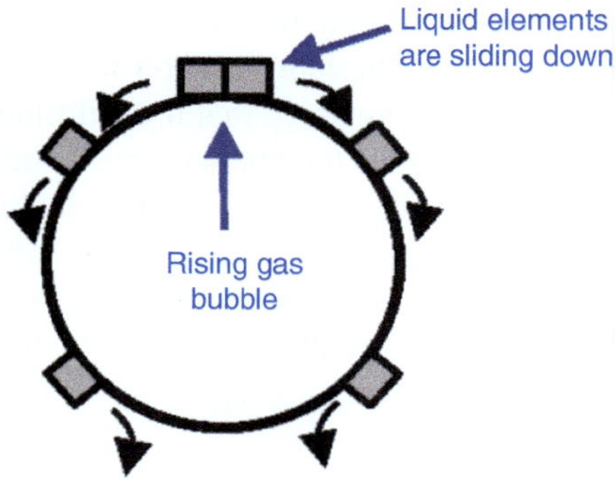

Figure 1.1 Schematic of the penetration model.

On solving the above partial differential equation, one obtains:

$$(c_{Ai} - c)/(c_{Ai} - c_{Ab}) = \mathrm{erf}[z/\{2\surd(D_{AB}t)\}] \tag{1.4}$$

If the process of mass transfer is a unidirectional diffusion and the surface concentration is very low: i.e. $c_{Ab} \approx 0$; then the mass flux of solute A, given by N_A (kg/m^2 s) may be estimated by the following equation:

$$N_A = [\{-\rho D_{AB}/(1 - c_{Ab})\}](\partial c/\partial z)_{z=0}$$
$$\approx -\rho(\partial c/\partial z)_{z=0} \tag{1.5}$$

From the above two expressions, the rate of mass transfer at time t is given by the following equation:

$$N_A(t) = [\surd(D_{AB}/\pi t)](c_{Ai} - c_{Ab}) \tag{1.6}$$

And the mass transfer coefficient is given by

$$k_L(t) = \surd(D_{AB}/\pi t) \tag{1.7}$$

Moreover, the average mass transfer coefficient during a time interval $t_c(t)$ may be obtained by integrating Eq. (1.4) as

$$k_{L,av} = (1/t_c)\int_0^{t_c} k(t)dt = 2\sqrt{(D_{AB}/\pi t_c)} \tag{1.8}$$

Thus, from the above equation, the mass transfer coefficient is ***proportional*** to the square root of the diffusivity. This was first proposed by R. Higbie in 1935 and the theory is called the ***Higbie penetration theory***.

1.2.3 Absorption into a Quiescent Liquid

First, consider the case in which no chemical reaction occurring between the dissolved gas and the liquid [2]. The surface of the liquid first contacts the gas at time $t = 0$, and it may be assumed that, from that instance onward, the concentration in the plane of the surface is uniformly equal to A^* – this concentration corresponds to the solubility of the gas at the prevailing partial pressure above the surface of the liquid – and is assumed to be constant. If this gas were mixed with another gas of different solubility, or if there were a resistance the passage of gas across the interface, then the concentration at the surface may vary with time.

Further, it is assumed that the diffusion of dissolved gas into the liquid does not appreciably affect the temperature, or other physical properties of the latter. This is likely to be true only when the solubility of the gas is not very great, so that A^* represents a mole fraction much less than unity.

Under these special circumstances, the variation in time and space of the concentration a of dissolved gas in the liquid in the absence of reaction is governed by the diffusion equation:

$$D_A \partial^2 a/\partial x^2 = \partial a/\partial t \tag{1.9}$$

and the rate of transfer of dissolved gas across unit area, R_x of any plane parallel to the surface is

$$R_x = -D_A \partial a/\partial x \tag{1.10}$$

Here, x is the distance measured from the surface, where $x = 0$, and D_A is the diffusivity or diffusion coefficient of the dissolved gas. Hence, the rate of absorption of gas at any time is

$$R = -D_A(\partial a/\partial x)_{x=0} \tag{1.11}$$

The term $(\partial a/\partial x)_{x=0}$ is the concentration gradient at the surface and is a function of time.

Let the initial concentration of A be uniformly equal to A^0, and its concentration remote from the surface remains A^0. Then the solution of Eq. (1.9) with boundary conditions:

$$\begin{aligned} a &= A^*, \quad x = 0, \quad t > 0 \\ a &= A^0, \quad x > 0, \quad t = 0 \\ a &= A^0, \quad x = \infty, \quad t > 0 \end{aligned} \tag{1.12}$$

is

$$\begin{aligned} a - A^0 &= (A^* - A^0)\text{erfc}\{x/2\sqrt{(D_A t)}\} \\ &= (A^* - A^0)[1 - \text{erf}\{x/2\sqrt{(D_A t)}\}] \end{aligned} \tag{1.13}$$

Giving the distribution of concentration in the case where the initial concentration is A_0, and the function

$$\text{erfc}\{x/2\sqrt{(D_A t)}\} = [1 - \text{erf}\{x/2\sqrt{(D_A t)}\}] \tag{1.14}$$

is the error function of $x/2\sqrt{(D_A t)}$, and is defined by

$$\text{erf}\{x/2\sqrt{(D_A t)}\} = (2/\sqrt{\pi})\int_0^{x/2\sqrt{(D_A t)}} \exp(-z^2)\,dz \tag{1.15}$$

Values of the error function may be found in standard mathematical tables, etc. From Eqs. (1.11) and (1.13), it follows that

$$R = (A^* - A^0)\sqrt{(D_A/\pi t)} \tag{1.16}$$

Thus, the rate of absorption is infinite when the gas and liquid are first in contact, decreasing with time, and the amount Q absorbed by a unit area of surface, in time t, is given by

$$\begin{aligned} Q &= \int_0^t R\,\mathrm{d}t \\ &= 2(A^* - A^0)\sqrt{(D_\mathrm{A}t/\pi)} \end{aligned} \tag{1.17}$$

1.2.3.1 Absorption Accompanied by Chemical Reactions

If the dissolved gas reacts with the liquid, or with a substance dissolved in the liquid [2], then Eq. (1.9) should be replaced by

$$D_\mathrm{A}\partial^2 a/\partial x^2 = \partial a/\partial t + r(x,t) \tag{1.18}$$

in which $r(x, t)$ is the rate *per unit volume of liquid* at which the reaction is using up the solute gas at time t and at a distance x below the surface. This rate will depend on the local concentration of the gas, and of any other solute with which it reacts. For some cases, numerical and/or analytical solutions of the diffusion reaction equations are available.

It is assumed throughout that the temperature and the values of such as physico-chemical quantities as diffusivities, reaction-rate constants, and solubilities remain constant and uniform. Moreover, enhancement factors E may be computed which are the ratios of the amount which would be absorbed if there were no reaction, viz.

$$2(A^* - A^0)\sqrt{(D_\mathrm{A}t/\pi)}$$

1.2.3.2 Irreversible Reactions

1.2.3.2.1 First-Order Reactions

Here

$$r = k_1 a \tag{1.19}$$

in which k_1 is the first-order rate constant for the reaction. The rate of reaction of dissolved gas at any point is directly proportional to its concentration. Under these circumstances, the solution to Eq. (1.18), with boundary conditions (1.12), and with $A^0 = 0$, is [2, 4]:

$$\begin{aligned} a/A^* = {} & 1/2\exp\{-x\sqrt{(k_1/D_\mathrm{A})}\}\mathrm{erfc}[x/\{2\sqrt{D_\mathrm{A}t}\} - \sqrt{(k_1 t)}] \\ & + 1/2\exp\{x\sqrt{(k_1/D_\mathrm{A})}\}\mathrm{erfc}[x/\{2\sqrt{D_\mathrm{A}t}\} + \sqrt{(k_1 t)}] \end{aligned} \tag{1.20}$$

so that

$$R = A^*\sqrt{(D_\mathrm{A}k_1)}[\mathrm{erf}\sqrt{(k_1 t)} + \exp(-k_1 t)/\sqrt{(\pi k_1 t)} \tag{1.21}$$

and

$$Q = A^*\sqrt{(D_\mathrm{A}/k_1)}\left[(k_1 t + 1/2)\mathrm{erf}\sqrt{(k_1 t)} + \sqrt{(k_1 t/\pi)}\exp(-k_1 t\right] \tag{1.22}$$

Thus, when k_1t is large, the distribution of concentration and absorption rate tend to limiting values and no longer change with time:

$$\text{for} \quad k_1t \gg 1$$

$$a/A^* = \exp\{-x\sqrt{(k_1/D_A)} \tag{1.23}$$

$$R = A^*\sqrt{(D_A k_1)} \tag{1.24}$$

and

$$Q = A^*\sqrt{(D_A k_1)}\{t + (1/2k_1)\} \tag{1.25}$$

The error in Eq. (1.25) is less than 3% when $k_1t > 2$.

When k_1t is large,

$$Q = tA^*\sqrt{(D_A k_1)} \tag{1.26}$$

to within 5% when $k_1t > 10$.

For short times of exposures, for $k_1t \ll 1$

$$R \approx A^*\sqrt{\{(D_A/\pi t)(1 + k_1t)} \tag{1.27}$$

$$Q \approx 2A^*\sqrt{\{(D_A t/\pi)(1 + k_1t/3)} \tag{1.28}$$

to within 5% when $k_1t < 0.5$.

The above equations form the basis of methods for measuring k_1 and $A^*\sqrt{D_A}$.

In practice, truly first-order irreversible reactions are seldom found. However, when the solute gas undergoes a reaction with a dissolved reactant, which is first order with respect to the concentration of the dissolved gas, then, under certain circumstances, the concentration of the reactant may be uniform and the reaction rate of the dissolved w order and the above equations do apply.

If the product of an irreversible first-order gas reaction has the same diffusivity as the dissolved gas, then its concentration P^* at the surface is given by Danckwerts (1967):

$$(P^*/yA^*) + 1 = \exp(-k_1t/2)[(1 + k_1t)I_0(k_1t/2) + k_1tI_1(k_1t/2)] = V_A \tag{1.29}$$

where y moles of product arise from the reaction of one mole of dissolved gas.

1.2.3.2.2 Instantaneous Reactions

In this case, the dissolved gas reacts instantaneously with a dissolved reactant. There is a plane underneath the surface where the concentration of both is zero, and the rate at which the two substances can diffuse to the reaction plane. The actual kinetics of the reaction are immaterial. The initial concentration of the reactant is uniformly B^0, and z moles of it react with each mole of dissolved gas.

The solution of the equations governing this case (and similar phenomena involving a moving boundary) has been given by Danckwerts [5]:

$$a/A^* = \left\{\text{erfc}[x/2\sqrt{(D_A t)}] - \text{erfc}\left(\beta/\sqrt{D_A}\right)\right\}/\text{erf}(\beta/\sqrt{D_A}), \quad 0 < x < 2\beta\sqrt{t} \tag{1.30}$$

$$a/A^* = 0, \quad x > 2\beta\sqrt{t} \tag{1.31}$$

$$b/B^0 = \left\{\text{erf}\left[x/2\surd(D_B t\right] - \text{erf}(\beta/\surd D_B)\right\}/\text{erfc}(\beta/\surd D_B), \quad x > 2\beta\surd t \tag{1.32}$$

$$b/B^0 = 0, \quad 0 < x < 2\beta\surd t \tag{1.33}$$

$$R = \{A^*/\text{erf}(\beta/\surd D_A)\}\surd(D_A/\pi) = E_i A^* \surd(D_A/\pi) \tag{1.34}$$

$$Q = \{2A^*/\text{erf}(\beta/\surd D_A)\}\surd(D_A t/\pi) = 2E_i A^* \surd(D_A t/\pi) \tag{1.35}$$

where

$$E_i = 1/\text{erf}(\beta/\surd D_A) \tag{1.36}$$

and β is defined by

$$\exp(\beta^2/D_B)\text{erfc}(\beta/\surd D_B) = (B^0/zA^*)\surd(D_B/D_A)\exp(\beta^2/D_A)\,\text{erf}(\beta/\surd D_A) \tag{1.37}$$

Thus, the factor E_i is a function of D_B/D_A and B^0/zA^*. Here a, b are the local concentrations of dissolved gas and reactant, respectively, and D_A, D_B *their* diffusivities. The reaction plane is at a depth of $2\beta\surd t$ beneath the surface. The quantity E_i is the factor by which the reaction increases the amount absorbed in a given time, as compared to absorption without reaction.

The concentration p of the product (assuming y moles to be formed from each mole of reacting gas) is

$$p = yA^*\sqrt{(D_A/D_P)}[\{\text{erfc}(\beta/\sqrt{D_P})\exp(-\beta^2/D_A)\}/ \{\text{erfc}(\beta/\sqrt{D_A})\exp(-\beta^2/D_P)\}], \quad 0 < x < 2\beta\sqrt{t} \tag{1.38}$$

$$p = yA^*\sqrt{(D_A/D_P)}[\{\text{erfc}(x/2\sqrt{D_P})\exp(-\beta^2/D_A)\}/ \{\text{erfc}(\beta/\sqrt{D_A})\exp(-\beta^2/D_P)\}], \quad x > 2\beta\sqrt{t} \tag{1.39}$$

where D_P is the diffusivity of the product.

When the diffusivities are equal:

$$p = yB^0/z, \quad 0 < x < 2\beta\surd t \tag{1.40}$$

If, in addition, one mole of reactant reacts with one mole of gas to produce one mole of product, then $y = z = 1$, and the concentration of product at the surface is the same as that of the reactant in the bulk.

When E_i is much greater than unity, then

$$E_i = \surd(D_A/D_B) + (B^0/zA^*)\surd(D_B/D_A) \tag{1.41}$$

the error being of order $1/2E_i$. When $D_A = D_B$,

$$E_i = 1 + B^0/zA^* \tag{1.42}$$

for all values of the variables.

Under these conditions, the rate of absorption is equal to that of physical absorption of a gas of solubility $(A^* + B^0/z)$.

Remarks: In many cases, the reactions of the dissolved gases are so fast that they may be considered as instantaneous under all circumstances, for example:

(1) $H_2S + OH^- \rightarrow HS^- + H_2O$
(2) $H_2S + RNH_2 \rightarrow RNH_3{}^+ + HS^-$
(3) $NH_3 + HCl \rightarrow NH_4{}^+ + Cl^-$

Other reactions may have finite rates which may control the rate of absorption under certain circumstances, for example:

(4) $CO_2 + H_2O \rightarrow HCO_3{}^- + H^+$
(5) $CO_2 + OH^- \rightarrow HCO_3{}^-$
(6) $CO_2 + 2RNH_2 \rightarrow RNHCOO^- + RNH_3{}^+$
(7) $4Cu^+ + 4H^+ + O_2 \rightarrow 4Cu^{++} + 2H_2O$

In all these reactions, in which the reaction rates may be controlled by diffusion under some circumstances, and then the chemical steps in the reaction may be treated as though they were instantaneous. In general, irreversible reactions may be treated as instantaneous if the following condition is fulfilled:

$$(Q_0/2A^*)\sqrt{(\pi/D_A t)} \gg 1 + (B^0/zA^*) \tag{1.43}$$

in which Q_0 is the amount of gas which would be absorbed in time t if there were no depletion of reactant B in the neighborhood of the surface; if the condition is fulfilled, then the reactant B is entirely depleted in the neighborhood of the surface and the rate of reaction is only controlled by diffusion.

Worked Example 1.1

H_2S at 1 atm is absorbed into quiescent water and into a 0.1 M solution of monoethanolamine (MEA) at 25 °C. The reaction

$$H_2S + RNH_2 \rightarrow HS^- + RNH_3{}^+$$

may be taken to be instantaneous and irreversible. Calculate the amount of H_2S absorbed per cm^2 of surface in 0.1 s in water.

The diffusivity of H_2S in water is 1.48×10^{-5} cm^2/s. The solubility of H_2S is 0.10 g-mol/1 atm.

Solution

In water:

$$Q = 2A^* \sqrt{(D_A t/\pi)} = 2 \times 0.1 \times 10^{-3} \times \sqrt{(1.48 \times 10^{-5} \times 0.1/\pi)}$$
$$= 1.373 \times 10^{-7} \text{ g-mol/cm}^2$$

1.2.3.2.3 *Simultaneous Absorption of Two Reacting Gases*

Consider the case of a mixture of two gases being absorbed simultaneously, in which each undergoes a second-order reaction with the same dissolved reactant [6]. As a typical example, this scenario may arise when a mixture of H_2S and CO_2 is absorbed by a solution of alkali amine. Here, the two solute gases are **A** and **B**, their concentrations at the surface, **A*** and **B***, will in general be different, as will their diffusivities $\boldsymbol{D}_{\mathbf{A}}$ and $\boldsymbol{D}_{\mathbf{B}}$, and their rate constants $\boldsymbol{k}_{\mathbf{A}}$ and $\boldsymbol{k}_{\mathbf{B}}$, governing their rates of reaction with the reactant **B**. The reactions are both supposed to be second order, so that the local rates of reaction are $\boldsymbol{k}_{\mathbf{A}}\boldsymbol{ab}$ and $\boldsymbol{k}_{\mathbf{B}}\boldsymbol{ab}$, respectively. Stoichiometrically:

$$\mathbf{A} + z\mathrm{B} \rightarrow \mathrm{P}$$

$$\mathbf{A} + z\mathrm{B} \rightarrow \mathrm{P}$$

The cases of a second-order reaction between A and B and an mth-, nth-order reaction between A and B will be considered in Chapter 3.

In the case of a second-order reaction, the necessary conditions will be discussed under Section 1.2.3.2.1 concentration where y moles of product arise from the reaction of one mole of dissolved gas. For large values of k_1t, it approaches $2\sqrt{(k_1t/\pi)}$.

1.2.4 Absorption into Agitated Liquids

In most commercial absorption mass transfer equipment such as plate or packed towers and columns [7], in almost all situations, one may and should consider the absorption of gases into liquids which are agitated vigorously so that the dissolved gases are transported from the surface to the interior by convective motion. Such agitations may occur in various ways, such as

(1) The liquid may be running in a layer over a vertical surface or an inclined plane, and the flow may be laminar or turbulent, or ripples may develop and enhance the absorption rate by convective motion. Discontinuities on the surface may cause periodic mixing of the liquid layer along its flow (as in a packed column).
(2) The gas may be blown through the liquid as streams of bubbles (as on a bubble plate).
(3) The liquid may be agitated by a mechanical stirrer, which may also entrain bubbles of gas into the liquid.
(4) The liquid may be sprayed.

The theory of diffusion and simultaneous chemical combination of several gases dissolving in a liquid and reacting near the interfaces had been discussed mathematically. The resultant set of nonlinear partial differential equations representing mass balances and corresponding to the penetration theory of mass transfer may be solved numerically. The chemical reaction rate expressions are second (or higher) order, corresponding to a bimolecular (or multi-molecular) mechanism. The results indicate that chemical reaction between two (or multiple) dissolved substances ***increases*** the rate of absorption of each by an amount (or amounts) that depend(s) on the ratio of solubilities, the reaction rate constant(s), and the diffusion coefficients. Mathematical solutions are presented for the simultaneous absorption of two gases in a liquid which contains a third component capable of reacting chemically with both of the dissolved gases. Based upon theoretical and experimental studies, one may express the rate of absorption of a gas in a liquid with which the gas reacts chemically in terms of diffusion and reaction kinetic mechanisms. The results are usually expressed by showing the ratio of absorption of the gas to the rate that would occur without the reaction, but with all other conditions held constant. This reaction rate enhancement ratio depends on:

- The reaction rate constant
- The time of exposure of the liquid surface to the gas
- The diffusion coefficients in the liquid
- Certain stoichiometric ratios.

The theoretical computation of such phenomena may be based on the assumption that diffusion and reaction take place in a laminar flow near the interface, the reaction rate per unit volume being so fast that only a small space in the liquid phase is sufficient to keep up with the diffusion of reactants to the reaction zone. With these simplifying assumptions, the resultant mathematical problem is that of solving the unsteady-state diffusion equation with a reaction term included. Thus, there will be as many different cases as there are orders of reaction and ways of bringing various numbers of reactants together.

The simplest problems are those in which linear partial differential equations are obtained, corresponding to first-order reactions of a single diffusing species after it dissolves in the liquid at the interface. Results of computations for both reversible and irreversible first-order reactions are available. Theoretical computations have been undertaken involving two dissolved substances: A and B, when A is initially in the gas, and B is a nonvolatile reagent initially present in the liquid. This situation is more complex than the linear problem; however, exact solutions have been obtained for fast reactions that are limited by diffusion of A and B to a reaction boundary that moves into the liquid as time elapses. When the reaction occurs at a finite rate, the reaction zone is diffuse, and no exact solutions have been obtained.

No theoretical calculations have been available for another physically interesting situation in which substances A and B are both gases that are brought into contact with a laminar layer of liquid in which both are soluble and within which they react with each other by a second-order chemical mechanism.

A typical practical example is the simultaneous absorption of CO_2 and NH_3 into water. These gases do ***not*** react in the vapor phase, but in the dissolved state they react with each other to form ammonium carbamate. The speed of the homogeneous reaction has been measured. As an additional example, gases containing both CO_2 and H_2S are sometimes washed with an alkaline liquid in which both gases dissolve and react, the two species competing for the same alkaline reagent.

However, it should not escape ones notice that these two examples are of fundamentally different types:

- In the former, a third substance is not involved, unless it is the solvent, which is present in a large excess and is not depleted by the reaction, or a homogeneous catalyst that is not consumed:
  - In the first example, the chemical equation is of the type: $A + B \rightarrow$
  - In the second example, a third substance is required for each dissolving gas to react, and its depletion from the solution near the interface ***limits*** the rate of absorption of *both* A *and* B. Then the chemical steps are typified by amounts that depend on the ratio(s) of solubilities, the various reaction rate constants, and the applicable diffusion coefficients. Mathematical solutions are presented for the simultaneous absorption of two gases in a liquid medium that contains a third component capable of reacting chemically with both of the dissolved gases.

    Based on the chemical steps beginning with

    $$A + C \rightarrow$$

    and, simultaneously:

    $$B + C \rightarrow$$

A mathematical model for the simultaneous diffusion and reaction of two dissolved gases results in the following model differential equations.

A material balance on a differential element having dimensions $dx \times dz$ in the direction perpendicular to the interface and the direction of flow perpendicular to the interface and the direction of flow parallel to the interface, respectively, leads to a pair of nonlinear partial differential equations:

$$D_A \partial^2 A/\partial x^2 = u\partial A/\partial z + kAB \quad (1.44A)$$

$$D_B \partial^2 B/\partial x^2 = u\partial B/\partial z + \nu kAB \quad (1.44B)$$

where the last terms on the right-hand side of the two equations represent the rate of the homogeneous chemical reaction between substances A and B in the solution per unit volume of solution. The reaction is assumed to have the stoichiometry corresponding to the chemical equation:

$$A + \nu B \rightarrow \quad (1.45)$$

and to be irreversible. Kinetically, it is assumed to be second order with respect to the two concentrations.

1.2.4.1 An Example of a First-Order Reaction

Consider a first-order irreversible reaction of a dissolved gas. It may be used to show how the effect of a chemical reaction may be calculated on the basis of three models. In this case, the local rate of reaction per unit volume is

$$r = k_1 a \quad (1.46)$$

k_1 being the first-order reaction-rate constant, and a is the local content of dissolved gas.

1.2.4.2 The Film Model

The equation for the film model now takes the form

$$D_A d^2a/dx^2 = k_1 a \quad \begin{cases} a = A^*, & x = 0 \\ a = A^0, & x = \delta \end{cases} \tag{1.47}$$

where $\delta = D_A/k_L$, k_L being the transfer coefficient the transfer for physical absorption. Thus,

$$a = (1/\sinh\sqrt{M})\{A^0 \sinh x\sqrt{(k_1/D_A)} + A^* \sinh[(D_A/k_L) - x]\sqrt{(k_1/D_A)}\} \tag{1.48}$$

where $M = D_A k_1/k_L{}^2$, and the absorption rate is

$$\underline{R} = -D_A(da/dx)_{x=0} = k_L\{A^* - (A^0/\cosh\sqrt{M})\}\{\sqrt{M}/\tanh\sqrt{M}\}\} \tag{1.49}$$

Remarks:

(1) $\underline{R}$ is ***not*** directly proportional to $(A^* - A^0)$, as in physical absorption. This is generally the case when chemical reactions occur, and the value of A^0 is nonzero. Moreover, it is not generally appropriate to define a transfer coefficient as $R/(A^* - A^0)$, except in case of physical absorption only.

(2) The result expressed in Eq. (1.12) was first obtained by Hatta [8].

(3) When $\sqrt{M} \gg 1$, A^0 tends to zero, and Eq. (1.12) becomes

$$\underline{R} = A^*\sqrt{(D_A k_1)} \tag{1.50}$$

giving

$$E = \sqrt{(D_A k_1)}/k_L = \sqrt{M} \tag{1.51}$$

Here, E is the factor by which the chemical reaction *increases* the rate of absorption compared with merely physical absorption (with $A^0 = 0$ in both cases).

When $\sqrt{M} \gg 1$, the dissolved gases all react in the film and none diffuses in the unreacted state into the bulk of the liquid. Thus, the film thickness, or the value of k_L, is irrelevant and does not appear in the expression for $\underline{R}$.

For the Higbie model, one finds the expression for transient absorption, accompanied by irreversible first-order reaction.

1.3 The Mathematical Theory of Simultaneous Mass Transfer and Chemical Reactions

Consider the absorption of gases into liquids which are in a state of agitation such that the dissolved gases are transported from the surface to the interior by agitated convective motion [2]. The agitation may occur in one or more of several ways, including:

(1) The liquid may be blown through the liquid as a moving stream of bubbles, as, for example, on a bubble plate or in a sparged vessel.

(2) The liquid may be running in a layer over a vertical or an inclined surface, and the liquid flow may be turbulent (as, for example, in a wetted-wall column operating at a sufficiently high Reynolds Number, or ripples may develop and, thus, enhance the absorption process by convective motion.) Discontinuities in the surface may result in periodic mixing of the liquid layer in the course of its flow – as, for example, in a packed column, strings of discs, or strings of spheres.
(3) The liquid may be sprayed, as drops or jets, through the gas, as in a spray column of chamber.
(4) The liquid may be agitated by a stirrer which may also entrain bubbles of gas into the liquid.
(5) The liquid and gases may be interfaced in any physical contacting configurations and equipment.

1.3.1 Physical Absorption

In physical absorption, the gas dissolves in the liquid without reacting. The rate of absorption, found by experiment, is given by

$$\underline{R}a = k_L a(A^* - A^0) \tag{1.52}$$

where

$\underline{R}$ is the average rate of transfer of gas per unit area
a is the area of interface between the gas and the liquid, per unit volume of the system
k_L is the "physical mass transfer coefficient"
A^* is the concentration of the dissolved gas corresponding to equilibrium with the partial pressure of the gas at the interface between gas and liquid
A^0 is the average concentration of dissolved gas in the bulk liquid.

In general, it is not possible to determine k_L and a separately by measurement of physical adsorption. For example, in a packed column, the fraction of the surface of packings which is effectively wetted is unknown; and in a system containing bubbles, the interfacial area is not known in general. Hence, the quantity that may be directly measurable by physical absorption measurements is the combined quantity $k_L a$.

However, Eq. (1.52) has been established in many experimental investigations so that this equation would be predicted from basic principles whenever certain conditions are met – the main conditions are that the diffusivity and temperature at the surface (where the concentration is A^*) should not be very different from those in the bulk of the liquid, and that no chemical reaction occurs – so that all molecules of the dissolved gas are in the same conditions.

1.3.2 Chemical Absorption

1.3.2.1 Preliminary Remarks on Simultaneous Mass Transfer (Absorption) with Chemical Reactions

It is difficult to ascertain whether a solute, in fact, chemically reacts with a liquid or just interacts with it physically!

For the present purpose, a "physical" solution means that the dissolved molecules are indistinguishable, viz. not distinguishable as "unreacted" and "reacted" portions!

It has not escaped one's attention that the effects of the hydrodynamics of agitated liquids may well have an important bearing on the issues at hand. Nevertheless, in the present context, it seems indeed useful to use some simplified models of the physical absorption process to describe and analyze the challenge at hand!

This approach will therefore proceed with a survey of some well-established models and proceed to describe the effect of chemical reactions on the mass transfer process!

In particular, the mass transfer models based upon:

(1) The Nernst one-film theory model,
(2) The Whitman two-film theory model,
(3) The Danckwerts surface renewal theory model,
(4) The Higbie penetration theory model,
(5) The boundary layer theory model.

A full presentation and discussion of these various approaches are presented in Chapter 3.

1.3.2.2 Some Solutions to the Mathematical Models of the Theory of Simultaneous Mass Transfer and Chemical Reactions

Some typical cases are:

(i) **Simultaneous Absorption of Two Reacting Gases** [2]:

Consider two gases being absorbed simultaneously, with each undergoing a second-order reaction with the same dissolved reactant. In practice, a typical example is the case of a mixture of H_2S and CO_2 being absorbed by a solution of amine or alkali.

Let A_1 and A_2 be the two solute gases, and their respective concentrations at the liquid surface be $A_1{}^*$ and $A_2{}^*$, respectively, and their diffusivities be D_{A1} and D_{A2}, respectively, the rate constants k_1 and k_2 be governing the rates of their reactions with reactant B.

The reactions are both assumed to be of second order, so that the local rates of reaction are k_1ab and k_2ab, respectively.

Now, stoichiometrically, one may write:

$$A_1 + z_1B \rightarrow P_1 \tag{1.53}$$

$$A_2 + z_2B \rightarrow P_2 \tag{1.54}$$

Danckwerts reported that

(a) As $\sqrt{(k_2Bt)}$ approached ∞, the system approaches that for infinitely fast reaction of both solute gases, and an analytical solution is possible.
(b) In general, the effect of the presence of A_1 is to reduce the rate of absorption of A_2, and vice versa.

(ii) **Irreversible Reaction of General Order:**
Here, one may assume that the reaction rate is given by

$$r = k_{mn}(a)^m(b)^n \tag{1.55}$$

Reactions of such type, including
(a) The zero-order case: $m = 0$, $n = 0$, so that $(m+n) = 0+0 = 0$
(b) The first-order case: $m = 1$, $n = 0$, so that $(m+n) = 1+0 = 1$
(c) The second-order case: $m = 1$, $n = 1$, so that $(m+n) = 1+1 = 2$.
Defining the dimensionless quantity M by

$$M = [\pi/\{2(m+1)\}]k_{mn}(A^*)^{m-1}\,(B^0)^n t \tag{1.56}$$

Then,
(1) If $\sqrt{M} \gg (B^0/zA^*)\sqrt{(D_B/D_A)}$, then the rate of absorption is governed by diffusion alone, viz. the reaction may be considered as "instantaneous," and

$$R = -D_A(\partial a/\partial x)_{x=0} \tag{1.57}$$

applies. The term $(\partial a/\partial x)_{x=0}$ is the concentration gradient at the surface and is a function of time,
(2) If $\sqrt{M} \ll (B^0/zA^*)\sqrt{(D_B/D_A)}$, then the concentration of the reactant is practically uniform up to the surface of the liquid and equal to B^0. The reaction is then pseudo-mth order with $k_m = k_{mn}(B^0)^n$, and $r = k_m(a)^m$.

(iii) For the system H_2S and CO_2, if k_L is so small that

$$\sqrt{(D_A k_2 B^0)/k_L} \gg 1 + B^0/\left(2A_1^* + A_2^*\right) \tag{1.58}$$

where A_1^* and A_2^* are the surface concentrations of H_2S and CO_2, and the factor 2 is the stoichiometric factor for the reaction of CO_2 with NH_3 and the enhancement factors are:

$$E_{i1} \approx \left[1 + \left\{D_B B^0/(zA^* D_A + zA^* D_A\right\}\right] \sqrt{(D_{1A}/D_B)} \tag{1.59}$$

$$E_{i2} \approx \left[1 + \left\{D_B B^0/(zA^* D_A + zA^* D_A\right\}\right] \sqrt{(D_{2A}/D_B)} \tag{1.60}$$

These expressions show that the presence of A_2 diminishes E_i, and hence the rate of absorption of A_1, and vice versa.

1.3.2.3 Approximate Closed Form Solutions

In the mathematical analysis of simultaneous gas absorption and chemical reaction, a number of closed form solutions for special cases have been reported [9].

Example 1.1 ***Theory of the Stagnant Film of Finite Thickness – Slow First-Order Reaction***

In general, for a particular problem, the approach of using numerical analysis to obtain a useful solution is recommended. Much has been written on this methodology [9]. A reference to the latest standard collection is readily available in the most recent edition of "Perry's Chemical Engineers' Handbook," which should be a fruitful starting point.

If the reaction is not very fast compared with the diffusion rate, then the zone of reaction will not be restricted to a thin region, but will be spread throughout the liquid film, but will be spread throughout the liquid film. Consider the following derivation for the absorption and simultaneous slow irreversible reaction of a solute A, with a large excess of the solvent.

In this system, the solute diffuses through the gas film under the influence of the driving force $(p - p_i)$. Upon entry into the liquid phase, it begins to diffuse toward the main body of the liquid and immediately starts to react with the dissolved substance B. Since the diffusing current of solute is being depleted as it diffuses into the liquid, the concentration gradient will be concave upward. The total rate of diffusion of reacted and unreacted forms of A is constant, since there is no accumulation of A in the film.

If the rate equation for the reaction is known, then the conditions of the process may be expressed mathematically. For example, assume that the reaction is of the first order and that the rate of elimination of A is proportional to the concentration of A at any point in time, viz.

$$-\mathrm{d}A/\mathrm{d}\theta = k_c cV \tag{1.61}$$

where

k_c is the specific reaction-rate constant,
c is the variable concentration of A, and
V is the liquid volume considered.

Next, consider a differential element, of unit cross-sectional area, and of thickness x_L.

The diffusion rate into this element will be

$$N_A = -D_A \mathrm{d}c/\mathrm{d}x \tag{1.62}$$

at $x = x$. The rate of diffusion out of this element, at $x = x + \mathrm{d}x$, will be

$$N_A = -D_A[\mathrm{d}c/\mathrm{d}x + (\mathrm{d}^2c/\mathrm{d}x^2)\mathrm{d}x] \tag{1.63}$$

The disappearance of A within the element, owing to reaction with B, is

$$\mathrm{d}A/\mathrm{d}\theta = -k_c c\,\mathrm{d}x \tag{1.64}$$

since the volume of the element is dx. The difference in diffusion rates in and out of the element must be equal to the rate of elimination of A by reaction with B, hence

$$D_A\,\mathrm{d}^2c/\mathrm{d}x^2 = k_c c \tag{1.65}$$

Now, assuming that k_c remains constant throughout the film, the solution of Eq. (1.65) may be expressed as

$$c = C_1 \exp(a_0 x) + C_2 \exp(-a_0 x) \tag{1.66}$$

where $a_0 = \sqrt{(k_c/D_A)}$.

Substituting the limits:

$$c = c_i \text{ at } x = 0, \quad \text{and} \quad c = c_L \text{ at } x = x_{La}$$

the constants C_1 and C_2 are obtained, and the solution takes the form:

$$c = [c_L \sinh(a_0 x) + c_i \sinh\{a_0(x_L - x)\}] / \sinh(a_0 x_L) \tag{1.67}$$

and for the case of $c_L = 0$, Eq. (1.67) reduces to

$$c = c_i \sinh\{a_0(x_L - x)\} / \sinh(a_0 x_L) \tag{1.68}$$

Remark I:

(1) Equation (1.68) is mathematically analogous to the description of heat conduction along a fin that is maintained at a constant temperature at the base, with heat dissipation along the fin in proportion to the temperature profile maintained on the fin.

(2) By differentiating Eq. (1.67), the slope of the concentration curve is obtained as follows:

$$dc/dx = [a_0 c_L \cosh(a_0 x) + a_0 c_i \cosh\{a_0(x_L - x)\}] / \sinh(a_0 x_L) \tag{1.69}$$

(3) The rate of diffusion into the liquid may be obtained by multiplying the slope at $x = 0$ by the diffusivity D_A, viz.

$$\begin{aligned} N_A &= -D_A(dc/dx)_{x=0} \\ &= D_A a_0 [c_i \cosh(a_0 x_L) - c_L] / \sinh(a_0 x_L) \end{aligned} \tag{1.70}$$

(4) Similarly, the rate of diffusion of A into the main body of the liquid may be obtained by substituting $x = x_L$:

$$\begin{aligned} N_A' &= -D_A(dc/dx)_{x=x_L} \\ &= D_A a_0 \{c_i - c_L \cosh(a_0 x_L - x)\} / \sinh(a_0 x_L) \end{aligned} \tag{1.71}$$

(5) With respect to the solute A entering the liquid phase, the fraction F arriving at the main body of the liquid *without* reacting is

$$\begin{aligned} F &= N_A'/N_A \quad \text{and upon substituting from Eqs. (1.70) and (1.71)} \\ &= [\{c_i - c_L \cosh(a_0 x_L - x)\} / \{c_i \cosh(a_0 x_L) - c_L\}]|_{x=0} \\ &= \{c_L \cosh(a_0 x_L) - c_i\} / \{c_L - c_i \cosh(a_0 x_L)\} \end{aligned} \tag{1.72}$$

Example 1.2 ***Theory of the Unsteady-State Absorption in a Stagnant Liquid – Slow First-Order Reaction***

At the ***initial*** exposure of a liquid to a gas during which the solute may accumulate in the liquid [9], an additional term should be added to Eq. (1.65)

$$D_A d^2c/dx^2 = k_c c$$

Remark II:

(1) Consider an elemental section make-up of the liquid phase near the interface, having a vertical side of unit cross section parallel near the interface, and a thickness dx. The rate of diffusion in and out of this liquid section are described

by the following two equations representing mass balances:

$$N_A = -D_A(dc/dx) \tag{1.73}$$

$$N_A = -D_A\{(dc/dx) + (d^2c/dx^2)\,dx\} \tag{1.74}$$

while the net rate of disappearance per unit volume owing to reaction is given by

$$k_c c_A - k_{c'} c_B \tag{1.75}$$

where k_c *and* $k_{c'}$ are the rate constants for the forward and reverse reactions:

$$A \leftrightarrow B \tag{1.76}$$

(2) Now, at equilibrium, the opposing rates are equal:

$$K_c' = k_c/K \tag{1.77}$$

where K is the chemical equilibrium constant.

(3) The rate of depletion of A from the elemental section is $-(\partial c_A/\partial\theta)\,dx$ in the unsteady state. Equating the rate of input of A by diffusion to the sum of the rates of output by diffusion, disappearance by reaction, and depletion, the result is

$$D_A(\partial^2 c_A/\partial x^2) = k_e c_A - (k_e/K)c_B + \partial c_A/\partial\theta \tag{1.78}$$

Similarly, a mass balance for B results in

$$D_A(\partial^2 c_A/\partial x^2) = (k_e/K)c_B - k_c c_A + \partial c_A/\partial\theta \tag{1.79}$$

(4) Should the concentration of B be eliminated and the diffusivities are assumed equal, the results is a *fourth-order* linear partial differential

$$\begin{aligned} D(\partial^4 c_A/\partial x^4) &- 2\partial^3 c_A/\partial\theta\partial x^2 - k_e\{1 + (1/K)\}(\partial^2 c_A/\partial x^2) \\ &+ (1/D)\partial^2 c_A/\partial\theta^2 + (k_e/D)\{1 + (1/K)\}(\partial c_A/\partial\theta) \end{aligned} \tag{1.80}$$

(5) It is assumed that
   (a) the product of the reaction is nonvolatile,
   (b) the interfacial concentration of A is equal to the value in equilibrium with the gas, where the partial pressure of A is constant,
   (c) the initial concentrations of A and B in the liquid before exposure to the gas are assumed to be uniformly equal to their equilibrium values,
   (d) the body of liquid is considered to be infinite effectively.

   Hence, correspondingly, the following boundary conditions are applicable:
   (A) At $x = 0$: $c_A = c_i = p_A/H$
   (B) At $x = 0$: $\partial c_B/\partial x = 0$
   (C) At $\theta = 0$: $c_B = Kc_A = \{K/(1 + K)\}c_0$, a constant
   (D) At $x = \infty$: $c_B = Kc_A = \{K/(1 + K)\}c_0$, a constant

   where c_0 represents the sum of the initial values of c_A and c_B, which are assumed to be in equilibrium with each other.

(6) Since Eq. (1.80) is linear (with respect to the derivative terms) with simple boundary conditions, the Laplace transform method may be used to obtain its solutions, with the following results:

$$\{(K+1)c_{\mathrm{av}} - c_0\}/\{(K+1)c_{\mathrm{i}} - c_0\} = (3/\sqrt{\pi})\left\{\sqrt{(D_{\mathrm{L}}\theta/B_{\mathrm{F}}{}^2)}\right\} \times \{f(K, k_{\mathrm{c}}\theta)/(K+1)\} \tag{1.81}$$

where

$$f(K, k_{\mathrm{c}}\theta) = K + 1 + [\pi K^5/\{4(K-1)(k_{\mathrm{c}}\theta)\}]^{1/2}\exp[k_{\mathrm{c}}\theta/\{K(K-1)\}] \times [\mathrm{erf}\{Kk_{\mathrm{c}}\theta/(K-1)\}]^{1/2} - \mathrm{erf}[k_{\mathrm{c}}\theta/\{K(K-1)\}^{1/2}] - [\{\pi K^3(K+1)\}/4k_{\mathrm{c}}\theta]^{1/2}\mathrm{erf}[\{(K+1)k_{\mathrm{c}}\theta\}/K]^{1/2} \tag{1.82}$$

where $(K+1)c_{\mathrm{av}}$ is the average total concentration of solute in the liquid in both the reacted and the unreacted forms and c_{av} represents the average concentration of A after absorption if A and B were in chemical equilibrium at every point.

(7) Similarly, $(K+1)c_{\mathrm{i}}$ represents the total concentration of unreacted and reacted solute in the liquid if it were allowed to come to chemical and physical equilibrium with the gas.

(8) Using the following functional notations:

$$\exp(x) = \mathrm{e}^x \quad \text{and} \quad \mathrm{erf}(x) = (2/\sqrt{\pi})\int_0^q \mathrm{e}^{-p^2}\,\mathrm{d}p \tag{1.83}$$

(9) the latter being the error function.

(i) For large values of $k_{\mathrm{c}}\theta$, the approximate equivalent of Eq. (1.81) is

$$f(K, k_{\mathrm{e}}\theta) = (1+K)\left[1 - (\pi^{1/2}/2)K\{K/(K+1)\}^{1/2}(k_{\mathrm{c}}\theta)^{-1/2} + [K^3/\{2(K+1)\}](k_{\mathrm{c}}\theta)^{-1}\right] \tag{1.84}$$

(ii) For small values of $k_{\mathrm{c}}\theta$ approximately, Eq. (1.84) simplifies to

$$f(K, k_{\mathrm{e}}\theta) = 1 + [\{K^{1/2}(K+1)^{3/2}(K-1)^2 - K^4 + 3K^2 - 2K\}/3(K-1)^2](k_{\mathrm{c}}\theta) \tag{1.85}$$

(iii) For the irreversible reaction, viz., $K = \infty$

$$f(\infty, k_{\mathrm{e}}\theta) = 1/2\,\exp(-k_{\mathrm{e}}\theta) + \sqrt{(\pi/4)}\left\{(1 + 2k_{\mathrm{e}}\theta)/(k_{\mathrm{e}}\theta)^{1/2}\right\}\mathrm{erf}(k_{\mathrm{c}}\theta)^{1/2} \tag{1.86}$$

When these equations are applied to absorption from a gas in a falling liquid film, the time of exposure is that of the liquid surface, which may be calculated from the surface velocity:

$$\theta = Z/u_{\mathrm{s}} = (2/3)Z/u_{\mathrm{av}} \tag{1.87}$$

where Z is the height of the surface. By means of a mass balance on the liquid layer of thickness B_ν, the liquid film absorption coefficient k_{L} is obtained as follows:

$$u_{\mathrm{av}}B_\nu[(K+1)c_{\mathrm{av}} - c_0] = k_{\mathrm{L}}Z[(K+1)c_{\mathrm{i}} - c_0] \tag{1.88}$$

and

$$k_L Z/u_{av}B_v = [(K+1)c_{av} - c_0]/[(K+1)c_i - c_0] \quad (1.89A)$$

$$= (3/\sqrt{\pi})(D_L\theta/Bv^2)^{1/2}\{f(K,k_c\theta)/(K+1)\} \quad (1.89B)$$

As before, in these equations:

- c_0 represents the concentration of total dissolved solute (both A and B forms) before exposure to the liquid surface and
- $(K+1)c_i$ represents the concentration of total dissolved solute, both A and B forms, in a liquid in phase equilibrium and reaction equilibrium with the gas – this latter quantity is not the actual solute concentration at the interface, where chemical equilibrium may not exist. It is the concentration which would be found in the usual static equilibrium experiment.

  The absorption coefficient k_L is defined in Eq. (1.88) on the basis of total dissolved-solute concentrations as a driving force. Comparing Eqs. (1.89A) and (1.89B) with the following equation (1.90), developed for pure physical adsorption:

$$k_L = \sqrt{(6/\pi)}\sqrt{(D_L u_{av}/Z)} \quad (1.90)$$

$$\text{viz.} \quad k_L Z/u_{av}B_r = (3/\sqrt{\pi})\sqrt{(D_L\theta/Br^2)} \quad (1.91)$$

  shows that $f(K, k_e\theta)/(K+1)$ is a multiplying factor which leads to the effect of chemical reaction on the coefficient for no reaction.
- Inspection of Eq. (1.85) shows that $f(K, k_e\theta)$ becomes equal to unity for the condition of a very unfavorable equilibrium ($K = 0$). Under this condition, the correction factor to the absorption coefficient becomes equal to 1, as expected.
- Also, $f(K, k_e\theta)$ becomes equal to unity when the equilibrium favors the production of B ($K > 0$), but the reaction is very slow ($k_e\theta \sim 0$).
- Equations (1.89A) and (1.89B) shows that under these conditions, the absorption coefficient is ***lower*** than that for purely physical absorption – owing to the fact that the solute dissolves at the interface in A form and does not react immediately: the interfacial concentration is only p_i/H, therefore, rather than $(K+1)(p_i/H)$, which is used to define k_L. The driving force causing diffusion is smaller, and the rate of absorption also smaller, owing to the reaction.
- When the reaction is fast: $k_e \to \infty$, Eq. (1.84) shows that $f(K, k_e\theta)$ approaches $(K+1)$ as a limiting value: the absorption coefficient then has its normal value, irrespective of K.

Example 1.3 ***Theory of the Unsteady-State Absorption in a Stagnant Liquid – Rapid Second-Order, Irreversible Reaction***

For a chemical reaction occurring in the liquid, after the absorption of gaseous substance A, it may be stated as

$$A + B \rightarrow AB \quad (1.92)$$

If the rate of absorption of A is sufficiently rapid, the rate may be limited by the rate at which B can diffuse from the liquid toward the gas–liquid interface – in order to react with A. Thus, the faster B is able to diffuse, the shorter the distance A will need to diffuse into the liquid to meet A. Thus, in contrast with the scenario for the first-order reaction:

$$A \rightarrow C \tag{1.93}$$

for which the absorption rate is independent of the concentration of B, in contrast to the second-order case wherein the absorption rate of A depends on the concentration of B.

When the reaction rate is extremely fast compared with the rate of diffusion of the reactants, the theory may be developed as shown here below.

In the absorption scenario, the following liquid layer has a uniform concentration of B when it is exposed to a gas containing A.

As A is absorbed, molecules of B which are near the interface are consumed in the reaction and are replenished by the diffusion of additional molecules of B from the main body of the liquid. As the reaction is assumed to take place instantaneously when A and B are brought together, regardless of the concentrations in the reaction zone, the point where A and B molecules meet each other will recede farther away from the interface. Hence, the absorption rate of A diminishes because the reacting molecules have to diffuse farther into the liquid to react! Thus, there is little or no opportunity for A to react until it arrives at the reaction zone at a distance x' from the interface. Hence, within the region $0 < x < x'$, the diffusion of A is represented by the second-order partial differential equation:

$$D_A \partial^2 c_A / \partial x^2 = \partial c_A / \partial \theta \tag{1.94}$$

Similarly, component B does not react with A until B reaches the point $x = x'$; and for $x > x'$, the diffusion of B is governed by a similar second-order partial differential equation:

$$D_B \partial^2 c_B / \partial x_2 = \partial c_B / \partial \theta \tag{1.95}$$

Moreover, for short times of exposure of the liquid surface, concentrations of A and B vary within a region near the interface, so that under these conditions, it may be assumed, without affecting the results significantly, that the liquid layer is infinitely deep: thus Eq. (1.95) applies within the region $x' < x < \infty$.

Sherwood and Pigford [10] showed, geometrically from the profiles of the concentration curves for A + B at times θ and $\theta + d\theta$, that solutions of Eqs. (1.94) and (1.95) will be suitable if they can be made to satisfy the necessary boundary conditions and are as follows:

$$c_A = A_1 + B_1 \operatorname{erf}[x/\{2\sqrt{(D_A \theta)}\}] \tag{1.96}$$

$$c_B = A_2 + B_2 \operatorname{erf}[x/\{2\sqrt{(D_B \theta)}\}] \tag{1.97}$$

where the functional notation of Eq. (1.83) is used.

1.3.3 Numerical Solutions

In practice, when seeking numerical solutions for similar system sets of partial differential equations, it is much more straightforward to use numerical techniques such as techniques available in numerical analysis and/or solving in the R domain.

Fully worked out examples are provided in Chapters 3 and 4.

1.4 Diffusive Models of Environmental Transport

If one considers only the diffusive element, leaving out the reactive element, in a similar problem, a comprehensive collection of formulations have been systematically worked out by Choy and Reible [11], including:

* Diffusive Models of Environmental Transport
* Equilibria Within Environmental Phases
* Diffusion in Semi-infinite Systems
* Diffusion in Finite Layers
* Diffusion in Two-Layer Composite Systems
* Diffusion in Three-Layer Composite Systems
* Advective Diffusive Models
* Diffusion with Volatile Liquid Evaporation
* Diffusion with Time-Dependent Partition Coefficients
* Diffusion with Constant Flux Liquid Evaporation
* Some Typical Numerical Examples of Diffusive Models of Environmental Transport.

It certainly should not escape one's attention that, should there be reactive processes involved in any one of these cases, suitable reactive terms may be added to any of these models to provide the concomitant model description to reflect the simultaneous mass transfer and chemical reactions processes.

Further Reading

Emmert, R.E. and Pigford, R.L. (1963). In: *Perry's Chemical Engineers' Handbook*, 4e (eds. R.H. Perry, C.H. Chilton, and S.D. Kirkpatrick). *Section 14 'Gas Absorption and Solvent Extraction'*. New York: McGraw-Hill Book Company.

Roper, Hatch, and Pigford have obtained a theoretical solution, using graphic presentations, for the case where two absorbing gases react with one another.

Hatch and Pigford (*Ind. Eng. Chem., Fund. Quarterly* 1: 209 (1962)) described an experimental in which CO_2 and NH_3 simultaneously dissolved in water and reacted with one another. Their results thus confirmed the foregoing theory for the pseudo first-order case.

2

Data Analysis Using R Programming

In an Internet on-line advertisement, there is a job vacancy advertisement for a statistician. The complete job description reads as follows:

Job Summary

Statistician I
Salary: Open
Employer: XYZ Research and Statistics
Location: City X, State Y
Type: Full Time – Entry Level
Category: Financial analyst/Statistics, Data analysis/processing, Statistical organization & administration
Required Education: Master's Degree preferred

XYZ Research and Statistics is a national leader in designing, managing, and analyzing financial data. XYZ partners with other investigators to offer respected statistical expertise supported by sophisticated web-based data management systems. XYZ services assure timely and secure implementation of trials and reliable data analyses.

Job Description

Position Summary: An exciting opportunity is available for a statistician to join a small but growing group focused on financial investment analysis and related translational research. XYZ, which is located in downtown City XX, is responsible for the design, management, and analysis of a variety of investment and financial, as well as the analysis of associated market data. The successful candidate will collaborate with fellow statistics staff and financial investigators to design, evaluate, and interpret investment studies.

Primary Duties and Responsibilities: Analyzes investment situations and associated ancillary studies in collaboration with fellow statisticians and other financial engineers. Prepares tables, figures, and written summaries of study

Simultaneous Mass Transfer and Chemical Reactions in Engineering Science, First Edition. Bertram K. C. Chan.

results; interprets results in collaboration with other financial; and assists in preparation of manuscripts. Provides statistical consultation with collaborating staff. Performs other job-related duties as assigned.

Requirements

Required Qualifications: Master's Degree in Statistics, Applied Mathematics, or related field. Sound knowledge of applied statistics. Proficiency in statistical computing in R.

Preferred Responsibilities/Qualifications: Statistical consulting experience.

S-Plus or R programming language experience. Experience with analysis of high-dimensional data. Ability to communicate well orally and in writing. Excellent interpersonal/teamwork skills for effective collaboration. Spanish language skills a plus.

In your cover letter, describe how your skills and experience match the qualifications for the position.

To learn more about XYZ, visit www.XYZ.org.

Clearly, one should be cognizant of the overt requirement of an acceptable level of professional proficiency in data analysis using R *programming!*
Even if one is not in such a job market, as a statistician working in the fields of Finance, Asset Allocations, Portfolio Optimization, etc., a skill set that would include R *programming would be helpful and interesting.*

2.1 Data and Data Processing

Data are facts or figures from which conclusions can be drawn. When the data have been recorded, classified, and organized, related or interpreted within a framework so that meaning emerges, they become **information**. There are several steps involved in turning data into information, and these steps are known as **data processing**. This section describes data processing and how computers perform these steps efficiently and effectively.

It will be indicated that many of these processing activities may be undertaken using R programming, or performed in an R environment with the aid of available R packages – where R functions and datasets are stored.

2.1.1 Introduction

The simplified flowchart below, shows how raw **data** are transformed into information [12]:

Data → Collection→ Processing →Information

Data processing takes place once all of the relevant data have been collected. They are gathered from various sources and entered into a computer where they can be processed to produce **information** – the output.

Data processing includes the following steps:

Data coding
Data capture
Editing
Imputation
Quality control
Producing results

2.1.2 Data Coding

First, before raw data can be entered into a computer, they must be coded. To do this, survey responses must be labeled, usually with simple, numerical codes.

This may be done by the interviewer in the field or by an office employee. The data coding step is important because it makes data entry and data processing easier. Surveys have two types of questions – closed questions and open questions. The responses to these questions affect the type of coding performed.

A **closed question** means that only a **fixed** number of predetermined survey responses are permitted. These responses will have already been coded.

The following question, in a survey on Sporting activities, is an example of a closed question:

To what degree is sport important in providing you with the following benefits?

(1) Very important
(2) Somewhat important
(3) Not important

An **open question** implies that **any** response is allowed, making subsequent coding more difficult. In order to code an open question, the processor must sample a number of responses, and then design a code structure that includes all possible answers.

The following code structure is an example of an open question:

What sports do you participate in?
Specify (28 characters)______________

In the Census and almost all other surveys, the codes for each question field are pre-marked on the questionnaire. To process the questionnaire, the codes are entered directly into the database and are prepared for data capturing. The following is an example of pre-marked coding:

What language does this person speak most often at home?

(1) English
(2) French
(3) Other – Specify____________

2.1.2.1 Automated Coding Systems

There are programs in use that will automate repetitive and routine tasks. Some of the advantages of an automated coding system are that the process increasingly becomes

- faster,
- more consistent, and
- more economical.

The next step in data processing is inputting the coded data into a computer database. This method is known as **data capture**.

2.1.3 Data Capture

This is the process by which data are transferred from a paper copy, such as questionnaires and survey responses, to an electronic file. The responses are then put into a computer. Before this procedure takes place, the questionnaires must be groomed (prepared) for data capture. In this processing step, the questionnaire is reviewed to ensure that all of the minimum required data have been reported, and that they are decipherable. This grooming is usually performed during extensive automated edits.

There are several methods used for capturing data:

Tally charts are used to record data such as the number of occurrences of a particular event and to develop frequency distribution tables.

Batch keying is one of the oldest methods of data capture. It uses a computer keyboard to type in the data. This process is very practical for high-volume entry where fast production is a requirement. No editing procedures are necessary, but there must be a high degree of confidence in the editing program.

Interactive capture is often referred to as intelligent keying. Usually, captured data are edited before they are imputed. However, this method combines data capture and data editing in one function.

Optical character readers or bar-code scanners are able to recognize alpha or numeric characters. These readers scan lines and translate them into the program. These bar-code scanners are quite common and often seen in department stores. They can take the shape of a gun or a wand.

Magnetic recordings allow for both reading and writing capabilities. This method may be used in areas where data security is important. The largest application for this type of data capture is the PIN number found on automatic bank cards. A computer keyboard is one of the best known input (or data entry) devices in current use. In the past, people performed data entry using punch cards or paper tape.

Some modern examples of data input devices are

- optical mark reader
- bar-code reader
- scanner used in desktop publishing
- light pen
- trackball
- mouse

Once data have been entered into a computer database, the next step is ensuring that all of the responses are accurate. This method is known as data editing.

2.1.4 Data Editing

Data should be edited before being presented as information. This action ensures that the information provided is accurate, complete, and consistent.

There are two levels of data editing – **micro-** and **macro-editing**.

Micro-editing corrects the data at the record level. This process detects errors in data through checks of the individual data records. The intent at this point is to determine the consistency of the data and correct the individual data records.

Macro-editing also detects errors in data, but does this through the analysis of aggregate data (totals). The data are compared with data from other surveys, administrative files, or earlier versions of the same data. This process determines the compatibility of data.

2.1.5 Imputations

Editing is of little value to the overall improvement of the actual survey results, if no corrective action is taken when items fail to follow the rules set out during the editing process. When all of the data have been edited using the applied rules and a file is found to have missing data, then imputation is usually done as a separate step.

Non-response and invalid data definitely impact the quality of the survey results.

Imputation resolves the problems of missing, invalid, or incomplete responses identified during editing, as well as any editing errors that might have occurred.

At this stage, all of the data are screened for errors because respondents are not the only ones capable of making mistakes; errors can also occur during coding and editing.

Some other types of imputation methods include:

hot deck uses other records as "donors" in order to answer the question (or set of questions) that needs imputation.

substitution relies on the availability of comparable data. Imputed data can be extracted from the respondent's record from a previous cycle of the survey, or the imputed data can be taken from the respondent's alternative source file (e.g. administrative files or other survey files for the same respondent).

estimator uses information from other questions or from other answers (from the current cycle or a previous cycle), and through mathematical operations, derives a plausible value for the missing or incorrect field.

cold deck makes use of a fixed set of values, which covers all of the data items. These values can be constructed with the use of historical data, subject-matter expertise, etc.

The donor can also be found through a method called **nearest-neighbor imputation**. In this case, some sort of criteria must be developed to determine which responding unit is "most like" the unit with the missing value in accordance with the predetermined characteristics. The closest unit to the missing value is then used as the donor.

Imputation methods can be performed automatically, manually, or in combination.

2.1.6 Data Quality

- Quality assurance
- Quality control
- Quality management in statistical agencies

Quality is an essential element at all levels of processing. To ensure the quality of a product or service in survey development activities, both quality assurance and quality control methods are used.

2.1.7 Quality Assurance

Quality assurance refers to all planned activities necessary in providing confidence that a product or service will satisfy its purpose and the users' needs. In the context of survey conducting activities, this can take place at any of the major stages of survey development: planning, design, implementation, processing, evaluation, and dissemination.

This approach anticipates problems prior to their unexpected occurrences, and uses all available information to generate improvements. It is not restricted to any specific quality the planning stage and is all-encompassing in its activities standards. It is applicable mostly at the planning stage and is all-encompassing in its activities.

2.1.8 Quality Control

Quality control is a regulatory procedure through which one may measure quality, with pre-set standards, and then act on any differences. Examples of this include controlling the quality of the coding operation, the quality of the survey interviewing, and the quality of the data capture.

Quality control responds to observed problems, using on-going measurements to make decisions on the processes or products. It requires a pre-specified quality for comparability. It is applicable mostly at the processing stage, following a set procedure that is a subset of quality assurance.

2.1.9 Quality Management in Statistical Agencies

The quality of the data must be defined and assured in the context of being "fit for use", which will depend on the intended function of the data and the fundamental characteristics of quality. It also depends on the users' expectations of what is considered to be useful information.

There is no standard definition among statistical agencies for the term "official statistics." There is a generally accepted, but evolving, range of quality issues underlying the concept of "fitness for use." These elements of quality need to be considered and balanced in the design and implementation of an agency's statistical program.

The following is a list of the elements of quality:

- Relevance
- Accuracy
- Timeliness
- Accessibility
- Interpretability
- Coherence

These elements of quality tend to overlap. Just as there is no single measure of accuracy, there is no effective statistical model for bringing together all these characteristics of quality into a single indicator. Also, except in simple or one-dimensional cases, there is no general statistical model for determining whether one particular set of quality characteristics provides higher overall quality than another.

2.1.10 Producing Results

After editing, data may be processed further to produce a desired output. The computer software used to process the data will depend on the form of output required. Software applications for word processing, desktop publishing, graphics (including graphing and drawing), programming, databases and spreadsheets are commonly used. The following are some examples of ways that software can produce data:

Spreadsheets are programs that automatically add columns and rows of figures, calculate means, and perform statistical analyses.

Databases are electronic filing cabinets. They systematically store data for easy access, and produce summaries, aggregates, or reports.

Specialized programs can be developed to edit, clean, impute, and process the final tabular output.

Review Questions for Section 2.1

1. In the Job Description for an entry-level statistician today, from the viewpoint of a prospective applicant for that position, what basic statistical computing languages are important in order to meet the requirement? Why?
2. For a typical MBA (Master of Business Administration) program in Business and Finance, should the core curriculum include the development of proficient skill in the use of R programming in statistics? Why?
3. (a) Contrast the concepts of data and information.
   (b) How does the process of data processing convert data to information?
4. In the steps which convert data into information, how are statistics and computing applied to the various data processing steps?

5. (a) Describe and delineate quality assurance and quality control in computer data processing.
   (b) In what way does statistics feature in these phases of data processing?

2.2 Beginning R

R is an open-source, freely available, integrated software environment for data manipulation, computation, analysis, and graphical display.

The R environment consists of

- a data handling and storage facility
- operators for computations on arrays and matrices
- a collection of tools for data analysis
- graphical capabilities for analysis and display
- an efficient, and continuing developing programming algebra-like programming language which consists of loops, conditionals, user-defined functions, and input and output capabilities.

The term "environment" is used to show that it is indeed a planned and coherent system [13, 14].

2.2.1 R and Statistics

R was initially written by Robert Gentleman and Ross Ihaka of the Statistics Department of the University of Auckland, New Zealand, in 1997. Since then there has been the R-development core group of about 20 people with write-access authority for the R source code.

The original introduction to the R environment, evolved from the S/S-Plus languages, was **not** primarily directed toward statistics. However, since its development in the 1990s, it appeared to have been "hijacked" by many working in the areas of classical and modern statistical techniques, including many applications in financial engineering, econometrics, biostatistics with respect to epidemiology, public health, and preventive medicine!

As of this writing, the latest version of R is R-3.3.2, officially released on 31 October 2016. The primary source of R packages is the Comprehensive R Archive Network, CRAN, at http://cran.r-project.org/

Another source of R packages may be found in numerous publications, e.g. the *Journal of Statistical Software*, now at its 45th volume, is available at http://www.jstatsoft.org/v45.

Let us get started (the R-3.3.2 version environment is being used here).

Recall that the R environment was obtained as follows:

Here is R:
Let us download the open-source high-level program R from the Internet and take a first look at the R computing environment.

Remark: Access the Internet at the website of CRAN (The Comprehensive R Archive Network: http://cran.r-project.org/

To install R: R-3.3.2-win32.exe

http://www.r-project.org/

=> download R

=> Select: USA

http://cran.cnr.Berkeley.edu <http://cran.cnr.berkeley.edu/>

University of California, Berkeley, CA

=> http://cran.cnr.berkeley.edu/

=> Windows (95 and later)

=> base

=> R-3.3.2-win32.exe
AFTER the downloading:
=> Double-click on: R-3.3.2-win32.exe
(on the DeskTop) to un-zip & install R
=> An icon (Script R 3.3.2) will appear on one's Computer "desktop" as follows: Figure 2.1

On the computer "desktop" is the R icon:

Figure 2.1 The R icon on the computer desktop. (The R 3.3.2 looks *exactly* the same as that for R 2.9.1).

In this book, different fonts are used for all statements during the computational activities in the R environment, to clarify the various inputs to and outputs from the computational process:

1. Texts in this book (***STIXTwoText*** font)
2. Line Input in R code (Verdana font)
3. Line Output *in* R code (Verdana font)
4. ***Line Comment Statements in*** **R** ***code (Italicized STIXTwoText font)***

Note: The # sign is the Comment Character: all text in the line following this sign is treated as a comment by the R program, i.e. no computational action will be taken regarding such a statement. That is, the computational activities will proceed as though the comment statements are ignored. These comment statements help the programmer and user by providing some clarification of the purposes involved in the remainder of the R environment. The computations will proceed even if these comment statements are eliminated.

is known as the Number Sign, it is also known as the pound sign/key, the hash key, and, less commonly, as the octothorp, octothorpe, octathorp, octotherp, octathorpe, and octatherp!

To use R under Windows: **Double-click on the R 3.3.2 icon**

Upon selecting and clicking on R, the R-window opens, with the following declaration:

R version 3.3.2 (2016-10-31)
Copyright (C) 2016. The R Foundation for Statistical Computing
ISBN 3-900051-07-0
R is free software and comes with ABSOLUTELY NO WARRANTY.

You are welcome to redistribute it under certain conditions.

Type "license()" or "licence()" for distribution details.
R is a collaborative project with many contributors.
Type "contributors()" for more information and
"citation()" on how to cite R or R packages in publications.
Type "demo()" for some demos, "help()" for on-line help, or
"help.start()" for an HTML browser interface to help.
Type "q()" to quit R.

```
> # This is the R computing environment.
> # Computations may begin now!
>
> # First, use R as a calculator, and try a simple arithmetic
> # operation, say: 1 + 1
> 1+1
> [1] 2 # This is the output!
>          # WOW! It's really working!
> # The [1] in front of the output result is part of R's way of printing
> # numbers and vectors. Although it is not so useful here, it does
> # become so when the output result is a longer vector
```

****** From this point on, this book is most beneficially read with the R environment at hand. It will be a most effective learning experience if one practices each R command as one goes along the textual materials!***

2.2.2 A First Session Using R

This section introduces some important and practical features of the R Environment (Figure 2.2).

Login and start an R session in the Windows system of the computer:

```
>
> # This is the R environment.
> help.start() # Outputting the page shown in Figure 2.2
>                # Statistical Data Analysis Manuals
```

Statistical Data Analysis

Manuals

An Introduction to R	The R Language Definition
Writing R Extensions	R Installation and Administration
R Data Import/Export	R Internals

Reference

Search Engine & Keywords

Packages

Miscellaneous Material

About R	Authors	Resources
License	Frequently Asked Questions	Thanks
NEWS	User Manuals	Technical papers

Material specific to the Windows port

CHANGES	Windows FAQ

Figure 2.2 Output of the R Command.

starting httpd help server ... done
If nothing happens, you should open
'http://127.0.0.1:28103/doc/html/index.html' yourself
At this point, explore the HTML interface to on-line help
right from the desktop, using the mouse pointer to note
the various features of this facility available within
the R environment. Then, returning to the R environment:

```
> help.start()
```

Carefully read through each of the sections under "Manuals" – to obtain an introduction to the basic language of the R environment. Then look through the items under "Reference" to reach beyond the elementary level, including access to the available "R Packages" – all R functions and datasets are stored in packages.

For example, if one selects the Packages Reference, the following R Package Ind ex window will open up, showing Figure 2.3, listing a collection of R program packages under the R library: C:\Program Files\R\R-2.14.1\library

Packages in C:\ProgramFiles\R\R-2.14.1\library

base	*The R Base Package*
boot	*Bootstrap Functions (originally by Angelo Canty for S)*
class	*Functions for Classification*
cluster	*Cluster Analysis Extended Rousseeuw et al.*
codetools	*Code Analysis Tools for R*
compiler	*The R Compiler Package*
datasets	*The R Datasets Package*
foreign	*Read Data Stored by Minitab, S, SAS, SPSS, Stata, Systat, dBase, ...*
graphics	*The R Graphics Package*
grDevices	*The R Graphics Devices and Support for Colours and Fonts*
grid	*The Grid Graphics Package*
KernSmooth	*Functions for kernel smoothing for Wand & Jones (1995)*
lattice	*Lattice Graphics*
MASS	*Support Functions and Datasets for Venables and Ripley's MASS*
Matrix	*Sparse and Dense Matrix Classes and Methods*
methods	*Formal Methods and Classes*
mgcv	*GAMs with GCV/AIC/REML smoothness estimation and GAMMs by PQL*
nlme	*Linear and Nonlinear Mixed Effects Models*
nnet	*Feed-forward Neural Networks and Multinomial Log-Linear Models*
parallel	*Support for Parallel computation in R*
rpart	*Recursive Partitioning*
spatial	*Functions for Kriging and Point Pattern Analysis*
splines	*Regression Spline Functions and Classes*
stats	*The R Stats Package*
stats4	*Statistical Functions using S4 Classes*
survival	*Survival analysis, including penalized likelihood.*
tcltk	*Tcl/Tk Interface*
tools	*Tools for Package Development*
utils	*The R Utils Package*

Figure 2.3 Package index.

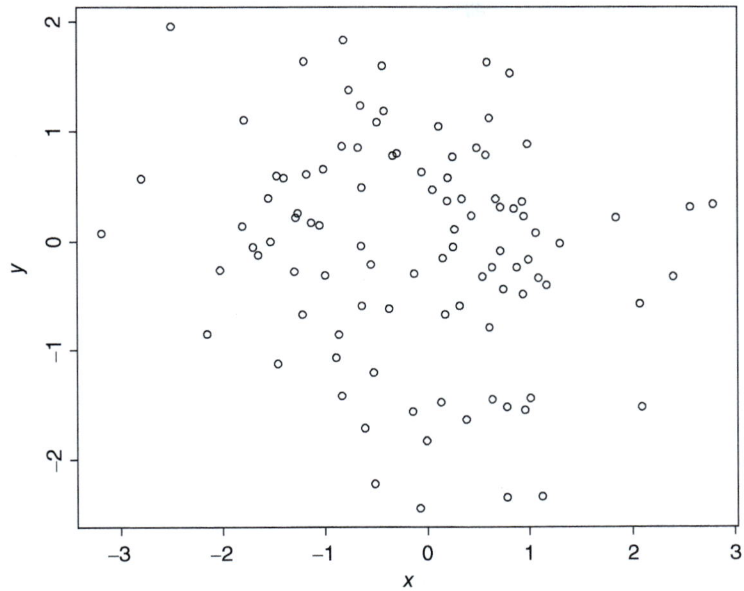

Figure 2.4 Graphical output for plot (x, y).

One may now access each of these R program packages, and use them for further applications as needed.

Returning to the R environment:

```
>
> x <- rnorm(100)
> # Generating a pseudo-random 100-vector x
> y <- rnorm(x)
> # Generating another pseudo-random 100-vector y
> plot (x, y)
> # Plotting x vs. y in the plane, resulting in a graphic
> # window: Figure 2.4.
```

Remark: For reference, Appendix 2.A contains the CRAN documentation of the R function plot(), available for graphic outputting, which may be found by the R code segment:

```
> ?plot
```

CRAN has documentations for many R functions and packages.

Again, returning to the R workspace, and enter:

```
>
>
> ls() # (This is a lower-case "L" followed by "s", viz., the 'list'
>      # command.)
```

```
>        # (NOT 1 = "ONE" followed by "s")
>        # This command will list all the R objects now in the
>        # R workspace:
>        # Outputting:
[1] "E" "n" "s" "x" "y" "z"
```

Again, returning to the R workspace, and enter:

```
>
> rm (x, y)  # Removing all x and all y from the R workspace
> x            # Calling for x
Error: object 'x' not found
>            # Of course, the xs have just been removed!
> y          # Calling for y
Error: object 'y' not found # Because the ys have also been
                            # removed!
>
> x <- 1:10  # Let x = (1, 2, 3, 4, 5, 6, 7, 8, 9, 10)
> x            # Outputting x (just checking!)
 [1]  1  2  3  4  5  6  7  8  9 10
> w <- 1 + sqrt(x)/2    # w is a weighting vector of
>                       # standard deviations
> dummy <- data.frame (x = x, y = x + rnorm(x)*w)
> # Making a data frame of 2 columns, x, and y, for inspection
> dummy     # Outputting the data frame dummy

    x         y
1   1  1.311612
2   2  4.392003
3   3  3.669256
4   4  3.345255
5   5  7.371759
6   6 -0.190287
7   7 10.835873
8   8  4.936543
9   9  7.901261
10 10 10.712029
>
> fm <- lm(y~x, data=dummy)
> # Doing a simple Linear Regression
> summary(fm)  # Fitting a simple linear regression of y on x,
>              # then inspect the analysis, and outputting:
Call:
lm(formula = y ~ x, data = dummy)

Residuals:
    Min       1Q   Median      3Q     Max
-6.0140  -0.8133  -0.0385  1.7291  4.2218
```

```
Coefficients:
             Estimate  Std. Error  t value  Pr(>|t|)
(Intercept)   1.0814    2.0604      0.525    0.6139
x             0.7904    0.3321      2.380    0.0445 *
---
Signif. codes:  0 '***' 0.001 '**' 0.01 '*' 0.05 '.' 0.1 ' ' 1

Residual standard error: 3.016 on 8 degrees of freedom
Multiple R-squared: 0.4146,  Adjusted R-squared: 0.3414
F-statistic: 5.665 on 1 and 8 DF,  p-value: 0.04453

> fm1 <- lm(y~x, data=dummy, weight=1/w^2)
> summary(fm1)  # Knowing the standard deviation, then doing a
>               # weighted regression and outputting:
Call:
lm(formula = y ~ x, data = dummy, weights = 1/w^2)

Residuals:
 Min  1Q Median 3Q  Max
-2.69867 -0.46190 -0.00072 0.90031 1.83202

Coefficients:
           Estimate Std. Error t value Pr(>|t|)
(Intercept)  1.2130   1.6294   0.744    0.4779
x            0.7668   0.3043   2.520   0.0358 *
---
Signif. codes: 0 '***' 0.001 '**' 0.01 '*' 0.05 '.' 0.1 ' ' 1

Residual standard error: 1.356 on 8 degrees of freedom
Multiple R-squared: 0.4424, Adjusted R-squared: 0.3728
F-statistic: 6.348 on 1 and 8 DF, p-value: 0.03583

> attach(dummy)  # Making the columns in the data
>                # frame as variables

The following object(s) are masked _by_ '.GlobalEnv': x

> lrf <- lowess(x, y)  # a non-parametric local
>                      # regression function lrf

> plot (x, y) # Making a standard point plot, outputting: Figure 2.5.
```

Remark: For reference, Appendix 2.A contains the CRAN documentation of the R function plot(), available for graphic outputting, which may be found by the R code segment:

```
> ?plot
> # CRAN has documentations for many R functions and packages.
```

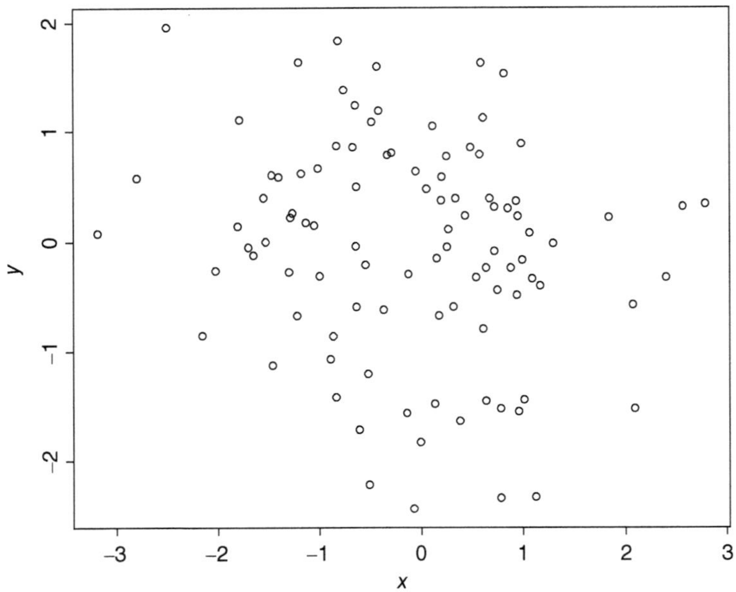

Figure 2.5 Graphical output for plot (x, y).

Again, returning to the R workspace, and enter:

```
>
> ls() # (This is a lower-case "L" followed by "s", viz., the 'list'
>      # command.)
>      # (NOT 1 = "ONE" followed by "s")
>      # This command will list all the R objects now in the
>      # R workspace:
>      # Outputting:

[1] "E" "n" "s" "x" "y" "z"
```

Again, returning to the R workspace, and enter:

```
>
> rm (x, y) # Removing all x and all y from the R workspace
> x          # Calling for x
Error: object 'x' not found
> # Of course, the xs have just been removed!
>
> y          # Calling for y
Error: object 'y' not found
> # Because the ys have been removed too!
```

```
>
> x <- 1:10  # Let x = (1, 2, 3, 4, 5, 6, 7, 8, 9, 10)
> x          # Outputting x (just checking!)

[1] 1 2 3 4 5 6 7 8 9 10
> w <- 1 + sqrt(x)/2  # w is a weighting vector of
>                     # standard deviations
> dummy <- data.frame (x = x, y = x + rnorm(x)*w)
> # Making a data frame of 2 columns, x, and y, for inspection

> dummy     # Outputting the data frame dummy

    x    y
1   1    1.311612
2   2    4.392003
3   3    3.669256
4   4    3.345255
5   5    7.371759
6   6    -0.190287
7   7    10.835873
8   8    4.936543
9   9    7.901261
10 10 10.712029

> fm <- lm(y~x, data=dummy)
>  # Doing a simple Linear Regression
> summary(fm)
> # Fitting a simple linear regression of y on x,
> # then inspect the analysis, and outputting:

Call:
lm(formula = y ~ x, data = dummy)

Residuals:
    Min      1Q  Median      3Q     Max
-6.0140 -0.8133 -0.0385  1.7291  4.2218

Coefficients:
            Estimate Std. Error t value Pr(>|t|)
(Intercept)  1.0814     2.0604   0.525   0.6139
x            0.7904     0.3321   2.380   0.0445 *

---
Signif. codes:  0 '***' 0.001 '**' 0.01 '*' 0.05 '.' 0.1 ' ' 1
```

```
Residual standard error: 3.016 on 8 degrees of freedom
Multiple R-squared: 0.4146,     Adjusted R-squared: 0.3414
F-statistic: 5.665 on 1 and 8 DF,  p-value: 0.04453

> fm1 <- lm(y~x, data=dummy, weight=1/w^2)
> summary(fm1) # Knowing the standard deviation,
>                # then doing a weighted
>                # regression and outputting:
Call:
lm(formula = y ~ x, data = dummy, weights = 1/w^2)

Residuals:
     Min       1Q   Median       3Q      Max
-2.69867 -0.46190 -0.00072  0.90031  1.83202

Coefficients:
            Estimate Std. Error t value Pr(>|t|)
(Intercept)   1.2130     1.6294   0.744   0.4779
x             0.7668     0.3043   2.520   0.0358 *
---
Signif. codes:  0 '***' 0.001 '**' 0.01 '*' 0.05 '.' 0.1 ' ' 1

Residual standard error: 1.356 on 8 degrees of freedom
Multiple R-squared: 0.4424,     Adjusted R-squared: 0.3728
F-statistic: 6.348 on 1 and 8 DF,  p-value: 0.03583
> attach(dummy)  # Making the columns in the data frame as
>                # variables
> lrf <- lowess(x, y)
> lrf
> plot (x, y) # Making a standard point plot,
>             # outputting:

> abline(0, 1, lty=3) # adding in the true regression line:
>                     # (Intercept = 0, Slope = 1),
>                     # outputting: Figure 2.6.

> abline(coef(fm)) # adding in the unweighted regression line:
>                  # outputting Figure 2.7

> abline(coef(fm1), col="red")
> # adding in the weighted regression line:
> # outputting Figure 2.8.

> detach() # Removing data frame from the search
>          # path
> plot(fitted(fm), resid(fm)),
> # Doing a standard diagnostic plot
```

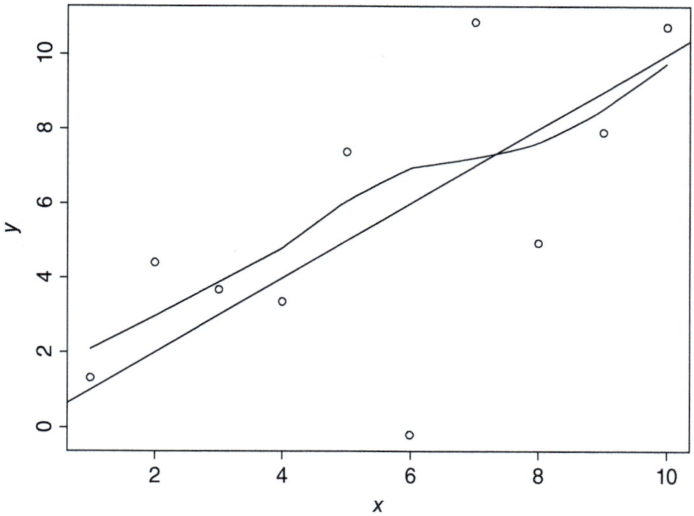

Figure 2.6 Adding in the true regression line (Intercept = 0, Slope = 1).

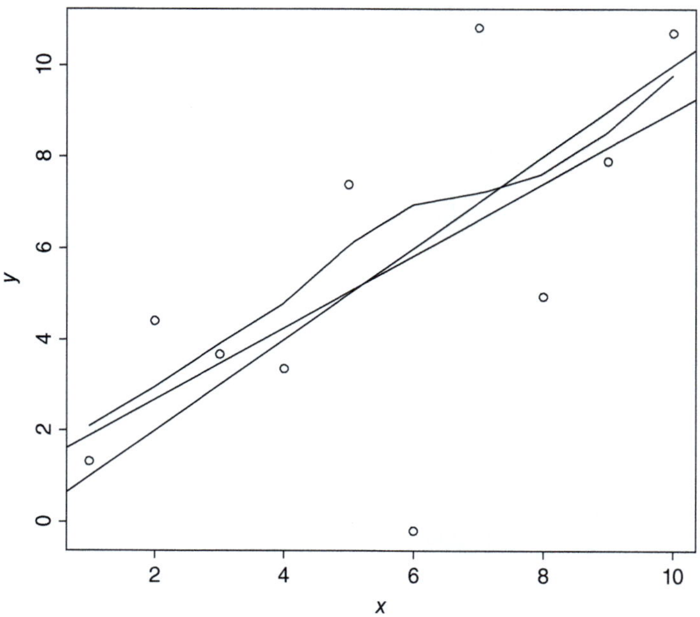

Figure 2.7 Adding in the unweighted regression line.

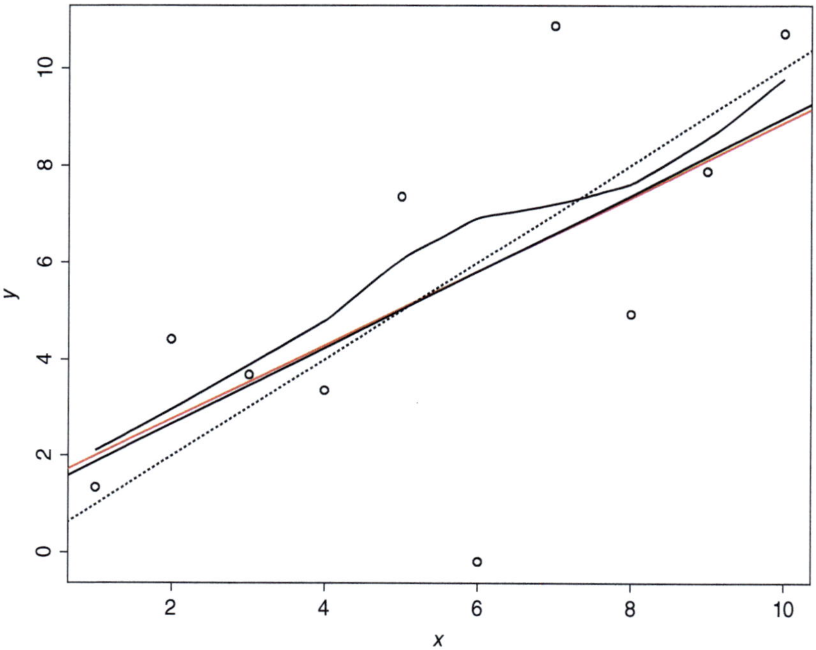

Figure 2.8 Adding in the weighted regression line.

```
+ xlab="Fitted values",  # to check for
+                        # heteroscedasticity**,

+ ylab="residuals",      # viz., checking for differing
+                        # variance.
+ main="Residuals vs Fitted")
# Outputting Figure 2.9.
```

**Heteroskedasticity occurs when the variance of the error terms differ across observations.

```
> qqnorm(resid(fm), main="Residuals Rankit Plot")
> # Doing a normal scores plot to check for
> # skewness, kurtosis, and outliers.
> # (Not very useful here.) Outputting Figure 2.10.

>
> rm(fm, fm1, lrf, x, dummy)
> # Removing these 5 objects
> fm

Error: object 'fm' not found  # Checked!
```

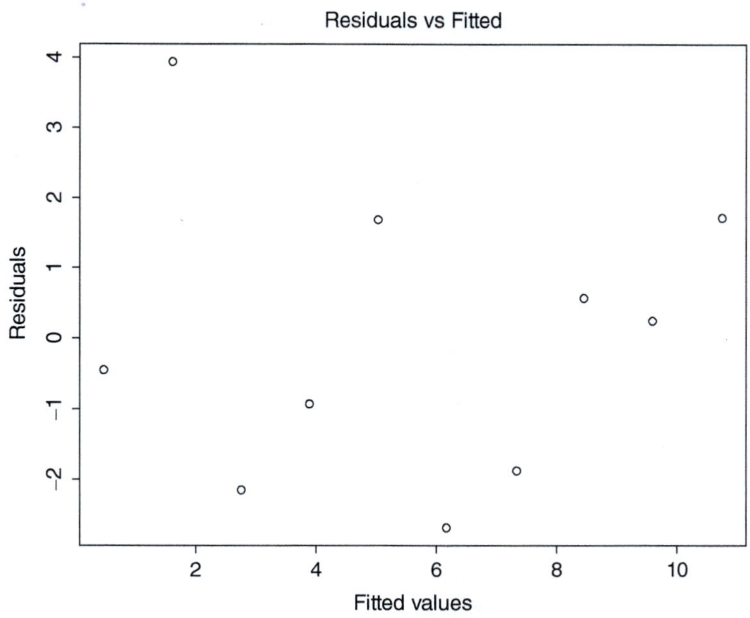

Figure 2.9 A standard diagnostic plot to check for heteroskedasticity.

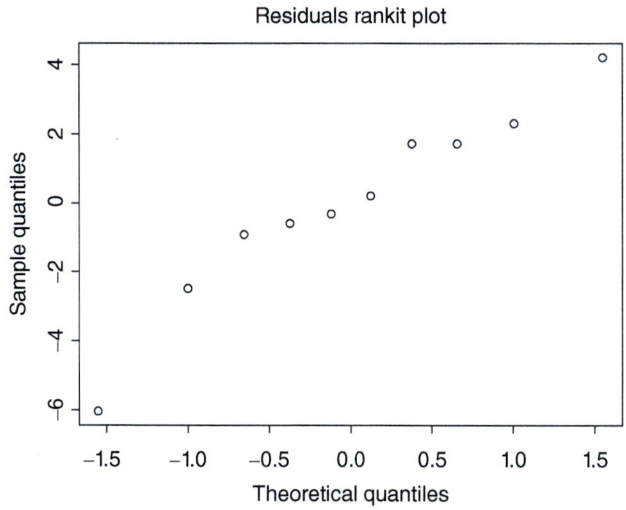

Figure 2.10 A normal scores plot to check for Skewness, Kurtosis, and Outliers.

```
> fm1
Error: object 'fm1' not found # Checked!
> lrf
Error: object 'lrf' not found # Checked!
> x
```

```
Error: object 'x' not found   # Checked!
> dummy
Error: object 'dummy' not found # Checked!
# END OF THIS PRACTICE SESSION!
```

2.2.3 The R Environment (This is Important!)

Getting through the First Session in Section 2.2.1, shows:

Technically, R is an expression language with a simple syntax which is almost self-explanatory. It is case sensitive: so x and X are different symbols and refer to different variables. All alphanumeric symbols are allowed, plus "." and "-", with the restriction that a name must start with "." or a letter, and if it starts with "." the second character must not be a digit. The command prompt > indicates when R is ready for input.

This is where one types commands to be processed by R, which will happen when one hit the ENTER key.

Commands consist of either expressions or assignments. When an expression is given as a command, it is immediately evaluated, printed, and the value is discarded. An assignment evaluates an expression and passes the value to a variable – but the value is not automatically printed. To printed the computed value, simple enter the variable again at the next command.

Commands are separated either by a new line, or separated by a semi-colon (";"). Several elementary commands may be grouped together into one compound expression by braces: ("{" and "}").

Comments, starting with a hashmark/number-sign ("#"), may be put almost anywhere: everything to the end of the line following this sign is a comment.

Comments may not be used in an argument list of a function definition or inside strings. If a command is not complete at the end of a line, R will give a different prompt, a "+" sign, by default:

On the second and subsequent lines, and continue to read input until the command is completed syntactically.

The result of a command is printed to the output device: if the result is an array, such as a vector or a matrix, then the elements are formatted with line break (wherever necessary) with the indices of the leading entries labeled in square brackets: [index].

For example, an array of 15 elements may be outputted as:

```
> array(8, 15)
      [1]  8 8 8 8 8 8 8 8 8 8
      [11] 8 8 8 8 8
```

The labels "[1]" and "[11]" indicate the 1st and 11th elements in the output. These labels are not part of the data itself!

Similarly, the labels for a matrix are placed at the start of each row and column in the output.

For example, for the 3 x 5 matrix M, it is outputted as:

```
>
> M <- matrix(1:15, nrow=3)
> M
        [,1] [,2] [,3] [,4] [,5]
   [1,]    1    4    7   10   13
   [2,]    2    5    8   11   14
   [3,]    3    6    9   12   15
>
```

Note that the storage is a column-major, viz. the elements of the first column are printed out first, followed by those of the second column, etc. To cause a matrix to be filled in a row-wise manner, rather than the default column-wise fashion, the additional switch byrow=T will cause the matrix to be filled row-wise rather than by column-wise:

```
>
> M <- matrix(1:15, nrow=3, byrow=T)
> M

     [,1] [,2] [,3] [,4] [,5]
[1,]    1    2    3    4    5
[2,]    6    7    8    9   10
[3,]   11   12   13   14   15
>
```

The First Session also shows that there is a host of helpful resources imbedded in the R environment that one can readily access, using the on-line help provided by CRAN.

Review Questions for Section 2.2

1. Let us get started!

   Please follow the step-by-step instructions given in the opening paragraphs of Section 2.2 to set up an R environment. The R window should look like this:

   ```
   >
   ```

   Great!

   Now enter the following arithmetic operations: press "*Enter*" after each entry:

   (a) 2 + 3 *<Enter>*
   (b) 13 – 7 *<Enter>*
   (c) 17 * 23 *<Enter>*
   (d) 100/25 <Enter>

(e) Did you obtain the following results:
 5, 6, 391, 4?

2. Here is a few more: The <*Enter*> prompt will be omitted from now on!
   (a) 2∧4
   (b) sqrt(3)
   (c) 1i [1i is used for the complex unit i, where $i^2 = -1$.]
   (d) (2 + 3i) + (4 + 5i)
   (e) (2 + 3i) * (4 + 5i)
3. Here is a short session on using R to do complex arithmetic: just enter the following commands into the R environment, and report the results:

```
> th <- seq(-pi, pi, len=20)
> th
```

   (a) How many numbers are printed out?

```
> z <- exp(1i*th)
> z
```

   (b) How many complex numbers are printed out?

```
> par(pty="s")
```

   (c) Along the menu-bar at the top of the R environment:

   * Select and left-click on "Window", then
   * Move downwards and select the 2nd option:
   R Graphic Device 2 (ACTIVE)
   * Go to the "R Graphic Device 2 (ACTIVE) Window"

   (d) What is there?

```
> plot(z)
```

   (e) Describe what is in the Graphic Device 2 Window.

2.3 R as a Calculator

2.3.1 Mathematical Operations Using R

To learn to do statistical analysis and computations, one may start by considering the R programming language as a simple calculator! [14, 16].

Start from here:

Just enter an arithmetic expression, press the **<Enter>** key, and the answer from the machine us found in the next line –

```
>
> 2 + 3
[1] 5
>
```

OK! What about other calculations? Such as:

13 - 7, 3 x 5, 12 /4, 72, $\sqrt{2}$, e3, $e^{i\pi}$,
ln 5 = $\log_e 5$, $(4 + \sqrt{3})(4 - \sqrt{3})$,
$(4 + i\sqrt{3})(4 - i\sqrt{3})$, ... and so on.

Just try:

```
>
> 13 - 7
[1] 6
>
> 3*5
[1] 15
>
> 12/4
[1] 3
>
> 7^2
[1] 49
>
> sqrt(2)
[1] 1.414214
>
> exp(3)
[1] 20.08554
>
> exp(1i*pi)
```

1i is used for the complex number i = $\sqrt{-1}$.

```
[1] -1-0i
```

[This is just the famous Euler's identity equation: $e^{i\pi} + 1 = 0$.]

```
> log(5)
[1] 1.609438
> (4 + sqrt(3))*(4 - sqrt(3))
[1] 13
```

[Checking: (4+$\sqrt{3}$)(4-$\sqrt{3}$) = 42 - ($\sqrt{3}$)2 =
16 - 3 = 13 (Checked!)

```
> (4 + 1i*sqrt(3))*(4 - 1i*sqrt(3))
[1] 19+0i
```

[Checking: (4+i$\sqrt{3}$)(4-i$\sqrt{3}$) = 42 -(i$\sqrt{3}$)2 = 16 - (-3) = 19 (Checked!)]

Remark: The [1] in front of the computed result is R's way of outputting numbers. It becomes useful when the result is a long vector. The number N in the brackets [N] is the index of the first number on that line. For example, if one generated 23 random numbers from a normal distribution:

```
>
> x <- rnorm(23)
> x
 [1] -0.5561324 0.2478934 -0.8243522 1.0697415 1.5681899
 [6] -0.3396776 -0.7356282 0.7781117 1.2822569 -0.5413498
[11] 0.3348587 -0.6711245 -0.7789205 -1.1138432 -1.9582234
[16] -0.3193033 -0.1942829 0.4973501 -1.5363843 -0.3729301
[21] 0.5741554 -0.4651683 -0.2317168
>
```

Remark: After the random numbers have been generated, there is no output until one calls for x, viz. x has become a vector with 23 elements, call that a 23-vector!

The [11] on the third line of the output indicates that 0.3348587 (colored in red here for emphasis!) is the 11th element in the 23-vector x. The numbers of outputs per line depend on the length of each element as well as the width of the page.

2.3.2 Assignment of Values in R, and Computations Using Vectors and Matrices

R is designed to be a Dynamically typed Language, viz. at any time one may change the data type of any variable. For example, one can first set x to be numeric as has been done so far, say: x = 7; next one may set x to be a vector, say: x = c (1, 2, 3, 4); then again one may set x to a word object, such as "Hi!". Just watch the following R environment:

```
>
> x <- 7
> x
[1] 7
> x <- c(1, 2, 3, 4)    # x is assigned to be a 4-vector.
> x
[1] 1 2 3 4
> x <- c("Hi!")  # x is assigned to be a character string.
> x
[1] "Hi!"
> x <- c("Greetings & Salutations!")
> x
[1] "Greetings & Salutations!"
> x <- c("The rain in Spain falls mainly on the
+          plain.")
> x
 [1] "The rain in Spain falls mainly on the plain."
> x <- c("Biostatistics", "Human", "Genetic",
+         "Epidemiology")
> x
[1] "Biostatistics", "Human", "Genetic",
```

```
+ "Epidemiology"
>
```

2.3.3 Computations in Vectors and Simple Graphics

The use of arrays and matrices was introduced in Section 2.2.2.

In finite mathematics, a matrix is a two-dimensional array of elements, which are usually numbers. In R, the use of the matrix extends to elements of any type, such as a matrix of character strings. Arrays and matrices may be represented as vectors with dimensions.

In statistics in which most variables carry multiple values, therefore computations are usually performed between vectors of many elements.

These operations among multivariates result in large matrices. To demonstrate the results, often graphical representations are useful.

The following simple example illustrates these operations being readily accomplished in the R environment

```
>
> weight <- c(73, 59, 97)
> height <- c(1.79, 1.64, 1.73)

> bmi <- weight/height^2
> bmi  # Read the BMI Notes below
[1] 22.78331  21.93635  32.41004
> # To summarize the results proceed to compute
> # as follows:
> cbind(weight, height, bmi) # Outputting:
     weight  height      bmi
[1,]    73     1.79   22.78331
[2,]    59     1.64   21.93635
[3,]    97     1.73   32.41004
>
> rbind(weight, height, bmi) # Outputting:
              [,1]        [,2]        [,3]
weight 73.00000    59.00000    97.00000
height  1.79000     1.64000     1.73000
bmi    22.78331    21.93635    32.41004
>
```

Clearly, the functions cbind and rbind bind (viz. join, link, glue, concatenate) by column and by row, respectively, the vectors to form new vectors or matrices.

2.3.4 Use of Factors in R Programming

In the analysis of, for example, health science datasets, categorical variables are often needed. These categorical variables indicate subdivisions of the origin dataset into various classes, e.g. age, gender, disease stages, and degrees of diagnosis. Input of the original dataset is generally delineated into several categories using a numeric code: 1 = age, 2 = gender, 3 = disease stage, etc. Such variables are specified as factors in R, resulting in a data structure that enables one to assign specific names to the various categories. In certain analyses, it is necessary for R to distinguish among categorical codes and variables whose values have direct numerical meanings.

A factor has 4 levels, consisting of 2 items:

(1) a vector of integers between 1 and 4, and
(2) a character vector of length four containing strings which describe the 4 levels.

Consider the following example:

** A certain type of cancer is being categorized into 4 levels: Levels 1, 2, 3, and 4.
** The corresponding pain levels consistent with these diagnoses are: none, mild, moderate, and severe, respectively.
** In the dataset, 5 case-subjects have been diagnosed in terms of their respective levels.

The following R code segment delineates the dataset:

```
> cancerpain <- c(1, 4, 3, 3, 2, 4)
> fcancerpain <- factor(cancerpain, level=1:4)

> levels(fcancerpain) <- c("none", "mild",
+                          "moderate", "severe")
```

The first statement creates a numerical vector cancerpain which encodes the pain levels of 6 case-subjects. This is being considered as a categorical variable for which, using the factor function, a factor fcancerpain is created. This may be called with one argument in addition to cancerpain, viz. levels = 1 to 4, which indicates that the input coding uses the values 1 – 4. In the final line, the pain level names are changed to the 4 specified character strings. The result is:

```
> fcancerpain
[1] none  severe  moderate  moderate mild  severe
Levels: none mild moderate severe
> as.numeric(fcancerpain)
[1] 1 4 3 3 2 4
> levels(fcancerpain)
[1] "none"    "mild"    "moderate"    "severe"
```

Remarks: The function as.numeric() outputs the numerical coding as numbers 1 to 4, and the function levels() outputs the names of the respective levels.

The original input coding in terms of the numbers 1 to 4 is no longer needed. There is an additional option using the function ordered which is similar to the function factor used here.

2.3.4.1 Body Mass Index

The body mass index (BMI) is a useful measure for human body fat based on an individual's weight and height – it does not actually measure the percentage of fat in the body [18]. Invented in the early 19th century, BMI is defined as a person's body weight (in kilograms) divided by the square of the height (in meters). The formula universally used in health science produces a unit of measure of kg/m^2:

$$\text{BMI} = \text{Body Mass (kg)}/\{\text{Height (m)}\}^2$$

A BMI chart may be used which, displaying BMI as a function of weight (horizontal axis) and height (vertical axis) with contour lines for different values of BMI or colors for different BMI categories: Figure 2.11.

2.3.5 Simple Graphics

Generating graphical presentations is an important aspect of statistical data analysis. Within the R environment, one may construct plots that allows production of plots and control of the graphical features. Thus, with the previous example, the

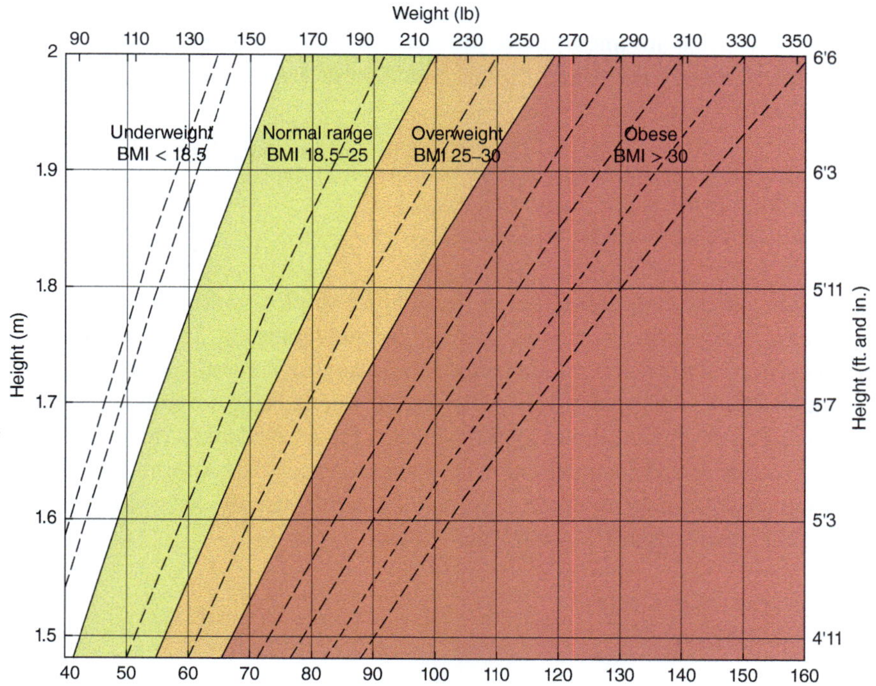

Figure 2.11 A graph of BMI (body mass index): the dashed lines represent subdivisions within a major class – the "Underweight" classification is further divided into "severe," "moderate," and "mild" subclasses. Source: BMI Notes 2012/Wikimedia Commons/Public Domain.

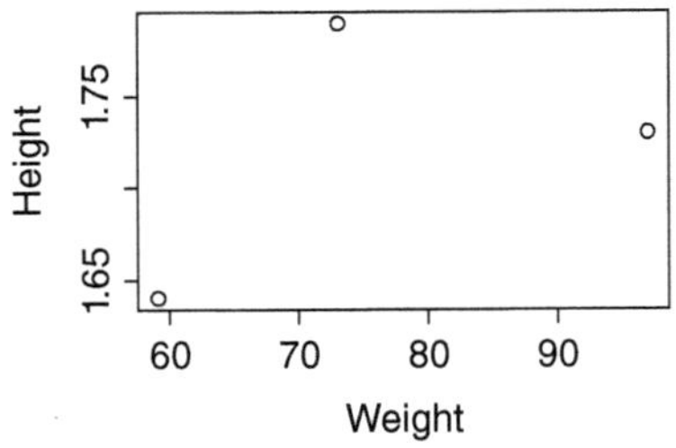

Figure 2.12 An *X*–*Y* plot for

> plot (weight, height)

relationship between Body Weight and Height may be considered by first plotting one versus the other by using the following R code segments:

```
>
> plot (weight, height)
> # Outputting: Figure 2.12.
> plot (weight, height)
```

Remarks:

(1) Note the order of the parameters in the plot (x, y) command: the first parameter is x (the independent variable – on the horizontal axis), and the second parameter is y (the dependent variable – on the vertical axis).
(2) Within the R environment, there are many plotting parameters that may be selected to modify the output. To get a full list of available options, return to the R environment and call for:

```
> ?plot # This is a call for "Help!" within the R environment.
> # The output is the R documentation for:
> plot {graphics} # Generic X-Y plotting
```

This is the official documentation of the R function plot, within the R package graphics – note the special notations used for plot and {graphics}. To fully make use of the provisions of the R environment, one should carefully investigate all such documentations. (R has many available packages, each containing a number of useful functions.) This document shows all the plotting options available with the R environment. A copy of this documentation is shown in Appendix 2.A for reference.

For example, to change the plotting symbol, one may use the keyword pch (for "plotting character") in the following R command:

```
> plot (weight, height, pch=8)
> # Outputting: Figure 2.13.
```

Note that the output is the same as that shown in Figure 2.12, except that the points are marked with little "8-point stars", corresponding to Plotting Character pch = 8.

In the documentation for pch, a total of 26 options are available, providing different plotting characteristics for points in R graphics. They are shown in Figure 2.14.

Figure 2.13 An X-Y plot for

> plot (weight, height, pch=8)

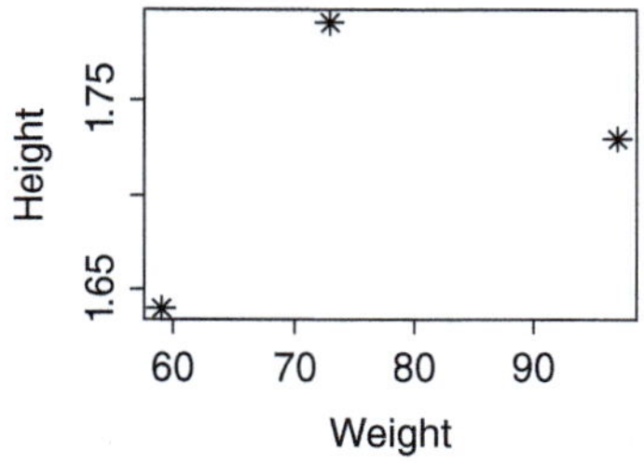

The parameter BMI was chosen in order that this value should be independent of a person's height, thus expressing as a single number or index indicative of whether a case-subject is overweight, and by what relative amount.

Of course, one may plot "height" as the abscissa (viz. the horizontal "x-axis"), and "weight" as the ordinate (viz., the vertical "y-axis"), as follows:

```
> plot(height, weight, pch=8)
```
Outputting: Figure 2.15.

Since a normal BMI is between 18.5 and 25, averaging (18.5 + 25)/2 = 21.75. For this BMI value, then the weight of a typical "normal" person would be (21.75 x height^2). Thus, one can superimpose a line-of-"expected"-weights at BMI = 21.75 on Figure 2.15. This line may be accomplished in the R environment by the following code segments:

```
> ht <- c(1.79, 1.64, 1.73)
> lines(ht, 21.75*ht^2)
```
Outputting: Figure 2.16.

In the last plot, a new variable for heights (ht) was defined instead of the original (height) because:

(i) The relation between height and weight is a quadratic one, and hence non-linear. Although it may not be obvious on the plot, it is preferable to use

Figure 2.14 Plotting symbols in R: pch = n, n = 0, 1, 2, …, 25.

Figure 2.15 An X-Y plot for

> plot (height, weight, pch=8)

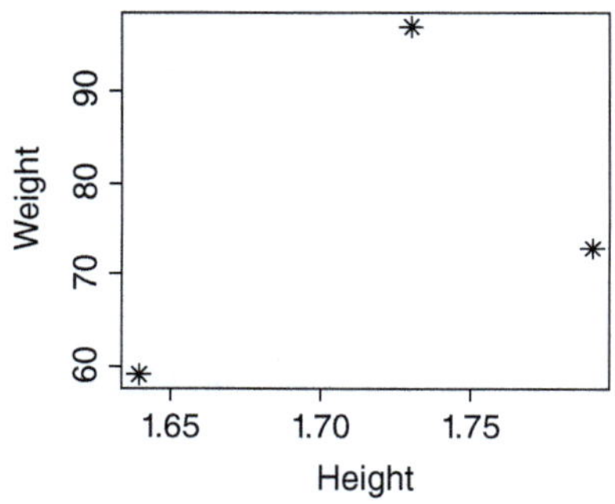

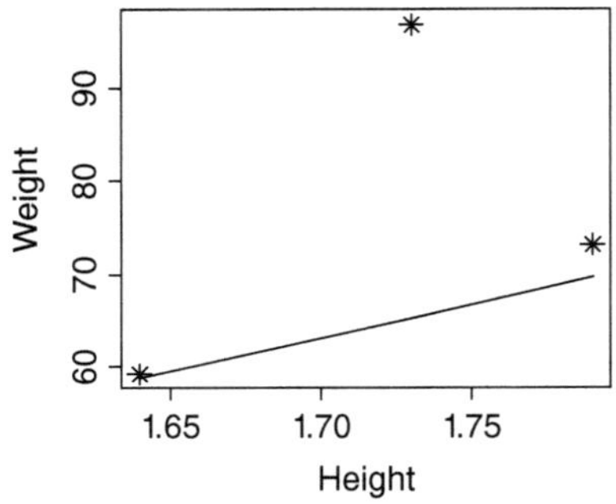

Figure 2.16 Superimposed reference curve using line:

```
(ht, 21.75*ht^2)
```

points that are spread evenly along the x-axis than to rely on the distribution of the original data.

(ii) As the values of height are not sorted, the line segments would not connect neighboring points but would run back and forth between distant points.

Remarks:

1. In the final example above, R was actually doing the arithmetic of vectors.
2. Notice that the two vectors weight and height are 3-vectors, making it reasonable to perform the next step.
3. The cbind statement, used immediately after the computations have been completed, forms a new matrix by binding together matrices horizontally, or column-wise. It results in a multivariate response variable. Similarly, the rbind statement does a similar operation vertically, or row-wise.
4. But, if for some reason (such as mistake in one of the entries) the two entries weight and height have different number of elements, then R will output an error message. For example:

```
>
> weight <- c(73, 59, 97) # a 3-vector
> height <- c(1.79, 1.64, 1.73, 1.48) # a 4-vector !
> bmi <- weight/height^2 # Outputting:
```

Warning message: # An Error message!

```
In weight/height^2 :
```

longer object length is not a multiple of shorter object length

>

2.3.6 x as Vectors and Matrices in Statistics

It has just been shown that a variable, such as x or M, may be assigned as

(1) a number, such as x = 7
(2) a vector or an array, such as x = c(1, 2, 3, 4)
(3) a matrix, such as

```
x =

        [,1] [,2] [,3] [,4] [,5]
[1,]     1    4    7   10   13
[2,]     2    5    8   11   14
[3,]     3    6    9   12   15
```

(4) a character string, such as

```
x = "The rain in Spain falls mainly on the plain."
```

(5) In fact, in R, a variable x may be assigned a complete dataset which may consist of a multiple-dimensional set of elements, each of which may in turn be anyone of the above kinds of variables.

For example, besides being a numerical vector, such as in (2) above, x may be:

(a) a character vector, which is a vector of text strings whose elements are expressed in quotes, using double-, single-, or mixed-quotes:

```
> c("one", "two", "three", "four", "five")
> # Double-quotes
  [1] "one"   "two"   "three" "four"  "five"
>
> c('one', 'two', 'three', 'four', 'five')
> # Single-quotes
[1] 'one'  'two'  'three'  'four'  'five'
>
> c("one", 'two', "three", 'four', "five")
> # Mixed-quotes
[1] "one"  'two'  "three" 'four' "five"
```

However, if there is a mixed pair of quotes such as "xxxxx', it will not be accepted! For example:

```
> c("one", "two", "three", "four", "five')
```

(a) a logical vector, which takes the value TRUE or FALSE (or NA). For inputs, one may use the abbreviations T or F. These vectors are similarly specified using the c function:

```
> c(T, F, T, F, T)
[1] TRUE  FALSE  TRUE  FALSE  TRUE
```

In most cases, there is no need to specified logical vectors repeated. It is acceptable to use a single logical value to provide the needed options as vectors of more than one value will respond in terms of relational expressions. Observe:

```
> weight <- c(73, 59, 97)
> height <- c(1.79, 1.64, 1.73)
> bmi <- weight/height^2
> bmi # Outputting:
[1] 22.78331 21.93635 32.41004
> bmi > 25 # A single logical value will suffice!
[1] FALSE FALSE TRUE
>
```

2.3.7 Some Special Functions that Create Vectors

Three functions that create vectors are: c, seq, and rep

(1) c, for "concatenate", or, the joining of objects end-to-end (this was introduced earlier) – for example:

```
> x <- c(1, 2, 3, 4) #x is assigned to be a 4-vector.
> x
[1] 1 2 3 4
```

(2) seq, for "sequence", for defining an equidistant sequence of numbers – for example:

```
> seq(1, 20, 2)
# To output a sequence from 1 to 20, in steps of 2
[1]  1  3  5  7  9 11 13 15 17 19
> seq(1, 20)
# To output a sequence from 1 to 20, in steps of 1,
# (which may be omitted)
[1]  1  2  3  4  5  6  7  8  9 10 11 12 13 14 15 16 17
[2] 18 19 20
> 1:20
# This is a simplified alternative to writing
# seq(1, 20)
[1]  1  2  3  4  5  6  7  8  9 10 11 12 13 14 15 16 17
[2] 18 19 20
> seq(1, 20, 2.5)
# To output a sequence from 1 to 20, in steps of
# 2.5.
[1]  1.0  3.5  6.0  8.5 11.0 13.5 16.0 18.5
```

(3) rep, for "replicate", for generating repeated values. This function takes two forms, depending on whether the second argument is a single number or a vector – for example:

```
> rep(1:2, c(3,5))
# Replicating the first element (1) 3 times, and
# then replicating the second element (2) 5 times
[1] 1 1 1 2 2 2 2 2          # This is the output.
> vector <- c(1, 2, 3, 4)

> vector   # Outputting the vector:
[1] 1 2 3 4
> rep(vector, 5)   # Replicating vector 5 times:
 [1] 1 2 3 4 1 2 3 4 1 2 3 4 1 2 3 4 1 2 3 4
```

2.3.8 Arrays and Matrices

In finite mathematics, a matrix M is a two-dimensional array of elements, generally numbers, such as

```
M  = 1   4   7   10   13
     2   5   8   11   14
     3   6   9   12   15
```

and the array is usually placed inside parentheses (), or some brackets {}, [], etc. In R, the use of a matrix is extended to elements of many types: numbers as well as character strings. For example, in R, the above matrix M is expressed as:

```
      [,1] [,2] [,3] [,4] [,5]
[1,]   1    4    7   10   13
[2,]   2    5    8   11   14
[3,]   3    6    9   12   15
```

2.3.9 Use of the Dimension Function dim() in R

In R, the above 3×5 matrix may be set up as vectors with dimension `dim(x)` using the following code segment:

```
> x <- 1:j
> x
 [1]  1  2  3  4  5  6  7  8  9 10 11 12 13 14 15
> dim(x) <- c(3, 5)   # a dimension of 3 rows by 5 columns
```

Remark: Here a total of 15 elements, 1 through 15, are set to be the elements of the matrix x. Then the dimension of x is set as `c(3, 5)`, making x to become a `3 X 5` matrix. The assignment of the 15 elements follows a column-wise procedure, viz. the elements of the first column are allocated a first, followed by those of the second column, then the third column, etc.

2.3.10 Use of the Matrix Function matrix() in R

Another way to generate a matrix is using the function matrix(). The above 3 X 5 matrix may be created by the following 1-line code segment:

```
> matrix (1:15, nrow=3)
> matrix # Outputting:

       [,1] [,2] [,3] [,4] [,5]
[1,]     1    4    7   10   13
[2,]     2    5    8   11   14
[3,]     3    6    9   12   15
```

However, if the 15 elements should be allocated by row, then the following code segment should be used:

```
> matrix (1:15, nrow=3, byrow=T)
> matrix # Outputting:

     [,1] [,2] [,3]  [,4] [,5]
[1,]   1    2    3    4    5
[2,]   6    7    8    9   10
[3,]  11   12   13   14   15
```

2.3.11 Some Useful Functions Operating on Matrices in R: colnames, rownames, and t (for transpose)

Using the previous example:

(i) the 5 columns of the 3 x 5 matrix x are first assigned thee names C1, C2, C3, C4, and C5, respectively, then
(ii) the transpose is obtained, and finally
(iii) one take the transpose of the transpose to obtain the original matrix x:

```
> matrix (1:15, nrow=3, byrow=T)
> matrix # Outputting:
      [,1]  [,2]  [,3]  [,4]  [,5]
[2,]    6     7     8     9    10
[3,]   11    12    13   1 4    15

> colnames(x) <- c("C1", "C2", "C3", "C4", "C5")
> x # Outputting:
      C1  C2  C3  C4  C5
[1,]   1   4   7  10 .gf
[2,]   2   5   8  11  14
[3,]   3   6   9  12  15
> t(x)
```

```
     [,1] [,2] [,3]
C1    1    2    3
C2    4    5    6
C3    7    8    9
C4   10   11   12
C5   13   14   15
> t(t(x)) # which is just x, as expected!

     C1 C2 C3 C4 C5
[1,]  1  4  7 10 13
[2,]  2  5  8 11 14
[3,]  3  6  9 12 15
```

Yet another way is to use the function LETTERS, which is a built-in variable containing the capital letters A through Z. Other useful vectors include letters, month.name, and month.abb for lower-case letters, month names, and abbreviated names of months, respectively. Take a look:

```
> X <-LETTERS
> X # Outputting:
 [1] "A" "B" "C" "D" "E" "F" "G" "H" "I" "J" "K" "L" "M" "N" "O"
[16] "P" "Q" "R" "S" "T" "U" "V" "W" "X" "Y" "Z"
> M # Outputting:
 [1] "January" "February" "March"     "April"     "May"
 [6] "June"    "July"    "August"  "September" "October"
[11] "November"  "December"
> m <- month.abb
> m # Outputting:
[1] "Jan" "Feb" "Mar" "Apr" "May" "Jun" "Jul" "Aug"
[9] ""Sep" "Oct"  Nov" "Dec"
```

2.3.12 NA "Not Available" for Missing Values in Datasets

NA is a logical constant of length 1 which contains a missing value indicator.

NA may be forced to any other vector type except raw. There are also constants NA integer, NA real, NA complex, and NA character of the other atomic vector types which support missing values: all of these are reserved words in the R language.

The generic function is .na indicates which elements are missing. The generic function is .na<- sets elements to NA.

The reserved words in R's parser are:

if, else, repeat, while, function, for, in next, break, NA complex, NA character, ... and ..1, ..2, etc., which are used to refer to arguments passed down from an enclosing function.

Reserved words outside quotes are always parsed to be references to the objects linked to in the foregoing list, and are not allowed as syntactic names. They are allowed as non-syntactic names.

2.3.13 Special Functions that Create Vectors

There are three useful R functions which are often used to create vectors:

(1) c for "concatenate", which was introduced in Section 2.3.2 for joining items together end-to-end, for example:

```
> c(2, 3, 5, 7, 11, 13, 17, 19, 23, 29)
> # The first 10 prime numbers
 [1] 2 3 5 7 11 13 17 19 23 29
```

(2) seq for "sequence", is used for listing equidistant sequences of numbers, for example:

```
> seq(1, 20)  # Sequence from 1 to 20
 [1]  1  2  3  4  5  6  7  8  9 10 11 12 13 14 15 16 17
     18 19 20
> seq(1, 20, 1)  # Sequence from 1 to 20, in steps of 1
 [1]  1  2  3  4  5  6  7  8  9 10 11 12 13 14 15 16 17
     18 19 20
> 1:20  # Sequence from 1 to 20, in steps of 1
[1]  1  2  3  4  5  6  7  8  9 10 11 12 13 14 15 16 17
    18 19 20
> seq(1, 20, 2)  # Sequence from 1 to 20, in steps of 2
[1]  1  3  5  7  9 11 13 15 17 19
> seq(1, 20, 3)  # Sequence from 1 to 20, in steps of 3
[1]  1  4  7 10 13 16 19
> seq(1, 20, 10)  # Sequence from 1 to 20, in steps of 10
[1]  1  11
> seq(1, 20, 20)  # Sequence from 1 to 20, in steps of 20
[1]  1
> seq(1, 20, 21)  # Sequence from 1 to 20, in steps of 21
[1] 1
>
```

(3) rep for "replicate", is used to generate repeated values, and may be expressed in 2 ways, for example

```
> x <- c(3, 4, 5)
> rep(x, 4)  # Replicate the vector x 4-times.
 [1] 3 4 5 3 4 5 3 4 5 3 4 5
> rep(x, 1:3)  # Replicate the elements of x:
# the first element once,
```

```
# the second element twice, and
# the third element three times.
[1] 3 4 4 5 5 5
> rep(1:3, c(3,4,5))  # For the sequence (1, 2, 3), replicate its
>                     # elements 3-, 4-, and 5-times, respectively
[1] 1 1 1 2 2 2 2 3 3 3 3 3
```

Review Questions for Section 2.3

1. A Tower of Powers – by computations using R. There is an interesting challenge in arithmetic which goes like this:

$$\sqrt{2}^{\sqrt{2}^{\cdots}}$$

What is the value of $\sqrt{2}^{\sqrt{2}^{\cdots}}$? viz. an infinity of ascending tower of powers of the square root of 2.
Solution: Let x be the value of this "Tower of Powers", then it is easily seen that $\sqrt{2}^{x} = x$ itself ! Agree?
Watch the lowest $\sqrt{2}$.
And clearly it follows that x = 2, because $\sqrt{2}^{2} = 2$.
This shows that the value of this "Infinite Tower of Powers of $\sqrt{2}$" is just 2.
Now use the R environment to verify this interesting result:
(a) Compute $\sqrt{2}$

```
> sqrt(2)
```

(b) Compute $\sqrt{2}^{\sqrt{2}}$

```
> sqrt(2)^sqrt(2)
```
[a 2-Towers of $\sqrt{2}$-s]

(c)
```
> sqrt(2)^sqrt(2)^sqrt(2)
```
[a 3-Towers of $\sqrt{2}$-s]

(d)
```
> sqrt(2)^sqrt(2)^sqrt(2)^sqrt(2)
```
[a 4-Towers of $\sqrt{2}$-s]

(e)
```
> sqrt(2)^sqrt(2)^sqrt(2)^sqrt(2)^sqrt(2)
```
[a 5-Towers of $\sqrt{2}$-s]

(f) Now try the following computations of 10-, 20-, 30-, and finally 40-Towers of Powers of $\sqrt{2}$, and finally reach the result of 2 (accurate to 6 places of decimal!).

```
> sqrt(2)^sqrt(2)^sqrt(2)^sqrt(2)^sqrt(2)^sqrt(2)^sqrt(2)^sqrt(2)^
sqrt(2)^sqrt(2)
[1] 1.983668
```
[a 10-Towers of Powers of $\sqrt{2}$-s]

```
> sqrt(2)^sqrt(2)^sqrt(2)^sqrt(2)^sqrt(2)^sqrt(2)^
+ sqrt(2)^sqrt(2)^sqrt(2)^sqrt(2)^sqrt(2)^sqrt(2)^
```

```
+ sqrt(2)^sqrt(2)^sqrt(2)^sqrt(2)^sqrt(2)^sqrt(2)^
+ sqrt(2)^sqrt(2)
[1] 1.999586
```
[a 20-Towers of Powers of $\sqrt{2}$-s]

```
> sqrt(2)^sqrt(2)^sqrt(2)^sqrt(2)^sqrt(2)^sqrt(2)^
+ sqrt(2)^sqrt(2)^sqrt(2)^sqrt(2)^sqrt(2)^sqrt(2)^
+ sqrt(2)^sqrt(2)^sqrt(2)^sqrt(2)^sqrt(2)^sqrt(2)^
+ sqrt(2)^sqrt(2)^sqrt(2)^sqrt(2)^sqrt(2)^sqrt(2)^
+ sqrt(2)^sqrt(2)^sqrt(2)^sqrt(2)^sqrt(2)^sqrt(2)
[1] 1.999989
```
[a 30-Towers of Powers of $\sqrt{2}$-s]

```
> sqrt(2)^sqrt(2)^sqrt(2)^sqrt(2)^sqrt(2)^sqrt(2)^
+ sqrt(2)^sqrt(2)^sqrt(2)^sqrt(2)^sqrt(2)^sqrt(2)^
+ sqrt(2)^sqrt(2)^sqrt(2)^sqrt(2)^sqrt(2)^sqrt(2)^
+ sqrt(2)^sqrt(2)^sqrt(2)^sqrt(2)^sqrt(2)^sqrt(2)^
+ sqrt(2)^sqrt(2)^sqrt(2)^sqrt(2)^sqrt(2)^sqrt(2)^
+ sqrt(2)^sqrt(2)^sqrt(2)^sqrt(2)^sqrt(2)^sqrt(2)^
+ sqrt(2)^sqrt(2)^sqrt(2)^sqrt(2)
[1] 2
```
(Accurate to 6 decimal places!)

[a 40-Towers of Powers of $\sqrt{2}$-s]

Thus, this R computation verifies the solution.

2. (a) What are the equivalents in R for the basic mathematical operations of:
   +, -, x, / (division), $\sqrt{}$, squaring of a number
   (b) Describe the use of factors in R programming. Give an example.
3. If x = (0, 1, 2, 3, 4, 5), and y = (0, 1, 4, 9, 16, 25), in R, plot:
   (a) y vs x
   (b) x vs y
   (c) $\sqrt{y}$ vs x
   (d) y vs $\sqrt{x}$
   (e) $\sqrt{y}$ vs $\sqrt{x}$
   (f) $\sqrt{x}$ vs $\sqrt{y}$.
4. Explain, given an example, how the following functions may be used to combine matrices to form new ones:
   (a) `cbind`
   (b) `rbind`.
5. (a) Describe is the R function `factor()`.
   (b) Give an example of using `factor()` to create new arrays.
6. Using examples, illustrate two procedures for creating:
   (a) a vector
   (b) a matrix
7. Describe, using examples, the following 3 functions for creating vectors:
   (a) `c`

(b) seq
(c) rep.

8. (a) Use the function dim() to set up a matrix. Give an example.
   (b) Use the function matrix() to set up a matrix. Give an example.
9. Describe, using an example, the use of the following functions operating on a matrix in R: t(), colnames(), and rownames().
10. (a) What is reserved word in the R environment?
    (b) In R, how is the logical constant NA used? Give an example.

Exercises for Section 2.3

Enter the R environment, and do the following exercises using R programming [15, 17]:

1. Perform the following elementary arithmetic exercises:
   (a) 7 + 31
   (b) 87 − 23
   (c) 3.1417 X (7)2
   (d) 22/7; (e) e$\sqrt{2}$
2. Body mass index (BMI) is calculated from your weight in kilograms and your height in meters:

$$\text{BMI} = \text{kg}/\text{m}^2$$

   Using 1 kg ≈ 2.2 lb, and 1m ≈ 3.3 ft ≈ 39.4 in
   (a) Calculate your BMI.
   (b) Is it in the "normal" range $18.5 \leq \text{BMI} \leq 25$?
3. In the MPH program, 5 graduate students taking the class in Introductory Epidemiology measured their weight (in kg) and height (in meters). The result is summarized in following matrix:

	John	Chang	Michael	Bryan	Jose
Weight	69.1	62.5	74.3	70.9	96.6
Height	1.81	1.46	1.69	1.82	1.74

   (a) Construct a matrix showing their BMI as the last row.
   (b) Plot:
      (i) WEIGHT (on the y-axis) vs HEIGHT (on the x-axis)
      (ii) HEIGHT vs WEIGHT
      (iii) Assuming that the weight of a typical "normal": is (21.75 x HEIGHT^2), superimpose a line-of-"expected"-weight at BMI = 21.75 on the plot in (i).
4. (a) To convert between temperatures in degrees Fahrenheit (F) and Celsius (C), the following conversion formulas are used:

$$F = (9/5)C + 32$$

$$C = (5/9) \times (F - 32)$$

At standard temperature and pressure, the freezing and boiling points of water are 0 and 100 °C, respectively. What are the freezing and boiling points of water in degrees Fahrenheit?

(b) For C = 0, 5, 10, 15, 20, 25, ..., 80, 85, 90, 95, 100, compute a conversion table that shows the corresponding F temperatures.

Note: To create the sequence of Celsius temperatures use the R function seq(0, 100, 5).

5. Use the data in Table A [14, 19]. Assume a person is initially HIV-negative. If the probability of getting infected per act is p, then the probability of not getting infected per act is $(1 - p)$.

The probability of not getting infected after 2 consecutive acts is $(1 - p)^2$, and the probability of not getting infected after 3 consecutive acts is $(1 - p)^3$. Therefore, the probability of not getting infected after n consecutive acts is $(1 - p)^n$, and the probability of getting infected after n consecutive acts is $1 - (1 - p)^n$.

(a) For the non-blood transfusion transmission probability (per act risk) in Table A, calculate the risk of being infected after one year (365 days) if one carries out the Needle-sharing Injection-Drug Use (IDU) once daily for one year.

(b) Do these cumulative risks seem reasonable? Why? Why not?

Table A. Estimated per act risk (transmission probability) for acquisition of HIV, by exposure route to an infected source.

Exposure route	Risk per 10,000 exposures
Blood Transfusion (BT)	9,000
Needle-sharing Injection-Drug Use (IDU)	67

Source: CDC [19].

Solution:

```
> p <- 67/10000
> p
[1] 0.0067
> q <- (1 - p)
> q
[1] 0.9933
> q365 <- q^365
> q365
[1] 0.08597238
> p365 <- 1 - q365
> p365
[1] 0.9140276
```

=> ***Probability of being infected in a year = 91.40%.***

A high risk, indeed!

2.4 Using R in Data Analysis in Human Genetic Epidemiology

In epidemiology, after preparing the collected datasets to undertake biostatistical analysis, the first step is to enter the datasets into the R environment. Once the datasets are placed within the R environment, analysis will process the data to obtain results leading to creditable conclusions, and likely recommendations for definitive courses of actions to improve pertinent aspects of public and personal data. Several methods for dataset entry will be examined.

2.4.1 Entering Data at the R Command Prompt

- **Data Frames and Datasets** [15]: For many biostatistical investigators, the terms data frame and dataset may be used interchangeably.
- **Datasets:** In many applications, a complete dataset contain several data frames, including the real data that have been collected.
- **Data Frames**: Rules for data frames are similar to those for arrays and matrices, introduced earlier. However, data frames are more complicated than arrays. In an array, if just one cell is a character, then all the columns will be characters. On the other hand, a data frame can consist of:
- a column of "IDnumber", which is numeric,
- a column of "Name", which is character.

  In a data frame:

(a) each variable can have long variable descriptions,
(b) a factor can have "levels" or value levels.

These properties can be transferred from the original dataset in other software formats (such as SPS and Stata). They can also be created in R.

2.4.1.1 Creating a Data-Frame for R Computation Using the EXCEL Spreadsheet (on a Windows Platform)

As an example using a typical set of real case-control epidemiologic research data, consider the dataset in Table B, from a clinical trial to evaluate the efficacy of maintenance chemotherapy for case-subjects with acute myelogenous leukemia (AML), conducted at Stanford University, California, U.S.A., in 1977. After reaching a status of remission through treatment by chemotherapy, the patients who entered the study were assigned randomly to two groups:

Maintained: this group received maintenance chemotherapy.
Non-maintained: this group did not, & is the Control Group.

The clinical trial was to ascertain whether maintenance chemotherapy prolonged the time until relapse (= "death").

Procedure:

(1) To create an acute myelogenous leukemia (AML) data file, called AML.csv, in Windows

(2) To input it into R as a data file AML
(1) Creating a Data-frame for R computation
   1. Data Input, Using EXCEL:
      (a) Open the EXCEL spreadsheet
      (b) Type in data such that the variable names are in the row 1 of the EXCEL spreadsheet.
      (c) Consider each row of data as an individual in the study.
      (d) Start with column A.
   2. Save as a .csv file:
      (a) Click: "File" → "Save as" → and then, in the file name box (the upper box at the bottom) type: AML
      (b) In the "Save in:" Box (at the top), choose: "Local Disc (C:)" The file AML will be then be saved in the top level of the C:Drive, but another level may also be chosen.

   In the "Save as Type" Box (the lower box at the bottom), scroll down, select, & click on: CSV (Comma delimited = Comma Separated Values)
   To close out of EXCEL using the big "X" at the top- right-hand corner: Click X.
   Table B**. Data for the AML maintenance Clinical Study (a + indicates a censored value).

Group	Duration for Complete Remission (weeks)
1=Maintained (11) 1=Uncensored	9,13,13+,18,23,28+,31,34,45+,48,161+
0=Non-maintained (12) 0=Censored (+)	5, 5, 8, 8, 12, 16+,23,27,30,33,43,45

NB: The non-maintained group may be considered as MBD.***

Source: The AML Clinical Study Data: Tableman and Kim (2004). Table B-1: 23 data points, taken from – "Survival Analysis Using S: Analysis of Time-to-Event Data" by Mara Tableman and Jong Sung Kimz, published by Chapman & Hall/CRC, Boca Raton, 2004
***The cancer epigenome is characterized by specific DNA methylation and chromatin modification patterns. The proteins that mediate these changes are encoded by the epigenetics genes here defined as: DNA methyltransferases (DNMT), methyl-CpG-binding domain (MBD) proteins, histone acetyltransferases (HAT), histone deacetylases (HDAC), histone methyltransferases (HMT), and histone demethylases.

3. In Windows, check the C:Drive for the AML.csv file., viz. C:AML
4. Read AML into R:
   (a) Open R
   (b) Use the read.csv() function:

```
> aml <- read.csv("C:\\AML.csv", header = T,
+                   sep= ",")
```

(c) Actually, it can be also be done by:

```
> aml <- read.csv("C:\\AML.csv")
> # Read in the AML.csv file from the C:Drive of
> # the Computer, and call it aml
```

5. Output the AML.csv file for inspection

```
> aml # Outputting:

 1    9  1  1
 2   13  1  1
 3   13  1  0
 4   18  1  1
 5   23  1  1
 6   28  1  0
 7   31  1  1
 8   34  1  1
 9   45  1  0
10   48  1  1
11  161  1  0
12    5  0  1
13    5  0  1
14    8  0  1
15    8  0  1
16   12  0  1
17   16  0  0
18   23  0  1
19   27  0  1
20   30  0  1
21   33  0  1
22   43  0  1
23   45  0  1
>
```

2.4.1.2 Obtaining a Data Frame from a Text File

Data from various sources are often entered using many different software programs.

They may be transferred from one format to another through the ASCII file format.

For example, in Windows, a text file is the most common ASCII file, usually having a ".txt" extension. There are other files in ASCII format, including the ".R" command file.

Data from most software programs can be outputted or saved as an ASCII file. From EXCEL, a very common spreadsheet program, the data can be saved as ".csv" (comma separated values) format. This is an easy way to interface between EXCEL spreadsheet files and R. Open the EXCEL file and "save as" the csv format.

- Files with field separators:

As an example, suppose the file "csv1.xls" is originally an EXCEL spreadsheet. After "save as" into csv format, the output file is called "csv1.csv", the contents of which is:

```
"name","gender","age"
   "A",     "F",   20
   "B",     "M",   30
   "C",     "F",   40
```

The characters are enclosed in quotes and the delimiters (variable separators) are commas. Sometimes the file may not contain quotes, as in the file "csv2.csv".

```
name, gender, age
  A,       F,  20
  B,       M,  30
  C,       F,  40
```

For both files, the R command to read in the dataset is the same.

```
> a <- read.csv("csv1.csv", as.is=TRUE)
> a

  name gender age
1  A     F     20
2  B     M     30
3  C     F     40
```

The argument 'as.is=TRU ' keeps all characters as they are, otherwise the characters would have been coerced into factors. The variable "name" should not be factored but "gender" should. The following command should therefore be entered:

```
> a$gender <- factor(a$gender)
```

Note that the object "a" has class data frame and that the names of the variables within the data frame "a" must be referenced using the dollar sign notation $. Otherwise, R will state that the object "gender" cannot be found.

For files with white space (spaces and tabs) as the separator, such as in the file "data1.txt", the command to use is read.table():

```
> a <- read.table("data1.txt", header=TRUE,
+                  as.is=TRUE)
```

- Files without field separators:

Consider the file "data2.txt" which in fixed field format without field separators.

```
     name  gender  age
1     A       F    20
2     B       M    30
3     C       F    40
```

To read in such a file, use the function read.fwf():

(i) Skip the first line, which is the header.
(ii) The width of each variable and the column names must be specified.

```
> a <- read.fwf("data2.txt", skip=1, width=c(1,1,2),
+ col.names = c("name", "gender", "age"),
+ as.is=TRUE)
```

2.4.1.3 Data Entry and Analysis Using the Function data.entry()

The previous section deals with creating data frames by reading in data created from programs outside R, such as EXCEL. It is also possible to enter data directly into R by using the function data.entry(). However, if the data size is large (say more than 15 columns and/or more than 25 rows), the chance of human error is high with the spreadsheet or text mode data entry. A software program specially designed for data entry, such as Epidata, is more appropriate.

The web site of Epidata is: http://www.epidata.dk.

2.4.1.4 Data Entry Using Several Available R Functions

The dataset [14], in Table C, listing deaths among subjects who received a dose of tolbutamide or a placebo in the Unversity Group Diabetes Program (1970), stratifying by age:

Table C[a]. Deaths Among Subjects Who Received Tolbutamide or a Placebo in the Unversity Group Diabetes Program (1970).

	Age < 55		Age ≥ 55		Combined	
	Tolbutamide	Placebo	Tolbutamide	Placebo	Tolbutamide	Placebo
Deaths	8	5	22	16	30	21
Survivors	98	115	76	69	174	184

[a]Available at http://www.medepi.net/data/ugdp.txt

The R functions that can be used to import the data frame have been previously introduced in Sections 2.3.3–2.3.13.

A convenient way to enter data at the command prompt is to use the R functions :

```
c(), matrix(), array(), apply(), list(), data.frame(), and odd.ratio(),
```

as shown by the following examples and using the data in Table C.

```
> #Entering data for a vector
> vector1 <- c(8, 98, 5, 115) # Using data from
+ Table C.
> vector1
[1]  8  98  5  115
>
> vector2 <- c(22, 76, 16, 69); vector2
> # Data from Table C.
[1] 22 76 16 69
>
> # Entering data for a matrix
> matrix1 <- matrix(vector1, 2, 2)
> matrix1
     [,1]  [,2]
[1,]    8     5
[2,]   98   115
> matrix2 <- matrix(vector2, 2, 2); matrix2
     [,1]  [,2]
[1,]   22    16
[2,]   76    69
>
> # Entering data for an array
> udata <- array(c(vector1, vector2), c(2, 2, 2))
> udata
, , 1
      [,1]  [,2]
[1,]     8     5
[2,]    98   115

, , 2
     [,1]   [,2]
[1,]  22     16
[2,]  76     69

> apply(udata, c(1, 2), sum); udata.tot

      [,1]   [,2]
[1,]  30     21
[2,] 174    184
>
> # Entering a list
> x <- list(crude.data = udata.tot, stratified.data =
+            udata)
> x$crude.data
```

```
     [,1] [,2]
[1,]  30   21
[2,] 174  184
> x$stratified
, , 1
     [,1] [,2]
[1,]    8    5
[2,]   98  115
, , 2
     [,1]  [,2]
[1,]   22    16
[2,]   76    69
>
> # Entering a simple data frame
> subjectname <- c("Peter", "Paul", "Mary")
> subjectnumber <- 1:length(subjectname)
> age <- c(26, 27, 28) # These are their true ages,
>                      # respectively, in 1964!
> gender <- c("Male", "Male", "Female")
> data1 <- data.frame(subjectnumber, subjectname,
+                     age, gender)
> data1
  subjectnumber subjectname age gender
1             1       Peter  26   Male
2             2        Paul  27   Male
3             3        Mary  28 Female
>
> # Entering a simple function
> odds.ratio <- function(aa, bb, cc, dd){ aa*dd /
+ (bb*cc)}
> odds.ratio(30, 174, 21, 184) # Data from Table C.
[1] 1.510673
```

2.4.1.5 Data Entry and Analysis Using the Function scan()

The R function scan() is taken from the CRAN package base. This function reads data into a vector or list from the console or file. This function takes the following usage form [20]:

```
scan(file = '', what = double(), nmax = -1, n = -1,
          sep = "",
          quote = if(identical(sep, "\n")) "" else
          "'\"", dec = ".",
          skip = 0, nlines = 0, na.strings = "NA",
          flush = FALSE, fill = FALSE, strip.white =
          FALSE,
          quiet = FALSE, blank.lines.skip = TRUE,
          multi.line = TRUE,
```

```
comment.char = "", allowEscapes =
FALSE,
fileEncoding = "", encoding = "unknown",
text)
```

2.4.1.5.1 Argument

what — The type of what gives the type of data to be read. The supported types are logical, integer, numeric, complex, character, raw, and list. If what is a list, it is assumed it is assumed that the lines of the data file are records each containing length (what) items ("fields") and the list components components should have elements that are one of the first six types six types listed or NULL.

The what argument describes the tokens that scan() should expect in the input file.

For a detailed description of this function, execute:

```
> ?scan
```

The methodology of applying scan() is similar to c(), as described in Section 2.4.1.4, except that it does not matter the numbers are being entered on different lines, it will still be a vector.

- Use scan() when accessing data from a file that has an irregular or a complex structure.
- Use scan() to read individual tokens and use the argument what to describe the stream of tokens in the file.
- scan() converts tokens into data, and then assemble the data into records.
- Use scan() along with readLines(), especially when one attempts to read an unorthodox file format. Together, these two functions will likely result in a successful processing the individual lines and tokens of the file!

The function readLines() reads lines from a file, and returns them to a list of character strings:

```
> lines <- readLines("input.text")
```

One may limit the number of lines to be read, per pass, by using the n parameter which gives the maximum number of lines to be read:

```
> lines <- readLines("input.text", n=5)
> # read 5 lines and stop
```

The function scan() reads one token at a time, and handles it accordingly as instructed.

An example:

Assume that the file to be scanned and read contains triplets of data (like the Dates, and the corresponding daily Highs & Lows of financial markets):

15-Oct-1987	2439.78	2345.63
16-Oct-1987	2396.21	2207.73
19-Oct-1987	2164.16	1677.55
20-Oct-1987	2067.47	1616.23
21-Oct-1987	2087.07	1951.76

Use a list to operate scan() that it should expect a repeating, 3-token sequence:

```
> triplets <- scan("triples.txt, what=list(character(0),
+                     numeric(0), numeric(0)))
```

Give names to the list elements, and scan() will assign those names to the data:

```
> triplets <- scan("triples.txt,
+                    what=list(date=character(0),
+                    high=numeric(0), low=numeric(0)))
```

Reads 5 records.

```
> triples # Outputs:
$date
[1] "15-Oct-1987" "15-Oct-1987" "19-oct-1987" "20-Oct-1987" "21-oct-1987"
$high
[1] 2439-78 2396.21 2164.16 2067.47 2081.07
$low
[1] 2345.63 2207.73 1677.55 1616.21 1951.76
```

2.4.1.6 Data Entry and Analysis Using the Function Source()

The R function source() is also taken from the CRAN package base [13, 14, 20]. This function reads data into a vector or list from the console or file. It takes the following usage form:

source() causes R to accept its input from the named file or URL or connection.

Input is read and parsed from that file until the end of the file is reached, then the parsed expressions are evaluated sequentially in the chosen environment:

```
source(file, local = FALSE, echo = verbose, print.eval
        = echo,
        verbose = getOption("verbose"),
        prompt.echo = getOption("prompt"),
        max.deparse.length = 150, chdir = FALSE,
        encoding = getOption("encoding"),
        continue.echo = getOption("continue"),
        skip.echo = 0, keep.source =
         getOption("keep.source"))
```

For commands which are stored in an external file, such as 'commands.R' in the working directory "work," they can be executed in an R environment with the command

```
> source("command.R")
```

The function source() instructs R to read the text and execute its contents. Thus, when one has a long, or frequently used, piece of R code, one may capture it inside a text file. This allows one to rerun the code without having to re-type it, and use the function source() to read and execute the code.

For example, suppose the file howdy.R contains the familiar greeting:

```
Print ("Hi, My Friend!")
```

Then by sourcing the file, one may execute the content of the file, as in the follow R code segment:

```
> source("howdy.R")
[1] "Hi, My Friend!"
```

Setting echo-TRUE will echo the same script lines before they are executed, with the R prompt shown before each line:

```
> source("howdy.R", echo=TRUE)
> Print("Hi, My Friend!")
[1] "Hi, My Friend!"
```

2.4.1.7 Data Entry and Analysis Using the Spreadsheet Interface in R

This method consists of the following R functions in the package Utils [14].

2.4.1.7.1 Spreadsheet Interface for Entering Data

This is a spreadsheet-like editor for entering or editing data, with the following R functions:

```
data.entry(..., Modes = NULL, Names = NULL)
dataentry(data, modes)
de(..., Modes = list(), Names = NULL)
```

The arguments of these R functions are:

A list of variables: currently these should be numerals or character vectors or a list containing such vectors.

Modes	The modes to be used for the variables.
Names	The names to be used for the variables.
data	A list of numeric and/or character vectors.
modes	A list of length up to that of data giving the modes of (some of) the variables. list() is allowed.

The function data.entry() edits an existing object, saving the changes to the original object name.

However, the function edit() edits an existing object but not saving the changes to the original object name so that one must assign it to an object name (even if it is the original name). To enter a vector one needs to initialize a vector and then use the function data.entry(). For example:

Start by entering the R environment, and set

```
> x <- c(2, 4, 6, 8, 10)
> # X is initially defined as an array of 5 elements.
> x # Just checking - to make sure!
[1] 2 4 6 8 10 #x is indeed set to be an array of 5 elements
>
> data.entry(x) # Entering the Data Editor:

> # The Data Editor window pops up, and looking at
> # the first
> # column: it is now named "x", with the first 5
> # rows (all on first
> # column) filled, respectively, by the numbers 2,
> # 4, 6, 8, 10
> # One can now edit this dataset by, say, changing
> # all the
> # entries to 2, then closing the Data Editor window,
> # and
> # returning to the R console window:
> x
[1] 2 2 2 2 2 # x is indeed changed!
> # Thus one can change the entries for x via the
> # Data Editor,
> # and save the changes.
```

When using the functions data.entry(x) and edit() for data entry, there are a number of limitations [14]:

(i) Arrays and non-tabular lists cannot be entered using a spreadsheet editor.
(ii) When using the function edit() to create a new data frame, one must assign an object name in order to save the data frame.
(iii) This approach is not a preferred method of entering data because one often prefers to have the original data to be in a text editor or available to be read in from a data file.

2.4.1.8 Human Genetic Epidemiology Using R: The CRAN Package Genetics

To illustrate the ease of use of R in Human Genetic Epidemiology, consider examples in the CRAN package Genetics.

2.4.2 The Function list() and the Construction of data.frame() in R

The Function list()

A list in R consists of an ordered collection of objects – its components, which may be of any modes or types [13, 16, 20]. For example, a list may consist of a matrix, a numeric vector, a complex vector, a logical value, a character array, a function, etc. Thus, some simple way to create a list would be:

Example 2.1 ***It is as easy as "1, 2, 3"!***

```
> x <- 1
> y <- 2
> z <- 3
> list1 <- list(x, y, z) # Forming a simple list
> list1 # Outputting:
[[1]]
[1] 1

[[2]]
[1] 2

[[3]]
[1] 3
```

Moreover, the components are always numbered, and may be referred to as such.

Thus if my.special.list is the name of a list with 4 components, they may be referred to, individually, as:

```
my.special.list[[1]], my.special.list[[2]],
my.special.list[[3]], and my.special.list[[4]].
```

If one defines my.special.list as:

```
> my.special.list <- list(name="John", wife="Mary",
+ number.of.children=3, children.age=c(2, 4, 6))
```

then

```
> my.special.list[[1]] # Outputting:
[1] "John"

> my.special.list[[2]]
[1] "Mary"
> my.special.list[[3]]
[1] 3
> my.special.list[[4]]
[1] 2 4 6
```

The Number of Components in a List: the number of (top-level) components in a list may be found by the function length(). Thus:

```
> length(my.special.list)
[1] 4
```

viz. the list my.special.list has 4 components.

To combine a set of objects into a larger composite collection for more efficient processing, the list function may be used to construct a list from its components.

As an example, consider

```
> odds <- c(1, 3, 5, 7, 9, 11,13,15,17,19)
> evens <- c(2, 4, 6, 8, 10, 12, 14, 16, 18, 20)
> mylist <- list(before=odds, after=evens)

> mylist
$before
 [1]  1  3  5  7  9 11 13 15 17 19
$after
 [1]  2  4  6  8  10 12 14 16 18 20
> mylist$before
 [1]  1  3  5  7  9 11 13 15 17 19
> mylist$after
 [1]  2  4  6  8  10 12 14 16 18 20
```

Components of a List: Components of a list may be named. In such a case, the component may be referred to either (1) by giving the component name as a character string in place of the number in double square brackets, or (2) by giving an expression of the form

> name$component_name
for the same object.

Example 2.2 *A family affair*

```
> my.special.list <- list(name="John", wife="Mary",
+ number.of.children=3, children.age=c(2, 4, 6))
> my.special.list # Outputting: $name
[1] "John"
$wife
[1] "Mary"
$number.of.children
[1] 3

$children.age
[1] 2 4 6
```

Thus, for this list:

```
my.special.list[[1]]
[1] "John"
> my.special.list$name
> # This is the same as my.special.list[[1]]
[1] "John"
> my.special.list[[2]]
[1] "Mary"
> my.special.list$wife
> # This is the same as my.special.list[[2]]
[1] "Mary"

> my.special.list[[2]]
[1] "Mary"
> my.special.list[[3]]
[1] 3
> my.special.list$number.of.children
> # This is the same as my.special.list[[3]]
[1] 3
> my.special.list[[4]]
[1] 2 4 6
> my.special.list$children.age
> # This is the same as my.special.list[[4]]
[1] 2 4 6
```

Extraction of a Variable from a List:

To extract the name of a component stored in another variable, one may use the names of the list components in double square brackets, viz., list1[["name"]]. The following R code segment may be used:

```
> x <- "name"; my.special.list[[John]]
[1] "John"
```

Constructing, Modifying, and Concatenating Lists:

New lists may be constructed from existing objects by the function list(). Thus, the form

```
> new.list <- list(name_1=object_1,... name-
+                  n=object_n)
```

will set up s list, list1, of n components using object_1,..., object_n for the components and giving then names as specified.

2.4.3 Stock Market Risk Analysis

2.4.3.1 Univariate, Bivariate, and Multivariate Data Analysis

A univariate dataset has only one variable: {x}, e.g. {patient name}.

A bivariate dataset has two variables: {x1, x2}, or {x, y}, e.g. {patient name, gender}.

A multivariate dataset has more than two, or many, variables: {x1, x2, x3, …, xn}, e.g. {case subject name, gender, age, weight, height, blood pressures, blood sugar level, heart rates, history of illnesses, etc.}

2.4.3.1.1 Univariate Data Analysis

As an example, enter the following code segments:

```
> x <- rexp(100); x
> # Outputting 100 exponentially-distributed
> # random numbers:
[1] 0.39136880 0.66948212 1.48543076 0.34692128 0.71533079 0.12897216
[7] 1.08455419 0.07858231 1.01995665 0.81232737 0.78253619 4.27512555
[13] 2.11839466 0.47024886 0.62351482 1.02834522 2.17253419 0.37622879
[19] 0.16456926 1.81590741 0.16007371 0.95078524 1.26048607 5.92621325
[25] 0.21727112 0.07086311 0.83858727 1.01375231 1.49042968 0.53331210
[31] 0.21069467 0.37559212 0.10733795 2.84094906 0.17899040 1.34612473
[37] 0.00290699 1.77078060 1.79505318 0.09763821 1.96568170 0.15911043
[43] 4.36726420 0.33652419 0.01196883 0.35657882 0.72797670 0.91958975
[49] 0.68777857 0.29100399 0.22553560 1.56909742 0.20617517 0.37169621

[55] 0.53173534 0.26034316 0.21965356 2.94355695 1.88392667 1.13933083
[61] 0.31663107 0.23899975 0.01544856 1.30674088 0.53674598 1.72018758
[67] 0.31035278 0.81074737 0.09104104 1.52426229 1.35520172 0.27969075
[73] 1.36320488 0.56317216 0.85022837 0.49031656 0.17158651 0.31015165
[79] 2.07315953 1.29566872 1.28955269 0.33487343 0.20902716 2.84732652
[85] 0.58873236 1.54868210 2.93994181 0.46520037 0.73687959 0.50062507
[91] 0.20275282 0.49697531 0.58578119 0.49747575 1.53430435 4.56340237
[97] 0.90547787 0.72972219 2.60686316 0.33908320
```

Note: The function rexp() is defined as follows:

```
rexp(n, rate = 1)
```

with arguments:

x vector

n number of observations. If length(n) > 1, the length is taken to be the number required.

The exponential distribution with rate λ has density:

$$f(x) = \lambda\, e^{-\lambda x}, \text{for}\, x \geq 0.$$

If the rate λ is not specified, it assumes the default value of 1.

Remark: The function rexp() is one of the functions in R under Exponential in the CRAN package stats.

To undertake a biostatistical analysis of this set of univariate data, one may call up the function univax(), in the package epibasix, using the following code segments:

```
> library(epibasix)
> univar(x) # Outputting:
Univariate Summary
Sample Size: 100
Sample Mean: 1.005
Sample Median: 0.646
Sample Standard Deviation: 1.067
>
```

Thus, for this sample, size 100 elements, the mean, median, and standard deviation have been computed.

For data analysis of univariate datasets, the R package epibasix may be used.

This CRAN [22] package covers many elementary financial functions for statistics and econometrics. It contains elementary tools for analysis of common financial problems, ranging from sample size estimation, through 2 x 2 contingency table analysis, and basic measures of agreement (kappa, sensitivity/specificity). Appropriate print and summary statements are also written to facilitate interpretation wherever possible. This work is appropriate for graduate financial engineering courses. This package is a work in progress.

To start, enter the R environment and use the code segment:

```
> install.packages("epibasix")
```

Installing package(s) into "C:/Users/bertchan/Documents/R/win-library/2.14"
(as "lib" is unspecified)

--- Please select a CRAN mirror for use in this session ---
> ***# Select CA1***
trying URL
"http://cran.cnr.Berkeley.edu/bin/windows/contrib/2.14/epibasix_1.1.zip"
Content type "application/zip" length 57 888 bytes (56 Kb)
opened URL
downloaded 56 Kb
package "epibasix" successfully unpacked and MD5 sums checked
The downloaded packages are in
C:\Users\bertchan\AppData\Local\Temp\RtmpMFOrEn\downloaded_packages

With epibasix loaded into the R environment, to learn more about this package, follow these steps:

1. Go to the CRAN website: http://cran.r-project.org/
2. Select (single-click) Packages, on the left-hand column
3. On the page: select E (for epibasix)
   Available CRAN Packages By Name

   A B C D E F G H I J K L M N O P Q R S T U V W X Y Z

4. Scroll down list of packages whose name starts with "E" or "e", and select: epibasix
5. When the epibasix page opens up, select: Reference manual: epibasix.pdf
6. The information is now on displayed, as follows:

Package	'epibasix'
	January 2, 2012
Version	1.1
Date	2009-05-13
Author	Michael A Rotondi <mrotondi@uwo.ca>
Maintainer	Michael A Rotondi mrotondi@uwo.ca
Depends	*R* (>=2.01)

For another example, consider the same analysis on the first one hundred Natural Numbers, using the following R code segments:

```
> x <-1:100; x # Consider, and then output, the first 100
>                       # Natural Numbers
 [1]  1  2  3  4  5  6  7  8  9 10 11 12 13 14 15 16 17 18
[19] 19 20 21 22 23 24 25 26 27 28 29 30 31 32 33 34 35 36
[37] 37 38 39 40 41 42 43 44 45 46 47 48 49 50 51 52 53 54
[55] 55 56 57 58 59 60 61 62 63 64 65 66 67 68 69 70 71 72
[73] 73 74 75 76 77 78 79 80 81 82 83 84 85 86 87 88 89 90
[91] 91 92 93 94 95 96 97 98 99 100
> # ANOVA Tables: Summarized in the following tables,
> # ANOVA is used for
> #   two different purposes:
> library(epibasix)
> univar(x) #  Performing a univariate data analysis
>           # on the vector x, and Outputting:
Univariate Summary
Sample Size: 100
Sample Mean: 50.5
Sample Median: 50.5
Sample Standard Deviation: 29.011
And that's it!
```

2.4.3.1.2 Bivariate and Multivariate Data Analysis

When there are two variables, (X, Y), one needs to consider two cases [21]:

Case I. In the classical regression model, only Y, called the dependent variable, is required to be random. X is defined as a fixed, non-random variable and is the independent variable. Under this model, observations are obtained by pre-selecting values of X and determining the corresponding value of Y.

Case II. If both X and Y are random variables, it is called the Correlation Model – under which sample observations are obtained by selecting a random sample of the units of association (such as persons, characteristics [age, gender, locations, points of time, specific events/actions/..., or elements on which the two measurements are based]) and by recording a measurement of X and of Y. In this case, values of X are not preselected but occurring at random, depending on the unit of association selected in the sample.

Regression Analysis

Case I. Correlation analysis cannot be meaningfully under this model.
Case II. Regression analysis can be performed under the Correlation Model.

Correlation for two variables implies a co-relationship between the variables, and does not distinguish between them as to which is the dependent or the independent variable. Thus one may fit a straight line to the data either by minimizing $\sum(x_i-x)^2$ or by minimizing $\sum(y_i-y)^2$. The fitted regression line will, in general, be different in the two cases – and a logical question arises as to which line to fit.

Two situation do exist, and should be considered:

1. If the objective is to obtain a measure of strength of the relationship between the two variables, it does not matter which line is fitted – the measure calculated will be the same in either case.
2. If one needs to use the equation describing the relationship between the two variables for the dependency of one upon the other, it does matter which line is to be fitted. The variable for which one wishes to estimate means or to make predictions should be treated as the depending variable. That is, this variable should be regressed with respect on the other variable.

Available R Packages for Bivariate Data Analysis:

Among the R packages for bivariate data analysis, a notable one available for sample size calculations for bivariate random intercept regression model there is the bivariate power.

An Example in Bivariate Data Analysis:

As an example, this package may be used to calculate necessary sample size to achieve 80% power at 5% alpha level for null and alternative hypotheses that correlation between RI is 0 and 0.2, respectively, across 6 time points. Other covariance parameter are set as follows:

Correlation between residuals = 0;
Standard deviations: 1st RI = 1, 2nd RI = 2, 1st residual = 0.5, 2nd residual = 0.75

The following R code segment may be used:

```
> library(bivarRIpower)
> bivarcalcn(power=.80,powerfor='RI',timepts=6,
+ d1=1,d2=2, p=0,p1=.2,s1=.5,s2=.75,r=0,r1=.1)
# Outputting:
```

Variance parameters

Clusters	= 209.2
Repeated measurements	= 6
Standard deviations	
1st random intercept	= 1
2nd random intercept	= 2
1st residual term	= 0.5
2nd residual term	= 0.75

Correlations

RI under H_o	= 0
RI under H_a	= 0.2
Residual under H_o	= 0
Residual under H_a	= 0.1
Con obs under H_o	= 0
Con obs under H_a	= 0.1831984
Lag obs under H_o	= 0
Lag obs under H_a	= 0.1674957

Correlation variances under H_o

Random intercept	= 0.005096138
Residual	= 0.0009558759
Concurrent observations	= 0.00358999
Lagged observations	= 0.003574277

Power (%) for correlations

Random intercept	= 80%
Residual	= 89.9%
Concurrent observations	= 86.4%
Lagged observations	= 80%

>

Bivariate Normal Distribution – Under the correlation model, the bivariates X and Y vary together in a joint distribution, which, if this joint distribution is a normal distribution, is called a bivariate normal distribution, from which inferences may be made based on the results of sampling properly from the population. If the joint distribution is known to be non-normal, or if the form is unknown, inferential procedures are invalid. The following assumptions must hold for inferences about the population to be valid when sampling from a bivariate distribution:

(i) For each value of X, there is a normally distributed sub-population of Y values.
(ii) For each value of Y, there is a normally distributed sub-population of X values.
(iii) The joint distribution of X and Y is a normal distribution called the Bivariate Normal Distribution.
(iv) The sub-population of Y values all have the same variance.
(v) The sub-population of X values all have the same variance.

Two random variables X and Y are said to be jointly normal if they can be expressed in the form

$$X = aU + bV \tag{2.1}$$

$$Y = cU + dV \tag{2.2}$$

where U and V are independent normal random variables.

If X and Y are jointly normal, then any linear combination

$$Z = s_1 X + s_2 Y \tag{2.3}$$

has a normal distribution. The reason is that if one has $X = aU + bV$ and $Y = cU + dV$ for some independent normal random variables U and V, then

$$Z = s_1(aU + bV) + s_2(cU + dV) = (as_1 + cs_2)U + (bs_1 + ds_2)V \tag{2.4}$$

Thus, Z is the sum of the independent normal random variables $(as_1 + cs_2)U$ and $(bs_1 + ds_2)V$, and is therefore normal.

A very important property of jointly normal random variables is that zero correlation implies independence.

Zero Correlation Implies Independence: If two random variables X and Y are jointly normal and are uncorrelated, then they are independent.

(This property can be verified using multivariate transforms.)

2.4.3.1.3 Multivariate Data Analysis

Two similar, but distinct, approaches used for multivariate data analysis are [21]:

1. The Multiple Linear Regression Analysis – assuming that a linear relationship exists between some variable Y, called the dependent variable, and n independent variables, $X_1, X_2, X_3, \ldots, X_n$, which are called explanatory or predictor variables because of their use.

The assumptions underlying multiple regression model analysis are:

(a) The X_i are non-random fixed variables, indicating that any inferences drawn from sample data apply only to the set of X values observed, but not to larger collections of X. Under this regression model, correlation analysis is not meaningful.

(b) For each set of X_i values, there is a sub-population of Y values. Usually, one assumes that these Y values are normally distributed.
(c) The variances of Y are all equal.
(d) The Y values are independent of the different selected set of X values.

For multiple linear regression, the model equation is:

$$y_j = \beta_0 + \beta_1 x_{1j} + \beta_2 x_{2j} + \beta_3 x_{3j} + \cdots + \beta_n x_{nj} + e_j \tag{2.5}$$

where y_j is a typical value from one of the sub-populations of Y values, and the β_i values are the regression coefficients.

$x_{1j}, x_{2j}, x_{3j}, \ldots, x_{nj}$ are, respectively, particular values of the independent variables $X_1, X_2, X_3, \ldots, X_n$, and e_j is a random variable with mean 0 and variance σ_2, the common variance of the sub-population of Y values. Generally, e_j is assumed normal and independently distributed.

When Eq. (2.1) consists of one dependent variable and two independent variables, the model becomes:

$$y_j = \beta_0 + \beta_1 x_{1j} + \beta_2 x_{2j} + e_j \tag{2.6}$$

A plane in three-dimensional space may be fitted to the data points. For models containing more than two variables, it is a hyperplane.

Let

$$f_j = \beta_0 + \beta_1 x_{1j} + \beta_2 x_{2j}$$

Now, $e_j = y_j - f_j$, forming a vector e.

If $\underline{y}$ is the mean of the observed data,

$$\text{viz.}, y = (1/n)\,\Sigma\, y_i, \quad \text{for } i = 1, 2, 3, \ldots, n$$

then the variability of the dataset may be measured using three sums of squares (proportional to the variance of the data):

(1) The Total Sum of Squares (proportional to the variance of the data):

$$\text{SS}_{\text{total}} = \Sigma\,(y_i - \underline{y})^2 \tag{2.7}$$

(2) The Regression Sum of Squares:

$$\text{SS}_{\text{reg}} = \Sigma\,(f_i - \underline{y})^2 \tag{2.8}$$

(3) The Sum of Squares of Residuals:

$$\text{SS}_{\text{res}} = \Sigma\,(y_i - f_i)^2 = \Sigma e_i{}^2 \tag{2.9}$$

The most general definition of the Coefficient of Multiple Determination is

$$R^2{}_{y,1,2,\ldots,n} \equiv 1 - (\text{SS}_{\text{res}}/\text{SS}_{\text{total}}) \tag{2.10}$$

The parameter of interest in this model is the Coefficient of Multiple Determination, $R^2{}_{y|1,2,3,\ldots n}$, obtained by dividing the explained sum of squares by the total sum of squares:

$$R^2{}_{y,1,2,\ldots,n} = \sum (y_i - y)^2 \Big/ \sum (y_i - y)^2 = \text{SSR/SSE} \tag{2.11}$$

where

$\sum(y_i - y)^2$ = the Explained Variation
= the original observed values from the calculated Y values
= the sum of squared deviation of the calculated values from the mean of the observed Y values, or
= the Sum of Squares due to Regression (SSR)

$\sum(y_i - y)^2$ = the Unexplained Variation
= the sum of squared deviations of the original observations from the calculated values
= the sum of squares about regression, or
= the Error Sum of Squares (SSE)

The total variation is the sum of squared deviations of each observation of Y from the mean of the observations:

$$\sum (y_j - y)^2 = \sum (y_i - y)^2 + \sum (y_i - y)^2 \tag{2.12}$$

viz.,

$$\text{SST} = \text{SSR} + \text{SSE} \tag{2.13}$$

or

$$\text{Total sum of squares} = \text{Explained (regression) sum.of squares} + \text{Unexplained (error) sum squares} \tag{2.14}$$

The Multiple Correlation Model Analysis – the object of this approach is to gain insight into the strength of the relationship between variables.

The Multiple Regression Correlation Model Analysis Equation is:

$$y_j = \beta_0 + \beta_1 x_{1j} + \beta_2 x_{2j} + \beta_3 x_{3j} + \cdots + \beta_n x_{nj} + e_j \tag{2.15}$$

where y_j is a typical value from one of the sub-populations of Y values, the β_i are the regression coefficients, $x_{1j}, x_{2j}, x_{3j}, \ldots, x_{nj}$ are, respectively, particular known values of the random variables $X_1, X_2, X_3, \ldots, X_n$, and e_j is a random variable with mean 0 and variance σ^2, the common variance of the sub-population of Y values. Generally, e_j is assumed normal and independently distributed.

This model is similar to model equation (2.5), with one important distinction: in Eq. (2.5), the x_i are non-random variables, but in Eq. (2.9), the x_i are random variables.

That is, in the correlation model, Eq. (2.9), there is a joint distribution of Y and the X_i that is called a Multivariate Distribution.

Under this model, the variables are no longer considered as being dependent or independent, because logically they are interchangeable, and either of the X_i may play the role of Y.

2. The Correlation Model Analysis

The Multiple Correlation Coefficient: To analyze the relationships among the variables, consider the multiple correlation coefficient, which is the square root of the coefficient of multiple determination and hence the sample value may be computed by taking the square root of Equation (2.12), viz.,

$$R^2 y, 12 \ldots n = \sqrt{R^2_{y,12\ldots n}} = \sqrt{\left\{\sum (y_i - y)^2 / \sum (y_i - y)^2\right\}} = \sqrt{(SSR/SSE)} \quad (2.16)$$

3. Analysis of Variance (ANOVA) [23]

In statistics, ANalysis Of VAriance (ANOVA) is a collection of statistical models in which the observed variance in a particular variable is partitioned into components from different sources of variation. ANOVA provides a statistical test of whether or not the means of several groups are all equal, and therefore generalizes *t*-test to more than two groups. Doing multiple two-sample *t*-tests would result in an increased chance of committing a Type I Error. For this reason, ANOVAs are useful in comparing two, three, or multiple means.

ANOVA Tables: Summarized in the following tables, ANOVA is used for two different purposes:

(1) To estimate and test hypotheses for simple linear regression about population variances, and
(2) To estimate and test hypotheses about population means.

ANOVA Table for Testing Hypotheses About Simple Linear Regression

Source	DF	Sum of squares	Mean squares	F-Value	P-Value
Model	1	$\Sigma(y_i - y)^2 =$ SSModel	SSM = MSM	MSG/MSE = $F_{1,n-2}$	Pr(F>F1,n-2)
Residual	$n-2$	$\Sigma e_i^2 =$ SSResidual	SSR/(n−2) = MSE		
Total	$n-1$	$\Sigma(y_i - y)^2 =$ SSTotal	SST/(n−1) = MST		

Residuals are often called errors since they are the part of the variation that the line could NOT explain, so

$$MSR = MSE = \text{sum of squared residuals}/df = \sigma$$
$$= \text{estimate for variance of the population regression line}$$

$$SSTot/(n-1) = MSTOT = s_y^{\,2} = \text{the total variance of the } y_s$$

$F = t^2$ for Simple Linear Regression.

The larger the F (the smaller the p-value) the more of y's variation the line explained so the less likely H_0 is true. One rejects a hypothesis when the p-value $< \alpha$.

R^2 = proportion of the total variation of y explained by the regression line

= SSM/SST

= 1–SSResidual/SST

ANOVA Table for Testing Hypotheses About Population Means.

Source	df	Sum of squares	Mean squares	F Value	p-value
Group (between)	$k-1$	$\sum n_i(\overline{x}_i \cdot \overline{x})^2$ = SSG	SSG/(k–1) = MSG	MSG/MSE Pr = $F_{k-1,\,N-k\,=\,(F\,>\,Fk-1,\,N-k)}$	$\Pr(F > F_{k-1,N-k})$
Error (within)	$N-k$	$\Sigma(n_i-1)s_i^{\,2}$ = SSE	SSE/(N–k) = MSE		
Total	$N-1$	$\sum (x_{ij} - \overline{x})^2$ = SSTot	SSTot/(N–1) = MST		

N = total number of observations = Σn_i,
where n_i = number of observations for group i

The F test statistic has two different degrees of freedom: the numerator = k–1, and the denominator = N–k, Fk–1, N–k

Note: SSE/(N – k) = MSE = sp2

$$= \text{(pooled sample variance)}$$

$$= \frac{(n_1 - 1)S_1^2 + \cdots + (n_k - 1)S_k^2}{(n_1 - 1) + \cdots + (n_k - 1)}$$

$$= \hat{\sigma}^2$$

= estimate for assumed equal variance

This is the "average" variance for each group.

SSTot/(N – 1) = MSTOT = s^2

= the total variance of the data (assuming NO groups)

F ≈ variance of the (between) samples means divided by the ∼average variance of the data, the larger the F (the smaller the p-value), the more varied the means are, so the less likely H_0 is true. It is rejected when the p-value < α.

R^2 = proportion of the total variation explained by the difference in means

$$= \frac{\text{SSG}}{\text{SSTot}}$$

Example 2.3

Package "fAssets"

February 19, 2015

Title Rmetrics – Analyzing and Modelling Financial Assets

Date 2014-10-30
Version 3011.83
Author Rmetrics Core Team,
Diethelm Wuertz [aut],
Tobias Setz [cre],
Yohan Chalabi [ctb]
Maintainer Tobias Setz <tobias.setz@rmetrics.org>
Description Environment for teaching
"Financial Engineering and Computational Finance."
Depends R (>= 2.15.1), timeDate, timeSeries, fBasics
Imports fMultivar, robustbase, MASS, sn, ecodist, mvnormtest, energy
Suggests methods, mnormt, RUnit
Note SEVERAL PARTS ARE STILL PRELIMINARY AND MAY BE CHANGED IN THE
FUTURE. THIS TYPICALLY INCLUDES FUNCTION AND ARGUMENT NAMES, AS
WELL AS DEFAULTS FOR ARGUMENTS AND RETURN VALUES.
LazyData yes
License GPL (>= 2)
URL https://www.rmetrics.org
NeedsCompilation no
Repository CRAN
Date/Publication 2014-10-30 13:38:28
R topics documented:
fAssets-package 2
assets-arrange 6
assets-distance 7
assets-lpm 9
assets-meancov 10
1
2 fAssets-package
assets-modeling 12
assets-outliers 14
assets-selection 15
assets-testing 16

```
> install.packages("fAssets")
```

Installing package into "C:/Users/Bert/Documents/R/win-library/3.2"

```
> library(fAssets)
```

https://www.rmetrics.org — Mail to: info@rmetrics.org
Warning messages:

1. package 'fAssets' was built under R version 3.2.5

2. package 'timeDate' was built under R version 3.2.5
3. package 'timeSeries' was built under R version 3.2.5
4. package 'fBasics' was built under R version 3.2.5

```
> ls("package:fAssets")
[1]  "abcArrange"             "assetsArrange"
[3]  "assetsBasicStatsPlot"   "assetsBoxPercentilePlot"
[5]  "assetsBoxPlot"          "assetsBoxStatsPlot"
[7]  "assetsCorEigenPlot"     "assetsCorgramPlot"
[9]  "assetsCorImagePlot"     "assetsCorTestPlot"
[11] "assetsCumulatedPlot"    "assetsDendrogramPlot"
[13] "assetsDist"             "assetsFit"
[15] "assetsHistPairsPlot"    "assetsHistPlot"
[17] "assetsLogDensityPlot"   "assetsLPM"
[19] "assetsMeanCov"          "assetsMomentsPlot"
[21] "assetsNIGFitPlot"       "assetsNIGShapeTrianglePlot"
[23] "assetsOutliers"         "assetsPairsPlot"
[25] "assetsQQNormPlot"       "assetsReturnPlot"
[27] "assetsRiskReturnPlot"   "assetsSelect"
[29] "assetsSeriesPlot"       "assetsSim"
[31] "assetsSLPM"             "assetsStarsPlot"
[33] "assetsTest"             "assetsTreePlot"
[35] "binaryDist"             "braycurtisDist"
[37] "canberraDist"           "corDist"
[39] "covEllipsesPlot"        "euclideanDist"
[41] "getCenterRob"           "getCovRob"
[43] "hclustArrange"          "jaccardDist"
[45] "kendallDist"            "mahalanobisDist"
[47] "manhattanDist"          "maximumDist"
[49] "minkowskiDist"          "mutinfoDist"
[51] "mvenergyTest"           "mvshapiroTest"
[53] "orderArrange"           "pcaArrange"
[55] "sampleArrange"          "sorensenDist"
[57] "spearmanDist"           "statsArrange"
```

assets-selection Selecting Assets from Multivariate Asset Sets

Description
Select assets from Multivariate Asset Sets based on clustering.

Usage
assetsSelect(x, method = c("hclust", "kmeans"), control = NULL, ...)

Arguments
x any rectangular time series object which can be converted by the function as.matrix() into a matrix object, e.g. like an object of class timeSeries, data.frame,

or mts. method a character string, which clustering method should be used? Either hclust for hierarchical clustering of dissimilarities, or kmeans for k-means clustering. control a character string with two entries controlling the parameters used in the underlying cluster algorithms. If set to NULL, then default settings are taken: For hierarchical clustering this is method=c(measure="euclidean", method="complete"), and for kmeans clustering this is method=c(centers=3, algorithm="Hartigan-Wong").

... optional arguments to be passed. Note, for the k-means algorithm the number of centers has to be specified!

Details

The function assetsSelect calls the functions hclust or kmeans from R's "stats" package. hclust performs a hierarchical cluster analysis on the set of dissimilarities hclust(dist(t(x))) and kmeans performs a k-means clustering on the data matrix itself.

The hierarchical clustering method has in addition a plot method.

Value

if use = "hclust" was selected then the function returns a S3 object of class "hclust," otherwise if

use = "kmeans" was selected then the function returns an object of class "kmeans."

For details we refer to the help pages of hclust and kmeans.

Author(s)

Diethelm Wuertz for the Rmetrics port.

Examples

```
## LPP -
  # Load Swiss Pension Fund Data:
  LPP <- LPP2005REC
  colnames(LPP)
  ## assetsSelect -
  # Hierarchical Clustering:
  hclust <- assetsSelect(LPP, "hclust")
  plot(hclust)
  ## assetsSelect -
  # kmeans Clustering:
  assetsSelect(LPP, "kmeans", control =
  c(centers = 3, algorithm = "Hartigan-Wong"))
  assets-
```

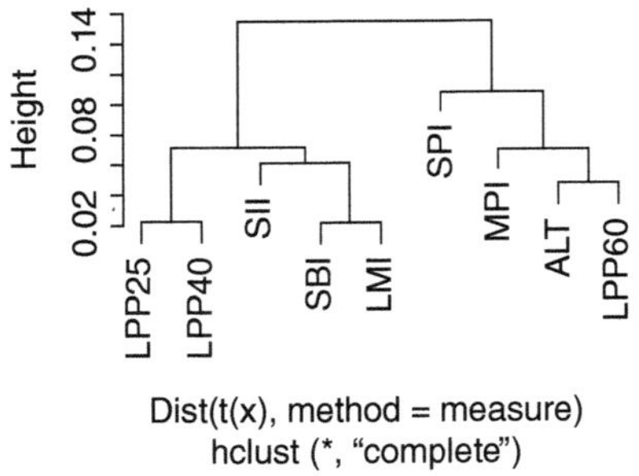

Figure 2.17 Cluster dendrogram.

```
> # LPP -
> # Load Swiss Pension Fund Data:
> LPP <- LPP2005REC
> colnames(LPP)
[1] "SBI" "SPI" "SII" "LMI" "MPI"
 "ALT" "LPP25" "LPP40" "LPP60"
>
> ## assetsSelect -
> # Hierarchical Clustering:
> hclust <- assetsSelect(LPP, "hclust")
> plot(hclust)
> # Outputting: Figure 2.17
>
> # assetsSelect -
> # kmeans Clustering:
> assetsSelect(LPP, "kmeans", control =
+ c(centers = 3, algorithm = "Hartigan-Wong"))
K-means clustering with 3 clusters of sizes 3, 1, 5
Cluster means:

2005-11-01    2005-11-02    2005-11-03    2005-11-04  2005-11-07
1 -7.174567e-05 -0.0004228193 0.0070806633 0.0086669900 0.003481978
2  8.414595e-03  0.0025193420 0.0127072920 -0.0007027570 0.006205226
3 -9.685162e-04 -0.0021474822 -0.0004991982 -0.0005400452 0.000752058

2005-11-08  2005-11-09  2005-11-10  2005-11-11   2005-11-14
1 0.0006761683  0.001840756 0.001091617 0.008291539 -0.0006469167
2 0.0003292600 -0.002378200 0.000922087 0.013334906 -0.0046930640
3 0.0014039486 -0.001694911 0.001119758 0.001929053 -0.0003708446

 2005-11-15   2005-11-16   2005-11-17   2005-11-18  2005-11-21
1 -0.0005408223  0.001686769 0.0048643273  0.005051224 0.001901881
2  0.0012668650 -0.007187498 0.0076581030  0.012527202 0.002659666
3 -0.0008278810  0.001879324 0.0003510514 -0.001005688 0.001511155
```

```
2005-11-22   2005-11-23   2005-11-24   2005-11-25   2005-11-28
1 0.0037938340 0.0024438233 -0.000394091 0.0022790387 -0.006316902
2 0.0021424940 0.0035671280 -0.002559544 0.0033748180 -0.009816739
3 0.0003204556 0.0008044326  0.001046522 0.0007771762 -0.001717004

2005-11-29   2005-11-30 2005-12-01   2005-12-02   2005-12-05
1 0.0031941750 -0.0037985373 0.011412029 0.0030362450 -0.003450339
2 0.0018864440 -0.0040370550 0.015977559 0.0070552900 -0.000268694
3 0.0001304264 -0.0001264086 0.002280999 0.0004551984 -0.001906796

2005-12-06   2005-12-07   2005-12-08   2005-12-09   2005-12-12
1 -0.001325485  0.0007240550 -0.0066049723 0.0031775860 -0.001708432
2  0.002864672 -0.0026064460 -0.0034829450 0.0007472420 -0.000278182
3  0.002119992 -0.0007569246  0.0008520178 0.0006191402  0.000879002
2005-12-13   2005-12-14   2005-12-15  2005-12-16   2005-12-19
1 0.0046313440 -0.004764346  0.0029376953 0.001213580 0.0004901883
2 0.0013586830 -0.007283576 -0.0073441080 0.004292528 0.0049806720
3 0.0008959768 -0.001095387  0.0007868516 0.001616794 0.0023614496

2005-12-20  2005-12-21   2005-12-22   2005-12-23  2005-12-26
1  0.0071593677 0.007886516  0.0001159380  0.001659755 0.001861593
2 -0.0016334200 0.004319502 -0.0036299400 -0.002234004 0.000000000
3  0.0009404428 0.002617884 -0.0002855498  0.000507245 0.000246350

2005-12-27   2005-12-28  2005-12-29   2005-12-30   2006-01-02
1 -0.001653545 -0.0005308263 0.004244265 -0.0020539910 0.0007404967
2  0.006571546  0.0016811710 0.007701353 -0.0039764730 0.0000000000
3  0.001191907  0.0015647450 0.001259883 -0.0007382012 0.0000960024
2006-01-03  2006-01-04   2006-01-05   2006-01-06   2006-01-09
1 -0.0013538470 0.001985783 -0.001291249 0.0015069157  0.006899025
2  0.0065956320 0.012163609 -0.002285923 0.0022278130 -0.000174350
3 -0.0001111704 0.001657156 -0.000790521 0.0005815596  0.000574890

2006-01-10   2006-01-11  2006-01-12   2006-01-13   2006-01-16
1 -0.001533005  0.0040728800 0.004428087 -0.001791852 -0.0005044543
2 -0.006484459  0.0109940950 0.005887780  0.000629145  0.0041673760
3 -0.000713456 -0.0008200066 0.001761519 -0.002024939 -0.0000711238

2006-01-17   2006-01-18  2006-01-19   2006-01-20   2006-01-23
1 -0.0064264427 -0.009959086 0.008732922 -0.005498331 -0.012300008
2 -0.0060610800 -0.007104074 0.004901245 -0.008712876  0.000080500
3 -0.0004887026 -0.000363750 0.001841324 -0.002770640 -0.000594795

2006-01-24   2006-01-25  2006-01-26  2006-01-27  2006-01-30   2006-01-31
1  0.0043293403  0.003928724 0.007596804 0.013018524 0.004510001
-0.000363935
```

```
2 -0.0009696750 -0.001245166 0.008348556 0.004847171 0.001426750
0.002736095
3 -0.0007514866 -0.001468543 0.001332284 0.002198388 0.000585831
0.000676559

2006-02-01   2006-02-02  2006-02-03  2006-02-06   2006-02-07
1  0.0025679990 -0.0033918090 0.001392298 0.003829826 -0.0016356553
2  0.0033390250 -0.0052315370 0.006229563 0.000010100 -0.0003661590
3 -0.0001937426 -0.0009799974 0.001561223 0.000789857 -0.0005261388

2006-02-08  2006-02-09   2006-02-10   2006-02-13   2006-02-14
1 -0.0034102587 0.004724740 -0.0002670790 -0.0043036747  0.0071958650
2 -0.0007797950 0.005323768 -0.0010290070  0.0030521410 -0.0005124970
3  0.0002705448 0.001218430 -0.0001116748 -0.0002136692  0.0009441608
2006-02-15  2006-02-16   2006-02-17  2006-02-20  2006-02-21
1 -0.0012863397 0.0045143087 -0.0008902317 -0.002237808 0.0050509303
2 -0.0099343840 0.0129171990  0.0012844190  0.004250758 0.0063638350
3  0.0002048986 0.0009410282  0.0012602358  0.001315374 0.0009478938

2006-02-22   2006-02-23   2006-02-24  2006-02-27   2006-02-28
1 0.007771075  0.0008127150  0.0050935397 0.0050390380 -0.0108695373
2 0.004433658 -0.0060669620 -0.0027457400 0.0053722110 -0.0119569840
3 0.002287663  0.0003148286 -0.0003851528 0.0009570004 -0.0006375348

2006-03-01   2006-03-02   2006-03-03   2006-03-06   2006-03-07
1 0.0038673107 -0.003960535 -0.004273008 0.0005488833 -0.001580974
2 0.0105047210 -0.004177747  0.001755361 0.0004602100 -0.007028455
3 0.0001534274 -0.003340033 -0.001016466 0.0001020288 -0.001403624
2006-03-08   2006-03-09   2006-03-10   2006-03-13   2006-03-14
1 -0.004631919 0.0054281930  0.0077513163 0.0026288817 -0.0011077267
2 -0.004230567 0.0035344700  0.0128746620 0.0082543960 -0.0005305960
3 -0.000989756 0.0001691144 -0.0002940006 0.0006133858  0.0006779516

2006-03-15   2006-03-16   2006-03-17   2006-03-20   2006-03-21
1  0.0039076300 -0.002874349  0.0035407443 -0.0002971253  0.0017513113
2 -0.0017679700  0.000666818  0.0043129660  0.0021624820 -0.0002753190
3  0.0001087012  0.000484471 -0.0001434976  0.0006474828 -0.0008360246

2006-03-22   2006-03-23   2006-03-24  2006-03-27  2006-03-28
1 0.0034417023  0.0042167183 -0.0005082437 -0.002477860 -0.005123098
2 0.0013273100 -0.0043185900  0.0027725900 -0.006945127 -0.000727865
3 0.0006235516  0.0002566278  0.0001903118 -0.000559792 -0.001748221
2006-03-29   2006-03-30   2006-03-31  2006-04-03   2006-04-04
1  0.0083719110 -0.0023536467  0.0032189740  0.002929918 -0.0053531483
2  0.0009252440  0.0062064890 -0.0009292510  0.006944399 -0.0026184930
3 -0.0004704764  0.0003112066 -0.0003227242 -0.001220841 -0.0005425296
```

```
2006-04-05  2006-04-06  2006-04-07  2006-04-10  2006-04-11
1 0.001672089 0.0035391920 0.0006688370 -0.0018577213 -0.0078612010
2 0.004752739 0.0015471340 0.0005119710 0.0000160000 -0.0095747430
3 0.000371034 0.0001308954 -0.0001606024 0.0004964256 0.0008645472

2006-04-12  2006-04-13  2006-04-14 2006-04-17  2006-04-18
1 -0.001794527 -6.576767e-05 0.0003451073 -0.01162293 0.0072958027
2 -0.002978612  3.927437e-03 0.0000000000 0.00000000 -0.0050312550
3 -0.001221078 -2.154967e-03 -0.0000179728 -0.00127895 0.0008801984
2006-04-19 2006-04-20  2006-04-21  2006-04-24  2006-04-25
1 0.005262631 0.004346695 -0.0001275733 -0.0055535810 -0.0008195537
2 0.010893639 0.004201724 0.0048380810 -0.0016947400 -0.0041756790
3 0.001021676 0.001559365 -0.0000182248 -0.0001482958 -0.0017556356

2006-04-26  2006-04-27  2006-04-28  2006-05-01 2006-05-02
1  0.001310372 -0.003174862 -0.0081970307 -0.0042204567 0.005113142
2  0.004699408  0.000883969 -0.0029437410  0.0000000000 0.003189327
3 -0.000656350  0.000053330  0.0002421188 -0.0007466388 0.001444051

2006-05-03  2006-05-04 2006-05-05  2006-05-08  2006-05-09
1 -0.001263695 -0.0006502133 0.006613750 0.0063954017 -0.003357079
2 -0.011943802  0.0030146350 0.011235363 0.0061756300  0.002517020
3 -0.001037496 -0.0000032026 0.002150292 0.0007181488 -0.001401839
2006-05-10  2006-05-11  2006-05-12  2006-05-15  2006-05-16
1 -0.0025826107 -0.008643966 -0.016678837 -0.006493649 -0.002878691
2 -0.0013872090 -0.001096609 -0.017989732 -0.013399941  0.003285150
3 -0.0005785598 -0.003397807 -0.004047129 -0.000274325  0.001044719

2006-05-17  2006-05-18 2006-05-19  2006-05-22 2006-05-23
1 -0.010999777 -0.0085399823 0.005933202 -0.021077161 0.007398834
2 -0.028406916 -0.0097114170 0.001848007 -0.025997761 0.018970677
3 -0.003777801  0.0001129756 0.002052023 -0.001940486 0.002266990

2006-05-24  2006-05-25 2006-05-26  2006-05-29  2006-05-30
1 -0.0004267907 0.0061746340 0.016697015 -0.0011805273 -0.017512962
2 -0.0111155890 0.0000000000 0.025842125  0.0017720070 -0.019842413
3 -0.0011767074 0.0006506848 0.002053482  0.0005608306 -0.003672446
2006-05-31  2006-06-01 2006-06-02  2006-06-05  2006-06-06
1 0.004247769  0.0048529913 0.002548541 -0.006215087 -0.007154035
2 0.009323326  0.0015364830 0.006993059  0.000000000 -0.022326545
3 0.001607127 -0.0008596914 0.002280503 -0.001082275 -0.001973220

2006-06-07  2006-06-08 2006-06-09  2006-06-12  2006-06-13
1 -0.0015405703 -0.008894756 0.004967781 -0.004911457 -0.017072989
2  0.0056383270 -0.027379310 0.012429170 -0.013895865 -0.023992295
3 -0.0003462346 -0.001317298 0.002478040 -0.001395419 -0.003113317
```

```
2006-06-14  2006-06-15   2006-06-16  2006-06-19  2006-06-20
1 -0.0008148080 0.0180758310  0.0023954587  0.001269003 -0.002988990
2  0.0022722680 0.0215693860 -0.0029262460  0.008192221  0.003859327
3 -0.0006585664 0.0004955194 -0.0000012182 -0.001387469 -0.001853536
2006-06-21  2006-06-22   2006-06-23   2006-06-26  2006-06-27
1  0.0019448093  0.005720239  0.0003236553  0.0022263750 -0.004784092
2  0.0033826850  0.007397751  0.0005459380 -0.0024934820 -0.006852493
3 -0.0002515638 -0.000396063 -0.0009850436 -0.0002276698 -0.001709434

2006-06-28  2006-06-29  2006-06-30  2006-07-03   2006-07-04  2006-07-05
1 0.002418113 0.013946565 0.0007082933 0.0051894390  0.0036357180
-0.003018747
2 0.002764602 0.014041765 0.0144738140 0.0087848690  0.0019620600
-0.008807793
3 0.000155729 0.003295715 0.0007134052 0.0007825544 -0.0005997304
-0.002343919

2006-07-06   2006-07-07 2006-07-10   2006-07-11  2006-07-12
1 0.0006826483 -0.0053623837 0.006527871 -0.0026637643 -0.000460215
2 0.0055633270 -0.0057979700 0.005982938 -0.0038346560  0.004470950
3 0.0008409888  0.0001182382 0.001124756  0.0001879154 -0.000366662
2006-07-13   2006-07-14   2006-07-17   2006-07-18  2006-07-19
1 -0.013003389 -0.0051940770 -0.0004672143  0.0002666863 0.010286439
2 -0.015294320 -0.0103727210 -0.0022662700 -0.0056064500 0.018419580
3 -0.002107831  0.0007863122  0.0008104944 -0.0013472198 0.001975698

2006-07-20   2006-07-21  2006-07-24  2006-07-25  2006-07-26  2006-07-27
1 0.001826095 -0.0080939067 0.013847674 0.0046651393 0.0005912200
-0.000995908
2 0.008023110 -0.0053387720 0.019239808 0.0006067870 0.0043343300
0.005560211
3 0.001140619 -0.0006294194 0.003391497 0.0009658202 0.0007748634
0.001358731

2006-07-28  2006-07-31   2006-08-01  2006-08-02   2006-08-03
1 0.0054530087 0.0009621367 -0.0016914860  0.001702776  3.388133e-05
2 0.0102982490 0.0002672360  0.0000000000 -0.003903872 -1.101953e-02
3 0.0008377254 0.0012288024 -0.0002328868 -0.000336470 -1.946893e-03
2006-08-04  2006-08-07  2006-08-08   2006-08-09  2006-08-10
1 0.0004074303 -0.005170347 0.001632508 0.0004397503  0.005330442
2 0.0095373900 -0.012098329 0.000553560 0.0107776200 -0.005122863
3 0.0025471316 -0.001348114 0.001391077 0.0003048744  0.000744234

2006-08-11   2006-08-14  2006-08-15  2006-08-16  2006-08-17
1  0.0003234277  0.0042420490 0.006281944 0.002505853 0.0022312940
2  0.0016952790  0.0080151810 0.014854679 0.004466055 0.0029039360
3 -0.0003461608 -0.0005048628 0.002816991 0.001751071 0.0003821246
```

```
2006-08-18   2006-08-21  2006-08-22   2006-08-23   2006-08-24
1  0.001322035 -0.0061511060 0.007819312 -0.0024164317 -0.0006088927
2 -0.003261951 -0.0030139970 0.003136533 -0.0002148440  0.0020208950
3  0.001275974 -0.0002004374 0.002229195 -0.0002547412 -0.0001443246
2006-08-25  2006-08-28   2006-08-29   2006-08-30   2006-08-31
1 0.001879964 0.0001150750  0.0039964717 -0.0013533323  0.003702581
2 0.000665318 0.0018015970  0.0060754350  0.0027585440 -0.001466397
3 0.000606633 0.0008119902 -0.0000166692  0.0008165558  0.002074951

2006-09-01   2006-09-04  2006-09-05  2006-09-06   2006-09-07
1 0.003825121  0.0022519797  0.002990688 -0.005576266 -0.0046239360
2 0.003126659  0.0045617030 -0.000424630 -0.005904139 -0.0061051260
3 0.001543151 -0.0000862626  0.000150264 -0.002215612 -0.0008681144

2006-09-08  2006-09-11  2006-09-12  2006-09-13   2006-09-14  2006-09-15
1 0.005100734 -0.006268633 0.0079822300 0.004665103 -0.0020439300
0.007261748
2 0.005038040 -0.008630266 0.0115443790 0.003047893 -0.0032925020
0.005426908
3 0.002150122 -0.002463850 0.0007386444 0.001504165 -0.0003697208
0.001212241
2006-09-18   2006-09-19  2006-09-20  2006-09-21   2006-09-22
1 -0.002313630 -0.0022077103 0.002879550 0.0017740240 -0.0120944313
2  0.003786049 -0.0025686540 0.012243951 0.0039715930 -0.0091896050
3 -0.001438723  0.0004212428 0.001333139 0.0009339414  0.0000184756

2006-09-25  2006-09-26   2006-09-27  2006-09-28  2006-09-29  2006-10-02
1  0.0015608587 0.007422705  0.0054981810 0.003993309 0.002248728
-0.004799167
2 -0.0027169680 0.012816618  0.0036014040 0.000706224 0.001409431
-0.005008288
3  0.0006968214 0.002279966 -0.0000253728 0.000284538 0.000532484
-0.001994171

2006-10-03  2006-10-04  2006-10-05   2006-10-06   2006-10-09
1  0.0011308367 0.007532294 0.007702274  0.0013383343 -1.842333e-06
2  0.0006241600 0.006657881 0.007166048  0.0010145910  3.046625e-03
3 -0.0000141456 0.001957197 0.001609076 -0.0009962644  2.434124e-04
2006-10-10  2006-10-11  2006-10-12   2006-10-13   2006-10-16
1  0.0058490050 -0.001032232 0.005763076  0.0057859347  0.0016738733
2  0.0089863210  0.001066884 0.004239901 -0.0020605700 -0.0018373360
3 -0.0004574944  0.000480683 0.001357373  0.0004827066 -0.0004780952

2006-10-17  2006-10-18  2006-10-19   2006-10-20   2006-10-23
1 -0.0072083463 0.0065625757 -0.004172731  0.0013755277  0.0067030360
2 -0.0105018630 0.0091540220  0.001017744  0.0026080710  0.0055130300
3 -0.0002134596 0.0004904236 -0.001158779 -0.0000334576 -0.0001258452
```

```
2006-10-24  2006-10-25  2006-10-26  2006-10-27  2006-10-30
1  0.0006909480 0.0007045297  0.0004360173 -0.006801505 -0.0013724143
2 -0.0035125790 0.0028470680 -0.0006172110  0.002649247 -0.0047902870
3  0.0001992902 0.0006721280  0.0005680150  0.000502184  0.0003469408
2006-10-31  2006-11-01  2006-11-02  2006-11-03  2006-11-06
1 -0.001518782 1.051133e-05 -0.0013112907  0.0043942913 0.006986784
2 -0.008238610 4.875862e-03  0.0034278750  0.0059986060 0.010778959
3  0.001557833 1.065105e-03 -0.0000054624 -0.0000061736 0.002111045

2006-11-07  2006-11-08  2006-11-09  2006-11-10  2006-11-13
1 -0.000492798  0.0010355810 -0.0042464430 -0.0022569473  0.0017332780
2  0.003869904 -0.0059257000 -0.0005821010 -0.0026741380  0.0015220670
3  0.002714624  0.0004917422  0.0007154908 -0.0007612248 -0.0003075366

2006-11-14  2006-11-15  2006-11-16  2006-11-17  2006-11-20
1  0.005042735  0.0039623527  0.000571091 -0.0041334430 -0.0008046817
2 -0.001419646  0.0067483290  0.000088600 -0.0046228550  0.0012852990
3  0.001633393 -0.0003903638 -0.000064044 -0.0004950332  0.0008012360
2006-11-21  2006-11-22  2006-11-23  2006-11-24  2006-11-27
1 0.0035024650 -0.001423817 -0.0022464847 -0.0074566157 -0.008884057
2 0.0015966630  0.000878824 -0.0028429630 -0.0120525280 -0.014190477
3 0.0000896774 -0.000362714 -0.0009733886 -0.0007358458 -0.001985401

2006-11-28 2006-11-29  2006-11-30  2006-12-01  2006-12-04
1 -0.0005720070 0.010011509 -0.0028207700 -0.0045097767 0.008035140
2 -0.0064998550 0.012901202 -0.0086468080 -0.0064254010 0.006552949
3  0.0004750374 0.001182122  0.0003041304 -0.0001392252 0.003841311

2006-12-05  2006-12-06  2006-12-07  2006-12-08  2006-12-11
1  0.0006129080  0.0012128337  0.0014398583 -0.001522731 0.006351745
2 -0.0028113180  0.0071741430  0.0074404230 -0.002915097 0.006216417
3 -0.0002819442 -0.0002501772 -0.0002762668  0.000266215 0.000853780
2006-12-12  2006-12-13  2006-12-14  2006-12-15  2006-12-18
1 0.0003822723  0.0032613870 0.0085999303 0.004852111  0.0007975687
2 0.0077836640  0.0016147920 0.0100852550 0.001294973  0.0045667980
3 0.0014112930 -0.0003111648 0.0006167068 0.002972998 -0.0018724352

2006-12-19  2006-12-20  2006-12-21  2006-12-22  2006-12-25
1 -0.007259400 0.0039956217 -0.001713398 -0.001404090 -0.0004215033
2 -0.006190852 0.0015531190  0.000940550 -0.005018284  0.0000000000
3 -0.001259345 0.0002665752  0.002535426 -0.002011179 -0.0000406544

2006-12-26 2006-12-27  2006-12-28  2006-12-29  2007-01-01
1 0.0026915907 0.009066652  0.0004164007 -0.001200784 -9.820667e-06
2 0.0000000000 0.010333820 -0.0020348850 -0.001101976  0.000000e+00
3 0.0004873476 0.002159424 -0.0010962306 -0.000159042 -1.140000e-05
```

```
2007-01-02  2007-01-03   2007-01-04  2007-01-05  2007-01-08
1 0.0016820867 0.004437189 -0.0002060123 -0.002683635 -0.001690284
2 0.0000000000 0.014834908  0.0001279650 -0.002580941 -0.004541686
3 0.0004246192 0.001879946  0.0006782908 -0.001927514 -0.001011925

2007-01-09   2007-01-10  2007-01-11   2007-01-12  2007-01-15
1 0.0044103520 -0.0006727953 0.007703170  0.003633612 0.005514158
2 0.0032808260 -0.0014696800 0.013481449  0.005427095 0.008135203
3 0.0001659808 -0.0012173928 0.001547265 -0.000644134 0.001385737

2007-01-16   2007-01-17   2007-01-18  2007-01-19  2007-01-22
1 -0.0002334037 -0.0009348027  0.0018915933 0.005485118 -0.001422125
2 -0.0023385150  0.0036009040 -0.0000306000 0.004899725 -0.004463529
3  0.0019940186 -0.0003052906 -0.0002513524 0.001870075  0.000533480
2007-01-23  2007-01-24   2007-01-25   2007-01-26   2007-01-29
1 -0.0024795987 0.010078597 -0.005617319  0.0003207103 0.0026204570
2 -0.0005070610 0.006027774 -0.001732155 -0.0095941300 0.0074250550
3 -0.0000391516 0.001921089 -0.001420198  0.0000189644 0.0006710216

2007-01-30   2007-01-31  2007-02-01  2007-02-02   2007-02-05   2007-02-06
1 0.002703326 -0.0017867277 0.005665622 0.005261889 1.452767e-05
0.0016964537
2 0.004207812 -0.0000649000 0.009002711 0.004474549 2.320000e-05
0.0009123640
3 0.002512929  0.0001274554 0.001459012 0.002004385 1.534503e-03
-0.0001289984

2007-02-07   2007-02-08   2007-02-09   2007-02-12   2007-02-13
1  0.001088440  0.0005737837 -0.0000244410 -0.003474908  0.0039971257
2  0.002937055 -0.0060861880  0.0065013820 -0.004102506 -0.0023852220
3 -0.001036504  0.0008835816 -0.0007572888 -0.001322240  0.0002060698
2007-02-14   2007-02-15   2007-02-16   2007-02-19   2007-02-20
1 0.003429143 0.0003378953 -0.0010496707  0.001237031  0.002481922
2 0.006839015 0.0014878740  0.0030582220  0.000894606 -0.003270610
3 0.001787848 0.0004864332 -0.0005123012 -0.002126213  0.001312554

2007-02-21   2007-02-22   2007-02-23   2007-02-26   2007-02-27
1 -0.0003544737  0.004133925 -0.003687659 -0.0006432653 -0.028181932
2 -0.0099202370  0.003907766 -0.000693145 -0.0035030230 -0.035746244
3  0.0008831304 -0.000198173  0.000224204  0.0003787182 -0.003557775

2007-02-28   2007-03-01   2007-03-02   2007-03-05   2007-03-06  2007-03-07
1 -0.006545974 -0.004681529 -0.007204097 -0.016965306 0.0131994283
0.001848668
2 -0.011946524 -0.001959064  0.002364749 -0.014462125 0.0116318040
0.011790712
```

```
3 -0.001276448 -0.000431924 -0.001183209 -0.000934475 0.0006908982
0.000013807
2007-03-08  2007-03-09   2007-03-12  2007-03-13  2007-03-14 2007-03-15
1 0.01238125 0.0054817963 -0.0019347513 -0.011489525 -0.016127064
0.010698674
2 0.01192565 0.0003297920 -0.0031733170 -0.007777588 -0.028205491
0.014756610
3 0.00211282 0.0003630802  0.0003176836 -0.001476343 -0.003026142
0.003330527

2007-03-16  2007-03-19  2007-03-20  2007-03-21  2007-03-22   2007-03-23
1 -0.0040807663 0.011023104 0.006051645 0.008317470 0.006371441
0.003735611
2  0.0011294700 0.014488916 0.004157206 0.007430739 0.013969587
0.001698465
3 -0.0004977044 0.001539199 0.001811865 0.002705776 0.002193543
-0.000066116

2007-03-26   2007-03-27  2007-03-28  2007-03-29   2007-03-30
1 -0.0038874940 -0.0024984800 -0.007625166 0.008954610  0.0028986417
2 -0.0075840220 -0.0047538740 -0.009870398 0.011153959  0.0004970940
3 -0.0006071596 -0.0008749648 -0.001164874 0.001407261 -0.0002237774
2007-04-02  2007-04-03   2007-04-04   2007-04-05   2007-04-06
1 -0.0013403807 0.008656804 0.0031424653 -0.0014326240 -0.0004997497
2 -0.0014515930 0.010335207 0.0013071590  0.0050004920  0.0000000000
3  0.0000685862 0.002129822 0.0004029036 -0.0001666756 -0.0000500682

2007-04-09   2007-04-10  2007-04-11
1 0.006564997  0.0004269093 -0.001272110
2 0.000000000  0.0063294250 -0.001044170
3 0.000570819 -0.0000510372 -0.000411593

Clustering vector:
  SBI  SPI  SII  LMI  MPI  ALT LPP25 LPP40 LPP60
    3    2    3    3    1    1    3     3     1

Within cluster sum of squares by cluster:
[1] 0.003806242 0.000000000 0.005432037
 (between_SS / total_SS =  75.3 %)
Available components:
[1] "cluster"    "centers"     "totss"       "withinss"     "tot.withinss"

[6] "betweenss"  "size"        "iter"        "ifault"
>
++++++++++++++++++++++++++++++++++++++
```

Example 2.1 in Multivariate Data Analysis (see the illustrative example in 4.2.1)
ANOVA Analysis:

ANOVA (analysis of variance) may be achieved using the R function anova(), which is in the CRAN package stats. The standard usage form of this function is

```
> anova(object, ...)
```

with arguments:

object ≡ an object for model fitting (e.g., lm or glm).
... ≡ additional objects of the same type.

Example 2.2 in Multivariate Data Analysis

2.A Appendix. Documentation for the Plot Function

plot {graphics} R **Documentation**

Generic X–Y Plotting

2.A.1 Description

Generic function for plotting of R objects.

For simple scatter plots, plot.default will be used. However, there are plot methods for many R objects, including functions, data.frames, density objects, etc. Use methods(plot) and the documentation for these.

2.A.2 Usage

plot(x, y, ...)

2.A.3 Arguments

X the x coordinates of points in the plot. Alternatively, a single plotting structure, function, or any R object with a plot method can be provided.

Y the y coordinates of points in the plot, optional if x is an appropriate structure.

2.A.4 Details

The two step types differ in their x-y preference: Going from (x1,y1) to (x2,y2) with x1 < x2, type = "s" moves first horizontal, then vertical, whereas type = "S" moves the other way around.

2.A.5 See Also

plot.default, plot.formula and other methods; points, lines, par.
For X–Y–Z plotting see contour, persp and image.

Examples

```
require(stats)
plot(cars)
lines(lowess(cars))
plot(sin, -pi, 2*pi) # see ?plot.function

## Discrete Distribution Plot:
plot(table(rpois(100,5)), type = "h", col = "red",
lwd=10, main="rpois(100,lambda=5)")

## Simple quantiles/ECDF, see ecdf() {library(stats)} for a better one:
plot(x <- sort(rnorm(47)), type = "s", main = "plot(x, type = \"s\")")

points(x, cex = .5, col = "dark red")
```

Further Reading

Adler, J. (2010). *R in a Nutshell – A Desktopbastopol*, CA: O'Reilly Media, Inc.

Chambers, J.M., Cleveland, W.S., Kleiner, B., and Tukey, P.A. (1983). *Graphical Methods for Data Analysis*. Belmont, CA: Wadsworth.

Chan, B.K.C. (2015). *Biostatistics for Epidemiology and Public Health Using* R. NY: Springer Publishing.

Mittal, H.V. (2011). *R Graphics Cookbook*. Birmingham, UK: PACKT Publiching Ltd.

Murrell, P. (2006). *R Graphics*, Computer Science and Data Analysis Series. Boca Raton, FL: Chapman & Hall/CRC.

Peng, R.D. and Domonici, F. (2008). *Statistical Methods for Environmental Epidemiology with R – A Case Study in Air Pollution and Health*. New York, NY: Springer Use R! Series, Springer Science+Business Media.

Schottenfeld, D. (2002). Epidemiology: an introduction. *Am. J. Epidemiol.* 156 (2): 188–190.

Seefeld, K. and Linder, L. (2005). "*Statistics Using R with Biological Examples*", Department of Mathematics & Statistics, University of New Hampshire, Durham, NH, *Also affiliated with the Dept. of Nephrology and the Biostatistics Research Center, Tufts-NEMC, Boston, MA. http://cran.rproject.org/doc/contrib/Seefeld_StatsRBio.pdf.

Statistician Job Search, http://jobs.amstat.org/jobs/4627784/biostatistician-1.

Teetor, P. (2011). *R Cookbook – Proven Recipes for Data Analysis, Statistics, and Graphics*. Sebastopol, CA: O'Reilly Media.

Verzani, J. (2005). *Using R for Introductory Statistics*. Boca Raton, FL: Chapman & Hall/CRC.

Wuertz, D., Chalabi, Y., Chen, W., and Ellis A. (2009). Portfolio Optimization with R/Rmetrics, Rmetrics eBook, Rmetrics Association and Finance Online, Zurich.

3

A Theory of Simultaneous Mass Transfer and Chemical Reactions with Numerical Solutions

3.1 Introduction

3.1.1 A Classical Experimental Study of Simultaneous Absorption of Carbon Dioxide and Ammonia in Water

Hatch and Pigford [24] studied the process of simultaneous gas absorption accompanied by chemical reaction, using the system CO_2–NH_3–Water. These two gases were mixed and brought into contact with a smooth laminar jet of water. After dissolution, they reacted with one another by a bimolecular reaction. The absorption rates were assessed by chemical analysis of the liquid phase leaving the jet absorption equipment. It was found that the rate of absorption of CO_2 was significantly affected by reaction with an excess of NH_3 in the liquid. (This result is in agreement with predictions based on the penetration theory for gas absorption of two gases followed by the second order and irreversible reaction.)

The high solubility of NH_3 in water which results in a large gas-side resistance to its absorption made it impossible to assess the effect of the rate of reaction on its rate of absorption owing to the difficulty in determining the NH_3 concentration at the gas–liquid interface. The confirmation of the diffusion-reaction theory was thus limited to the observations on CO_2 absorption, and its validity depends on the accuracy of the NH_3 gas-phase resistance measurements.

This classical experimental study illustrates the industrially important situations in which two chemically active gases are simultaneously dissolved in a liquid, diffusing and reacting rapidly near the interface immediately after the solution occurs.

Two important practical examples of this situation are

(i) In the production of synthesis gas: the simultaneous removal of CO_2 and H_2S from a gas mixture by washing with an alkaline liquid.
(ii) In the Solvay process: the simultaneous washing, with water, of a gas mixture containing both NH_3 and CO_2.

In these situations, the rate of absorption of each component depends on the rate of the other component, because the concentration distribution near the interface of one component is affected by its reaction with the other component. The experimental study seemed to show that the results closely followed the penetration

Simultaneous Mass Transfer and Chemical Reactions in Engineering Science, First Edition. Bertram K. C. Chan.

theory when the effect of reaction is allowed for with the aid of known reaction mechanisms.

The chemical system in this study involves NH_3 and CO_2 as the gaseous components and distilled water as the solvent.

The chemistry of this system consists of the gases reacting in the solution via the following steps:

$$CO_2 + NH_3 \rightarrow NH_2COOH \tag{R1}$$

$$NH_2COOH + NH_3 \rightarrow NH_4^+ + NH_2COO^- \tag{R2}$$

$$NH_2COO^- + H_2O \rightarrow NH_4CO_3^- \tag{R3}$$

$$NH_4CO_3^- \rightarrow NH_4^+ + CO_3^- \tag{R4}$$

For this chemical system:

- Reaction (R1) shows the formation of carbamic acid NH_2COOH.
- Reaction (R2) shows the carbamic acid picking up another NH_3 to form ammonium carbamate, which partially ionizes into the ammonium ion NH_4^+ and the carbamate ion NH_2COO^-.
- Reaction (R3) shows that the carbamate ions NH_2COO^- slowly react with water to form the ammonium bicarbonate ions $NH_4CO_3^-$.
- Reaction (R4) shows that the ammonium bicarbonate ions $NH_4CO_3^-$ may then dissociate into ammonium ions NH_4^+ and carbonate ions CO_3^-.

In this reaction system, ionic equilibria are present:

Reaction (R1) is the rate-determining step in the absorption process because Reaction (R2) is very ***fast*** and ***irreversible***. Taken together, these two reactions consist of the consumption of two molecules of NH_3 per molecule of CO_2 via the following net reaction:

$$CO_2 + 2NH_3 \rightarrow NH_4^+ + NH_2COO^- \tag{R5}$$

which follows the second-order reaction rate law: the rate is proportional to the ***product*** of the concentrations of CO_2 and NH_3.

Reaction (R3) is much slower than the others and occurs mostly outside the experimental apparatus for the absorption process.

3.1.2 Physical Absorption

For purely physical absorption into a laminar liquid phase, the well-known result from the penetration theory is

$$k^0{}_L = 2\sqrt{(D_A/\pi t)} \tag{3.1}$$

And for the experimental jet absorber, in which t is the time of flight of a liquid particle in the jet (from the nozzle exit to the receiver), is approximately given by:

$$t = h\pi d^2/4q_L \tag{3.2}$$

where

t = Elapsed time of exposure of liquid to gas
h = Jet length
d = Jet diameter
q_L = Volumetric flow rate of liquid stream through the nozzle for a jet of *perfectly* uniform diameter, having a rod-like velocity distribution

For such as ideal jet, the total rate of absorption for the whole jet surface is

$$\phi^0{}_A = 4\sqrt{(D_A q_L h)(A_i - A_1)} \tag{3.3}$$

where

$\phi^0{}_A$ = Absorption rate for the whole jet
D_A = Diffusivity of dissolved gas A
q_L = Volumetric flow rate of liquid stream through the nozzle
h = Jet length
A_i = Interfacial concentration of A (CO_2) in the liquid phase
A_1 = Initial concentration of A (CO_2) in the liquid phase

3.1.2.1 Results

The data for both NH_3–N_2 and NH_3–He, in which the NH_3 diffusion coefficient was about four times greater, had been satisfactorily correlated by a log–log plot, giving a straight line for an empirical equation:

$$[k*_{GB}\pi h d P_M/q_v(P/RT)] = 0.76(D_{vB}h/q_v)^{0.28} \tag{3.4}$$

where

$k*_{GB}$ = Average mass transfer coefficient for B
d = Jet diameter
P_M = Arithmetic-average partial pressure of the inert gas
q_v = Volumetric flow rate of gas mixture into the apparatus
P = Total pressure
R = D_B/D_A
T = Temperature
D_{vB} = Diffusivity of B in the gas phase
h = Jet length

Equation (3.4), the empirical correlation, showed that the Stanton number for diffusion in the gas phase was only slightly influenced by flow, diffusion, or jet length.

(The Stanton number for diffusion in the gas phase, $St = D_{vB}/\rho u c_p$, where ρ = density of the fluid, u = speed of the fluid, and c_p = specific heat of the fluid.)

Overall, it was found that the absorption rate for CO_2 seems to correspond closely to the penetration theory when the effect of the reaction is assumed by using known reaction mechanisms.

3.2 Biomolecular Reactions

3.2.1 Occurrences of Simultaneous Biomolecular Reactions and Mass Transfer Are Common in Many Biomedical Environments

Some typical examples are [1]:

(1) **Intestinal Drug Absorption Involving Bio-transporters and Metabolic Reactions with Enzymes**: The absorption of drugs via the oral route is a subject of ongoing and serious investigations in the pharmaceutical industry since good bio-availability implies that the drug is able to reach the systemic circulation via the oral path. Oral absorption depends on both the drug *properties* and the physiology of the gastrointestinal tract, or *patient properties*, including drug dissolution, drug interaction with the aqueous environment and membrane, permeation across membrane, and irreversible removal by organs such as the liver, intestines, and the lung.

(2) **Oxygen Transport via Metal Complexes**: On the average, an adult at rest consumes 250 ml of pure oxygen per minute to provide energy for all the tissues and organs of the body, even when the body is at rest. During strenuous activities, such as exercising, the oxygen needs to increase dramatically. The oxygen is transported in the blood from the lungs to the tissues where it is consumed.

However, only about 1.5% of the oxygen transported in the blood is dissolved directly in the blood plasma. Transporting the large amount of oxygen required by the body and allowing it to leave the blood when it reaches the tissues that demand the most oxygen, require a more sophisticated mechanism than simply dissolving the gas in the blood.

To meet this challenge, the body is equipped with a finely tuned transport system that centers on the metal complex ***heme***. The metal-ions bind and then release ligands in some processes and oxidize and reduce in other processes, making them ideal for use in biological systems. The most common metal used in the body is iron, and it plays a central role in almost all living cells. For example, iron complexes are used in the transport of oxygen in the blood and tissues. Metal-ion complexes consist of a metal ion that is bonded via "coordinate-covalent bonds" to a small number of anions or neutral molecules called ligands. For example, the ammonia (NH_3) ligand is a monodentate ligand; i.e. each monodentate ligand in a metal-ion complex possesses a single electron-pair-donor atom and occupies only one site in the coordination sphere of a metal ion.

Some ligands have two or more electron-pair-donor atoms that can simultaneously coordinate to a metal ion and occupy two or more coordination sites; these ligands are called ("claw") agents, because they appear to grasp the metal ion between two or more electron-pair-donor atoms. The ***coordination number*** for a metal refers to the total number of occupied coordination sites around the central metal ion (i.e. the total number of metal–ligand bonds in the complex). This process is another important example of biomolecular reaction and transport.

(3) **Carotenoid Transport in the Lipid Transporters SR-BI, NPC1L1, and ABCA1**: The intestinal absorption of carotenoids in vivo involves several crucial steps:
   (1) release from the food matrix in the lumen,
   (2) solubilization into mixed micelles,
   (3) uptake by intestinal mucosal cells,
   (4) incorporation into chylomicrons,
   (5) secretion into the lymph.

   Research has shown that:
   (A) EZ is an inhibitor of the intestinal absorption of carotenoids, an effect that decreased with increasing polarity of the carotenoid molecule;
   (B) SR-BI is involved in intestinal carotenoid transport;
   (C) EZ acts not only by interacting physically with cholesterol transporters as previously suggested, but also by down-regulating the gene expression of three proteins involved in cholesterol transport in the enterocyte, the transporters SR-BI, NPC1L1, and ABCA1.

   The intestinal transport of carotenoid is thus a facilitated process resembling that of cholesterol; therefore, carotenoid transport in intestinal cells may also involve more than one transporter.

   Hence, the study of biomolecular reaction and transport is an area of importance in biomedical processes and their occurrences in epidemiologic investigations. In this section, one applies the facilities available in the R environment to solve problems arisen from these processes. This study is being approached from two directions:
   - Using the R environment as a ***support*** to numerical analytical schemes that may be developed to solve this class of problems.
   - Applying the R functions in the CRAN package ReacTran.[24]

3.3 Some Examples in Chemical Engineering Sciences [25]

3.3.1 Simultaneous Chemical Reactions and Mass Transfer

A large number of highly significant industrial systems consist of simultaneous mass transfer operations and chemical reactions, some of these are:

(1) Absorption of carbon dioxide gas into a dilute aqueous solution of sodium hydroxide in a packed column, to remove the CO_2 gas from a vapor stream.
(2) Absorption of CS_2 in aqueous amine solutions for the manufacture of dithiocarbamates.
(3) Absorption of O_2 in aqueous solutions of CuCl for the conversion to $CuCl_2$ and copper oxychloride.
(4) Addition chlorination: reaction between Cl_2 and C_2H_4 in a $C_2H_4Cl_2$ medium.
(5) Substitution chlorination: chlorination of many organic compounds such as benzene, toluene, phenols, etc.

3.4 Some Models in the Diffusional Operations of Environmental Transport *Unaccompanied* by Chemical Reactions [11]

In a theoretical model analysis of environmental contamination of liquid media, Choy and Reible [11] provided a similar analysis of the transport of contaminants in a medium. However, in that study, *no chemical interactions were considered*, and only the movements of the contaminants were modeled. Two principal cases were analyzed.

3.4.1 Diffusion Models of Environmental Transport

The concentration of the contaminant A, c_A, may be found from the solutions of the following equation:

$$\partial c_A/\mathrm{d}t = (D_A/R_f)\partial^2 c_A/\partial x^2 \quad x \in [0, \infty)$$

in which

D_A = The diffusivity, or dispersion coefficient, of species A
R_f = The retardation factor of the solution medium

3.4.2 Advection–Diffusion Models of Environmental Transport

The concentration of the contaminant A, c_A, may be found from the solutions of the following equation:

$$\partial c_A/\mathrm{d}t = (D_A/R_f)\partial^2 c_A/x^2 - (v/R_f)\partial c_A/\partial x \quad x \in [0, \infty)$$

in which

D_A = The diffusivity, or dispersion coefficient, of species A
R_f = The retardation factor of the solution medium
v = The Darcy velocity of the mobile phase

For simple model geometries, analytical expressions were obtained for these cases.

Remarks:

(1) Should chemical reactions take place among the interactive species, and with the liquid medium itself, more complicated reactions will occur.
(2) Nevertheless, for such case, solutions may be obtained in the manner discussed in the rest of this book, to provide the appropriate and applicable analytical solutions. These solutions may then be used to guide a full understanding of the systems considered.

3.5 The Concept of Diffusion

3.5.1 Publishers' Remarks

Now in its 85th Anniversary Edition, this industry-standard resource has equipped generations of engineers and chemists with vital information, data, and insights.

Thoroughly revised to reflect the latest technological advances and processes, ***Perry's Chemical Engineers' Handbook, Ninth Edition***, provides unsurpassed coverage of every aspect of chemical engineering. You will get comprehensive details on chemical processes, reactor modeling, biological processes, biochemical and membrane separation, process and chemical plant safety, and much more.

This fully updated edition covers
"Unit Conversion Factors and Symbols • Physical and Chemical Data including Prediction and Correlation of Physical Properties • Mathematics including Differential and Integral Calculus, Statistics, Optimization • Thermodynamics • Heat and Mass Transfer • Fluid and Particle Dynamics *Reaction Kinetics • Process Control and Instrumentation• Process Economics • Transport and Storage of Fluids • Heat Transfer Operations and Equipment • Psychrometry, Evaporative Cooling, and Solids Drying • Distillation • Gas Absorption and Gas–Liquid System Design • Liquid–Liquid Extraction Operations and Equipment • Adsorption and Ion Exchange • Gas–Solid Operations and Equipment • Liquid–Solid Operations and Equipment • Solid–Solid Operations and Equipment •Chemical Reactors • Bio-based Reactions and Processing • Waste Management including Air, Wastewater and Solid Waste Management• Process Safety Including Inherently Safer Design • Energy Resources, Conversion and Utilization* Materials of Construction."

3.5.2 Fick's Laws of Diffusion

3.5.2.1 Fick's First Law of Diffusion (Steady-State Law)

Fick (1955) first described the molecular diffusion in an isothermal, and isobaric binary system of components A and B.

For molecular diffusion, the molar flux of a species relative to an observer moving with molar average velocity is ***proportional*** to the concentration gradient in a given direction:

$$J_A \propto (dC_A/dZ) \tag{3.5A}$$

or

$$J_A = -D_{AB}(dC_A/dZ) \tag{3.5B}$$

in which

J_A = The molar flux of component A in the Z direction
C_A = The concentration of A
Z = The distance of diffusion
D_{AB} = The constant of proportionality, D_{AB}, is the ***diffusion coefficient, or*** the ***diffusivity,*** of the molecule A in B. This is valid only at steady-state condition of diffusion. Equation (3.5B) is called ***Fick's first law of diffusion.***

3.5.2.1.1 Illustrating Fick's Laws of Diffusion

The technique use to measure cardiac output is illustrated in Figure 3.1 and shows molecular diffusion from a microscopic and macroscopic point of view.

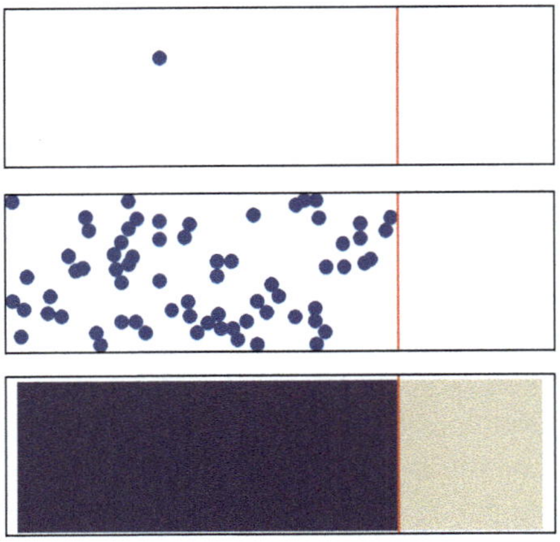

Figure 3.1 A demonstration of the process of diffusion. Source: Sbyrnes321/Wikimedia Commons/Public Domain.

Initially, there are *solute* molecules on the left side of a barrier (purple line) and ***none*** on the right. The barrier is then removed, and the solute diffuses to fill the whole container.

Top: A single molecule moves around randomly.
Middle: With more molecules, there is a clear trend where the solute fills the container more and more uniformly.
Bottom: With an enormous number of solute molecules, randomness becomes undetectable: The solute appears to move smoothly and systematically from high-concentration areas to low-concentration areas. This smooth flow is described by Fick's laws.

Fick's Laws of Diffusion describe ***diffusion*** and was derived by Fick in 1855. They can be used to solve for the diffusion coefficient, D.

Fick's first law may be used to derive his second law which in turn is identical to the diffusion equation.

3.5.2.1.2 Some Remarks on Fick's First Law of Diffusion

Fick's first law relates the diffusive *flux* to the concentration under the assumption of *steady state*. It postulates that the flux goes from regions of high concentration to regions of low concentration, with a magnitude that is proportional to the concentration gradient (spatial derivative), or in simplistic terms the concept that a solute will move from a region of high concentration to a region of low concentration across a concentration gradient. In one (spatial) dimension, the law is

$$J = -D\,\mathrm{d}\varphi/\mathrm{d}x \tag{3.6}$$

where

- J is the "diffusion flux," of which the dimension is the amount of material per unit area per unit time, so it is expressed in such units as mol/m^2 s. J measures the amount of material that will flow through a unit area during a unit time interval.
- D is the **diffusion coefficient** or **diffusivity**. Its dimension is area per unit time, so typical units for expressing it would be m^2/s.

- φ (*for* ideal mixtures) is the concentration, of which the dimension is amount of material per unit volume. It may be expressed in units of mol/m^3.
- x is the position coordinate, the dimension of which is length. It may be expressed in the unit m.

D is proportional to the velocity-squared of the diffusing particles, which depends on the temperature, viscosity of the fluid, and the size of the particles according to the Stokes–Einstein relation. In dilute aqueous solutions, the diffusion coefficients of most ions are similar and have values that at room temperature are in the range of 0.6×10^{-9} to 2×10^{-9} m^2/s. For biological molecules, the diffusion coefficients normally range from 10^{-11} to 10^{-10} m^2/s.

In two or more dimensions, one may use ∇, the *del* or *gradient* operator, which generalizes the first derivative, obtaining

$$J = -D\nabla\varphi \tag{3.7}$$

where J denotes the diffusion flux vector.

The driving force for the one-dimensional diffusion is the quantity $-\partial\phi/\partial x$, which for ideal mixtures is the concentration gradient. In biological and chemical systems other than ideal solutions or mixtures, the driving force for diffusion of each species is the gradient of chemical potential of this species. Then Fick's first law (one-dimensional case) may be expressed as:

$$J_i = -(D_{ci}/RT)(\partial\mu_i/\partial x) \tag{3.8}$$

where

the index i denotes the ith species,
c is the concentration (mol/m^3),
R is the Universal Gas Constant (J/(K mol)),
T is the absolute temperature (K), and
μ is the chemical potential (J/mol).

If the primary variable is mass fraction (y_i, given, for example, in kg/kg), then the equation changes to:

$$J_i = -\rho D\nabla y_i \tag{3.9}$$

where ρ is the fluid density (for example, in kg/m^3).

Note that the density ρ is outside the gradient operator ∇.

3.5.2.2 Fick's Second Law of Diffusion

Fick's second law predicts how diffusion causes the concentration to change with time.

It is a *partial differential equation* which, in one dimension, it reads

$$\partial\varphi/\partial t = D\partial^2\varphi/\partial x^2 \tag{3.10}$$

where

φ is the concentration in dimensions of ([amount of substance] length^{-3}), for example: mol/m^3,

$\varphi = \varphi(x, t)$ is a function that depends on location x and time t,
t is time (s),
D is the diffusion coefficient in dimensions of (length^2 time^{-1}), example: m^2/s, and
x is the position (length), example: m

In two or more dimensions, one should use the Laplacian $\Delta = \nabla^2$, which generalizes the second derivative, obtaining the equation:

$$\partial\varphi/\partial t = D\Delta\varphi \tag{3.11}$$

3.5.3 Derivation of Fick's Laws of Diffusion

3.5.3.1 Remarks: Additional Remarks on Fick's Laws of Diffusion

3.5.3.1.1 Fick's First Law of Diffusion

Derivation of Fick's First Law of Diffusion: In one dimension, the following derivation is based on a similar argument made in Berg [26].

Consider a collection of particles performing a random walk in one dimension with length scale Δx and timescale Δt. Let $N(x, t)$ be the number of particles at position x at time t. At a given time step, half of the particles would move left and half would move right. Since half of the particles at point x move right, and half of the particles at point $x + \Delta x$ move left, the net movement to the right is

$$[-1/2\{N(x + \Delta x, t) - N(x, t)\}]$$

The flux, J, is this ***net*** movement of particles across some area element of area a, normal to the random walk during a time interval Δt. Hence, one may write

$$\begin{aligned} J &= [-1/2\{N(x + \Delta x, t) - N(x, t)\}]/a\Delta t \\ &= -1/2[\{N(x + \Delta x, t)/a\Delta t\} - \{N(x, t)/a\Delta t\}] \end{aligned} \tag{3.12}$$

Multiplying the top and bottom of the right-hand side by $(\Delta x)^2$ and rewriting, one obtains

$$J = \{(\Delta x)^2/2\Delta t\}[\{N(x + \Delta x, t)/a(\Delta x)^2\} - \{N(x, t)/a(\Delta x)^2\}] \tag{3.13}$$

Defining concentration as the number of particles per unit volume, viz.

$$\varphi(x, t) = N(x, t)/a\Delta x \tag{3.14}$$

Moreover, $(\Delta x)^2/2\Delta t$ is the definition of the diffusion constant in one dimension, D. Thus, the expression simplifies to

$$J = D[\{\varphi(x + \Delta x, t)/\Delta x\} - \{\varphi(x, t)/\Delta x\}] \tag{3.15}$$

In the limit when Δx approaches an infinitesimal limit, the right-hand side becomes a space derivative:

$$J = -D(\partial\varphi/\partial x) \tag{3.16}$$

3.5.3.1.2 Fick's Second Law of Diffusion

Fick's second law of diffusion may be obtained from Fick's first law of diffusion together with the property of mass conservation *in the absence of any chemical reactions*:

$$\partial\varphi/\partial t + \partial J/\partial x = 0 \quad \Rightarrow \quad \partial\varphi/\partial t - \partial/\partial x(D\partial\varphi/\partial x) = 0 \tag{3.17}$$

Assuming the diffusion coefficient D to be constant, one may exchange the orders of the differentiation and multiply by the constant:

$$\partial/\partial x(D\partial\varphi/\partial x) = D(\partial/\partial x)(\partial/\partial x)\varphi = D(\partial^2\varphi/\partial x^2) \tag{3.18}$$

and thus receive the form of the Fick's equations as was stated above.

For the case of diffusion in two or more dimensions, Fick's second law becomes

$$\partial\varphi/\partial t = D\nabla^2\varphi \tag{3.19}$$

which is analogous to the well-known ***heat conduction equation***.

If the diffusion coefficient is ***not*** a constant, but depends on the coordinate or concentration, Fick's second law yields

$$\partial\varphi/\partial t = \nabla \cdot (D\nabla\varphi) \tag{3.20}$$

An important example is the case where φ is at a steady state, i.e. the concentration does not change with time, so that the left part of the above equation is identically zero. In one dimension with constant D, the solution for the concentration will be a linear change of concentrations along x. In two or more dimensions, one obtains

$$\nabla^2\varphi = 0 \tag{3.21}$$

which is the Laplace equation, the solutions to which are known as ***harmonic functions***.

A Derivation of Fick's Second Law of Diffusion Fick's second law of diffusion is a special case of the convection–diffusion equation in which there is no advective flux and no net volumetric source, viz., in the absence of any chemical reactions. It may be derived from Fick's first law of diffusion ***and*** the mass conservation continuity equation:

$$\partial\varphi/\partial t + \partial J/\partial x = 0 \quad \Rightarrow \quad \partial\varphi/\partial t - (\partial/\partial x)(D\partial\varphi/\partial x) = 0 \tag{3.22}$$

Assuming the diffusion coefficient D to be constant, one may exchange the orders of the differentiation and multiply by the constant:

$$(\partial/\partial x)(D\partial\varphi/\partial x) = D(\partial/\partial x)(\partial/\partial x)\varphi = D(\partial^2\varphi/\partial x^2) \tag{3.23}$$

and thus resulted in the form of the equation for Fick's second law of diffusion:

$$\partial\varphi/\partial t = D\nabla^2\varphi \tag{3.24}$$

(This is analogous to the heat conduction equation.) If the flux were the result of both diffusive and advective fluxes, then the convection–diffusion equation is the result.

3.5.3.2 Example Solution in One Dimension: Diffusion Length

A simple case of diffusion with time t in one dimension (taken as the x-axis) from a boundary located at position $x = 0$, where the concentration is maintained at the value n_0 is

$$n(x,t) = n_0 \operatorname{erfc}\{x/2\sqrt{(Dt)}\} \tag{3.25}$$

where erfc{ } is the complementary ***Error Function***. This models the case when corrosive gases diffuse through the oxidative layer toward the metal surface (if it is assumed that concentration of gases in the environment is constant and the diffusion space (i.e. corrosion product layer) is *semi-infinite* – starting at 0 at the surface and spreading infinitely deep in the material). If the diffusion space is *infinite*:

$$n(x,0) = 0, \quad x > 0 \tag{3.26}$$

and

$$n(x,0) = n_0, \quad x \leq 0 \tag{3.27}$$

then the solution is amended only with coefficient 1/2 in front of n_0 (this is quite obvious, as the diffusion now occurs in both positive and negative directions).

This case is valid when some solution with concentration n_0 is put in contact with a layer of pure solvent.

The length $2\sqrt{(Dt)}$ is called the *diffusion length* and provides a measure of how far the concentration has propagated in the x-direction by diffusion in time t.

As an approximation of the error function, the first two terms of the Taylor series may be used:

$$n(x,t) = n_0[1 - 2\{x/2\sqrt{(Dt\pi)}\}] \tag{3.28}$$

If D is time-dependent, the diffusion length becomes $2\sqrt{\{\int D(t')\mathrm{d}t'\}}$

This idea is useful for estimating a diffusion length over a heating and cooling cycle, where D varies with temperature.

3.6 The Concept of the Mass Transfer Coefficient

Movement of the bulk fluid particles in the turbulent condition is, as in any turbulent fluid dynamics environment, highly complex.

For gases, it is well known as a molecular diffusion since it is described in terms of the kinetic theory of gases. The rate of mass transfer from the interface to the turbulent zone in the same manner can be useful for molecular diffusion. The term F is called mass transfer coefficient. The value of F depends on the local nature of the fluid motion. It is a local mass transfer coefficient defined for a particular location on the interface.

3.7 Theoretical Models of Mass Transfer

3.7.1 Nernst *One-Film* Theory Model and the Lewis–Whitman *Two-Film* Model

In chemical engineering analysis of mass transfer, the film theory is based on the following assumptions.

Mass transfer occurs by molecular diffusion through a fluid layer at the phase boundary. Mass transfer occurs by molecular diffusion through a fluid layer at phase boundary (like a solid wall). Beyond this film, concentration is homogeneous.

3.7.1.1 Gas Transfer Rates

If either phase concentration cannot be predicted by Henry's law, then there will be a transfer of mass across the interface until equilibrium is reached. The mechanisms and rate expressions for this transfer process have been conceptualized in a variety of ways so that quantitative descriptions are possible.

One common conceptualization method is the ***film theory***, or the **one-film theory** or the **two-film theory**.

3.7.1.2 The Nernst One-Film Model

A simple model of the gas–liquid transfer process is attributed to Nernst (1904): Nernst ***postulated*** that near the interface there exists a *stagnant film*. This stagnant film is ***hypothetical***, since the details of the velocity profile near the interface are unknown. In this film, mass transport is governed by *molecular diffusion*.

Therefore, Fick's law describes flux J through the film:

$$J = -D\partial C/\partial X \tag{3.29}$$

(typical units: mg/cm^2 s), where D is the diffusivity or the diffusion coefficient.

If the thickness of the stagnant film is given by δ_n, then the gradient may be approximated by:

$$\partial C/\partial X \approx (C_\mathrm{b} - C_\mathrm{i})/\delta_n \tag{3.30}$$

in which C_b and C_i are concentrations in the bulk and at the interface, respectively. At steady state, *if* there are no reactions in the stagnant film, then there will be no accumulation in the film (assuming that D remains constant). Hence, the gradient must be linear and the approximation is appropriate, and:

$$J = -D(C_\mathrm{b} - C_\mathrm{i})/\delta_n \tag{3.31}$$

3.7.1.3 Mass Transfer Coefficients

To proceed with calculations of the mass transfer coefficients, define a mass transfer coefficient for either the liquid or gas phase as k_l or k_g (with dimensions = $L/T = LT^{-1}$):

$$k_{\mathrm{l,g}} = D/\delta_n \tag{3.32}$$

Calculation of C_i is done by assuming that equilibrium (Henry's law) is attained instantly at the interface (i.e. use Henry's law based on the bulk concentration of the other bulk phase). This assumes that the other phase does not have a "film." This problem will be addressed later. Thus, at this stage of the analysis:

$$C_i = C_g/H_c \tag{3.33}$$

(If the film side is liquid and the opposite side is the gas phase).

3.7.1.4 The Lewis-Whitman Two-Film Model

In cases with gas–liquid transfer, one may have transfer considerations from both sides of the interface. In such cases, one may use the **Lewis–Whitman** (1923) **two-film model** as described here.

Calculation of C_i is done by assuming that equilibrium (viz., Henry's law) is attained ***instantly*** at the interface (i.e. using Henry's law based on the bulk concentration of the other bulk phase). This assumes that the other phase does not have a "film." (This problem will be addressed later.)

For the moment:

$$C_i = C_g/H_c \tag{3.34}$$

(If the film side is liquid and the opposite side is the gas phase.)

3.7.1.5 The Two-Film Model

In many cases with gas–liquid transfer, one has transfer considerations from *both* sides of the interface. Therefore, one may use the Lewis–Whitman two-film model as described in Figure 3.2.

The same assumptions will be applied to the two-film model as apply in the single Nernst film model. The problem is that one may now have difficulty in finding the interface concentrations, C_{gi} or C_{li}. However, one may assume that equilibrium will

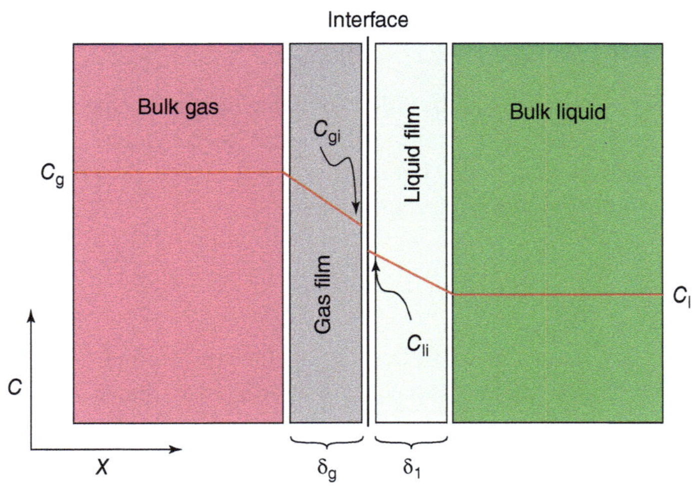

Figure 3.2 The Lewis–Whitman two-film model for mass transfer with chemical reactions.

be attained at the interface (since, generally, gas solubilization reactions occur rather fast), so that:

$$C_{li} = C_{gi}/H_c \tag{3.35}$$

A steady-state flux balance (assumed for thin films) through each film may now be performed. The fluxes are then given by:

$$J = k_l(C_l - C_{li}) \tag{3.36A}$$

and

$$J = k_g(C_{gi} - C_g) \tag{3.36B}$$

assuming that the flux is in the X-direction, viz., from bulk gas phase to bulk liquid phase.

Since the concentrations at the interface ***cannot*** be measured, so overall mass transfer coefficients are defined. These coefficients are based on the difference between the bulk concentration in one phase and the concentration that would be in equilibrium with the bulk concentration in the other phase. Now, define

$$J = K_l\left(C_l - C_l^*\right) \tag{3.37A}$$

$$J = K_g\left(C_g^* - C_g\right) \tag{3.37B}$$

where

K_l = ***Overall*** mass transfer coefficient based on liquid-phase concentration
K_g = ***Overall*** mass transfer coefficient based on gas-phase concentration

Note: Both K_l and K_g have dimensions L/T, or LT^{-1}

C_l^* = Liquid-phase concentration that would be in equilibrium with the bulk gas concentration
= C_g/H_c (typical dimensions are mol/m^3)

C_g^* = Gas-phase concentration that would be in equilibrium with the bulk liquid concentration
= H_cC_l (typical dimensions are mol/m^3)

Expanding the liquid-phase overall flux equation to include the interface liquid concentration:

$$J = K_l\left\{[C_l - C_{li}] + \left[C_{li} - C_l^*\right]\right\} \tag{3.38}$$

Then substitute

$$C_{li} = C_{gi}/H_c \tag{3.39A}$$

and

$$C_l^* = C_g/H_c \tag{3.39B}$$

to obtain

$$J = K_l\{(C_l - C_{li}) + (C_{gi} - C_g)/H_c\} \tag{3.40}$$

In the steady state, fluxes through all films must be equal.

Let all these fluxes be equal to J.

On an individual film basis:

$$(C_l - C_{li}) = J/k_l \tag{3.41A}$$

and

$$(C_{gi} - C_g) = J/k_g \tag{3.41B}$$

Since all the Js are equal:

$$J = K_l\{(J/k_l) + (J/H_c \cdot k_g)\} \tag{3.42}$$

which may can be rearranged to give

$$(1/K_l) = (1/k_l) + (1/H_c \cdot k_g) \tag{3.43A}$$

A similar analysis starting with the overall flux equation based on gas-phase concentration will give

$$(1/K_g) = (H_c/k_l) + (1/k_g) \tag{3.43B}$$

These last two equations (3.43A) and (3.43B) may be viewed as "resistance" expressions, where $1/K_g$ or $1/K_l$ represent total resistance to mass transfer based on gas- or liquid-phase concentration, respectively.

Thus, the total resistance to mass transfer consists of three resistances in series:

(1) Liquid film,
(2) Interface,
(3) Gas film.

In this model analysis, it was assumed that instant equilibrium at the interface, so there is no transfer limitation here.

Remark: It should be noted that model selection (film, surface renewal, or penetration) does ***not*** influence the outcome of this analysis.

3.7.1.6 Single-Film Control

It is possible that one of the films exhibits relatively high resistance and therefore dominates the overall resistance to transfer. This, of course, depends on the relative magnitudes of k_l, k_g, and H_c. Thus, the solubility of the gas and the hydrodynamic conditions which establish the film thickness or renewal rate (in either phase) determine if a film controls.

In general, highly soluble gases (low H_c) have transfer rates controlled by gas film (or renewal rate) and vice versa. For example,

(i) Oxygen (slightly soluble) transfer is usually controlled by liquid film.
(ii) Ammonia (highly soluble) transfer is usually controlled by gas-phase film.

3.7.1.7 Applications

Transfer of gas across a gas–liquid interface may be accomplished by bubbles or by creating large surfaces (interfaces).

3.7.2 Higbie's Penetration Theory Model

Many of the industrial processes of mass transfer with chemical reactions (such as gas absorption) are *unsteady-state processes* where the gas–liquid contact time is too short to achieve steady-state condition. For example, in the absorption of gases from bubbles or absorption by wetted wall columns, the mass transfer surface is formed instantaneously and transient diffusion of the material takes place, see Figure 3.3.

Assume that a bubble is rising in a pool of liquid (as in a bubble column) where the liquid elements are swept on its surface, remain in contact with it during their motion, and finally get detached at the bottom.

The basic assumptions of the penetration theory of mass transfer are

- Unsteady-state mass transfer occurs to a liquid element as long as it is in contact with the gas bubbles
- Equilibrium exists at gas–liquid interface
- Each liquid elements stays in contact with the gas for same period of time (The liquid elements are moving at the same rate and there is not velocity gradient within the liquid.)

Under these assumption, *the convective terms in the diffusion can be neglected* and the unsteady-state mass transfer of gas (penetration) into the liquid element may be obtained from Fick's second law for unsteady-state diffusion as:

$$\partial c/\partial t = D_{AB}(\partial^2 c/\partial z^2) \tag{3.44}$$

subject to the following boundary conditions:

$$t = 0, z > 0: \quad c = c_{Ab}, \text{ and}$$
$$t > 0, z = 0: \quad c = c_{Ai}$$

where

c_{Ab} = The concentration of solute A at infinite distance from the surface (viz., the bulk concentration)

c_{Ai} = The interfacial concentration of solute A at the surface

On solving the above partial differential equation, the result is

$$(c_{Ai} - c)/(c_{Ai} - c_{Ab}) = \text{erf}\{z/2\sqrt{(D_{AB}t)}\} \tag{3.45}$$

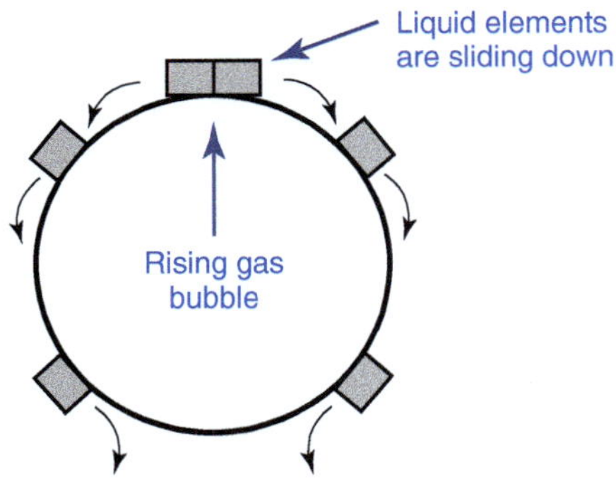

Figure 3.3 The penetration model for mass transfer with chemical reactions.

in which erf{} is the error function.

- Some Remarks on the error function erf(x)

  In mathematics, the **Error Function** (also known as the **Gaussian error function**) is a nonelementary special function of sigmoid shape that occurs in the theory of probability, statistics, and partial differential equations when describing diffusion.

  It is defined as follows:[1, 2]

  - In statistics, for nonnegative values of x, the error function has the following interpretation: for a *random variable* Y that is normally distributed with mean 0 and variance 1/2, erf(x) describes the probability of Y falling in the range $[-x, x]$ (Figure 3.4).

Remarks:

(1) The name "**error function**" and the abbreviation for the error function (and the error function complement) were developed on account of its connection with "the theory of Probability, and notably the theory of errors," for the "law of facility of errors," viz., the normal distribution whose density is given by:

$$f(x) = (c/\pi)^{1/2} \exp(-cx^2) \tag{3.46}$$

The chance of an error lying between p and q is

$$(c/\pi)^{1/2} \int_p^q \exp(-cx^2)dx = 1/2 \left[\mathrm{erf}(q\sqrt{c}) - \mathrm{erf}(p\sqrt{c})\right] \tag{3.47}$$

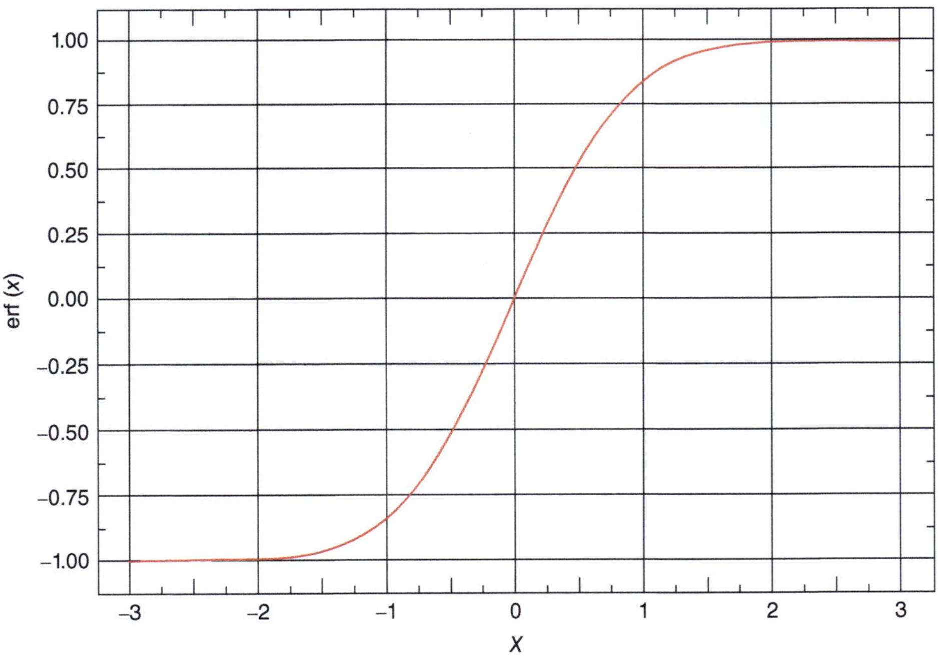

Figure 3.4 A plot of the error function erf(x) vs. the function → −1 and → +1, asymptotically, for $x \to -\infty$ and $+\infty$, respectively, and passes through the point (0, 0).

(2) **Complementary Error Function**
The **complementary error function**, denoted as erfc(x), is defined as:

$$\begin{aligned}\mathrm{erfc}(x) &= 1 - \mathrm{erf}(x)\\ &= \frac{2}{\sqrt{\pi}}\int_x^{\infty} \mathrm{e}^{-t^2}\,\mathrm{d}t\\ &= \mathrm{e}^{-x^2}\,\mathrm{erfcx}(x)\end{aligned} \tag{3.48}$$

Now, if the process of mass transfer is a unidirectional diffusion and the surface concentration is very low, viz.,

$$C_{\mathrm{Ab}} \approx 0 \tag{3.49}$$

then the mass flux of solute A, viz., N_{A} (kg/m^2 s) may be estimated by the following equation:

$$N_{\mathrm{A}} = \{-\rho D_{\mathrm{AB}}/(1 - c_{\mathrm{Ab}})\}(\partial c/\partial z)_{z=0} \approx -\rho(\partial c/\partial z)_{z=0} \tag{3.50}$$

(3) **The Mass Transfer Coefficient k_{L}:**
From the above two expressions, the rate of mass transfer at time t is given by the following equation:

$$N_{\mathrm{A}}(t) = \{\sqrt{(D_{\mathrm{AB}}/\pi t)}\}(c_{\mathrm{Ai}} - c_{\mathrm{Ab}}) \tag{3.51}$$

And the ***instantaneous mass transfer coefficient*** k_{L} is given by:

$$k_{\mathrm{L}}(t) = \sqrt{(D_{\mathrm{AB}}/\pi t)} \tag{3.52}$$

since the ***instantaneous*** mass transfer coefficient k_{L} is ***defined*** by the following equation:

$$N_{\mathrm{A}}(t) = k_{\mathrm{L}}(t)(c_{\mathrm{Ai}} - c_{\mathrm{Ab}}) \tag{3.53}$$

The **time-averaged mass transfer coefficient** during a time interval t_{c} is obtainable by integrating Eq. (3.53) over one time period from $t = 0$ to $t = t_{\mathrm{c}}$ as follows:

$$k_{\mathrm{L,av}} = (1/t_{\mathrm{c}})\int_0^{t_{\mathrm{c}}} k(t)\mathrm{d}t = 2\sqrt{(D_{\mathrm{AB}}/\pi t_{\mathrm{c}})} \tag{3.54}$$

Thus, from the above equation, the time-averaged mass transfer coefficient is proportional to the square root of the diffusivity. This was first proposed by Higbie in 1935 and the theory is also known as the **Higbie's penetration theory**.

3.7.3 Danckwerts' Surface Renewal Theory Model

The surface renewal theory model for mass transfer is an unsteady-state theory where molecules of solute are in random motion. It assumes that these solute molecules form a cluster which goes to the interphase where they remain there for some unspecified time which basically depends on experiments. Some of solute then pass through this interphase, while the rest mixes back to the bulk. Here, the specific term is ***uncertain time of contact***.

The surface renewal theory has been conceived so that the deficiencies of the first two theories can be rectified through incorporation of some statistical components into the description of interphasal mass or heat transport.

Numerous variants of this theory, giving rise to various mathematical models, have been proposed; still, the acceptance or popularity of the surface renewal theory appears to lag behind the other two theories. This is probably attributable to the fact that while the mathematical formulation of the theory is abundantly couched in statistical or stochastic terms, the methodologies and procedures followed are those of continuum mechanics and deterministic mathematics, thereby rendering the model or theory less acceptable – from a practical, or utilitarian, viewpoint.

The present work attempts to derive a self-consistent mathematical model of the surface renewal theory of interphase mass transfer by resorting to the theories and methodologies of stochastic processes based on the Markovian assumption. Specifically, the expression for the contact time distribution of fluid elements or solid particles participating in the interphase transport has been derived from the stochastic population balance of these elements or particles. Moreover, the expression for the dynamic rate of transfer of molecules or microscopic components across the interface has been derived as the continuous limit of the probability balance equation of the random walks of these entities around the interface. Proper coupling of the two expressions constitutes the desired model. By analogy, this model is applicable to the turbulent interphase heat transport and plausibly to the momentum transport under certain circumstances.

For mass transfer in the liquid phase, Danckwerts (1951) modified the Higbie's penetration theory as follows: assume that a portion of the mass transfer surface is replaced with a new surface by the motion of eddies near the surface and applied the following assumptions:

(1) The liquid elements at the interface are being randomly swapped by fresh elements from bulk liquid;
(2) At any moment, each of the liquid elements at the surface has the same probability of being substituted by fresh elements;
(3) Unsteady-state mass transfer occurs at the element during its sojourn at the interface.

Hence, the average molar flux, $N_{A,av}$, is

$$N_{A,av} = (C_{Ai} - C_{Ab})\sqrt{(sD_{AB})} \tag{3.55}$$

Comparing Eq. (3.55) with Eq. (3.53), one obtains Eq. (3.56):

$$k_{L,av} = \sqrt{(sD_{AB})} \tag{3.56}$$

where s is **fraction of the surface renewed in unit time**, i.e. ***the rate of surface renewal*** (s^{-1}).

3.7.4 Boundary Layer Theory Model

Boundary layer theory takes into account the hydrodynamics or the flow field that characterizes a system and gives a realistic picture of the occurrence of mass transfer at an interfacial boundary.

3.7.4.1 Fluid–Fluid Interfaces

Fluid–fluid interfaces are typically not fixed and are strongly affected by the flow leading to heterogeneous systems that make it difficult to develop a general theory to support the mass transfer correlations.

3.7.4.2 Fluid–Solid Interfaces

Solids typically have fixed and well-defined surfaces, allowing one to develop theoretical models for the empirical mass transfer correlations. In contrast to fluid–fluid interfaces, the models are far more rigorous but, unfortunately, computationally intensive as shown in Section 3.3.

3.7.4.3 Example: Prandtl's Experimental Mass Transfer from a Flat Plate

As an example, consider the classical case of mass transfer from a "Flat Plate" configuration: a sharp-edged, flat plate that is sparingly dissolvable is immersed in a rapidly flowing solvent [27, 28] (Figure 3.5).

From experimental data, using dimensionless parameters, the following mass transfer correlation was established (Figures 3.6 and 3.7):

$$N_{Sh} = 0.646(N_{Re})^{1/2}(N_D)^{1/3} \tag{3.57}$$

where

N_{Sh} = The Sherwood number = kL/D
N_{Re} = The Reynolds number = LU_∞/v
N_D = The diffusivity number = v/D
L = The plate length
U_∞ = The bulk fluid velocity
D = The diffusivity of the material from the plate into the bulk fluid
v = The diffusivity/viscosity of the fluid

Figure 3.5 Prandtl: flat plate experiment – the concept of boundary layers.

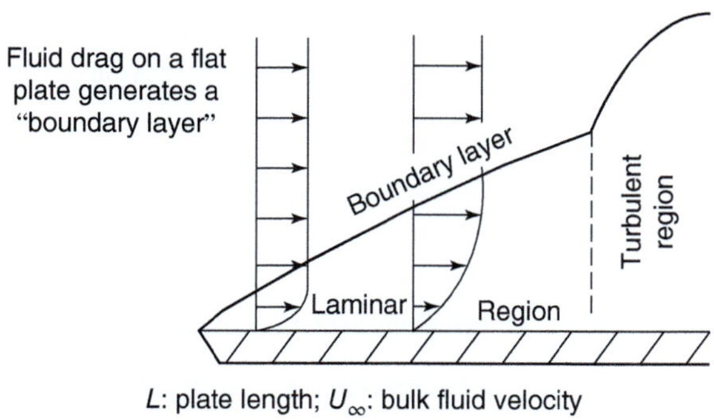

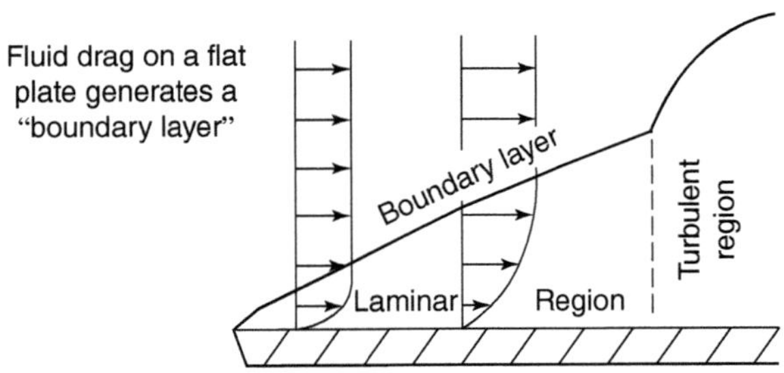

Figure 3.6 The flat plate experiment for boundary layer mass transfer for a flat plate: 1 of 2.

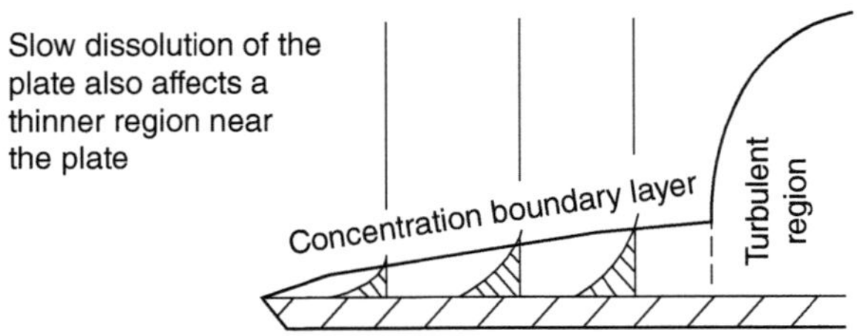

Figure 3.7 The flat plate experiment for boundary layer mass transfer for a flat plate: 2 of 2.

3.7.5 Mass Transfer Under Laminar Flow Conditions

Mass transfer coefficient does not play a big role in laminar flow conditions as molecular diffusion exists there. In the laminar flow regime, the average liquid-phase mass transfer coefficient, $k_{L,av}$, is correlated with Sherwood number (N_{Sh}) and D_{AB} as follows:

$$k_{L,av} = 3.41(D_{AB}/\delta) \tag{3.58}$$

$$N_{Sh} = k_{L,av}\delta/D_{AB} = 3.41 \tag{3.59}$$

3.7.6 Mass Transfer Past Solids Under Turbulent Flow

Mass transfer under flow past solid is important in practical applications. Several theories have attempted to clarify the behavior of the corresponding mass transfer coefficients. All the theories have some assumptions and some drawbacks. Hence, they are revised frequently. In turbulent flow medium, small fluid in element of different sizes, called eddies, move randomly. These eddies form and interact among others and disappear in the flow path. The total molar flux of a solute "A" owing to molecular diffusion and eddy diffusion, J_A, is as follows:

$$J_A = -(D_{AB} + D_E)dC_A/dZ \tag{3.60}$$

where D_E is the eddy diffusivity. (Eddy diffusivity depends on the intensity of local turbulence.)

3.7.7 Some Interesting Special Conditions of Mass Transfer

3.7.7.1 Equimolar Counter-Diffusion of A and B ($N_A = -N_B$)

Now, for gas-phase diffusion, it is known that

$$N_A = k'_G(p_{A1} - p_{A2})$$

$$= k'_y(y_{A1} - y_{A2})$$
$$= k'_c(C_{A1} - C_{A2})$$
$$\rightarrow \text{Gas phase} \tag{3.61}$$

$$N_A = k'_x(x_{A1} - x_{A2})$$
$$= k'_L(C_{A1} - C_{A2})$$
$$\rightarrow \text{Liquid phase} \tag{3.62}$$

For gas-phase diffusion, it is known that

$$N_A = D_{AB}(p_{A1} - p_{A2})/RT\delta \tag{3.63}$$

Equating Eqs. (3.61) and (3.63), the result is

$$k'_G = D_{AB}/RT\delta \tag{3.64}$$

Again,

$$N_A = D_{AB}(p_{A1} - p_{A2})/RT\delta = D_{AB}P(y_{A1} - y_{A2})/RT\delta \tag{3.65}$$

Equating Eqs. (3.61) and (3.65),

$$k'_G = D_{AB}/RT\delta y \tag{3.66}$$

Also,

$$N_A = D_{AB}(p_{A1} - p_{A2})/RT\delta = D_{AB}(C_{A1} - C_{A2})/\delta \tag{3.67}$$

Equating Eqs. (3.61) and (3.67),

$$k'_c = D_{AB}/\delta \tag{3.68}$$

3.7.7.2 For Liquid-Phase Diffusion

The flux may be expressed as:

$$N_A = (D_{AB}/\delta x_{BM})(\rho/M)_{av}(x_{A1} - x_{A2}) \tag{3.69}$$

in which,

$$x_{BM} = (x_{B2} - x_{B1})/\ln(x_{B2}/x_{B1}) \tag{3.70}$$

Equating Eqs. (3.62) and (3.69), one obtains

$$k'_x = (D_{AB}/\delta x_{BM})(\rho/M)_{av} \tag{3.71}$$

Also,

$$N_A = (D_{AB}/\delta)(\rho/M)_{av}(x_{A1} - x_{A2})$$
$$= (D_{AB}/\delta)(C_{A1} - C_{A2}) \tag{3.72}$$

Also,

$$N_A = (D_{AB}/\delta x_{BM})(\rho/M)_{av}(x_{A1} - x_{A2})$$
$$= (D_{AB}/\delta x_{BM})(C_{A1} - C_{A2}) \tag{3.73}$$

Equating Eqs. (3.61) and (3.73)

$$k_L = D_{AB}/\delta x_{BM} \tag{3.74}$$

3.7.7.3 Conversions Formulas for Mass Transfer Coefficients in Different Forms

$$k_c = RTk_{\mathrm{G}} \tag{3.75A}$$

$$k_y = Pk_{\mathrm{G}} \tag{3.75B}$$

$$k_x = (\rho/M)_{\mathrm{av}}k_{\mathrm{L}} \tag{3.75C}$$

3.7.8 Applications to Chemical Engineering Design

3.7.8.1 Designing a Packed Column for the Absorption of Gaseous CO_2 by a Liquid Solution of NaOH, Using the Mathematical Model of Simultaneous Gas Absorption with Chemical Reactions

Abstract

This paper present a research work on the modeling of a gas absorption packed column with the aim of formulating a mathematical model and simulation of the model using a computer software to obtain the rate of absorption and the amount of absorbed carbon dioxide CO_2 into dilute sodium hydroxide NaOH. Arm field gas absorption column titration techniques were used for the analysis. The total concentration of carbonate and hence the amount of carbon dioxide absorbed was determined using gravimetric methods. The comparison between the experimentally obtained and simulated results shows that the formulated model is a good representation of the system. (A program was written in Basic for simulation of any gas absorption packed column, and the model and computer program could therefore be used to determining the condition in a counter-current continuous column at any given time, for any chemically reacting absorption system.)

Introduction

Chemical reactor design is based on the modeling of reactors and of the reactions that take place in them (Levinspiel 1972). A mathematical model is a simplified image of the processes that take place in a reactor. It retains the most essential properties of the actual process and presents them in mathematical form. Depending on the objective, a mathematical model may involve a varying number of properties of the prototype, and so it may be wide or narrow. Mathematical modeling is so important that it is now being used in computer simulations (Levinspiel 1972).

Simulation represents the application of modeling techniques to real systems, thus enabling information on plant characteristics to be gained without either constructing or operating the full-scale plant or the system under consideration (Levinspiel 1972). Simulation methods in chemical reactor designs are of two types: digital and analog. Digital simulation involves the use of codes and programs that are more in use, since they can be implemented on modern computers with exceptional speed (Levinspiel 1972). Most reactions utilized in the chemical process industries involve substances, which exist in different phases. There may be two-phase (binary) and three-phase (tertiary) systems (Coulson 1996). For all the differences between them, they have one thing in common. Before a chemical reaction can take place, the reactants must be transported from the bulk of the stream and carry one phase to the interphase boundary or into bulk of the other phase. Gas–liquid absorption is a heterogeneous process, which involves the transfer of a soluble component of a gas phase into a relatively nonvolatile liquid absorbent (Franks 1967). Gas–liquid absorption could be employed in purification of waste gases (Coulson 1996). In this work, a mathematical model was developed and simulated for a gas–liquid absorption packed column operating under a continuous mode for the absorption of carbon dioxide into dilute caustic soda via determination of the overall performance of the absorption column.

Description and Operation of the Gas Absorption Column

A gas absorption experiment is usually carried out in vertical counter-current packed column, about 75 mm in diameter, in which there are two lengths of branching. The liquid solvent is fed at the top of the column and is distributed over the surface of the packing either by nozzle or distribution plates. Pressure tappings are provided at the base, center, and top of the column to determine pressure drops across the column.

Sampling points are also provided for the gas at these same three points. The liquid outlet stream and feed solution are also equipped with sampling point. Suitable manometer measurement is included. Water/solvent is taken from a sump tank and pumped to the column via a calibrated flow meter. Gas is taken from a pressure cylinder through a calibrated flow meter, mixed with air, and supplied and monitored from a small compressor in a predetermined (but variable) mixed ration. The mixture is fed to the base of the tower, in which a liquid seal is provided. The effluent gas leaves the top of the column and is intended to be exhausted to atmosphere outside the laboratory building. The apparatus is designed to absorb CO_2–Air mixture into an aqueous solution flowing down the column. Gas analysis is provided for this system (see Figure 3.8).

Application of Mass Transfer Theories

The analysis of the process of absorption so far given is based on the two-film theory of Whitman (1923). It is supposed that the two films have negligible capacity, but

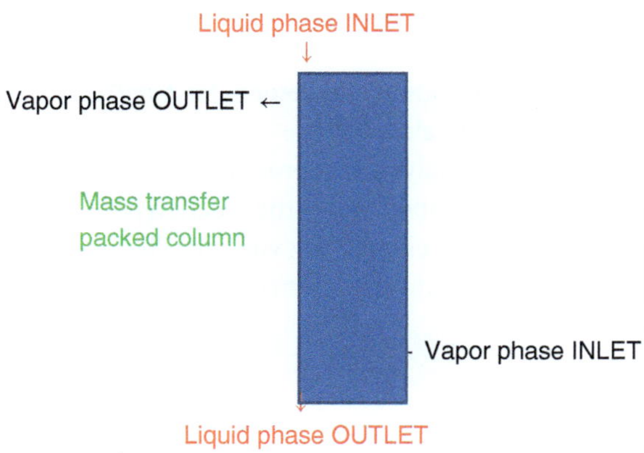

Figure 3.8 Gas absorption packed column model of a gas-absorption packed column for the carbon dioxide–sodium hydroxide system.

offer all the resistance to mass transfer, and thus the flow is considered to be laminar and parallel to the surface. An alternative theory was proposed by Higbie (1935) and later extended by Danckwerts and Kennedy (1954) which gave mass transfer rate (N_A) to be directly proportional to the concentration difference ($C_{AI} - C_{AL}$).

However, Higbie–Danckwerts also used $N'_A \propto VD_L$. Whereas $N'_a \propto VD_L$ in the two-film theory, Danckwerts applied this theory to the problem of absorption coupled with chemical reaction, but although in this case the three proposals give somewhat different results, it has not been possible to distinguish between them.

Also, the application of penetration theory to the interpretation of experimental results obtained in wetted wall columns has been studied by Lynn et al. (1955).

An Example: The CO_2–NaOH Reaction System

The reaction is the absorption of carbon dioxide into dilute caustic soda, termed as chemisorption, is similar to the basic gas–liquid reactions widely used in chemical process industries as auxiliary steps in the purification of gas mixtures:

$$CO_2 + NaOH \rightarrow Na_2CO_2 + H_2O \tag{3.76}$$

Mathematical Modeling for *Arm Field* Gas Absorption Column

The process of interest is the absorption of carbon dioxide into a dilute aqueous solution of caustic soda using the arm field gas absorption column (packed with ranching rings). For proper understanding of the behavior of the system, prior knowledge of rate of reaction mass and energy balances together with simple chemical reaction kinetics is required.

The following assumptions are taken into consideration in developing a model for gas absorption packed column and in the analysis of the predicted result of the model:

1. The absorption process is carried out in a series of contacts with the recurrence of unidirectional mass transfer and occurs counter-currently using diluted absorbent.
2. The column operates at steady-state conditions, hence there are negligible heat effects in the column and the absorption process is accompanied by pseudo-first-order reaction.
3. There exist constant molal flow rates owing to the occurrence of perfect mixing.

Derivation of a Mathematical Model for *Arm Field* Gas Absorption Column: Overall Material Balance

Mathematical analysis of a process is based on the material balance of the process for one of the components of the reaction mixture and may be written as differential equations for the elementary volume. The material balance equation for the main component contains the following terms:

$$G_t = G_d + G_r + G_a \tag{3.77}$$

where

G_t is the amount of the component fed to the elementary volume per unit time;
G_d is the amount of the component out from the elementary volume per unit time;
G_r is the rate of consumption of the component in the chemical reaction in the elementary volume; and
G_a is the rate of accumulation of the component in the elementary volume.

This may be written for the whole reactor if there is uniform distribution of materials over the reactor volume.

Differential Material Balance

The material balance of "A" (CO) and "B" (NaOH) is obtained by noting that of the two reactants, A alone is present in the main body of gas and B alone is present in the main liquid. For each mole of "A" reacted, "*b*" moles of "B" are consumed. In this case, for each mole of CO_2 reacted, 2 mol of NaOH are consumed:

$$CO_2 + 2NaOH \rightarrow Na_2CO_3 + H_2O \tag{3.78A}$$

given by $A + bB \rightarrow$ Products, $b = 2$. Referring to the diagram below (see Figure 3.9),

$$A_{(gas)} + bB_{(liquid)}(\text{fast}) \rightarrow \text{Product} \tag{3.78B}$$

However, a material balance about the end of the packed column gives composition at any point in the column.

The differential material balance for the counter-current flow is as shown below:

$$[A' \text{ lost by gas}] = (1/b)\,[B' \text{ lost by liquid}] \tag{3.79}$$

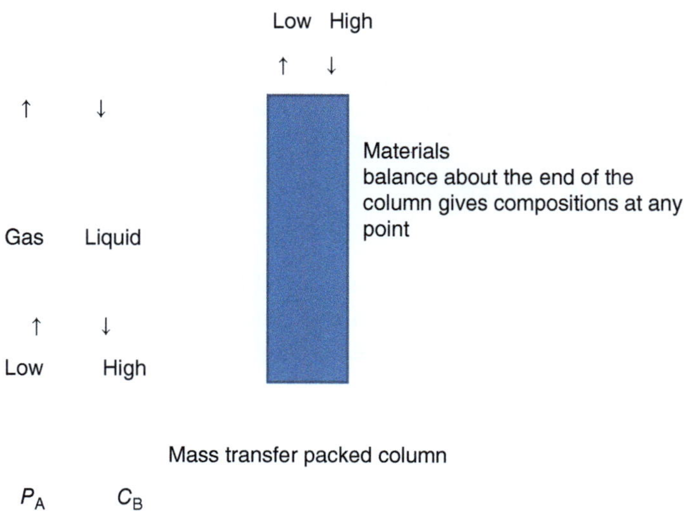

Figure 3.9 Material balances.

or

$$\begin{aligned} GdY_A &= (-L/b)dX_B \\ &= -Ld[C_B/C_U] \\ &= d[G'PA/\pi] \\ &= (-1/b)d[L'C_B/C_T] \end{aligned} \tag{3.80}$$

Compositions at any point in the tower are found in terms of the end conditions by integrating the material balance equation. Thus,

$$\begin{aligned} G[Y_A - Y_{AI}] &= -L[(X_B - X_{BI})/b] \\ &= G[(P_A/P_U) - (P_{AI}/P_{UI})] \\ &= -(L/b)[(C_B/C_U) - (C_{BI}/C_{UI})] \\ &= (G''P_A/\pi) - (G''P_{AI}/\pi) \\ &= (-I/b) - [(L'C_B/C_T) - (L'C_{BI}/C_{TI})] \end{aligned} \tag{3.81}$$

The above expressions give the concentrations of reactants A and B in the phases throughout the tower:

$$P_U \approx \pi C_U \approx C_T \tag{3.82}$$

Equations (3.81) and (3.82) for differential material balance become

$$GDP_A/bC_T = -LdC_B \tag{3.83}$$

and for point conditions.

$$G(P_A - P_{AI}) = -(L/bC_T)(C_B - C_{BI}) \tag{3.84}$$

Height of Tower Equation

The height of the tower (column) is found by combining rate equation and material balance for a differential element of tower and is given by:

$$h = \gamma_A = (1/K_{Ag}) + HA/\sqrt{(ALKC_B)} \tag{3.85}$$

rearranging and integrating, in terms of A and B.

Rate of absorption of a gas which diffuses into the liquid film stagnant medium in which it has a limited solubility and with which it undergoes an irreversible first-order or pseudo-first-order reaction is assumed to obey ideal diffusion law:

$$\mathrm{d}Q/\mathrm{d}t = -D\partial C/\partial X \tag{3.86}$$

This holds for the diffusion of unreacted solute through the medium, and also that the mass of solvent per unit volume of the medium is virtually constant throughout.

Let C^* be the saturated concentration of the solute reigning at the surface ($X = 0$); C = concentration at a distance X below the surface; t = time; and k = velocity constant of the reaction between solute and medium.

Initial concentration of solute in the medium is uniformly equal to zero.

Hence, the rate at which the solute crosses unit area of any plane of c constant X in the direction of increasing X is

$$(\mathrm{d}Q/\mathrm{d}t) = -D\partial C/\partial X \tag{3.87}$$

Consider the element of volume of unit cross-sectional area between planes X and $X + dX$. The following changes in its content of solute occur in timidity:

$$\text{Diffusion (in)} = -(D\partial C/\partial X)\mathrm{d}t \tag{3.88}$$

$$\text{Diffusion (out)} = [-D\,\mathrm{d}t\{(\partial C/\partial X) + (\partial^2 C/\partial X^2)\,\mathrm{d}X\}] \tag{3.89}$$

$$\text{Reacting} = -kc\,\mathrm{d}t \cdot \mathrm{d}x \tag{3.90}$$

$$\text{Net increase} = [D\partial^2 C/\partial X^2 - kC]\,\mathrm{d}t \cdot \mathrm{d}x \tag{3.91}$$

If the increase in concentration is dc, then the net increase above may be equated to dc/dx, giving

$$\mathrm{d}c/\mathrm{d}t = D\partial^2 C/\partial x^2 - KC \tag{3.92}$$

With the boundary conditions:

$$\begin{aligned} &C = C^*, \quad x = 0, \quad t > 0,\\ &C = 0, \quad x > 0, \quad t = 0,\\ &C = 0, \quad x = \infty, \quad t > 0 \end{aligned} \tag{3.93}$$

the solution to equation [17] is

$$C/C^* = 1/2\ \exp[-X\sqrt{(K/D)}]^{\mathrm{erfc}} \tag{3.94}$$

where

$$\mathrm{erfc} = 1 - \mathrm{erf}(z) \tag{3.95}$$

and

$$z = 1 - (2/\sqrt{\pi})\int_0^z \exp(-y^2)\mathrm{d}y \tag{3.96}$$

Differentiating Eq. (3.94),

$$\partial c/\partial x(\partial c/\partial x)_{x=0} = -C^*\sqrt{(D/k)}[\mathrm{erf}\sqrt{(Kt)} + \{\exp(-kt)/\sqrt{(\pi kt)}\}] \tag{3.97}$$

The rate of absorption $\mathrm{d}Q/\mathrm{d}t$ per unit area of surface becomes

$$\mathrm{d}Q/\mathrm{d}t = -D(\partial c/\partial x)_{x=0} \tag{3.98}$$

$$= C^*\sqrt{(D/k)}[\mathrm{erf}\sqrt{(Kt)} + \{\exp(-kt)/\sqrt{(\pi kt)}\}] \tag{3.99}$$

The quantity Q absorbed in time "t" is

$$Q = C^*\sqrt{\{(D/K)[(Kt + 1/2)\mathrm{erf}\sqrt{(kt)} + \sqrt{\{(kt/\pi)\mathrm{e}^{-kt})\}]\}} \tag{3.100}$$

when Kt is sufficiently large, erf $= \sqrt{(Kt)} \approx 1$, and

$$(C/C^*) \approx \exp\{-x\sqrt{(K/D)}\} \tag{3.101}$$

Then,

$$(\partial Q/\partial x)\,dQ \approx C^* \sqrt{(DK)} \tag{3.102}$$

and

$$Q = C^* \sqrt{(D/K)}(Kt + 1/2) \tag{3.103}$$

Determination of the Absorption Acceleration Factor in the Case of a Chemical Reaction "ε"

If $aA_{(g)} + bB_{(l)} \rightarrow$ Product, the true rate equation is

$$\omega_{rA} = KC_A C_B \tag{3.104}$$

which could also be written as:

$$\omega_{rA} = K^1 C_A \tag{3.105}$$

where

$$K^1 = KC^n{}_B \tag{3.106}$$

The concentration of reactant A in the boundary film decreases not only owing to diffusion but also because it is being consumed in the reaction, thus leading to an increased concentration gradient of the reactant at the interface and the numerical value of the slope of the tangent of the $C_A Z$ curve at point M equal to grad C_A increases. As a result, the individual mass transfer coefficient from the gas to the liquid phase is increased:

$$W_A = D \operatorname{grad} C_A \tag{3.107}$$

which be expressed as:

$$W_A = \beta_i\{C_{Ab} - C_{A1} \cosh(M^{1/2}) M^{1/2} \tanh(M^{1/2})\} \tag{3.108}$$

where

$$\beta_i = D/\delta = \varepsilon \tag{3.109}$$

which describes the absorption accompanied by a chemical reaction and it differs from the equation:

$$W_A = \beta_i(C_A - C_{AL}) \tag{3.110}$$

which describes physical absorption.

When the fast chemical reaction takes place in the film and $C_{AL} = 0$, Eq. (3.110) takes a simpler form:

$$\omega A = \beta C_{AB}\{M^{1/2}/\tanh M^{1/2}\} \tag{3.111}$$

The absorption accelerator factor ε may be found by equating Eqs. (3.111) and (3.106) at $C_{AL} = 0$. Here, ε is the factor in the equation describing absorption accompanied by a chemical reaction. For the physical absorption equation at $C_{AL} = 0$,

$$\varepsilon = (\beta C_{Ab}/\beta_1 C_{Ab})M^{1/2}\tanh(M^{1/2})$$
$$= M^{1/2}\tanh(M^{1/2}) \quad (3.112)$$

or at $M^{1/2} \gg 1$ (because then $\tanh M^{1/2}$ tends to unity),

$$\varepsilon = M^{1/2}$$
$$= (DK/B^2L)^{1/2}$$
$$= (KD)^{1/2}/(\beta_2 L) \quad (3.113)$$

Thus, it is seen that an increase in the reaction rate constant K leads to greater absorption acceleration factors ε.

The predictive model equation for the quantity of absorbed CO_2 in a given time (t) by diffusion accompanied by chemical reaction is given by:

$$Q = C^* \sqrt{(D/K)}[(Kt + 1/2)\text{erf}\sqrt{(Kt)} + \sqrt{\{Kt/\pi\}}\exp(-kt)\}] \quad (3.114)$$

$$Q = C^* \sqrt{(D/K)}(Kt + 1/2) \quad (3.115)$$

From these expressions, one would expect that the $Q_{v,t}$ plot to be linear with a slope of $C^*\sqrt{(KD)}$ until the depletion of the OH^- near the surface which disturbs the first-order character of the reaction and makes it impossible to deduce the velocity reaction constant from the observation.

Conclusions

Mathematical modeling of a gas absorption packed column carbon dioxide sodium hydroxide system had been performed, and a model equation was developed and tested. The result obtained shows that it can be used to predict the quantity of CO_2 absorbed in a given time. And $Q_{v,t}$ plot cannot be linear as expected due to the depletion of OH near the surface which disturbed the first-order characteristics of the reaction and makes it impossible to deduce the reaction velocity constant.

3.7.8.2 Calculation of Packed Height Requirement for Reducing the Chlorine Concentration in a Chlorine–Air Mixture

3.7.8.2.1 Packed Tower Design

Methods for calculating the height of the ***active*** section of counter-current differential contactors such as packed towers, spray towers, and falling-film absorbers are based on rate expressions representing mass transfer at a point on the gas–liquid interface and on material balances representing the changes in bulk composition in the two phases that flow past each other. The rate expressions are based on the interphase mass transfer principles. Combination of such expressions leads to an integral expression for the Number of Transfer Units (NTU) or to equations related closely to the Number of Theoretical Plate. The following discussions describe a direct

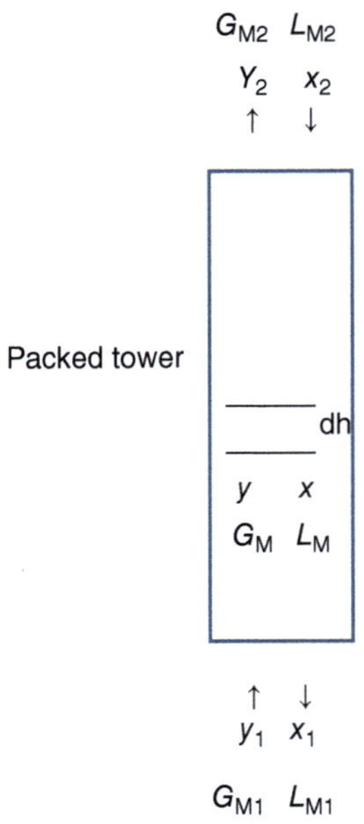

Figure 3.10 Mass balances in a packed tower stripper or absorber.

method for using such equations, first in a general case, and then for cases in which simplifying assumptions are valid. Use of mass transfer rate expression Figure 3.10 shows a section of a packed absorption tower together with the nomenclature that will be used in developing the equations which follow. In a differential section dh, one may equate the rate at which solute is lost from the gas phase to the rate at which it is transferred through the gas phase to the interface as follows:

$$-d(G_M y) = -G_M\,\mathrm{d}y - y\,\mathrm{d}G_M = N_A a\,\mathrm{d}h \tag{3.116}$$

When only one component is transferred,

$$\mathrm{d}G_M = -N_A a\,\mathrm{d}h \tag{3.117}$$

Substitution of this relation into Eq. [41] and rearranging yields

$$\mathrm{d}h = -G_M\,\mathrm{d}y/[N_A a(1-y)] \tag{3.118}$$

For this derivation, one use the gas-phase rate expression:

$$N_A = kG(y - y_i) \tag{3.119}$$

and integrate over the tower to obtain

$$h_T = \int_0^{y_1} [G_M/\{k_G a(1-y)(y-y_i)\}]\mathrm{d}y \tag{3.120}$$

Multiplying and dividing by y_{BM} place Eq. (3.120) into the $H_G N_G$ format:

$$h_T = \int_{y_2}^{y_1} [G_M/k_G a y_{BM}]\{y_{BM}/(1-y)(y-y_i)\}\mathrm{d}y \tag{3.121A}$$

$$= H_{G,av} \int_{y_2}^{y_1} \{y_{BM}/(1-y)(y-y_i)\} dy \tag{3.121B}$$

$$= H_{G,av} N_G \tag{3.121C}$$

The general expression given by Eqs. (3.121A)–(3.121C) is more complex than generally is required, but it should be used when the mass transfer coefficient varies from point to point, as may be the case when the gas is not dilute or when the gas velocity varies as the gas dissolves. The values of y_i to be used in Eqs. (3.121A)–(3.121C) depend on the local liquid composition x_i and on the temperature. This dependency is best represented by using the operating and equilibrium lines as one moves along the packed column. The following worked example illustrates the use of Eqs. (3.121A)–(3.121C) for scrubbing chlorine from air with aqueous caustic solution. For this case, one may make the simplifying assumption that y_i, the interfacial partial pressure of chlorine over the aqueous caustic solution, is zero owing to the rapid and complete reaction of the chlorine after it dissolves. Note that the feed gas is not dilute.

Example 3.1 ***Packed Height Requirement Calculations***

The Problem

One may compute the height of packing needed to reduce the chlorine concentration of 0.537 kg/(s m^2), or 396 lb/(h ft^2), of a chlorine–air mixture containing 0.503 mol fraction chlorine to 0.0403 mol fraction. On the basis of test data described by Sherwood and Pigford [29], the value of $k_G a y_{BM}$ at a gas velocity equal to that at the bottom of the packing is equal to 0.1175 kmol/(s m^3), or 26.4 lb mol/(h ft^3). The equilibrium back pressure y_i may be assumed to be negligible.

The Solution

By assuming that the mass transfer coefficient varies as the 0.8 power of the local gas mass velocity, one may derive the following relation:

$$\underline{K}_{Ga} = k_G a y_{BM}$$
$$= 0.1175[\{71y + 29(1-y)\}/\{71y_1 + 29(1-y_1)\}\{1 - y_1/(1-y)\}]^{0.8} \tag{3.122}$$

where 71 and 29 are the molecular weights of chlorine and air, respectively. Noting that the inert gas (air) flow rate is given by:

$$G'_M = G_M(1-y) \tag{3.123A}$$

$$= 5.34 \times 10^{-3} \text{ kmol/(s m}^2) \tag{3.123B}$$

or

$$= 3.94 \text{ lb mol/(h ft}^2) \tag{3.123C}$$

Upon introducing these expressions into the integral gives

$$h_T = 1.82 \int_{0.0403}^{0.503} \left\{ \left[(1-y)/(29+42y)\right]^{0.8} / \left[(1-y)^2 \ln\{1/(1-y)\}\right] \right\} dy \tag{3.124}$$

This definite integral may be evaluated numerically by the use of Simpson's rule to obtain

$$h_T = 0.305 \text{ m} \tag{3.125}$$

(or, approximately 1 ft).

For other conditions, such as the case in which the back pressure cannot be neglected, other design methods are available, such as those listed in Perry [30], and they should be consulted.

3.8 Theory of Simultaneous Bimolecular Reactions and Mass Transfer in Two Dimensions

(To facilitate referencing, some of the **R**eaction **T**ransport modeling equations in this section is predesignated by **RT**-.)

3.8.1 Numerical Solutions of a Model in Terms of Simultaneous Semi-linear Parabolic Differential Equations

In formulating a mathematical model representing a typical **bimolecular reaction and transport** such as the examples cited at the beginning of this chapter, etc., consider the two principal reactants A and B being brought into contact with a medium for reaction. There are physically interesting situations in which substances A and B are both gases that are brought into contact with a liquid in which both are soluble and within which they react with each other. Such theoretical and experimental results have been reported by Roper et al.[32]

This analysis presents useful and efficient numerical methods for solving the systems equations arising from such situations in which A and B are brought into contact with a laminar layer of liquid in which they react by a general $(n_1 + n_2)$-order biochemical or chemical mechanism. Some aspects of the nature of the mathematics of these equations will be discussed as their solutions are sought. Some numerical results are obtained and are presented.

The R environment will be used to present the theoretical results graphically.

3.8.1.1 Theory of Simultaneous Bimolecular Reactions and Mass Transfer in Two Dimensions

3.8.1.1.1 A Mathematical Model for Simultaneous Biochemical, or Chemical, Reaction and Mass Transport of Two Species in a Medium

NB: All equations in Sections 3.10–3.14 are designated in the form of (**RT***n*- …), where $n = 1, 2, 3$, and 4, where **RT** denotes **R**eaction and **T**ransport.

The reaction is assumed to have the stoichiometry corresponding to the irreversible biochemical or chemical equation

$$A + \upsilon B \rightarrow \text{Products}$$

Let y and z be two mutually perpendicular axes in space.

Consider a material balance on a differential element having dimensions $dy \times dz$ in the direction perpendicular to the interface and the direction of flow parallel to the interface.

Let $A = A(y, z)$ and $B = B(y, z)$ be the concentrations of biochemical or chemical species A and B at the point (y, z).

Let the absorbing medium flow with velocity u in the direction of positive z-axis, with diffusion occurring in the direction of positive y-axis.

Consider the mass balance of A and B over a volume element $dV = (dy)(dz)(1)$, with A and B reacting in the medium phase after absorption, according to the irreversible biochemical mechanism indicated, viz., $A + \upsilon B \rightarrow$ Products.

Now for species A:

Net Rate of Diffusion = Net Rate of Accumulation + Net Rate of Reaction

viz.,

[(Rate of Diffusion ***into*** dV) – (Rate of Diffusion ***out of*** dV)]
= [(Rate of flow ***into*** dV) – (Rate of Flow ***out of*** dV)]
+ [Rate of Biochemical, or Chemical Reaction]

Hence,

$$\begin{aligned}&[-(dz \cdot 1)D_A(\partial A/\partial y)] - [-(dz \cdot 1)D_A\{(\partial A/\partial y) + (\partial^2 A/\partial y^2)dy\}]\\&\quad = \{(dy \cdot 1)uA\} - \{(dy \cdot 1)u(A + (\partial A/\partial z)dz)\}\\&\quad\quad + kA^{n1}B^{n2}(dy \cdot dz \cdot 1)\end{aligned}$$

which results in Eq. (RT1-1A).

For species B, a similar mass balance results in Eq. (RT1-1B).

$$D_A \partial^2 A/\partial y^2 = u\,\partial A/\partial z + kA^{n1}B^{n2} \tag{RT1-1A}$$

$$D_B \partial^2 B/\partial y^2 = u\,\partial B/\partial z + \vartheta kA^{n1}B^{n2} \tag{RT1-1B}$$

with the following boundary conditions on $A(y, z)$ and $B(y, z)$:

$$A(0, z) = A_i, \quad B(0, z) = B_i$$

$$A(y, 0) = 0, \quad B(y, 0) = 0 \tag{RT1-1C}$$

Now, let $A(y, z)$ and $B(y, z)$ are bounded as y and z tend to infinity.

The three terms in each of the above pair of equations (RT1-1A) and (RT1-1B), represent the diffusion, convection, and reaction terms, respectively.

3.8.1.1.2 Transformation of the Mathematical Model into Condensed and/or Dimensionless Forms

For the simple case of reaction for the second-order biochemical, or chemical, mechanism:

Let

$$n_1 = 1 = n_2$$

Let

$$a = A/A_i, \quad b = B/B_i$$

$$x = (kB_i/DA)^{1/2}y, \quad t = (kB_i/u)z \tag{RT1-2}$$

then Eqs. (RT1-1A) and (RT1-1B) become

$$\partial^2 a/\partial x^2 = \partial a/\partial t + ab \tag{RT1-3A}$$

$$R\partial^2 b/\partial x^2 = \partial b/\partial t + mab \tag{RT1-3B}$$

with the boundary conditions on $a(x, t)$ and $b(x, t)$ given by:

$$a(0, t) = 1 = b(0, t)$$

$$a(x, 0) = 0 = b(x, 0)$$

$$a(x, z) \text{ and } b(x, t) \text{ are bounded as } x, t \to \infty \tag{RT1-3C}$$

and the following dimensionless parameters:

$$R = D_B/D_A, \quad m = \upsilon A_i/B_i \tag{RT1-4}$$

In general, the model system equations ((RT1-1A)–(RT1-1C)) may be transformed into a condensed form by the following substitution of new variables.

Let the variables A, B, y, and z be represented by the following condensed variables:

$$a = A/A_i^p$$

$$b = B/B_i^p$$

$$x = \left(kA_i^r B_i^{\underline{s}}/D_A\right)^{1/2} y$$

$$t = \left(kA_i^r B_i^{\underline{s}}/u\right)^{1/2} z$$

respectively, where p, q, r, and s are exponents *to be determined.*

Substituting these new condensed variables into Eqs. (RT1-1A) and (RT1-1B), the result is

$$\left(kA_i^{p+r}B_i^{s}\right) \partial^2 a/\partial x^2 = \left(kA_i^{p+r}B_i^{s}\right) \partial a/\partial t + \left(kA_i^{pn1}B_i^{qn2}\right) a^{n1}b^{n2}$$

$$\left(kA_i^{r}B_i^{q+s}\right) R\partial^2 b/\partial x^2 = \left(kA_i^{r}B_i^{q+s}\right) \partial b/\partial t + \left(kA_i^{pn1-1}B_i^{qn2+1}\right) ma^{n1}b^{n2}$$

where

$$a = A/A_i, \quad b = B/B_i$$

$$x = \left(kA_i^{n1-1}B_i^{n2}/D_A\right)^{1/2} y, \quad t = \left(kA_i^{n1-1}B_i^{n2}/u\right) z \tag{RT1-5}$$

resulting in

$$\partial^2 a/\partial x^2 = \partial a/\partial t + a^{n1}b^{n2} \tag{RT1-6A}$$

$$R\partial^2 b/\partial x^2 = \partial b/\partial t + ma^{n1}b^{n2} \tag{RT1-6B}$$

together with boundary conditions given by Eq. (RT1-3C) and dimensionless parameters given by Eq. (RT1-4) as before.

In fact, the condensed equations ((RT1-6A) and (RT1-6B)), are obtained if the respective exponents of A_i and B_i are equated, viz.,

$$p + r = pn_1, \quad s = qn_2, \quad r = pn_1 - 1, \quad q + s = qn_2 + 1$$

Choosing $p = 1 = q$, one obtains: $r = n_1 - 1$, and $s = n_2$. This choice of p and q makes the condensed variables a and b dimensionless, and the boundary condition equation (RT1-1C) is reduced to Eq. (RT1-3C).

Hence, the new condensed variables are defined by Eq. (RT1-2).

3.8.1.1.3 Solutions of the System Equations

For somewhat different boundary conditions, Roper et al.[32] obtained some approximate solutions using the Method of Moments in which concentration profiles for A and B were assumed to be similar cubic polynomials. The approximate solutions were forced to satisfy the boundary conditions exactly but to satisfy the differential equations only on the average. Empirical factors were used to adjust the solutions to known asymptotic solutions, with a reported error of about 3%. Since the boundary conditions appear to be incorrect, as shown at the end of this section, it seems that any success of their work may be fortuitous. However, in this work, a versatile scheme will be developed to solve the system of equations. The remainder of this section will discuss this scheme.

3.8.1.1.4 Numerical Schemes

Several finite difference discretization schemes are used to solve the equation system (RT1-1A)–(RT1-1C):

(a) Fully explicit scheme
(b) Semi-explicit scheme
(c) Semi-implicit schemes, for which the following variations are possible for solving the corresponding system of linear equations:
  (i) Calculating the inverse
  (ii) For triangular systems, solution by elimination
  (iii) Choleski's method, with and without iterations
(d) Fully implicit scheme
(e) Crank–Nicholson method

Fully Explicit Scheme Now, Eqs. (RT1-3A) and (RT1-3B) may be re-written as:

$$\partial a/\partial t = \partial^2 a/\partial x^2 - ab \qquad \text{(RT1-7A)}$$

$$\partial b/\partial t = R\partial^2 b/\partial x^2 - mab \qquad \text{(RT1-7B)}$$

Let $a \sim A$, and $b \sim B$, also $A_i^j = a(x_i, t_j)$ and $B_i^j = b(x_i, t_j)$,

Then the boundary conditions become

$$a(0, t) = A_1^j = 1 = B_1^j = b(0, t)$$

$$a(x,0) = A_i^{\ 1} = 0 = B_i^{\ 1} = b(x,0)$$

$$a(x,t) \text{ and } b(x,t) \text{ are bounded as } x,t \to \infty \quad \text{(RT1-7C)}$$

The corresponding finite difference equations, obtainable by discretizing Eqs. (RT1-7A) and (RT1-7B), are, respectively:

$$\left(A_i^{j+1} - A_i^j\right)/\Delta t = \left(A_{i-1}{}^j - 2A_i^j + A_{i+1}{}^j\right)/(\Delta x)^2 - A_i^j B_i^j \quad \text{(RT1-8A)}$$

$$\left(B_i^{j+1} - B_i^j\right)/\Delta t = R\left(B_{i-1}{}^j - 2B_i^j + B_{i+1}{}^j\right)/(\Delta x)^2 - mA_i^j B_i^j \quad \text{(RT1-8B)}$$

with the restriction that $\Delta t/(\Delta x)^2 \leq 1/2$.

Equations (RT1-8A) and (RT1-8B) may now be solved explicitly for A_i^{j+1} and B_i^{j+1} to yield

$$A_i^{j+1} = \left[1 - 2\Delta t/(\Delta x)^2 - \Delta t B_i^j\right] A_i^j + [\Delta t/(\Delta x)^2]\left(A_{i+1}{}^j + A_{i-1}{}^j\right) \quad \text{(RT1-9A)}$$

$$B_i^{j+1} = \left[1 - 2R\Delta t/(\Delta x)^2 - m\Delta t A_i^j\right] B_i^j + [R\Delta t/(\Delta x)^2]\left(B_{i+1}{}^j + B_{i-1}{}^j\right) \quad \text{(RT1-9B)}$$

which may be used for direct computation of A_i^{j+1} and B_i^{j+1}, given $A_n^{\ j}$ and $B_n^{\ j}$, for $n = i-1, i, i+1$.

This algorithm is programmed for digital computer calculation, and convergent results are obtained.

Semi-explicit Scheme In this scheme, the finite difference equations, discretizing Eqs. (RT1-7A) and (RT1-7B), are

$$\left(A_i^{j+1} - A_i^j\right)/\Delta t = \left(A_{i-1}{}^j - 2A_i^j + A_{i+1}{}^j\right)/(\Delta x)^2 - A_i^{j+1} B_i^j \quad \text{(RT1-10A)}$$

$$\left(B_i^{j+1} - B_i^j\right)/\Delta t = R\left(B_{i-1}{}^j - 2B_i^j + B_{i+1}{}^j\right)/(\Delta x)^2 - mA_i^{j+1} B_i^{j+1} \quad \text{(RT1-10B)}$$

Solving for A_i^{j+1} from Eq. (RT1-10A), then B_i^{j+1} from Eq. (RT1-10B), one obtains

$$A_i^{j+1} = \left[\{\Delta t/(\Delta x)^2\}\left(A_{i-1}{}^j - 2A^j + A_{i+1}{}^j\right) + A_i^j\right] / \left[1 + \Delta t B_i^j\right] \quad \text{(RT1-11A)}$$

$$B_i^{j+1} = \left[\{R\Delta t/(\Delta x)^2\}\left(B_{i-1}{}^j - 2B_i^j + B_{i+1}{}^j\right) + B_i^j\right] / \left[1 + m\Delta t A_i^j\right] \quad \text{(RT1-11B)}$$

For this procedure, it is seen that A_i^{j+1} must first be evaluated, and then B_i^{j+1} is obtained. The rest of the computation may proceed as in the fully explicit case. This algorithm is programmed for digital computer calculation, and convergent results are obtained.

Semi-implicit Scheme In this scheme, the finite difference equations, discretizing Eqs. (RT1-7A) and (RT1-7B), are

$$\left(A_i^{j+1} - A_i^j\right)/\Delta t = \left(A_{i-1}{}^{j+1} - 2A_i^{j+1} + A_{i+1}{}^{j+1}\right)/(\Delta x)^2 - A_i^{j+1} B_i^j \quad \text{(RT1-12A)}$$

$$\left(B_i^{j+1} - B_i^j\right)/\Delta t = R\left(B_{i-1}{}^{j+1} - 2B_i^{j+1} + B_{i+1}{}^{j+1}\right)/(\Delta x)^2 - mA_i^{j+1} B_i^{j+1} \quad \text{(RT1-12B)}$$

Equations (RT1-12A) and (RT1-12B) may be rewritten in matrix form:

$$\underline{\mathbf{C}}\underline{A}^{j+1} = \underline{A}^{j} \qquad \text{(RT1-13A)}$$

$$\underline{\mathbf{D}}\underline{B}^{j+1} = \underline{B}^{j} \qquad \text{(RT1-13B)}$$

where $\underline{\mathbf{C}}$ is a tri-diagonal matrix given by:

$$\underline{\mathbf{C}} = [\underline{\mathrm{I}} - \{\Delta t/(\Delta x)^2\}\underline{\mathrm{A}} + \Delta t\underline{\mathrm{B}}^{j}]$$

$$= \begin{bmatrix} c_1 & u_1 & 0 & 0 & 0 & 0 & 0 & \cdots & 0 & 0 \\ u_1 & c_2 & u_2 & 0 & 0 & 0 & 0 & \cdots & 0 & 0 \\ 0 & u_2 & c_3 & u_3 & 0 & 0 & 0 & \cdots & 0 & 0 \\ 0 & 0 & u_3 & c_4 & u_4 & 0 & 0 & \cdots & 0 & 0 \\ \cdots & \cdots & \cdots & \cdots & \cdots & \cdots & \cdots & \cdots & \cdots & \cdots \\ 0 & 0 & 0 & - & - & - & - & - & \cdots & 0 \\ 0 & 0 & 0 & - & - & - & - & 0 & c_{n-1} & u_n \\ 0 & 0 & 0 & - & - & - & - & 0 & u_{n-1} & c_n \end{bmatrix}$$

with $c_i = 1 - 2\Delta t/(\Delta x)^2 + \Delta t B_i^{\,j}$, and $u_i = -\Delta t/(\Delta x)^2$.

$\underline{\mathbf{C}}$ is a tri-diagonal matrix given by:

$$\underline{\mathbf{C}} = \begin{bmatrix} -2 & +1 & 0 & 0 & 0 & 0 & 0 & 0 & 0 \\ +1 & -2 & +1 & 0 & 0 & 0 & 0 & 0 & 0 \\ 0 & +1 & -2 & +1 & 0 & 0 & 0 & 0 & 0 \\ 0 & 0 & +1 & -2 & +1 & 0 & 0 & 0 & 0 \\ \cdots & \cdots & \cdots & \cdots & \cdots & \cdots & \cdots & \cdots & \cdots \\ 0 & 0 & 0 & - & - & \cdots & +1 & -2 & +1 \\ 0 & 0 & 0 & - & - & - & \cdots & +1 & -2 \end{bmatrix}$$

$\underline{\mathbf{B}}^{j}$ is a diagonal matrix given by:

$$\underline{\mathbf{B}}^{j} = \begin{bmatrix} B_1^{\,j} & 0 & 0 & 0 & 0 & 0 & 0 & 0 & 0 \\ 0 & B_2^{\,j} & 0 & 0 & 0 & 0 & 0 & 0 & 0 \\ 0 & 0 & B_3^{\,j} & 0 & 0 & 0 & 0 & 0 & 0 \\ 0 & 0 & 0 & B_4^{\,j} & 0 & 0 & 0 & 0 & 0 \\ \cdots & \cdots & \cdots & \cdots & \cdots & \cdots & \cdots & \cdots & \cdots \\ 0 & 0 & 0 & 0 & 0 & 0 & 0 & 0 & B_n^{\,j} \end{bmatrix}$$

$\underline{\mathbf{D}}$ is a tri-diagonal matrix given by:

$$\underline{\mathbf{D}} = [\underline{\mathrm{I}} - R\{\Delta t/(\Delta x)^2\}\underline{\mathrm{A}} + m\Delta t\underline{\mathrm{A}}^{j}]$$

$$= \begin{bmatrix} d_1 & v_1 & 0 & 0 & 0 & 0 & 0 & \cdots & 0 & 0 \\ v_1 & d_2 & v_2 & 0 & 0 & 0 & 0 & \cdots & 0 & 0 \\ 0 & v_2 & d_3 & v_3 & 0 & 0 & 0 & \cdots & 0 & 0 \\ 0 & 0 & v_3 & d_4 & v_4 & 0 & 0 & \cdots & 0 & 0 \\ \cdots & \cdots & \cdots & \cdots & \cdots & \cdots & \cdots & \cdots & \cdots & \cdots \\ 0 & 0 & 0 & - & - & - & - & - & - & 0 \\ 0 & 0 & 0 & - & - & - & - & 0 & d_{n-1} & v_{n-1} \\ 0 & 0 & 0 & - & - & - & - & 0 & v_{n-1} & d_n \end{bmatrix}$$

with $d_i = 1 - R2\Delta t/(\Delta x)^2 + \mathrm{m}\Delta t A_i^{\,j+1}$, and $v_i = -R\Delta t/(\Delta x)^2$.

$\underline{\mathbf{A}}^{j+1}$ is the diagonal matrix given by:

$$\underline{\mathbf{A}}^{j+1} = \begin{bmatrix} A_1^{j+1} & 0 & 0 & 0 & 0 & 0 & 0 & 0 & 0 \\ 0 & A_2^{j+1} & 0 & 0 & 0 & 0 & 0 & 0 & 0 \\ 0 & 0 & A_3^{j+1} & 0 & 0 & 0 & 0 & 0 & 0 \\ 0 & 0 & 0 & A_4^{j+1} & 0 & 0 & 0 & 0 & 0 \\ \cdots & \cdots & \cdots & \cdots & \cdots & \cdots & \cdots & \cdots & \cdots \\ 0 & 0 & 0 & 0 & 0 & 0 & 0 & 0 & A_n^{j+1} \end{bmatrix}$$

The coupled matrix equations (RT1-13A) and (RT1-13B) may be solved numerically by several procedures, including:

(i) Calculating the inverse

$$\underline{\mathbf{C}}\underline{A}^{j+1} = \underline{A}^j \quad \Rightarrow \quad \underline{A}^{j+1} = \underline{\mathbf{C}}^{-1}\underline{A}^j$$
$$\underline{\mathbf{D}}\underline{B}^{j+1} = \underline{B}^j \quad \Rightarrow \quad \underline{B}^{j+1} = \underline{\mathbf{D}}^{-1}\underline{B}^j$$

This method is generally not used owing to computational inefficiency and large data storage requirements.[25]

(ii) For triangular systems, solution by elimination[24]

This algorithm is programmed for digital computer calculation using one subroutine, and convergent results are obtained. In order to properly include the boundary condition of $a(0, 1) = 1 = b(0, t)$, one may set

$$A_0^{j+1} = 1 = B_0^{j+1}$$

so that, for calculations, Eqs. (RT1-14A), (RT1-14B) and (RT1-22A) become

$$\begin{bmatrix} \left[1 + \left\{\frac{2\cdot\Delta t}{(\Delta x)^2}\right\} + \Delta t \cdot B_1^j\right] & \frac{-\Delta t}{(\Delta x)^2} & 0 & 0 & 0 \\ \frac{-\Delta t}{(\Delta x)^2} & \left[1 + \left\{\frac{2\cdot\Delta t}{(\Delta x)^2}\right\} + \Delta t \cdot B_2^j\right] & \frac{-\Delta t}{(\Delta x)^2} & 0 & 0 \\ 0 & \frac{-\Delta t}{(\Delta x)^2} & \left[1 + \left\{\frac{2\cdot\Delta t}{(\Delta x)^2}\right\} + \Delta t \cdot B_3^j\right] & \frac{-\Delta t}{(\Delta x)^2} & 0 \\ 0 & 0 & \cdots & \ddots & 0 \\ 0 & 0 & 0 & \frac{-\Delta t}{(\Delta x)^2} & \left[1 + \frac{2\cdot\Delta t}{(\Delta x)^2}\right] + \Delta t \cdot B_n^j \end{bmatrix}$$
$$\times \begin{bmatrix} A_1^{j+1} \\ A_2^{j+1} \\ A_3^{j+1} \\ \vdots \\ A_n^{j+1} \end{bmatrix} = \begin{bmatrix} \{\Delta t/(\Delta x)^2\}A_0^{j+1} + A_1^j \\ \{\Delta t/(\Delta x)^2\}A_0^{j+1} + A_2^j \\ \{\Delta t/(\Delta x)^2\}A_0^{j+1} + A_3^j \\ \vdots \\ \{\Delta t/(\Delta x)^2\}A_0^{j+1} + A_n^j \end{bmatrix} \tag{RT1-14A}$$

and

$$\begin{bmatrix} \left[1 + \left\{\frac{R\cdot 2\cdot\Delta t}{(\Delta x)^2}\right\} + m\Delta t \cdot A_1^j\right] & \frac{-\Delta t}{(\Delta x)^2} & 0 & 0 & 0 \\ \frac{-\Delta t}{(\Delta x)^2} & \left[1 + \left\{\frac{R\cdot 2\cdot\Delta t}{(\Delta x)^2}\right\} + m\Delta t \cdot A_2^j\right] & \frac{-\Delta t}{(\Delta x)^2} & 0 & 0 \\ 0 & \frac{-\Delta t}{(\Delta x)^2} & \left[1 + \left\{\frac{R\cdot 2\cdot\Delta t}{(\Delta x)^2}\right\} + m\Delta t \cdot A_3^j\right] & \frac{-\Delta t}{(\Delta x)^2} & 0 \\ 0 & 0 & \cdots & \ddots & \cdots \\ 0 & 0 & 0 & \frac{-\Delta t}{(\Delta x)^2} & \left[1 + \left\{\frac{R\cdot 2\cdot\Delta t}{(\Delta x)^2}\right\} + m\Delta t \cdot A_n^j\right] \end{bmatrix}$$

$$\times \begin{bmatrix} B_1^{j+1} \\ B_2^{j+1} \\ B_3^{j+1} \\ \vdots \\ B_n^{j+1} \end{bmatrix} = \begin{bmatrix} \{R\Delta t/(\Delta x)^2\}B_0^{j+1} + B_1^j \\ \{R\Delta t/(\Delta x)^2\}B_0^{j+1} + B_2^j \\ \{R\Delta t/(\Delta x)^2\}B_0^{j+1} + B_3^j \\ \vdots \\ \{R\Delta t/(\Delta x)^2\}B_0^{j+1} + B_n^j \end{bmatrix} \tag{RT1-14B}$$

(iii) Computation using Choleski's method

Since the matrices $\underline{\mathbf{C}}$ and $\underline{\mathbf{D}}$, in Eqs. (RT1-13A) and (RT1-13B), respectively, are each symmetric and positive definite, one can write

$$\underline{C} = \underline{LL}^{\mathrm{T}} \tag{RT1-15A}$$

where $\underline{L}$ is a lower triangular matrix. Now,

$$\underline{CA}^{j+1} = \underline{A}^j$$

$$\Rightarrow (\underline{LL}^T)\underline{A}^{j+1} = \underline{A}^j$$

$$\Rightarrow \underline{L}(\underline{L}^T\underline{A}^{j+1}) = \underline{A}^j$$

$$\Rightarrow \underline{L}(\underline{g}) = \underline{A}^j$$

where $\underline{L}^T\underline{A}^{j+1} = \underline{g}$ and $\underline{LL}^T = \underline{C}$

$$\Rightarrow \begin{bmatrix} \alpha_1 & 0 & 0 & 0 & & & & \\ \beta_1 & \alpha_2 & 0 & 0 & 0 & & & \\ 0 & \beta_2 & \alpha_3 & 0 & 0 & 0 & & \\ 0 & 0 & & & & & & \\ 0 & 0 & 0 & 0 & & & \beta_{n-1} & \alpha_n \end{bmatrix} \begin{bmatrix} \alpha_1 & \beta_1 & 0 & 0 & & & & \\ 0 & \alpha_2 & \beta_2 & 0 & 0 & & & \\ 0 & 0 & \alpha_3 & \beta_3 & 0 & 0 & 0 & \\ 0 & 0 & & & & & & \\ 0 & 0 & 0 & & 0 & 0 & 0 & \alpha_n \end{bmatrix}$$

$$= \begin{bmatrix} c_1 & u_1 & 0 & 0 & 0 & & \\ u_1 & c_2 & u_2 & 0 & 0 & & \\ 0 & u_2 & c_3 & u_3 & 0 & 0 & \\ 0 & 0 & 0 & 0 & & & \\ 0 & 0 & 0 & 0 & & u_{n-1} & c_n \end{bmatrix}$$

$$\Rightarrow \alpha_1 = \sqrt{c_1}$$

$$\alpha_i = \sqrt{(c_i - \beta_{i-1}{}^2)}, \quad i = 2, 3, 4, \dots, n$$

$$\beta_i = u_i/\alpha_i, \quad i = 1, 2, 3, \dots, n-1$$

which forms the algorithm for $\underline{\mathrm{L}}$ and $\underline{\mathrm{L}}^{\mathrm{T}}$, ***the subroutine for which is denoted by*** CHOL. Also,

$$\underline{Lg} = \underline{A}^j$$

$$\Rightarrow \begin{bmatrix} \alpha_1 & 0 & 0 & 0 & & & & 0 \\ \beta_1 & \alpha_2 & 0 & 0 & 0 & & & 0 \\ 0 & \beta_2 & \alpha_3 & 0 & 0 & 0 & & 0 \\ 0 & 0 & & & & & & 0 \\ 0 & 0 & 0 & 0 & & & \beta_{n-1} & \alpha_n \end{bmatrix} \begin{bmatrix} g_1 \\ g_2 \\ g_3 \\ \vdots \\ g_n \end{bmatrix} = \begin{bmatrix} \gamma_1 \\ \gamma_2 \\ \gamma_3 \\ \vdots \\ \gamma_n \end{bmatrix}$$

$$\Rightarrow g_1 = \gamma_1/\alpha_1$$

$$\Rightarrow g_i = (\gamma_i - \beta_{-1} g_{i-1})/\alpha_i \quad i = 2, 3, 4, \dots, n$$

which forms the algorithm for g, ***the subroutine for which is named*** SOLV.

Further

$$\underline{L}^T \underline{A}^{j+1} = g$$

$$\Rightarrow \begin{bmatrix} \alpha_1 & \beta_1 & 0 & 0 & & & & 0 \\ 0 & \alpha_2 & \beta_2 & 0 & 0 & & & 0 \\ 0 & 0 & \alpha_3 & \beta_3 & 0 & 0 & 0 & 0 \\ 0 & 0 & & & & & & 0 \\ 0 & 0 & 0 & & 0 & 0 & 0 & \alpha_n \end{bmatrix} \begin{bmatrix} \delta_1 \\ \delta_2 \\ \delta_3 \\ \vdots \\ \delta_n \end{bmatrix} = \begin{bmatrix} g_1 \\ g_2 \\ g_3 \\ \vdots \\ g_n \end{bmatrix}$$

$$\Rightarrow \delta_n = g_n/\alpha_n$$

$$\delta_i = (g_i - \beta_i \delta_{i+1})/\alpha_i, \quad i = n-1, \dots 3, 2, 1$$

which forms the algorithm for $\underline{A}^{j+1}$ and is computed using the subroutine SOLV.

Similarly,

$$\underline{D} = \underline{MM}^T \tag{RT1-15B}$$

and

$$\underline{D}\underline{B}^{j+1} = \underline{B}^j \Rightarrow (\underline{M}\,\underline{M}^T)\underline{B}^{j+1} = \underline{B}^j$$

$$\Rightarrow \underline{M}(\underline{M}^T \underline{B}^{j+1}) = \underline{B}^j$$

$$\Rightarrow \quad \underline{M}\,\underline{h} = \underline{B}^j$$

where $\underline{M}^T \underline{B}^{j+1} = \underline{h}$, in which $\underline{M}$ is a lower-triangular matrix given by:

$$\underline{M}\ \underline{M}^T = \underline{D}$$

$$\Rightarrow \begin{bmatrix} a_1 & 0 & 0 & 0 & & & & \\ b_1 & a_2 & 0 & 0 & 0 & & & \\ 0 & b_2 & a_3 & 0 & 0 & 0 & & \\ 0 & 0 & & & & & & \\ 0 & 0 & & 0 & 0 & & b_{n-1} & a_n \end{bmatrix} \begin{bmatrix} a_1 & b_1 & 0 & 0 & & & & \\ 0 & a_2 & b_2 & 0 & 0 & & & \\ 0 & 0 & a_3 & b_3 & 0 & 0 & 0 & \\ 0 & 0 & & & & & & \\ 0 & 0 & 0 & & 0 & 0 & 0 & a_n \end{bmatrix}$$

$$= \begin{bmatrix} d_1 & v_1 & 0 & 0 & 0 & & & \\ v_1 & d_2 & v_2 & 0 & 0 & & & \\ 0 & v_2 & d_3 & v_3 & 0 & 0 & & \\ 0 & 0 & 0 & 0 & & & & \\ 0 & 0 & 0 & 0 & & & v_{n-1} & d_n \end{bmatrix}$$

$$\Rightarrow a_1 = \sqrt{d_1}$$
$$a_i = \sqrt{(d_i - b_{i-1}{}^2)}, \quad i = 2, 3, 4, \ldots, n$$
$$b_i = v_i/a_i, \quad i = 1, 2, 3, \ldots, n-1$$

which may be computed again by the subroutine CHOL. Again,

$$\underline{M}\,\underline{h} = \underline{B}^j$$

$$\Rightarrow \begin{bmatrix} a_1 & 0 & 0 & 0 & & & & 0 \\ b_1 & a_2 & 0 & 0 & 0 & & & 0 \\ 0 & b_2 & a_3 & 0 & 0 & 0 & & 0 \\ 0 & 0 & & & & & & 0 \\ 0 & 0 & 0 & & / & 0 & b_{n-1} & a_n \end{bmatrix} \begin{bmatrix} h_1 \\ h_2 \\ h_3 \\ \vdots \\ h_n \end{bmatrix} = \begin{bmatrix} w_1 \\ w_2 \\ w_3 \\ \vdots \\ w_n \end{bmatrix}$$

$$\Rightarrow h_1 = w_1/a_1$$
$$h_i = (w_i - b_{i-1}h_{i-1})/a_i, \quad i = 1, 2, 3, \ldots, n$$

which again may be computed by the forms the subroutine SOLV.

Finally,

$$\underline{M}^{\mathrm{T}}\underline{B}^{j+1} = \underline{h}$$

$$\Rightarrow \begin{bmatrix} a_1 & b_1 & 0 & 0 & \cdots & \cdots & \cdots & 0 \\ 0 & a_2 & b_2 & 0 & 0 & \cdots & \cdots & 0 \\ 0 & 0 & a_3 & b_3 & 0 & 0 & 0 & 0 \\ 0 & 0 & \cdots & \cdots & \cdots & \cdots & \cdots & 0 \\ 0 & 0 & 0 & \cdots & 0 & 0 & 0 & a_n \end{bmatrix} \begin{bmatrix} s_1 \\ s_2 \\ s_3 \\ \vdots \\ s_n \end{bmatrix} = \begin{bmatrix} h_1 \\ h_2 \\ h_3 \\ \vdots \\ h_n \end{bmatrix}$$

$$\Rightarrow s_n = h_n/a_n$$
$$s_i = (h_i - b_i s_{i+1})/a_i, \quad i = n-1, \ldots, 3, 2, 1$$

which again may be computed using the subroutine SOLV.

Thus, this scheme consists of the following algorithm:

$$\underline{A}^j, \underline{B}^j \;\underset{\text{SOLV}}{\overset{\text{CHOL}}{\Rightarrow}}\; \underline{A}^{j+1} \;\underset{\text{SOLV}}{\overset{\text{CHOL}}{\Rightarrow}}\; \underline{B}^{j+1}$$

$$\underset{\text{SOLV}}{\Rightarrow} \;\overset{\text{CHOL}}{}\; \underline{A}^{j+2} \;\underset{\text{SOLV}}{\overset{\text{CHOL}}{\Rightarrow}}\; \underline{B}^{j+2}, \text{ etc.}$$

This procedure is programmed for digital computer calculations, using these two subroutines, and results are obtained. As in the calculation for (ii), viz., solution by elimination for tri-diagonal systems, two vectors $\underline{A}^j$ and $\underline{B}^j$ were modified so as to properly include the boundary conditions: $a(0, t) = 1 = b(0, t)$.

Fully Implicit Scheme In this scheme, the finite difference equations, discretizing Eqs. (RT1-7A) and (RT1-7B), are

$$\left(A_i^{j+1} - A_i^j\right)/\Delta t = \left(A_{i-1}{}^{j+1} - 2A_i^{j+1} + A_{i+1}{}^{j+1}\right)/(\Delta x)^2 - A_i^{j+1}B_i^{j+1} \tag{RT1-16A}$$

$$\left(B_i^{j+1} - B_i^j\right)/\Delta t = R\left(B_{i-1}{}^{j+1} - 2B_i^{j+1} + B_{i+1}{}^{j+1}\right)/(\Delta x)^2 - mA_i^{j+1}B_i^{j+1} \tag{RT1-16B}$$

which may be written in matrix form as:

$$[\underline{\mathrm{I}} - \Delta t/(\Delta x)^2\underline{\mathbf{A}}]\underline{A}^{j+1} = \underline{A}^j - \Delta t\underline{\mathrm{B}}^{j+1}\underline{A}^{j+1} \tag{RT1-17A}$$

$$[\underline{\mathrm{I}} - R\Delta t/(\Delta x)^2\underline{\mathbf{A}}]\underline{B}^{j+1} = \underline{B}^j - m\Delta t\,\underline{\mathrm{B}}^{j+1}\underline{B}^{j+1} \tag{RT1-17B}$$

where

$$\underline{\mathbf{A}} = \begin{bmatrix} -2 & +1 & & & 0 & \\ +1 & -2 & +1 & & 0 & \\ & +1 & -2 & +1 & 0 & \\ & & & & & \\ & 0 & & & +1 & -2 \end{bmatrix}$$

$$A_1{}^{j+1} = \begin{bmatrix} A_1{}^{j+1} & & & 0 & \\ & A_2{}^{j+1} & & 0 & \\ & & A_3{}^{j+1} & 0 & \\ \cdots\cdots & \cdots\cdots & \cdots\cdots & \cdots\cdots & \cdots\cdots \\ 0 & & & & A_n{}^{j+1} \end{bmatrix}$$

and

$$\underline{B}_1{}^{j+1} = \begin{bmatrix} B_1{}^{j+1} & & & 0 & \\ & B_2{}^{j+1} & & 0 & \\ & & B_3{}^{j+1} & 0 & \\ \cdots\cdots & \cdots\cdots & \cdots\cdots & \cdots\cdots & \cdots\cdots \\ 0 & & & & B_n{}^{j+1} \end{bmatrix}$$

As before, to incorporate the boundary conditions: $a(0, t) = 1 = b(0, t)$, the matrices for these schemes become

$$\begin{bmatrix} [1 + \{2\Delta t/(\Delta x)^2\} & -\Delta t/(\Delta x)^2 & 0 & \\ -\Delta t/(\Delta x)^2 & [1 + \{2\Delta t/(\Delta x)^2\} & -\Delta t/(\Delta x)^2 & 0 \\ 0 & -\Delta t/(\Delta x)^2[1 + \{2\Delta t/(\Delta x)^2\} & -\Delta t/(\Delta x)^2 & 0 \\ 0 & 0 & \cdots\cdots\cdots\cdots & \cdots \\ 0 & 0 & -\Delta t/(\Delta x)^2 & [1 + \{2\Delta t/(\Delta x)^2\} \end{bmatrix} \cdot \begin{bmatrix} A_1{}^{j+1} \\ A_2{}^{j+1} \\ A_3{}^{j+1} \\ \cdots \\ A_n{}^{j+1} \end{bmatrix}$$

$$= \begin{bmatrix} [\{\Delta t/(\Delta x)^2\}1 + A_1{}^j - \Delta tB_1{}^{j+1}A_1{}^{j+1}] \\ A_2{}^j - \Delta tB_2{}^{j+1}A_2{}^{j+1}] \\ A_3{}^j - \Delta tB_3{}^{j+1}A_3{}^{j+1} \\ \cdots\cdots\cdots\cdots\cdots\cdots \\ A_n{}^j - \Delta tB_n{}^{j+1}A_n{}^{j+1}] \end{bmatrix} \tag{RT1-18A}$$

and

$$\begin{bmatrix} [1 + R\{2\,\Delta t/(\Delta x)^2\}] & & -R\Delta t/(\Delta x)^2 & 0 \\ -R\Delta t/(\Delta x)^2 & [1 + R\{2\,\Delta t/(\Delta x)^2\} & -R\Delta t/(\Delta x)^2 & 0 \\ 0 & -R\Delta t/(\Delta x)^2 & [1 + R\{2\,\Delta t/(\Delta x)^2\}] - R\Delta t/(\Delta x)^2 & 0 \\ 0 & 0 & \cdots\cdots & \cdots \\ 0 & 0 & -R\Delta t/(\Delta x)^2 & [1 + R\{2\,\Delta t/(\Delta x)^2 \end{bmatrix}$$

$$= \begin{bmatrix} B_1^{j+1} \\ B_2^{j+1} \\ B_3^{j+1} \\ \cdots\cdots \\ B_n^{j+1} \end{bmatrix} \begin{bmatrix} R\{\Delta t/(\Delta x)^2\}1 + B_1^{j} - m\Delta t A_1^{j+1} B_1^{j+1} \\ B_2^{j} - m\Delta t A_2^{j+1} B_2^{j+1} \\ B_3^{j} - m\Delta t\, A_3^{j+1} \\ \cdots\cdots\cdots\cdots\cdots\cdots \\ B_n^{j} - m\Delta t A_n^{j+1} \end{bmatrix} \tag{RT1-18B}$$

Clearly, the implicit relations between A_i^{j+1} and A_i^j, and between B_i^{j+1} and B_i^j, mean that an iterative scheme is to be used. The symmetric tri-diagonal matrices on the left-hand side of Eqs. (RT1-18A) and (RT1-18B) can be readily obtained and factored by the subroutine CHOL. Since it is independent of A_i^j and B_i^j, this part of the calculation need to be done only once per increment in time or space.

Given the vector A_i^j, the next incremental vector A_i^{j+1} is obtained by iteration, using Eq. (RT1-18A) and SOLV, to a predetermined accuracy. Similarly, B_i^{j+1} can be obtained from B_i^j, using Eq. (RT1-18B). This algorithm is programmed for computer calculations, and results are obtained.

Crank–Nicholson Semi-implicit Scheme In this scheme, the finite difference equations, discretizing Eqs. (RT1-7A) and (RT1-7B), are

$$\left(A_i^{j+1} - A_i^j\right)/\Delta t = \frac{1}{2}\left[\frac{\left(A_{i-1}{}^{j+1} - 2A_i^{j+1} + A_{i+1}{}^{j+1}\right)}{(\Delta x)^2} + \frac{\left(A_{i-1}{}^{j} - 2A_i^{j} + A_{i+1}{}^{j}\right)}{(\Delta x)^2}\right] - \frac{1}{2}\left(A_i^{j+1}B_i^{j+1} + A_i^jB_i^j\right) \tag{RT1-19A}$$

$$\left(B_i^{j+1} - B_i^j\right)/\Delta t = \frac{1}{2}R\left[\frac{\left(B_{i-1}{}^{j+1} - 2B_i^{j+1} + B_{i+1}{}^{j+1}\right)}{(\Delta x)^2} + \frac{\left(B_{i-1}{}^{j} - 2B_i^{j} + B_{i+1}{}^{j}\right)}{(\Delta x)^2}\right] - \frac{1}{2}m\left(A_i^{j+1}B_i^{j+1} + A_i^jB_i^j\right) \tag{RT1-19B}$$

which, in matrix form, are

$$\underline{C}_1\underline{A}^{j+1} = \underline{C}_2\underline{A}^{j} \tag{RT1-20A}$$

$$\underline{D}_1\underline{B}^{j+1} = \underline{D}_2\underline{B}^{j} \tag{RT1-20B}$$

where

$$\underline{C}_1 = [\underline{I} - 1/2\{\Delta t/(\Delta x)^2\}\underline{A} + 1/2\Delta t\underline{B}^{j+1}]$$

$$\underline{C}_2 = [\underline{I} + 1/2\{\Delta t/(\Delta x)^2\}\underline{A} + 1/2\Delta t\underline{B}^{j}]$$

$$\underline{D}_1 = [\underline{I} - 1/2\{\Delta t/(\Delta x)^2\}\underline{A} + 1/2m\Delta t\underline{A}^{j+1}]$$

$$\underline{D}_2 = [\underline{I} - 1/2\{\Delta t/(\Delta x)^2\}\underline{A} + 1/2m\Delta t\underline{A}^{j}]$$

with

$$\underline{A}^j = \begin{bmatrix} A_1^j & & & 0 & \\ & A_2^j & & 0 & \\ & & A_3^j & 0 & \\ \cdots\cdots & \cdots\cdots & \cdots\cdots & \cdots\cdots & \cdots\cdots \\ 0 & & & & A_n^j \end{bmatrix}$$

$$\underline{B}^j = \begin{bmatrix} B_1^{\ j} & & & 0 & \\ & B_2^{\ j} & & 0 & \\ & & B_3^{\ j} & 0 & \\ \cdots\cdots & \cdots\cdots & \cdots\cdots & \cdots\cdots & \cdots\cdots \\ 0 & & & & B_n^{\ j} \end{bmatrix}$$

and $\underline{A}^j, \underline{A}^{j+1}$, and $\underline{B}^{j+1}$ are defined as before. Clearly, this required an iterative scheme.

Again, to incorporate the boundary conditions: $a(0, t) = 1 = b(0, t)$, one may first rewrite Eqs. (RT1-20A) and (RT1-20B) as follows:

$$[\mathrm{I} - 1/2\{\Delta t/(\Delta x)^2\}\mathrm{A}]A^{j+1} = [\mathrm{I} + 1/2\{\Delta t/(\Delta x)^2\} - 1/2\Delta t\mathrm{B}^j]A^j - 1/2\Delta t\mathrm{B}^{j+1}A^{j+1} \tag{RT1-21A}$$

$$[\underline{\mathrm{I}} - \frac{1}{2}R\{\Delta t/(\Delta x)^2\}\underline{\mathrm{A}}]\underline{B}^{j+1} = [\underline{\mathrm{I}} - \frac{1}{2}R\{\Delta t/(\Delta x)^2\} - \frac{1}{2}m\Delta t\underline{A}^j]\underline{B}^j - \frac{1}{2}m\Delta t\underline{\mathrm{A}}^{j+1}B^{j+1} \tag{RT1-21B}$$

Upon inclusion of the boundary conditions, these matrix equations become

$$\begin{bmatrix} a & m & 0 & 0 & 0 & 0 & \cdots & 0 \\ m & a & m & 0 & 0 & 0 & \cdots & 0 \\ 0 & m & a & m & 0 & \cdots & \cdots & 0 \\ \cdots & \cdots & \cdots & \cdots & \cdots & \cdots & \cdots & \cdots \\ 0 & 0 & 0 & 0 & \cdots & \cdots & m & a \end{bmatrix} \begin{bmatrix} A_1^{\ j+1} \\ A_2^{\ j+1} \\ A_3^{\ j+1} \\ \vdots \\ A_n^{\ j+1} \end{bmatrix} = \begin{bmatrix} \{\Delta t/(\Delta x)^2\} \\ 0 \\ 0 \\ \vdots \\ 0 \end{bmatrix}$$
$$+ \begin{bmatrix} p_1 & n & 0 & 0 & 0 & 0 & \cdots & 0 \\ n & p_2 & n & 0 & 0 & 0 & \cdots & 0 \\ 0 & n & p_3 & n & 0 & 0 & \cdots & 0 \\ \cdots & \cdots & \cdots & \cdots & \cdots & \cdots & \cdots & \cdots \\ 0 & 0 & 0 & 0 & \cdots & \cdots & n & p_n \end{bmatrix} \begin{bmatrix} A_1^{\ j} \\ A_2^{\ j} \\ A_3^{\ j} \\ \vdots \\ A_n^{\ j} \end{bmatrix}$$
$$- 1/2\Delta t \begin{bmatrix} B_1^{\ j+1} & 0 & 0 & 0 & \cdots & \cdots & \cdots & 0 \\ 0 & B_2^{\ j+1} & 0 & 0 & 0 & \cdots & \cdots & 0 \\ 0 & 0 & B_3^{\ j+1} & 0 & \cdots & \cdots & \cdots & 0 \\ \cdots & \cdots & \cdots & \cdots & \cdots & \cdots & \cdots & \cdots \\ 0 & 0 & 0 & \cdots & \cdots & \cdots & 0 & B_n^{\ j+1} \end{bmatrix} \begin{bmatrix} A_1^{\ j+1} \\ A_2^{\ j+1} \\ A_3^{\ j+1} \\ \vdots \\ A_n^{\ j+1} \end{bmatrix} \tag{RT1-22A}$$

where

$$a = 1 + \{\Delta t/(\Delta x)^2\}$$

$$m = -(1/2)\{\Delta t/(\Delta x)^2\}$$

$$p_i = 1 - \{\Delta t/(\Delta x)^2\} - (1/2)\Delta t B_i^{\ j}$$

$$n = (1/2)\{\Delta t/(\Delta x)^2\}$$

and

$$\begin{bmatrix} c & r & 0 & 0 & 0 & 0 & \cdots & 0 \\ r & c & r & 0 & 0 & 0 & \cdots & 0 \\ 0 & r & c & r & 0 & \cdots & \cdots & 0 \\ \cdots & \cdots & \cdots & \cdots & \cdots & \cdots & \cdots & \cdots \\ 0 & 0 & 0 & 0 & \cdots & \cdots & r & c \end{bmatrix} \begin{bmatrix} B_1^{j+1} \\ B_2^{j+1} \\ B_3^{j+1} \\ \vdots \\ B_n^{j+1} \end{bmatrix} = \begin{bmatrix} R\{\Delta t/(\Delta x)^2\} \\ 0 \\ 0 \\ \vdots \\ 0 \end{bmatrix}$$

$$+ \begin{bmatrix} q_1 & s & 0 & 0 & 0 & 0 & \cdots & 0 \\ s & q_2 & s & 0 & 0 & 0 & \cdots & 0 \\ 0 & s & q_3 & s & 0 & 0 & \cdots & 0 \\ \cdots & \cdots & \cdots & \cdots & \cdots & \cdots & \cdots & \cdots \\ 0 & 0 & 0 & 0 & \cdots & \cdots & s & q_n \end{bmatrix} \begin{bmatrix} B_1^{j} \\ B_2^{j} \\ B_3^{j} \\ \vdots \\ B_n^{j} \end{bmatrix}$$

$$- 1/2m\Delta t \begin{bmatrix} A_1^{j+1} & 0 & 0 & 0 & \cdots & \cdots & \cdots & 0 \\ 0 & A_2^{j+1} & 0 & 0 & 0 & \cdots & \cdots & 0 \\ 0 & 0 & A_3^{j+1} & 0 & \cdots & \cdots & \cdots & 0 \\ \cdots & \cdots & \cdots & \cdots & \cdots & \cdots & \cdots & \cdots \\ 0 & 0 & 0 & \cdots & \cdots & \cdots & 0 & A_n^{j+1} \end{bmatrix} \begin{bmatrix} B_1^{j+1} \\ B_2^{j+1} \\ B_3^{j+1} \\ \vdots \\ B_n^{j+1} \end{bmatrix} \quad \text{(RT1-22B)}$$

where

$$c = 1 + R\{\Delta t/(\Delta x)^2\}$$

$$r = -(1/2)R\{\Delta t/(\Delta x)^2\}$$

$$q_i = 1 - R\{\Delta t/(\Delta x)^2\} - (1/2)m\Delta t A_i^j$$

$$s = (1/2)R\{\Delta t/(\Delta x)^2\}$$

It is easily seen that the implicit relations between A_i^{j+1} and A_i^j and between B_i^{j+1} and B_i^j require the use of an iterative procedure. This procedure for computation is the same as that for the fully implicit scheme, viz., using Choleski's method. This algorithm is programmed for machine computation, and results are obtained.

3.8.1.1.5 Biomedical, Biopharmaceutical, and Chemical Applications

An important application of this study is the effects of biochemical, biopharmaceutical, and chemical reactions on the rate of transport into the media, e.g. the enhancement of rates of mass transport with reactions *over* the rate of physical mass transport. A typical measure of these absorption rates may be expressed in terms of the ***transport coefficients***, defined as the rates of transport of dissolved reactants *per unit of interfacial area*.

For physical mass transport, this mass transfer coefficient is

$$\text{for Reactant A:} \quad k_{\text{LA}} = -\sqrt{(kB_i D_{\text{A}})} \quad \text{(RT1-23A)}$$

$$\text{for Reactant B:} \quad k_{\text{LB}} = -\sqrt{(kB_i D_{\text{B}})} \quad \text{(RT1-23B)}$$

The mass transfer coefficient, in general, will be different for A and for B – each dependent on the reaction rates between the reacting substances:

$$k_{\mathrm{LA}}(t) = -(D_{\mathrm{A}}/A_i)(\partial A/\partial z)_{z=0} = -\surd(kB_iD_{\mathrm{A}})(\partial a/\partial x)_{x=0} \quad \text{(RT1-24A)}$$

$$k_{\mathrm{LB}}(t) = -(D_{\mathrm{B}}/B_i)(\partial B/\partial z)_{z=0} = -\surd(kB_iD_{\mathrm{A}})R(\partial b/\partial x)_{x=0} \quad \text{(RT1-24B)}$$

After Eqs. (RT1-7A) and (RT1-7B) have been solved, subject to the boundary conditions in Eq. (RT1-7C), then the derivatives required in Eqs. (RT1-24A) and (RT1-24B) may be determined.

The **Mass Transport Rate Enhancement Factor** F_{E} may be expressed as the ratio:

$$\frac{\text{Mass Transfer Coefficient for Mass Transport with Biochemical Reaction}}{\text{Mass Transfer Coefficient for Physical Mass Transport } \textbf{\textit{only}}}$$

Thus, for Reactant A:

$$F_{\mathrm{EA}} = k_{\mathrm{LA}}/k_{\mathrm{LA}}(t)|_{\mathrm{mean}} = (1/T)\int_0^T (\partial a/\partial x)_{x=0}\mathrm{d}T \quad \text{(RT1-25A)}$$

where $k_{\mathrm{LA}}(t)|_{\mathrm{mean}}$ = time-mean mass transfer coefficient for mass transport with biochemical reaction, for Reactant Component A,

$$= -\sqrt{(kB_iD_{\mathrm{A}})}\left[\left\{\int_0^T (\partial a/\partial x)_{x=0}\mathrm{d}T\right\}\Big/T\right]$$

Similarly, for Reactant B:

$$k_{\mathrm{LB}}(t)|_{\mathrm{mean}} = -\sqrt{(kB_iD_{\mathrm{B}})}\left[\left\{\int_0^T (\partial b/\partial x)_{x=0}\mathrm{d}T\right\}\Big/T\right]$$

and

$$F_{\mathrm{EB}} = k_{\mathrm{LB}}/k_{\mathrm{LB}}(t)|_{\mathrm{mean}} = (1/T)\int_0^T (\partial b/\partial x)_{x=0}\mathrm{d}T \quad \text{(RT1-25B)}$$

From the results of the numerical solutions of Eqs. (RT1-7A) and (RT1-7B), the gradients at $x = 0$ may be obtained, and using Eqs. (RT1-25A) and (RT1-25B), the Enhancement Factors, (owing to biochemical reaction) of the mass transport rates may be evaluated for each reactant component.

3.8.1.1.6 Some Numerical Results

The Steady-State Solution[25] The following steady-state solution may be used for comparison:

$$a = b = 1/(1 + x\surd 6)^2$$

Fully Implicit Scheme For $R = 1$, and $m = 1$, using the Choleski method, with iterations and double precision calculations, numerical results for two cases have been obtained, see Appendix 3.A (in which an outline of the FORTRAN IV computer code is shown, together with some pertinent outputs of the calculated results):

Case I NX (number of distance steps) = 15
NT (number of time steps) = 100

Case II NX (number of distance steps) = 20
NT (number of time steps) = 100

The convergence within an error less than $1/2 \times 10^{-4}$ unit.

Previously observed instabilities no longer appeared, even for large time steps:

$\Delta t/\Delta x = 3/20$ to $1/5$, and
$\Delta t/(\Delta x)^2 = 9/40$ to $2/5$

Results, truncated to four significant digits, are obtained, listed below, and presented graphically using the open-source R program, showing the time-dependent ***penetration*** of the mass transport under the second-order irreversible process: (Figures A–D).

Case I: $R = 1$, $m = 1$, $X = 10$, NX = 15, $T = 10$, NT = 100
Case II: $R = 1$, $m = 1$, $X = 10$, NX = 20, $T = 10$, NT = 100

Evaluation of the Reaction Rate Enhancement Factor F_E From the FORTRAN program output, the integrals in Eqs. (RT1-25A) and (RT1-25B) may be evaluated, by a

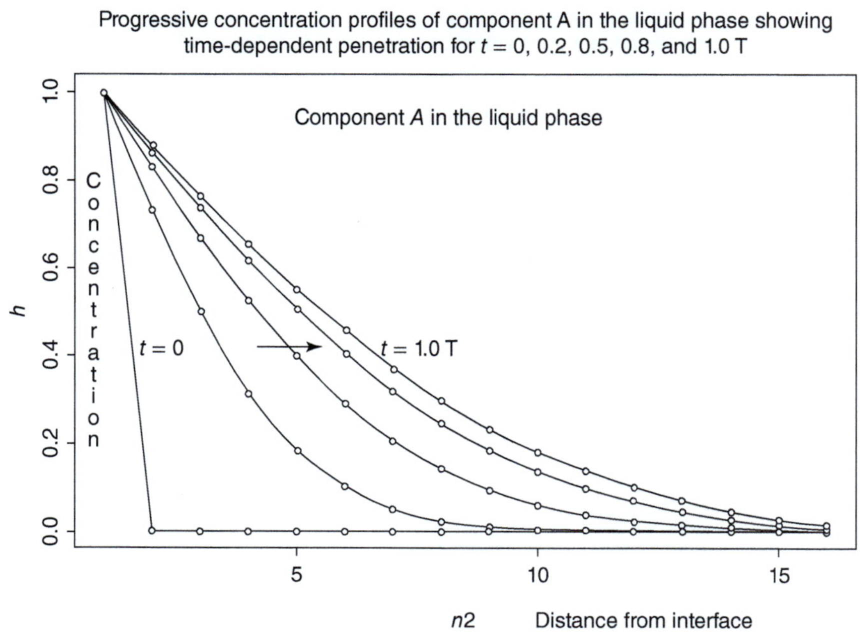

Figure A Using R to provide graphical output for the fully implicit scheme, Case I, Component A: progressive concentration profiles of Component A in the liquid phase showing time-dependent penetration for $t = 0, 0.2, 0.5, 0.8$, and 1.0 T.

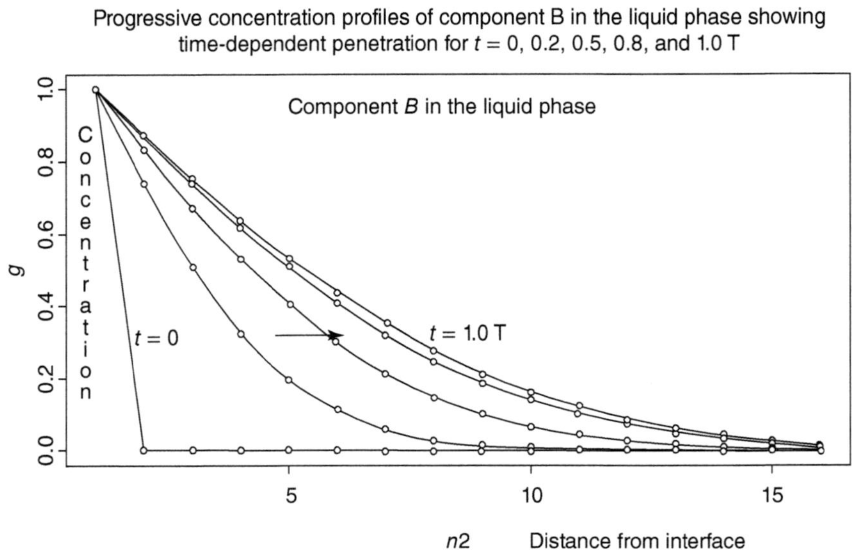

Figure B Using R to provide graphical output for the fully implicit scheme, Case I, Component B: progressive concentration profiles of Component B in the liquid phase showing time-dependent penetration for $t = 0, 0.2, 0.5, 0.8$, and 1.0 T.

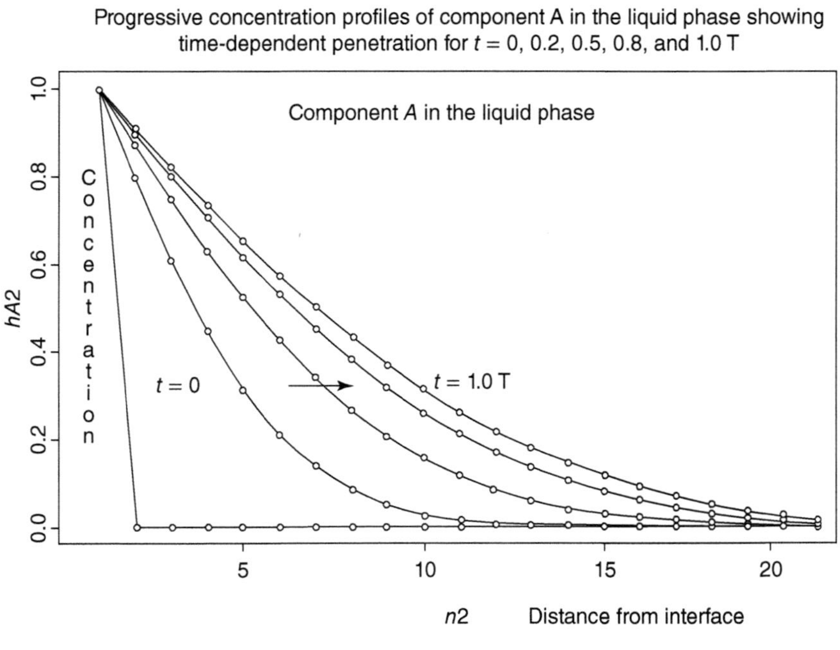

Figure C Using R to provide graphical output for the fully implicit scheme, Case II, Component A: progressive concentration profiles of Component A in the liquid phase showing time-dependent penetration for $t = 0, 0.2, 0.5, 0.8$, and 1.0 T.

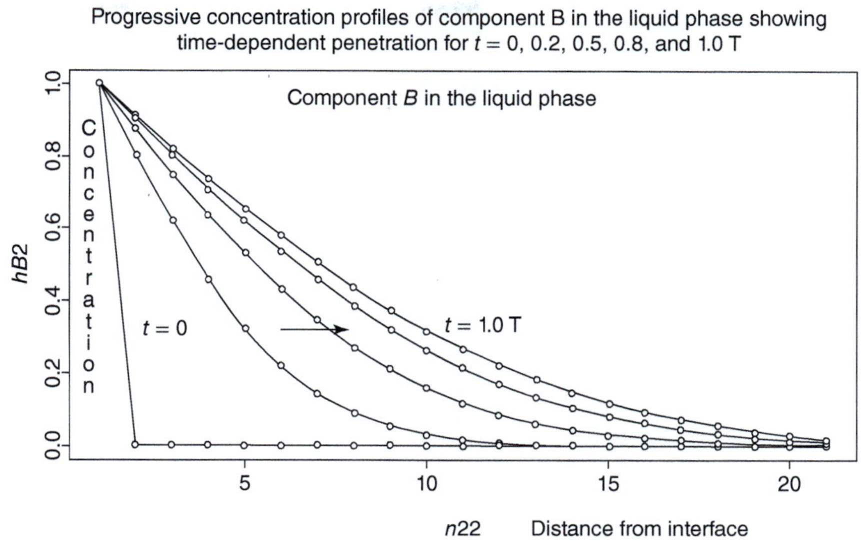

Figure D Using R to provide graphical output for the fully implicit scheme, Case II, Component B: progressive concentration profiles of Component B in the liquid phase showing time-dependent penetration for $t = 0, 0.2, 0.5, 0.8$, and 1.0 T.

summation approximation, viz.:

$$F_{\mathrm{EA}} = (1/T)\int_0^T (\partial a/\partial x)_{x=0}\mathrm{d}T = (1/T)\sum_0^T\{(\Delta A/\Delta X)_{x=0}\Delta T\}$$

$$= (\Delta T/T)\sum_0^T\{(\Delta A/\Delta X)_{x=0}\} \qquad \text{(RT1-26)}$$

for a fixed ΔT, where $(\Delta A/\Delta X)_{x=0}$ = gradient of the A vs. x profile, at $x = 0$.

In order to avoid the error owing to the abruptness of the *numerical* concentration gradient at the interface, viz at $T = 02$, a more realistic value of F_{E} may be obtained by evaluating the gradient integral by leaving out the value at $T = 02$, viz., the integral is taken from $T = 22$ to $T = 102$, or over the final 8 units of time T (out of the total of 10 units of T):

$$\Delta T = T/4 = 8/4 = 2, \quad \text{where } T = 8, \text{and } \Delta T/T = 2/8 = 0.25$$

Case I: For Gas A, $\Delta X = 1/15 = 0.0667$

At $T = 22$: $(\Delta A/\Delta X)_{x=0} = -(0.7331 - 1.0000)/0.0667 = 4.0015$
At $T = 52$: $(\Delta A/\Delta X)_{x=0} = -(0.8315 - 1.0000)/0.0667 = 2.5262$
At $T = 82$: $(\Delta A/\Delta X)_{x=0} = -(0.8669 - 1.0000)/0.0667 = 1.9955$
At $T = 102$: $(\Delta A/\Delta X)_{x=0} = -(0.8809 - 1.0000)/0.0667 = 1.7856$

and $\sum_0^T\{(\Delta A/\Delta X)_{x=0}\} = 10.3088$

Hence,

$$F_{\mathrm{EA}} = (\Delta T/T)\sum_0^T\{(\Delta A/\Delta X)_{x=0}\}$$

$$= (0.25)(10.3088)$$

$$= 2.5772$$

Similarly, for Gas B,

$$F_{EB} = (\Delta T/T)\sum_{0}^{T} \{(\Delta A/\Delta X)_{x=0}\}$$
$$= (0.25)(10.1648)$$
$$= 2.5412$$

Case II: Similarly, for Gas A, $\Delta X = 1/20 = 0.0500$

At $T = 22$: $(\Delta A/\Delta X)_{x=0} = -(0.7984 - 1.0000)/0.0500 = 4.0320$
At $T = 52$: $(\Delta A/\Delta X)_{x=0} = -(0.8733 - 1.0000)/0.0500 = 2.5340$
At $T = 82$: $(\Delta A/\Delta X)_{x=0} = -(0.9000 - 1.0000)/0.0500 = 2.0000$
At $T = 102$: $(\Delta A/\Delta X)_{x=0} = -(0.9106 - 1.0000)/0.0500 = 1.7880$
and $\sum_{0}^{T} \{(\Delta A/\Delta X)_{x=0}\} = 10.3540$

Hence,

$$F_{EA} = (\Delta T/T)\sum_{0}^{T} \{(\Delta A/\Delta X)_{x=0}\}$$
$$= (0.25)(10.3540)$$
$$= 2.5885$$

Similarly, for Gas B,

$$F_{EB} = (\Delta T/T)\sum_{0}^{T} \{(\Delta A/\Delta X)_{x=0}\}$$
$$= (0.25)(10.2080)$$
$$= 2.5520$$

Remarks: In the work of Roper et al.[32], the results were presented as F_E vs. $\sqrt{T}$, for various parametric values of R and m. For $T = 10$, and for $R = 1$, and $m = 1$, at $\sqrt{T} = \sqrt{10} = 3.1623$, the value obtained by Roper et al. was: $F_E = 2.42$.

In this work, the four cases of computed values of F_E are 2.5772, 2.5412, 2.5885, and 2.5520, averaging 2.5572, which are in satisfactory agreement with Roper's results (to within about 5%).

Appendix 3.A

On the Boundary Conditions of $a(x, t)$ and $b(x, t)$ at Large Values of x and t: Equation (3C)

Roper et al. [32] considered only the case when $n_1 = 1 = n_2$: instead of taking a and b to be bounded as x approached infinity, for all t, they ***assumed*** the condition that

$$\lim_{x\to\infty} a(x, t) = 0 = \lim_{x\to\infty} b(x, t)$$

However, this assumption seems to lead to a contradiction except for the special case when $R = 1 = m$, for which they obtained the explicit steady-state solution mentioned in this work.

To demonstrate this contradiction, let the steady solution be

$$\alpha(x) = \lim_{x\to\infty} a(x,t) \quad \text{and} \quad \beta(x) = \lim_{x\to\infty} b(x,t)$$

so that one obtains:

$$\alpha''(x) = \alpha^{n1}\beta^{n2} \quad \text{and} \quad R\beta''(x) = m\alpha^{n1}\beta^{n2}$$

Combining the last two equations by eliminating $\alpha^{n1}\beta^{n2}$, one obtains

$$\partial^2/\partial x^2\{\alpha - (R/m)\beta\} = 0$$

Hence on integrating twice,

$$\alpha(x) - (R/m)\beta(x) = c_1x + c_2$$

Now, $c_1 = 0$, for otherwise $\{\alpha - (R/m)\beta\} \to \infty$ as $x \to \infty$.

Moreover, at $x = 0$, $\alpha(0) = (R/m)\beta(0) = 1 - (R/m) = c_2$. Hence,

$$\alpha(x) - (R/m)\beta(x) = 1 - (R/m)$$

and

$$\lim_{x\to\infty}[\alpha(x) - (R/m)\beta(x)] = 1 - (R/m) \neq 0, \quad \text{unless } R = m$$

Thus, the boundedness condition as $x \to \infty$ seems to be at least more mathematically consistent. Physically, this assumption of boundedness also appears to be realistic and consistent. In other words, while both a and b approaches zero at large values of x and t, ***they do not become zero at the same point in space and time.***

Appendix 3.B

FORTRAN IV Program for the Fully Implicit Scheme, Section 3 (d), for Solving the System Equations (RT1-7A)–(RT1-7C)

The discretizing equations for this system, using the fully implicit scheme, are given by Eqs. (RT1-16A) and (RT1-16B), which may be written in matrix form as Eqs. (RT1-17A) and (RT1-17B). Setting the Choleski iterative error to be $\leq 1/2 \times 10^{-4}$, for double precision computations, the following parameters are chosen:

Case I: $R = 1, m = 1, X = 10$, NX = 15, $T = 10$, NT = 100
Case II: $R = 1, m = 1, X = 10$, NX = 20, $T = 10$, NT = 100

The following computer program, written in FORTRAN IV, was used for running in a mainframe computer system. Results, truncated to four significant digits, are obtained, listed below, and presented graphically using the open-source R program, showing the time-dependent penetration of the mass transfer process, under the second-order irreversible process:

FORTRAN IV Program

```
C    Bertram K. C. Chan
C
C    bertchan1@gmail.com
C
C    SIMULTANEOUS ABSORPTION AND CHEMICAL REACTION
C    OF TWO GASES IN A LIQUID
```

```
C
C     PENETRATION THEORY MODEL
C
C     NUMERICAL SOLUTION BY FINITE DIFFERENCE METHOD
C
C     IMPLICIT SCHEME, CHOLESKI METHOD WITH ITERATIONS

C
C
C
        IMPLICIT REAL*8(A-H, O-Z)
        DIMENSION A(51,102), B(51, 102)
        DIMENSION C(51), U(50), ALPHA(51), BETA(50), CL(51)
        DIMENSION GAMMA(51), G(51)
        DIMENSION DELTA(51)
        DIMENSION D(51), V(50), AA(51), BB(50), DM(50)
        DIMENSION W(51), H(51)
        DIMENSION S(51)
        DIMENSION A1(50), A2(50)
        DIMENSION B1(50), B2(50)
C
  100   READ (1,1000) N
        WRITE (3,1001) N
 1000   FORMAT (I3)
 1001   FORMAT (I4)
C
        NN = -1
        IF (N.EQ.NN) GO TO 999
        CONTINUE
C
C  TO INPUT THE PARAMETERS AND VARIABLES

C

        READ (1,1002) R
        WRITE (3,1003) R
 1002   FORMAT (F16.6)
 1003   FORMAT (5H R   , F16.6)

C
        READ (1,1002) XM
        WRITE (3,1004) XM
 1004   FORMAT (5H M   ,F16.6)
C
        READ (1,1002) X
        WRITE (3,1005) X
```

```
 1005   FORMAT (5H X    ,F16.6)
C
        READ (1,1002) T
        WRITE (3,1006) T
 1006   FORMAT (5H T    ,F16.6)
C
        READ (1,1000) NX
        WRITE (3,1007) NX
 1007   FORMAT (6H NX    ,I4)
C
        READ (1,1000) 1008   FORMAT (6H NT    ,I4)
C
       DX = X/NX
       DT = T/NT
C
       NX1 = NX+1
       NT1 = NT+1
       NT2 = NT+2
C
C  TO INPUT THE BOUNDARY CONDITIONS AND INITIAL
C  CONDITIONS
C
       READ (1,1002) AIJ
       READ (1,1002) BIJ

       READ (1,1002) ANXIJ
       READ (1,1002) BNXIJ
       READ (1,1002) AI1
       READ (1,1002) bi1
C
       WRITE (3,1009) AIJ, BIJ, ANXIJ, BNXIJ, AI1, BI1
  1009 FORMAT (6F10.4)
C
C  TO PRE-SET THE FIELDS
C
      DO 1     I = 1,NX1,1
      DO 2     J = 1,NT2,1
      A(I,J) = AI1
      B(I,J) = BI1
    2  CONTINUE
 1. CONTINUE
C
C  TO OUTPUT THE INITIAL CONDITIONS
C
      J = 1
      WRITE (3,1001) J
```

```
      WRITE (3,1010)  (A(I,J),  I = 1,NX)
      WRITE (3,1010)  (B(I,J),  I = 1,NX)
 1010 FORMAT (5E16.9)
      WRITE (3,99)
   99 FORMAT (/)
C
C TO SET THE BOUNDARY CONDITIONS
C
      DO 3    J = 2,NT2,1
      A(I,J) = A1J
      B(I,J) = B1J
      A(NX1,J) = ANX1J
      B(NX1,J) = BNX1J


    3 CONTINUE
C
C TO OUTPUT THE FIRST BOUNDARY CONIN
C
      J = 2
      WRITE (3,1001) J
      WRITE (3,1010)  (A(I,J),  I = 1,NX)
      WRITE (3,1010)  (B(I,J),  I = 1,NX)
C
C TO START THE COMPUTATION
C
      DO 500    I = 2,NX1,1
      A(I,3) = A(I,2)
      B(I,3) = B(I,2)

500 CONTINUE
      DO 101    J = 2,NT2,1
C
C
C TO CALCULATE A(I,J+1) FROM A(I,J) AND B(I,J)
C
C ITERATING ON  A  AND  B
C
C TO OBTAIN THE VECTOR A(J)
      GAMMA(2)=(DT/(DX**2))+A(2,J)-DT*B(2,J + 1)*A(2,J + 1)
      D0 105    I = 3,NX,1
      GAMMA(I) = A(I,J) - DT*B(I,J+1)*A(I,J+1)
  105 CONTINUE
C
C TO OBTAIN THE VECTOR B(J)
      W(2) = R*(DT/(DX**2))+B(2,J)-XM*DT*A(2,J+1)*B(2,J+1)
```

```
      DO 205    I = 3,NX1,1
      W(I) = B(I,J) -XM*DT*A(I,J+1)*B(I,J+1)
 205  CONTINUE
C
      DO 700    IT = 1, 10
C  TO OBTAIN THE POSITIVE DEFINITE SYMMETRIC
C  TRI-DIAGONAL MATRIX C
C(2)0

      DO 102    I = 2,NX1,1
      C(I) = 1.0 + (2.0*(DT/(DX**2)))
 102  CONTINUE
C
      DO 103    I = 2,NX,1
      U(I) = - (DT/(DX**2))
 103  CONTINUE
C
C  TO COMPUTE THE LOWER TRO-DIAGONAL MATRIX L

C
C
      ALPHA(2) = DSQRT(C(2))
      DO 104    I = 2,NX,1
      BETA(I) = U(I)/A;PHA(I)
      CL(I+1) = C(I+1) - ((BETA(I))**2)
     ALPHA(I+1) = DSQRT(CL(I+1))
 104  CONTINUE
C
C  COMPUTE THE VECTOR G
      G(2) = GAMMA(2)/ALPHA(2)
C
      DO 106    I = 3,NX1,1
      G(I) = (GAMMA(I) - (BETA(I-1)*G(I-1)))/APLHA(I)
 106  CONTINUE
C
C  TO COMPUTE THE VECTOR A(J+1)
C
      DELTA(NX1) = G(NX1)/ALPHA(NX1)
C
      I = NX1-1
 107  CONTINUE
      DELTA(I) = (G(I) - (BETA(I)*DELTA(I+1)))/ALPHA(I)
      I = I-1
      IF (I.GE.2)  GO TO 107
      CONTINUE
C
C  TO OBTAIN A(I,J+1)
```

```
C
          DO 108    I = 2,NX1,1
       A(I,J+1) = DELTA(I)
108  CONTINUE
C
C
C  TO CALCULATE B(I,J+1) FROM B(I,J) AND A(I,J+1)
C
C
        DO 202    I = 2,NX1,1
        D(I) = 1.0+(2.0*(DT/DX**2))*R)
 202  CONTINUE
C
        DO 203    I = 2,NX,1
        V(I) = -((DT/(DX**2))*R)
 203  CONTINUE
C

C  TO COMPUTE THE LOWER TRI-ANGULAR MATRIX A
C
C         AA(2) = DSQRT(D(2))
        DO 204    I = 2,NX,1
        BB(I) = V(I)/AA(I)
        DM(I+1) = D(I+1) - ((BB(I)**2)
        AA(I+1) = DSQRT(DM(I+1))
 204  CONTINUE
C
C  TO COMPUTE THE VECTOR H
C
        H(2) = W(2)/AA(2)
C
        DO 206    I = 3,NX1,1
        H(I) = (W(I) -BB(I-1)*H(I-1))/AA(I)
 206  CONTINUE
C
C  TO COMPUTE THE VECTOR B(J+1)
C
        S(NX1) = H(NX1)/AA(NX))
C
        I = NX1-1
 207  CONTINUE
        S(I) = (H(I) - (BB(I)*S(I+1)))/AA(I)
        I = I-1
        IF (I.GE.2)  GO TO 207
         CONTINUE
C
C  TO OBTAIN B(I,J+1)
```

```
C
      DO 208    I = 2,NX,1
     B(I,J+1) = S(I)
208 CONTINUE
C
      IF (IT.GT.1)  GO TO 707
      CONTINUE
      DO 800    I=2,NX1
800 AI(I) = A(I,J+1)
      DO 900    I = 2,NX1
900 BI(I) = B(I,J+1)
      GO TO 706
707 CONTINUE
      DO 801    I = 2,NX1
801 A2(I) = A(I,J+1)
      DO 901    I = 2,NX1
901 B2(I) = B(I,J+1)
      ERA = DABS(A1(2) = A2(2))
      DO 802    I = 3,NX1

      ERA1 = DABS(A1(I) - a2(I))
      IF  (ERA1.LT.ERA)  GO TO 802
      ERA = ERA1
802 CONTINUE
      ERB = DABS(B1(2) - B2(2))
      DO 902    I = 3,NX1

     ERB1  = DABS(B1(I) - B2(I))
      IF (ERB1.LT.ERB)  GO TO 902
      ERB = ERB1
902 CONTINUE
      ERROR = 0.00005
      ERR = ERA
      IF (E.GT.ERB)  GO TO 705
      ERR = ERB
705 CONTINUE
799 FORMAT (3E16.9)
      IF (ERR.LE.ERROR)  GO TO 101
      CONTINUE
706 CONTINUE
      DO 803    I = 2,NX1
803 A1(I) = A(I,J+1)
      DO  903    I = 2,NX1
903 B1(I) = B(I,J+1)
      DO 701    I = 2,NX1
701 B(I,J) = B(IJ+1)
700 CONTINUE
```

```
101 CONTINUE
C
C

C TO PRINT OUT THE FINAL RESULTS OF THIS RUN
C
C
      DO 4    J = 2,NT2,10
      WRITE (3,1001) J
      WRITE (3,1010) (A(I,J), I = 1,NX1))
      WRITE (3,1010) (B(I,J), I = 1,NX1))
    4 CONTINUE
C
C
C  TO OUTPUT THE RESULTS FOR THE FIRST 10 TIME INTERVALS
C
C
      DO 5    J = 1,I2,1
      WRITE (3,1001) J
      WRITE (3,1010)  (A(I,J), I = 1,NX1)
      WRITE (3,1010)  (B(I,J), I = 1,NX1)
  2. CONTINUE
      C
      C  END OF CALCULATIONS FOR THIS SET OF DATA
      C
      C  IS THERE ANOTHER SET OF DATA TO BE PROCESSED?
      C
        GO TO 100
   C
      C  END OF DATA
      C
      999  CONTINUE
     STOP
     END
```

Appendix 3.C

Using the R software for plotting the results of the numerical solutions of the system equations (RT1-7A)–(RT1-7C), to obtain:

Figures A and B (for Case I)
Figures C and D (for Case II)

In the computed numerical solutions for each of the two cases:

Case I: $R = 1$, $m = 1$, $X = 10$, NX = 15, $T = 10$, NT = 100
Case II: $R = 1$, $m = 1$, $X = 10$, NX = 20, $T = 10$, NT = 100

there are two independent variables:

(i) Displacement points, x, and
(ii) Time points, t.

(1) **Displacement points**: All displacement points are selected for the presentation
  Case I: There are (NX + 1) = 15 + 1 = 16 displacement points.
  Case II: There are (NX + 1) = 20 + 1 = 21 displacement points.
(2) There are (NT + 1) = 100 + 1 = 101 time points.
  For each case, five discrete time points are chosen for the presentation: at time points: 1, 21, 51, 81, and 101.

The requisite R code for processing the output data sets (from the FORTRAN program computations) are as follows:

```
R version 2.10.1 (2009-12-14)
Copyright (C) 2009 The R Foundation for Statistical Computing
ISBN 3-900051-07-0

R is a free software and comes with ABSOLUTELY NO WARRANTY.
You are welcome to redistribute it under certain conditions.
Type 'license()' or 'licence()' for distribution details.
Natural language support but running in an English locale
R is a collaborative project with many contributors.
Type 'contributors()' for more information and
'citation()' on how to cite R or R packages in publications.

Type 'demo()' for some demos, 'help()' for on-line help, or
'help.start()' for an HTML browser interface to help.
Type 'q()' to quit R.
[Previously saved workspace restored]
```

Case I: R = 1, m = 1, X = 10, NX = 15, T = 10, NT = 100

```
# Defining the Displacement Points
> x <- c(1, 2, 3, 4, 5, 6, 7, 8, 9, 10, 11, 12, 13, 14, 15)
# Defining the Corresponding Component Concentration Values in the Liquid Phase
> A2 <- c(1.0, 0,0,0,0,0,0,0,0,0,0,0,0,0,0,0)
> B2 <- c(1.0, 0,0,0,0,0,0,0,0,0,0,0,0,0,0,0)
# Inputting the Selected Data from the FORTRAN Program Output
> A22 <- c(1.0, 0.7331, 0.4971, 0.3118, 0.1812, 0.0981, 0.0497, 0.0237,
+          0.0107,0.0046, 0.0019, 0.0007, 0.0003, 0.0001, 0.000034, 0.000010)
> B22 <- c(1.0, 0.7396, 0.5078, 0.3235, 0.1916, 0.1059, 0.0548, 0.0267, 0.0123,
+           0.0054, 0.0023, 0.0009, 0.0003, 0.0001, 0.000046, 0.000014)
> A52 <- c(1.0, 0.8315, 0.6706, 0.5239, 0.3961, 0.2897, 0.2049, 0.1401,
+           0.0927, 0.0594, 0.0368, 0.0221, 0.0128, 0.0071, 0.0037, 0.0016)
> B52 <- c(1.0, 0.8332, 0.6737, 0.5281, 0.4009, 0.2945, 0.2094, 0.1440,
+           0.0959, 0.0618, 0.0386, 0.0233, 0.0136, 0.0077, 0.0040, 0.0017)
```

```
> A82 <- c(1.0, 0.8669, 0.7375, 0.6152, 0.5030, 0.4028, 0.3158, 0.2423,
+            0.1819, 0.1336, 0.0958, 0.0670, 0.0455, 0.0295, 0.0175, 0.0081)
> B82 <- c(1.0, 0.8677, 0.7391, 0.6175, 0.5057, 0.4057, 0.3188, 0.2452,
+            0.1846, 0.1359, 0.0978, 0.0687, 0.0467, 0.0304, 0.0181, 0.0084)
> A102 <- c(1.0, 0.8809, 0.7646, 0.6533, 0.5494, 0.4544, 0.3695, 0.2954,
+             0.2319, 0.1786, 0.1349, 0.0994, 0.0710, 0.0483, 0.0299, 0.0142)
> B102 <- c(1.0, 0.8752, 0.7535, 0.6376, 0.5302, 0.4329, 0.3469, 0.2728,
+             0.2104, 0.1590, 0.1176, 0.0850, 0.0595, 0.0397, 0.0242, 0.0114)
>
# Collating the Data for Component A
> h <- c(A2, A22, A52, A82, A102)
> xx2 <- c(x, x, x, x, x)
# Plotting the Data for Component A
> plot (xx2, h)
> lines (x, A2)
> lines (x, A22)
> lines (x, A52)
> lines (x, A82)
> lines (x, A102)
> title ("Progressive Concentration Profiles of Component A in the
Liquid Phase Showing Time-dependent Penetration for t = 0,
+ 0.2T, 0.5T, 0.8T, and 1.0T")
# Outputting: Figure 36.
# Collating the Data for Component B
> g <- c(B1, B22, B52, B82, B102)
# Plotting the Data for Component B
> plot (xx2, g)
> lines (x, B2)
> lines (x, B22)
> lines (x, B52)
> lines (x, B82)
> lines (x, B102)
> title ("Progressive Concentration Profiles of Component B in the
# Liquid Phase Showing Time-dependent Penetration for
# t = 0, 0.2T, 0.5T, 0.8T, and 1.0T")
>
>
```

Case II: R = 1, m = 1, X = 10, NX = 20, T = 10, NT = 100

```
# Defining the Displacement Points
> x2 <-
+ c(1,2,3,4,5,6,7,8,9,10,11,12,13,14,15,16,17,18,19,20,21)
# Defining the Corresponding Component Concentration Values in the Liquid Phase
> A22 <- c(1.0,0,0,0,0,0,0,0,0,0,0,0,0,0,0,0,0,0,0,0,0)
> B22 <- c(1.0,0,0,0,0,0,0,0,0,0,0,0,0,0,0,0,0,0,0,0,0)
# Inputting the Selected Data from the FORTRAN Program Output
> A222 <- c(1.0, 0.7984, 0.6102, 0.4459, 0.3116, 0.2083, 0.1334, 0.0820,
```

```
+             0.0484, 0.0276, 0.0152, 0.0081, 0.0042, 0.0021, 0.0010, 0.000484,
+             0.000224, 0.000102, 0.000045, 0.000019, 0.000007)
> B222 <- c(1.0, 0.8034, 0.6191, 0.4572, 0.3234, 0.2193, 0.1426, 0.0892,
+             0.0536,0.0311,0.0174, 0.0095, 0.0050, 0.0025, 0.0012, 0.000615,
+             0.000291, 0.000135, 0.000061, 0.000026, 0.000010)
> A252 <- c(1.0, 0.8733, 0.7498, 0.6326, 0.5241, 0.4263, 0.3402, 0.2664, 0.2047,
+         0.1542, 0.1140, 0.0827, 0.0588, 0.0411, 0.0281, 0.0189, 0.0124, 0.0080,
+             0.0049, 0.0028, 0.0013)
> B252 <- c(1.0, 0.8746, 0.7523, 0.6360, 0.5283, 0.4309, 0.3450, 0.2712,
+         0.2091, 0.1583, 0.1175, 0.0857, 0.0613, 0.0430, 0.0297, 0.0200, 0.0133,
+             0.0086, 0.0053, 0.0031, 0.0014)
> A282 <- c(1.0, 0.9000, 0.8016, 0.7062, 0.6154, 0.5301, 0.4514, 0.3798,
+         0.3158, 0.2593, 0.2104, 0.1685, 0.1332, 0.1040, 0.0800, 0.0605, 0.0449,
+             0.0323, 0.0222, 0.0138, 0.0066)
> B282 <- c(1.0, 0.9006, 0.8028, 0.7080, 0.6176, 0.5327, 0.4542, 0.3828,
+         0.3188, 0.2623, 0.2132, 0.1711, 0.1356, 0.1061, 0.0818, 0.0620, 0.0461,
+             0.0333, 0.0229, 0.0143, 0.0066)
> A2102 <- c(1.0, 0.9106, 0.8223, 0.7363, 0.6534, 0.5747, 0.5008, 0.4323,
+         0.3695, 0.3128, 0.2621, 0.2174, 0.1783, 0.1445, 0.1156, 0.0909, 0.0700,
+             0.0523, 0.0370, 0.0236, 0.0115)
> B2102 <- c(1.0, 0.9110, 0.8232, 0.7375, 0.6550, 0.5766, 0.5029, 0.4345,
+         0.3719, 0.3152, 0.2645, 0.2196, 0.1804, 0.1464, 0.1173, 0.0924, 0.0713,
+             0.0533, 0.0378, 0.0241, 0.0118)
>
# Collating the Data for Component A
> hA2 <- c(A22, A222, A252, A282, A2102)
> xx22 <- c(x2, x2, x2, x2, x2)
>
# Plotting the Data for Component A
> plot (xx22, hA2)
> lines (x2, A22)
> lines (x2, A222)
> lines (x2, A252)
> lines (x2, A282)
> lines (x2, A2102)
> title ("Progressive Concentration Profiles of Component A in the
# Liquid Phase Showing Time-dependent Penetration for
# t = 0, 0.2T, 0.5T, 0.8T, and 1.0T")
>


# Collating the Data for Component
> hB2 <- c(B22, B222, B252, B282, B2102)
>
# Plotting the Data for Component B
> plot (xx22, hB2)
> lines (x2, B22)
```

```
> lines (x2, B222)
> lines (x2, B252)
> lines (x2, B282)
> lines (x2, B2102)
> title ("Progressive Concentration Profiles of Component B in the
# Liquid Phase Showing Time-dependent Penetration for t = 0,
# 0.2T, 0.5T, 0.8T, and 1.0T")
>
```

3.8.2 Existence and Uniqueness Theorems of First-Order Linear Ordinary Differential Equations [31]

3.8.2.1 Differential Equations

Recall that in differential equations I, one learned how to solve certain types of differential equations. A typical example one would have encountered is the first-order differential equation (DE) such as

$$d/dt[y(t)] = t^2 + ty^3(t) \tag{RT2-1}$$

with some initial conditions such as $y(0) = 5$. More generally, the first-order DE is of the form $y'(t) = f[t, y(t)]$ for some function f, with initial condition $y(t_0) = y_0$. The goal is to find $y(t)$.

Only certain classes of first-order DEs may be solved explicitly. The two standard classes are linear equations and separable equations.

Many ordinary differential equations (such as the one above) are neither linear nor separable. Nevertheless, there is a useful theorem which asserts that many ODEs, including the one above, have unique solutions, even if one cannot express them explicitly. The proof of this theorem is a classical application of pure mathematics, with a brief introduction to the concept of a Banach Space: see Appendix 3.D.

3.8.2.2 Contraction Mappings on a Banach Space

Let V denotes a Banach space and let T denote a mapping or operator from V to V. One may say that T is a *contraction* if for all $x, y \in V$, one has

$$||T(x) - T(y)|| \leq C||x - y|| \tag{RT2-2}$$

for some constant $0 \leq C < 1$. A contraction mapping "draws" any two points closer together.

Example 3.2 Let $V = C([0, 1/2])$ with the supremum norm and $T(f)(s) = \int_0^s f(u)du$.

Then for all $s \in [0, 1/2]$ and $f, g \in V$, one has

$$|T(f)(s) - T(g)(s)| = \left| \int_0^s (f - g) \right| \leq \int_0^s ||f - g||_\infty = s||f - g||_\infty \leq 1/2||f - g||_\infty \tag{RT2-3}$$

Since this holds for all $s \in [0, 1/2]$, we clearly have $\|T(f) - T(g)\|_\infty \le 1/2\|f - g\|_\infty$, so T is a contraction with constant $C = 1/2$.

Example 3.3 The operator $T(f) = f^2$ is ***not*** a contraction on $C([0, 1/2])$, for, e.g. if $f \equiv 1$ and $g \equiv 2$, then $\|f - g\|_\infty = 1$, while $\|T(f) - T(g)\|_\infty = 3$.

Remark: The contraction mappings are continuous.

In real analysis, one may have seen that a contraction mapping f from R to R (that is, a function f for which $|f(x) - f(y)| \le c|x - y|$ where $c < 1$) has a unique fixed point, i.e. there is one and only one solution to the equation $x = f(x)$. The same fact is true in any Banach space.

Theorem 3.1 ***(The Contraction Mapping Principle)*** *Let T be a contraction on a Banach space V. Then T has a unique fixed point in V, i.e. there is one and only one solution $x \in V$ to the equation: $x = T(x)$.*

Proof: Let us do uniqueness first:

Suppose one has two solutions x_1 and x_2 to $x_i = T(x_i)$. Then one has

$$\|x_1 - x_2\| = \|T(x_1) - T(x_2)\| \le C\|x_1 - x_2\| \tag{RT2-4}$$

But since $C < 1$ this gives an immediate contradiction unless $\|x_1 - x_2\| = 0$, that is $x_1 = x_2$. This proves uniqueness.

Now one may show the existence of a fixed point. Let x_0 be any element of V. Define a sequence recursively by $x_{n+1} = T(x_n)$. The claim is that x_n converges to a fixed point. To show this, one will first show that the sequence x_n is Cauchy, and hence must converge. One has

$$\begin{aligned}\|X_{k+1} - x_k\| &= \|T(x_k) - T(x_{k-1})\|\\ &\le c\|x_k - x_{k-1}\|\\ &\le c\|T(x_{k-1}) - T(x_{k-2})\|\\ &\le c^2\|x_{k-1} - x_{k-2}\|\\ &\le c^k\|x_1 - x_0\|,\end{aligned}$$

so

$$\|x_{k+1} - x_k\| \le c^k\|x_1 - x_0\| \tag{RT2-5}$$

As a result one has, for $m < n$, that

$$\begin{aligned}\|x_m - x_n\| &\le \|x_m - x_{m+1}\| + \|x_{m+1} - x_{m+2}\| + \cdots + \|x_{n-1} - x_n\|\\ &\le \|x_1 - x_0\|(c^m + c^{m+1} + \cdots + c^{n-1})\\ &= c^m\|x_1 - x_0\|(1 + c + c^2 + \cdots + c^{n-m-1})\\ &< \{c^m/(1 - c)\}\|x_1 - x_0\|\end{aligned} \tag{RT2-6}$$

It is clear that since $c < 1$, one has, for any $\varepsilon > 0$, that $c^{m-1}/(1 - c) < \varepsilon$, for $m \ge N$ sufficiently large. Thus if $m, n \ge N$, one has $\|x_m - x_n\| < \varepsilon$. The sequence f_n is Cauchy, and since V is complete, x_n converges to some limit x.

Of course, x must be the fixed point one wants. This is easy to prove, as

$$\begin{aligned}\|x - T(x)\| &= \|x - x_n + x_n - T(x)\| \\ &\leq \|x - x_n\| + \|x_n - T(x)\| \\ &= \|x - x_n\| + \|T(x_{n-1}) - T(x)\| \\ &\leq \|x - x_n\| + \|x_{n-1} - x\|\end{aligned} \tag{RT2-7}$$

The first inequality above came from the triangle inequality, and the last inequality uses that T is a contraction. Both norms on the right above limit to zero and one may conclude that $\|x - T(x)\| = 0$, so $x = T(x)$.

Example 3.4 Let $V = C([0, 1/2])$ and define T as

$$T(f)(s) = 1 + \int_0^s f(u)du \tag{RT2-8}$$

This is similar to **Example 3.3**, and one may show that T is a contraction. As a result, the equation $f = T(f)$, which is just the integral equation:

$$f(s) = 1 + \int_0^s f(u)du \tag{RT2-9}$$

must have a unique continuous solution, at least for $0 \leq s \leq 1/2$. The integral equation above is an example of a Volterra integral equation (in which the upper or lower limit of the integral depends on the independent variable). Actually, one may find the solution explicitly in this case by differentiating both sides of $f(s) = 1 + \int_0^s f(u)du$ with respect to s, to obtain $f'(s) = f(s)$. Thus, $f(s) = Ce^s$ for some C. Also by putting $s = 0$ in $f(s) = 1 + \int_0^s f(u)du$, one finds that $f(0) = 1$, so $f(s) = e^s$.

Example 3.5 Let T be defined on $C([0, 1])$ as:

$$T(f)(x) = x^2 + (x/2)\int_0^1 y \sin(f(y))dy \tag{RT2-10}$$

It's easy to check that T is a contraction (with respect to the sup norm), for

$$\begin{aligned}|T(f)(x) - T(g)(x)| &= (x/2)\left|\int_0^1 y[\sin(f(y)) - \sin(g(y))dy]\right| \\ &\leq 1/2\left|\int_0^1 y|f(y) - g(y)|dy\right| \\ &\leq 1/2\left|\int_0^1 y\|f - g\|_\infty dy\right| \\ &\leq 1/2\|f - g\|_\infty \int_0^1 ydy \\ &= 1/4\|f - g\|_\infty\end{aligned} \tag{RT2-11}$$

so that $\|T(f) - T(g)\|_\infty \leq 1/4\|f - g\|_\infty$. In obtaining the above estimates, one used the inequality:

$$|\sin(a) - \sin(b)| \leq |a - b| \tag{RT2-12}$$

which is easy to prove using the mean value theorem. Since T is a contraction, there is thus one and only one continuous function on [0, 1] which satisfies

$$f(x) = x^2 + (x/2)\int_0^1 y\sin(f(y))dy \quad \text{(RT2-13)}$$

An integral equation like this in which the limits in the integral are constant is called a Fredholm integral equation.

3.8.2.3 Application to Differential Equations

The general first-order differential equation has a general form:

$$y'(t) = f\{t, y(t)\} \quad \text{(RT2-14)}$$

where $f(t, y)$ is some function in two variables. For example, if $f(t, y) = t + t^2y^2$ then the differential equation is $y'(t) = t + t^2y^2(t)$. One is also given an initial condition $y(t_0) = y_0$.

The object of the exercise is to find a function $y(t)$ which satisfies the differential equation and the initial condition. The function should be defined and continuous on an interval $(t_0, t_0 + \delta)$ and differentiable on the open interval $(t_0, t_0 + \delta)$ for some $\delta > 0$. The theorem below shows that one can, under the right conditions, assert that a differential equation has a unique solution, even if the solution cannot be written down ***in closed form***.

In what follows, one will consider only functions $f(t, y)$ which satisfy the following property.

It will require that f be differentiable in y and that

$$|\partial f(t, y)/\partial y| \leq C \quad \text{(RT2-15)}$$

for all t in some interval $[t_0, t_0 + \delta_1]$ and all y in $[y_0 - \delta_2, y_0 + \delta_2]$, where δ_1 and δ_2 are some positive constants. The constant C must be independent of t and y. The simplest situation in which this is guaranteed is that $\partial f/\partial y\ (t, y)$ is continuous for $(t, y) \in [t_0, t_0 + \delta_1] \times [y_0 - \delta_2, y_0 + \delta_2]$ (because a continuous function is bounded on a compact set). Under these conditions, one can prove the following theorem.

Theorem 3.2 ***(Existence-Uniqueness Theorems)*** Let the function $f(t, y)$ be continuous and satisfy the bound (RT-40). Then the differential equation (RT-39) with initial condition $y(t_0) = y_0$ has a unique solution which is continuous on some interval $[t_0, t_0 + \delta)$ and differentiable on $(t_0, t_0 + \delta)$, where $\delta > 0$.

Before proving the theorem, apply it to the DE $y'(t) = t^2 + ty^3(t)$ with initial condition $y(0) = 2$ that appeared at the start of this discussion. One may compute that $\partial f/\partial y = 3ty^2$, which is continuous for all t and y, and hence is bounded by a constant on any compact set of the form $0 \leq t \leq \delta_1$, $2 - \delta_2 \leq y \leq 2 + \delta_2$. Take, for example, $\delta_1 = \delta_2 = 1$ and one has $|\partial f/\partial y| \leq 27$. For some $\delta > 0$, this differential equation must have a unique solution for $t \in (0, \delta)$ with $y(0) = 2$.

A result like that above in which one proves a differential equation has a solution on some (possibly very small) interval is called a local existence result, as opposed to a global existence result in which one proves that the differential equation has a solution for ***all time***, or at least all time up to some preassigned limit.

Proof of Existence-Uniqueness Theorems: First, recast the differential equation as an integral equation. Note that if $y(t)$ is continuous for $t_0 \leq t \leq t_0 + \delta_1$ then $f(t, y(t))$ is continuous, hence $y'(t) = f(t, y(t))$ must also be continuous for $t_0 \leq t \leq t_0 + \delta_1$. As a result one may integrate both sides of Eq. (RT-39) for $s = t_0$ to $s = t < t_0 + \delta_1$ and apply the Fundamental Theorem of Calculus to obtain

$$y(t) = y_0 + \int_{t_0}^{t} f(s, y(s))ds \tag{RT2-16}$$

Conversely, any continuous solution to Eq. (RT2-16) must be a solution to the original differential equation (RT-39), for the right side of Eq. (RT2-16) must be differentiable in t (by the Fundamental Theorem of Calculus) and so one may differentiate both sides of Eq. (RT-41) with respect to t to obtain the original differential equation.

Also, plugging $t = t_0$ into Eq. (RT2-17) produces $y(t_0) = y_0$. Any differentiable solution $y(t)$ to Eq. (RT2-15) with initial condition $y(t_0) = y_0$ is necessarily a continuous solution to equation (RT2-17) and vice versa. One may show that Eq. (RT2-17) has a unique continuous solution on some interval

$$t_0 \leq t < t_0 + \delta. \tag{RT2-17}$$

Now, consider a proof of the theorem in a simple special case. Suppose that the bound (RT2-15) is satisfied with "$\delta_2 = \infty$", i.e. for all y. For any real numbers a_1 and a_2, it then follows from the mean value theorem that

$$\{f(s, a_1) - f(s, a_2)\}/(a_1 - a_2) = \partial f(s, z)/\partial y \tag{RT2-18}$$

for some z between a_1 and a_2. Taking the absolute values of both sides of the above equation and using the bound (RT-40) shows that

$$|f(s, a_1) - f(s, a_2)| \leq C|a_1 - a_2| \tag{RT2-19}$$

for ALL real numbers a_1, a_2 and all $s \in [t_0, t_0 + \delta_1]$.

Define an operator $T(y)$ on $C([t_0, t_0 + \delta_1])$ as:

$$T(y)(t) = y_0 + \int_{t_0}^{t} f(s, y(s))ds \tag{RT2-20}$$

Note that by the Fundamental Theorem of Calculus, T really does turn continuous functions into continuous (in fact, differentiable) functions. From the discussion showing the equivalence of Eqs. (RT2-19) and (RT2-20), any fixed point for T yields a solution to $y' = f(t, y)$ with $y(t_0) = y_0$. It may be proved that there exists such a fixed point.

By using Eq. (RT2-20), one may estimate that for any two functions y_1 and y_2 in $C([t_0, t_0 + \delta_1])$, one has

$$|T(y_1)(t) - T(y_2)(s)| = \left| \int_{t_0}^{t} [f\{s, y_1(s)\} - f\{s, y_2(s)\}]ds \right|$$

$$\leq \int_{t_0}^{t} |f\{s, y_1(s)\} - f\{s, y_2(s)\}| ds$$

$$\leq C \int_{t_0}^{t} |y_1(s) - y_2(s)| ds \text{ by Eq. (RT2-19)}$$

$$\leq C\|y_1 - y_2\|_\infty \int_{t_0}^{t} ds$$

$$= C\|y_1 - y_2\|_\infty (t - t_0) \tag{RT2-21}$$

If one chooses t close enough to t_0 so that $C(t-t_0)<1$, i.e. $t<t_0+1/C$, then T becomes a contraction. Thus, let $\delta = \min(\delta_1, 1/C)$ and let $I = [t_0, t_0+\delta]$. On $C(I)$, the operator T is a contraction and hence Eq. (RT2-21) must have a unique fixed point $y(t)$. By the Fundamental Theorem of Calculus, the right of Eq. (RT2-21) is differentiable in t, hence $y(t)$ is also differentiable. By putting $t = t_0$ into the integral equation version, it follows that $y(t_0) = y_0$, and differentiating both sides shows that $y'(t) = f\{t, y(t)\}$. This proves the theorem if the bound (RT2-19) holds for all y.

What does one do in the more general case in which Eq. (RT2-19) holds only for $y \in [y_0 - \delta_2, y_0 + \delta_2]$? First, one may define a new function $g(s, y)$ for $s \in [t_0, t_0 + \delta_1]$ and $y \in \boldsymbol{R}$ as:

$$g(s,y) = \begin{cases} f(s,y), & y_0 - \delta_2 < y < y_0 + \delta_2 \\ f(s, y_0 - \delta_2), & y \leq y_0 - \delta_2 \\ f(s, y_0 + \delta_2), & y \geq y_0 + \delta_2 \end{cases} \tag{RT2-22}$$

The function g extends f continuously outside the rectangle $[t_0, t_0+\delta_1]\times[y_0-\delta_2, y_0+\delta_2]$ in such a way that the bound (RT2-22) holds. Next, replace the differential equation $y' = f(t, y)$ with the differential equation $y' = g(t, y)$, prove solvability of the latter, and use this to get (local) solvability of $y' = f(t, y)$.

Exercise

- Verify that the function g, as defined above, is continuous (although it may not be differentiable) and satisfies

$$|g(s, a_1) - g(s, a_2)| \leq C|a_1 - a_2| \tag{3.53}$$

where C is whatever constant works for f in Eq. (RT-40), a_1 and a_2 are any real numbers, and $s \in [t_0, t_0+\delta_1]$.

Given that g satisfies Eq. (RT-47), one can apply exactly the same argument as above to conclude that the equation:

$$y(t) = y_0 + \int_{t_0}^{t} g\{s, y(s)\} ds \tag{RT2-24}$$

has a unique solution on $C(I)$ where $I = [t_0, t_0+\delta]$ for some $\delta>0$. As before this shows that $y(t_0) = y_0$ and $y'(t) = g\{t, y(t)\}$. But, since $y(t)$ is continuous and $y(t_0) = y_0$, there must be some interval $[t_0, t_0+\delta_3]$ with $\delta_3>0$ on which $|y(t)-y_0|<\delta_2$, i.e. $y_0-\delta_2<y(t)<y_0+\delta_2$. But for $y(t)$ on such an interval, the functions $f\{t, y(t)\}$ and $g\{t, y(t)\}$ are identical. As a result, one has $y(t)$ satisfying $y(t_0) = y_0$ and $y'(t) = f\{t, y(t)\}$ for $t \in [t_0, t_0+\delta_3]$.

This completes the proof of the existence-uniqueness theorem.

By the way, although a bounded like (RT-40) is not the most general condition possible for existence and uniqueness, one cannot get away with dropping all conditions on the function $f(t, y)$, for then the differential equation may not have a unique solution.

For example, consider the differential equation $y'(t) = \sqrt[3]{\{y(t)\}}$ with initial condition $y(0) = 0$. One can check this differential equation and initial condition has three solutions:

$$y(t) = \pm 2\sqrt{6}t^{3/2}/9, \quad \text{and} \quad y(t) \equiv 0 \tag{RT2-25}$$

Of course this f does ***not*** satisfy the bound (RT2-20) near $y = 0$.

This proof may look rather abstract, but in fact it is quite constructive, for it not only tells us that a solution exists but also gives a procedure for approximating the solution. The solution is a fixed point for a contraction and one proved such points exist by making an initial guess y_0, then iterating at $y_{k+1} = T(y_k)$; the iterates converge to the solution. It is interesting to look at what happens when this procedure is applied to a specific differential equation.

Suppose one wants to solve $y'(t) = y^2(t) + t$ with initial condition $y(0) = 1$. The relevant integral operator is

$$T(y)(t) = 1 + \int_0^t \{y^2(s) + s\}\mathrm{d}s \tag{RT2-26}$$

The solution one seeks is a fixed point of T. First, make an initial guess $y_0(t) \equiv 1$.

Then one can easily compute

$$y_1(t) = T(y_0)(t) = 1 + \int_0^t \left\{y_0{}^2(s) + s\right\} \mathrm{d}s = 1 + t + 1/2t^2$$

$$y_2(t) = T(y_1)(t) = 1 + \int_0^t \left\{y_1{}^2(s) + s\right\} \mathrm{d}s = 1 + t + 3/2t^2 + 2/3t^3 + 1/4t^4 + 1/20t^5$$

etc., with the higher iterates becoming more complicated!

Appendix 3.D

Banach Space (From Wikipedia)

In mathematics, more specifically in functional analysis, a **Banach Space** (pronounced [banax]) is a complete normed vector space. Thus, a Banach space is a vector space with a metric that allows the computation of vector length and distance between vectors and is complete in the sense that a Cauchy sequence of vectors always converges to a well-defined limit that is within the space.

Banach spaces are named after the Polish mathematician Stefan Banach, who introduced this concept and studied it systematically in 1920–1922 along with Hans Hahn and Eduard Helly. Banach spaces originally grew out of the study of function spaces by Hilbert, Fréchet, and Riesz earlier in the century. Banach spaces play a central role in functional analysis. In other areas of analysis, the spaces under study are often Banach spaces.

Definition A Banach space is a vector space X over the field $\boldsymbol{R}$ of real numbers, or over the field **C** of complex numbers, which is equipped with a norm $||\cdot||_X$ and which is *complete* with respect to the distance function induced by the norm, that is to say, for every Cauchy sequence $\{x_n\}$ in X, there exists an element x in X such that

$$\lim_{n\to\infty} x_n = x \qquad \text{(RT2-27)}$$

or equivalently:

$$\lim_{n\to\infty} ||x_n - x||_X = 0 \qquad \text{(RT2-28)}$$

The vector space structure allows one to relate the behavior of Cauchy sequences to that of converging series of vectors. A normed space X is a Banach space if and only if each absolutely convergent series in X converges:

$$\sum_{n=1}^{\infty} ||v_n||_X < \infty \quad \text{implies that} \quad \sum_{n=1}^{\infty} v_n \quad \text{converges} \qquad \text{(RT2-29)}$$

Completeness of a normed space is preserved if the given norm is replaced by an equivalent one. All norms on a finite-dimensional vector space are equivalent. Every finite-dimensional normed space over $\boldsymbol{R}$ or $\boldsymbol{C}$ is a Banach space.

General Theory of a Banach Space

Linear Operators and Isomorphisms If X and Y are normed spaces over the same ground field **K**, the set of all continuous **K**-linear maps $T: X \to Y$ is denoted by $B(X, Y)$. In infinite-dimensional spaces, not all linear maps are continuous. A linear mapping from a normed space X to another normed space is continuous if and only if it is bounded on the closed unit ball of X. Thus, the vector space $B(X, Y)$ may be given the *operator norm*:

$$||T|| = \sup \{ ||T_x||_Y \mid x \in X, \ \ ||x||_X \leq 1 \} \qquad \text{(RT2-30)}$$

For Y a Banach space, the space $B(X, Y)$ is a Banach space with respect to this norm. If X is a Banach space, the space $B(X) = B(X, X)$ forms a unital Banach algebra; the multiplication operation is given by the composition of linear maps.

If X and Y are normed spaces, they are **isomorphic normed spaces** if there exists a linear bijection $T: X \to Y$ such that T and its inverse T^{-1} are continuous. If one of the two spaces X or Y is complete (or reflexive, separable, etc.), then so is the other space. Two normed spaces X and Y are ***isometrically isomorphic*** if, in addition, T is an isometry, i.e. $||T(x)|| = ||x||$ for every x in X. The Banach–Mazur distance $d(X, Y)$ between two isomorphic but not isometric spaces X and Y gives a measure of how much the two spaces X and Y differ.

Remarks:

(1) **Basic Notions**

Every normed space X can be isometrically embedded in a Banach space. More precisely, for every normed space X, there exist a Banach space Y and a mapping $T: X \to Y$ such that T is an ***isometric mapping*** and $T(X)$ is dense in Y. If Z is another Banach space such that there is an isometric isomorphism from X onto a dense subset of Z, then Z is isometrically isomorphic to Y.

(2) **Completion**
This Banach space Y is the ***completion*** of the normed space X. The underlying metric space for Y is the same as the metric completion of X, with the vector space operations extended from X to Y. The completion of X is often denoted by $\underline{\mathbf{X}}$.

(3) **Cartesian Product**
The Cartesian product $X \times Y$ of two normed spaces is not canonically equipped with a norm. However, several equivalent norms are commonly used, such as

$$\|(x, y)\|_1 = \|x\| + \|y\| \tag{RT2-31}$$

$$\|(x, y)\|_\infty = \max(\|x\|, \|y\|) \tag{RT2-32}$$

and give rise to isomorphic normed spaces. In this sense, the product $X \times Y$ (or the direct sum $X \oplus Y$) is complete if and only if the two factors are complete.

(4) **Quotient Space**
If M is a **closed linear subspace** of a normed space X, there is a natural norm on the **quotient space** X/M,

$$\|x + M\| = \inf_{m \in M} \|x + m\| \tag{RT2-33}$$

The quotient X/M is a Banach space when X is complete. The **quotient map** from X onto X/M, sending x in X to its class $x + M$, is linear, onto and has norm 1, except when $M = X$, in which case the quotient is the null space.

(5) **Complemented Subspace**
The closed linear subspace M of X is said to be a **complemented subspace of X** if M is the range of a bounded linear projection P from X onto M. In this case, the space X is isomorphic to the direct sum of M and Ker(P), the kernel of the projection P.

(6) **Canonical Factorization**
Suppose that X and Y are Banach spaces and that $T \in B(X, Y)$, then there exists a **canonical factorization** of T as

$$T = T_1 \circ \pi, \quad T : X \xrightarrow{\pi} X/\mathrm{Ker}(T) \xrightarrow{T_1} Y \tag{RT2-34}$$

where the first map π is the quotient map, and the second map T_1 sends every class $x + \mathrm{Ker}(T)$ in the quotient to the image $T(x)$ in Y. This is well defined because all elements in the same class have the same image. The mapping T_1 is a linear bijection from $X/\mathrm{Ker}(T)$ onto the range $T(X)$, whose inverse need not be bounded.

(7) **Classical Spaces**
Basic examples of Banach spaces include the L^p spaces and their special cases, the sequence spaces ℓ^p that consist of scalar sequences indexed by **N**; among them, the space ℓ^1 of absolutely summable sequences and the space ℓ^2 of square summable sequences; the space c_0 of sequences tending to zero and the space ℓ^∞ of bounded sequences; the space $C(K)$ of continuous scalar functions on a compact Hausdorff space K, equipped with the max norm.

3.8.3 An Existence Theorem of the Governing Simultaneous Semi-linear Parabolic Partial Differential Equations

(In Part 1, the system appears to be a well-posed initial and boundary value problem for a *system* of simultaneous semi-linear parabolic partial differential equations.)

Theorem 3.3 ***Existence Theorem*** *For a system defined by Eqs. (RT1-7A) and (RT1-7B), there exists a stable solution.*

Proof of the Existence Theorem: First, two Preliminary Lemmas [32] ***are stated as the first step in proving the Existence Theorem.***

Preliminary Lemmas: Existence and Uniqueness Lemmas for First-Order ODEs

The general first-order ODE is

$$dy/dx = F(x;y); \quad y(x_0) = y_0 \tag{*}$$

One is interested in the following questions:

(i) Under what conditions can one be sure that a solution to (*) exists?
(ii) Under what conditions can one be sure that there is a unique solution to (*)?

Here are the answers.

Preliminary Lemma A (Existence). *Suppose that $F(x, y)$ is a continuous function defined in some region*

$$R = \{(x,y) : x_0 - \delta < x < x_0 + \delta; y_0 - \varepsilon < y < y_0 + \varepsilon\}$$

containing the point $(x_0; y_0)$. Then there exists a number δ_1 (possibly smaller than δ) so that a solution $y = f(x)$ to (*) *is defined for $x_0 - \delta_1 < x < x_0 + \delta_1$.*

Preliminary Lemma 2 (Uniqueness). Suppose that both $F(x, y)$ and $(\partial F/\partial y)(x, y)$ are continuous functions defined on a region R. Then there exists a number δ_2 (possibly smaller than δ_1) so that the solution $y = f(x)$ to (*), whose existence was guaranteed by **Preliminary Lemma A (Existence)**, is the unique solution to (*) for $x_0 - \delta_2 < x < x_0 + \delta_2$.

The finite element method may be used for ***approximating*** the system solution.

The finite element procedure is simply the Rayleigh–Ritz–Galerkin (RRG) approximation applied to the spaces of piecewise polynomial functions.

Methodology in numerical analysis using this approach is, probably, better known as ***Spline Analysis*** in engineering sciences. It represents a unified and mathematically vigorous approach to the finite element method for solving continuous or infinite-dimensional problems [32, 33].

Basically, the Galerkin procedure is used to "discretize" the space variables of the semi-linear system:

$$a_{xx} = a_t + ab \tag{RT3-1A}$$

$$Rb_{xx} = b_t + mab \tag{RT3-1B}$$

where

$$a = a(x, t)m$$

having

$$b = b(x, t)$$

with

$$a(x, t) = 0 = b(x, t), \quad x > 0$$

$$a(0, t) = 1 = b(x, t), \quad t > 0$$

whose equivalent system having homogeneous boundaries is (dropping the suffices):

$$a_{xx} = a_t - ab + ab_s + a_s b \tag{RT3-2A}$$

$$Rb_{xx} = b_t - mab + mab_s + ma_s b \tag{RT3-2B}$$

or

$$a_t - a_{xx} - ab + ab_s + a_s b = 0 \tag{RT3-3A}$$

$$b_t - Rb_{xx} - mab + mab_s + ma_s b \tag{RT3-3B}$$

with

$$a(x, 0) = 0 = b(x, 0), \quad x > 0$$

$$b(0, t) = 0 = b(0, t), \quad t > 0$$

Using the notation $(f, g) \equiv \int_0^\infty f(x)g(x)\mathrm{d}x$ and multiplying Eqs. (RT-29A) and (RT-29B) by smooth but arbitrary functions v and w, one obtains the following weak formulations:

$$\int_0^\infty a_t v\,\mathrm{d}x - \int_0^\infty a_{xx} v\,\mathrm{d}x - \int_0^\infty abv\,\mathrm{d}x + \int_0^\infty b_s av\,\mathrm{d}x + \int_0^\infty a_s bv\,\mathrm{d}x = 0 \tag{RT3-4A}$$

$$\int_0^\infty b_t w\,\mathrm{d}x - R\int_0^\infty b_{xx} w\,\mathrm{d}x - m\int_0^\infty abw\,\mathrm{d}x + m\int_0^\infty ab_s w\,\mathrm{d}x + m\int_0^\infty a_s bw\,\mathrm{d}x = 0 \tag{RT3-4B}$$

for all $v, w \in C^\infty(0, \infty) \equiv$ the set of infinitely differentiable and continuous which are nonnegative. Now, let

$$\int_0^\infty a_t v\,\mathrm{d}x = (a_t, v)$$

$$-\int_0^\infty a_{xx} v\,\mathrm{d}x = -\int_0^\infty \partial^2 a/\partial x^2 v\,\mathrm{d}x$$

$$= -\int_0^\infty \{\partial(\partial a/\partial x)/\partial x\} v\,\mathrm{d}x$$

$$= -\int_0^\infty (\partial a_x/\partial x)v \, dx$$

$$= -\lceil a_x y \rceil_0 + \int_0^\infty a_x(\partial v/\partial x)dx$$

$$= 0 + \int_0^\infty a_x v_x \, dx$$

$$= (a_x, v_x)$$

$$-\int_0^\infty ab \, v \, dx = -(ab, v)$$

$$\int_0^\infty b_s a \, v \, dx = -(b_s a, v)$$

$$\int_0^\infty a_s b \, v \, dx = (a_s b, v)$$

and, similarly:

$$\int_0^\infty b_t w \, dx = (b_t, w)$$

$$-R\int_0^\infty b_{xx} w \, dx = R(b_x, w_x)$$

$$-m\int_0^\infty abw \, dx = -m(ab, w)$$

$$m\int_0^\infty b_s aw \, dx = -m(b_s a, w)$$

$$m\int_0^\infty a_s bw \, dx = -m(a_s b, w)$$

Substituting these into (RT-29A) and (RT-29B), the results are

$$(a_t, v) + (a_x, v_x) - (ab, v) + (b_s a, v) + (a_s b, v) = 0 \qquad \text{(RT3-5A)}$$

$$(b_t, w) + R(b_x, w_x) - m(ab, w) + m(b_{ss} a, w) + m(a_s b, w) = 0 \qquad \text{(RT3-5B)}$$

for all $t > 0$.

Now, let S_n be a finite-dimensional sum of elements defined in $C^\infty(0, \infty)$, viz., $S_n \subset C^\infty(0, \infty)$, that is, S_n is spanned by the basis of $\{\phi_i(x)\}$, $i = 1, 2, 3, \ldots, n$.

Defining a semi-discrete Galerkin approximation, let

$$a^h = a^h(t) = \sum_{i=1}^{n} c_i(t)\phi_i(x) = \sum_i c_i\phi_i \qquad \text{(RT3-6A)}$$

$$b^h = b^h(t) = \sum_{i=1}^{n} d_i(t)\phi_i(x) = \sum_i d_i\phi_i \qquad \text{(RT3-6B)}$$

Thus, if the generalized solution of (RT-30A and RT-30B) is a and b, then a weak formulation, or semi-discrete Galerkin approximation to a and b for each element i, is given by (RT-31A) and (RT-31B). The coefficients $\{c_i(t)\}_{i=1,2,3,\ldots,n}$ and $\{d_i(t)\}_{i=1,2,3,\ldots,n}$ are functions of time which are determined as the solution of the linear system of ordinary differential equations:

$$\left(a_t{}^h, v\right) + \left(a_x{}^h, v_x\right) - (a^h, b^h, v) + (b_s a^h, v) + (a_s b^h, v) = 0 \quad \text{(RT3-7A)}$$

$$\left(b_t{}^h, w\right) + R\left(b_x{}^h, w_x\right) - m(a^h, b^h, w) + m(b_s a^h, w) + m(a_s b^h, w) = 0 \quad \text{(RT3-7B)}$$

where

$$\begin{aligned} &1 \le t \le n, && \text{for all } t > 0 \\ &(a^h(0), v) = (0, v), && \text{for all } v \in S_n \\ &(b^h(0), w) = (0, w), && \text{for all } w \in S_n \end{aligned}$$

Now

$$\begin{aligned} \left(a^h{}_t, v\right) &= \int_0^\infty \left[\sum_{i=1}^n \phi_i(x)\{\mathrm{d}C_i(t)/\mathrm{d}t\}\right] \phi_k(x) \\ &= \sum_{i=1}^n \left\{\left[\int_0^\infty \phi_i(x)\phi_k(x)\mathrm{d}x\right]\right\} \mathrm{d}C_i(t)/\mathrm{d}t \\ &= \sum_{i=1}^n (\phi_i(x)\phi_k(x))\mathrm{d}C_i(t)/\mathrm{d}t \end{aligned} \quad \text{(RT3-8A)}$$

$$\begin{aligned} -\left(a^h{}_x, v_x\right) &= \int_0^\infty \left[\sum_{i=1}^n C_i(t)\phi_{ik}(x)\right] \phi_k(x)\mathrm{d}x \\ &= \sum_{i=1}^n \left\{\left[\int_0^\infty \phi_{ix}(x)\phi_{kx}(x)\mathrm{d}x\right] C_i(t)\right\} \\ &= \sum_{i=1}^n [\phi_i(x)\phi_k(x)]C_i \end{aligned} \quad \text{(RT3-8B)}$$

$$\begin{aligned} -(a^h b^h, v) &= -\left(\left[\sum_i c_i\phi_i\right]\left[\sum_j d_i\phi_j\right], \phi_k\right) \\ &= -\int_0^\infty \left[\sum_i c_i\phi_i\right]\left[\sum_j d_i\phi_j\right], \phi_k \,\mathrm{d}x \\ &= -\sum_{i,j}\left[\int_0^\infty \gamma_{ij}\phi_i\phi_j \,\mathrm{d}x\right] \end{aligned} \quad \text{(RT3-8C)}$$

$$(b_s a^h, v) = \sum_{i_1 j_1}\left[\int_0^\infty \gamma_{ij}\phi_{i1}\phi_{j1}\,\mathrm{d}x\right] \quad \text{(RT3-8D)}$$

$$(a_s b^h, v) = \sum_{i_2 j_2}\left[\int_0^\infty \gamma_{ij}\phi_{i2}\phi_{j2}\,\mathrm{d}x\right] \quad \text{(RT3-8E)}$$

Hence, the first equation, for c_i, viz., Eq. (RT3-8A):

$$\left(a_t{}^h, v\right) + \left(a_x{}^h, v_x\right) - (a^h, b^h, v) + (b_s a^h, v) + (a_s b^h, v) = 0 \quad \text{(RT3-9)}$$

becomes

$$\sum_i (\phi_i(x)\phi_k(x))\mathrm{d}C_i/\mathrm{d}t + \sum_i (\phi_i(x)\phi_k(x))C_i - \sum_{i,j} \int \gamma_{ij}\phi_i\phi_j \,\mathrm{d}x$$
$$+ \sum_{i_1,j_1} \left[\int_0^\infty \gamma_{1ij}\phi_{i1}\phi_{j1}\,\mathrm{d}x\right] + \sum_{i_2,j_2} \left[\int_0^\infty \gamma_{2ij}\phi_{i2}\phi_{j2}\,\mathrm{d}x\right] = 0 \qquad \text{(RT3-10)}$$

which is a set of ***linear ordinary*** differential equations for the c_i-s. A similar set of equations for the d_i-s may be obtained for Eq. (RT1-33B).

Moreover, the rest of Eq. (RT1-33), viz., Eqs. (RT1-33C), (RT1-33D), and (RT1-33E), become

$$\begin{aligned}(a^h(0), v) &= \int_0^\infty \left[\sum_i^n c_{i(0)}\phi_i(x)\right]\phi_k(x)\,\mathrm{d}x \\ &= \sum_i^n \left[\int_0^\infty \phi_{i(x)}\phi_{k(x)}\,\mathrm{d}x\right] c_i(0) \\ &= \sum_{i=1}^n (\phi_i, \phi_k)c_i(0) \\ &= 0\end{aligned}$$

and

$$(b^h(0), w) = \sum_{i=1}^n (\phi_i, \phi_k)d_i(0) = 0$$

Now let the vector

$$\underline{y} = \underline{y}(t) = \begin{matrix}(y_1) \\ (y_2) \\ \vdots \\ (y_n) \\ (y_{n+1}) \\ (y_{n+2}) \\ \vdots \\ (y_{2n})\end{matrix} \quad \begin{matrix}(c_1(t)) \\ (c_2(t)) \\ \vdots \\ (c_n(t)) \\ (d_1(t)) \\ (d_2(t)) \\ \vdots \\ (d_n(t))\end{matrix} = \begin{matrix}(c_1) \\ (c_2) \\ \vdots \\ (c_n) \\ (d_1) \\ (d_2) \\ \vdots \\ (d_n)\end{matrix}$$

then the system of ordinary differential equations equivalent to Eq. (RT1-34) may be written as:

$$\sum_{i,j=1}^n a_{ij}\mathrm{d}y_k(t)/\mathrm{d}t + \sum \beta_{ij}\,y_k(t) + G_k(y_1, y_2, y_3, \dots, y_n) = 0$$

with $\sum_{i,j=1}^n a_{ij}\,y_k(t) = 0$, for $k = 1, 2, 3, \dots, 2n$, which, in metric form, is

$$\begin{aligned}\underline{P}\mathrm{d}y(t)/\mathrm{d}t + \underline{Q}y(t) &= R, \quad t > 0 \\ Py(t) &= 0\end{aligned} \qquad \text{(RT3-11)}$$

Since P is symmetric, positive definite, and hence non-singular, it follows, from a standard theoretical result in the analysis of ordinary differential equations [32], that the equation system (RT3-11) has a unique solution $y^*(t)$ which may be expressed analytically as:

$$y^*(t) = \int_0^t \left[\exp\{(s-t)P^{-1}Q\}\right] P^{-1}Q\, \mathrm{d}s, \quad t > 0 \qquad \text{(RT3-12)}$$

Hence, the semi-discrete Galerkin method is well defined for the semi-linear systems (RT1-27A) and (RT1-27B).

This completes the proof of the ***Existence Theorem***.

3.8.4 A Uniqueness Theorem of the Governing Simultaneous Semi-linear Parabolic Partial Differential Equations

From physical considerations, $a(x, t)$ and $b(x, t)$ may be restricted to nonnegative, bounded, and continuous functions. The following uniqueness theorem for a and b will be established:

Theorem 3.4 ***Uniqueness Theorem*** *If a and b are the classical solutions of Eqs. (RT-3A), (RT-3B), and (RT-3C), then this system has at most one solution.*

Proof of the Theorem: The inhomogeneous boundary conditions in Eqs. (RT-3A) and (RT-3B) may be transformed to homogeneous boundaries by modifying the dependent variables a and b in Eq. (RT-3C).

Let a_s and b_s be the *steady-state* solutions of Eqs. (RT1-3A)–(RT1-3C), viz.,

$$\partial a_s/\partial t = 0 = \partial b_s/\partial t, \quad \text{and} \quad a_s = a_s(x), b_s = b_s(x)$$

Hence

$$\partial^2 a_s/\partial x^2 = a_s b_s \qquad \text{(RT4-1A)}$$

$$R\partial^2 b/\partial x^2 = m a_s b_s \qquad \text{(RT4-1B)}$$

Also

$$a_s(0) = 1 = b_s(0) \qquad \text{(RT4-1C)}$$

Now, let

$$a_1 = a_s - a$$

$$b_1 = b_s - b$$

$$\text{From physical considerations:} \quad a_s \geq a, \text{and } b_s \geq b \qquad \text{(RT4-2A)}$$

$$\text{hence} \quad a_1 \geq 0, \text{ and } b_1 \geq 0 \qquad \text{(RT4-2B)}$$

Writing $a = a_s - a_1$, and $b = b_s - b_1$, and substituting in Eqs. (RT4-1A) and (RT-3B), one obtains

$$\partial^2 a_s/\partial x^2 - \partial^2 a_1/\partial x^2 = \partial a_s/\partial t - \partial a_1/\partial t + (a_s - a_1)(b_s - b_1) \qquad \text{(RT4-3A)}$$

$$R\partial^2 b_s/\partial x^2 - R\partial^2 b_1/\partial x^2 = \partial b_s/\partial t - \partial b_1/\partial t + m(a_s - a_1)(b_s - b_1) \qquad \text{(RT4-3B)}$$

Similarly, substituting in Eq. (RT-3C), one obtains

$$a_s(0) - a_1(0,t) = 1 = b_s(0) - b_1(0,t), \quad t > 0 \quad \text{(RT4-4A)}$$

$$a_s(x) - a_1(x,0) = 0 = b_s(x) - b_1(x,0), \quad x > 0 \quad \text{(RT4-4B)}$$

Using Eqs. (RT-3A) and (RT-3B), the following transformed system is obtained from simplifying Eqs. (RT4-3A) and (RT4-3B):

$$\partial^2 a_1/\partial x^2 = \partial a_1/\partial t - a_1 b_1 + (a_1 b_s + a_s b_1) \quad \text{(RT4-5A)}$$

$$R\partial^2 b_1/\partial x^2 = \partial b_1/\partial t - m a_1 b_1 + m(a_1 b_s - a_s b_1) \quad \text{(RT4-5B)}$$

with ***homogeneous*** initial and boundary conditions

$$a_1(x,0) = 0 = b_1(x,0), \quad x > 0 \quad \text{(RT4-6A)}$$

$$a_1(0,t) = 0 = b_1(0,t), \quad t > 0 \quad \text{(RT4-6B)}$$

It is seen that homogeneity of the boundaries, Eqs. (RT4-6A) and (RT4-6B), has been obtained ***at the expense of*** inhomogeneity of the system: Eqs. (RT-29A) and (RT-29B), which may be written, using suffix notations for simplicity and convenience (where $f_{xx} \equiv \partial^2 f/\partial x^2$, and $f_t \equiv \partial f/\partial t$):

$$a_{1t} - a_{1xx} - a_1 b_1 + a_1 b_s + a_s b_1 = 0 \quad \text{(RT4-7A)}$$

$$b_{1t} - R b_{1xx} - m a_1 b_1 + m a_1 b_s - m a_s b_1 = 0 \quad \text{(RT4-7B)}$$

Let a' and b' be ***another*** solution of Eqs. (RT4-5A) and (RT4-5B), and let $a_2 = a_s - a'$, $b_2 = b_s - b'$ then

$$a_{2t} - a_{2xx} - a_2 b_2 + a_2 b_s + a_s b_2 = 0 \quad \text{(RT4-8A)}$$

$$b_{2t} - R b_{2xx} - m a_2 b_2 + m a_2 b_s - m a_s b_2 = 0 \quad \text{(RT4-8B)}$$

If

$$A = a_1 - a_2, \quad \text{and} \quad B = b_1 - b_2$$

then the theorem will be established if it can be shown that

$$A = 0 = B, \quad \text{for all } t.$$

Now, combining terms from Eqs. (RT4-7A), (RT4-7B), (RT4-8A) and (RT4-8B):

$$A_t - A_{xx} - a_1 b_1 + a_2 b_2 + a_1 b_s - a_2 b_s + a_s b_1 - a_s b_2 = 0$$

or,

$$A_t - A_{xx} - (a_1 b_1 + a_2 b_2) + (a_1 b_s - a_2 b_s) + (a_s b_1 - a_s b_2) = 0 \quad \text{(RT4-9A)}$$

$$B_t - R B_{xx} - m a_1 b_1 + m a_2 b_2 + m a_1 b_s - m a_2 b_s - m a_s b_1 + m a_s b_2 = 0$$

or

$$B_t - R B_{xx} - m(a_1 b_1 - a_2 b_2) + m(a_1 b_s - a_2 b_s) - m a_s (b_1 - b_2) = 0 \quad \text{(RT4-9B)}$$

Now,

$$\begin{aligned}-a_1b_1 + a_2b_2 &= -a_1b_1 + a_2b_1 - a_2b_1 + a_2b_2 \\ &= -(a_1 - a_2)b_1 - a_2(b_1 - b_2) \\ &= -Ab_1 - a_2B\end{aligned}$$

$$a_1b_s - a_2b_s = (a_1 - a_2)b_s = Ab_s \quad \text{(RT4-10)}$$

$$a_sb_1 - a_sb_2 = a_s(b_1 - b_2) = a_sB$$

Substituting the relationships expressed in (RT2-33) into Eqs. (RT2-32A) and (RT2-32B), the result is

$$A_t - A_{xx} + (-Ab_1 - a_2B) + (Ab_s) + (a_sB) = 0$$

$$B_t - RB_{xx} - m(-Ab_1 - a_2B) + m(Ab_s) - m(a_sB) = 0$$

or,

$$A_t - A_{xx} + (b_s - b_1)A + (a_s - a_2)B = 0 \quad \text{(RT4-11A)}$$

$$B_t - RB_{xx} - m(b_s - b_1)A + m(a_s - a_{s2}) = 0 \quad \text{(RT4-11B)}$$

with ***homogeneous*** initial and boundary conditions:

$$A(x, 0) = 0 = B(x, 0), \quad x > 0 \quad \text{(RT4-11C)}$$

$$A(0, t) = 0 = B(0, t), \quad t > 0 \quad \text{(RT4-11D)}$$

Multiplying Eq. (3-11A) by A, and integrating with respect to all possible values of x, one obtains

$$\int_0^{\infty} \{A_t - A_{xx} + (b_s - b_1)A + (a_s - a_2)B\}A \, dx = 0 \quad \text{(RT4-12)}$$

For simplicity, the limits notation of $(0, \infty)$ are omitted in the remaining analysis, so that Eq. (RT4-12) may be written as:

$$\int A_tA \, dx - \int A_{xx}A \, dx + \int (b_s - b_1)A^2 \, dx + \int (a_s - a_2)AB \, dx = 0 \quad \text{(RT4-13)}$$

Now,

(a)

$$\begin{aligned}\int A_tA \, dx &= \int (\partial A/\partial t)A \, dx = \int (A\partial A/\partial t) \, dx \\ &= \int (\partial 1/2A^2/\partial t) \, dx = 1/2(d/dt) \int A^2 \, dx \quad \text{(i)}\end{aligned}$$

and writing

$$\|A\| = (A, A)^{1/2} = \left[\int AA \, dx\right]^{1/2} = \left[\int A^2 \, dx\right]^{1/2}$$

hence

$$\int A^2\,\mathrm{d}x = ||A||^2 \qquad \text{(ii)}$$

and combining (i) and (ii), $\int A_t A\,\mathrm{d}x = 1/2(\mathrm{d}/\mathrm{d}t)||A||^2$.

(b)

$$\begin{aligned} -\int A_{xx}A\,\mathrm{d}x &= -[A_{xx}A]_0^\infty + \int A_x A_x\,\mathrm{d}x, \text{ integrating by parts} \\ &= 0 + \int [A_x]^2\mathrm{d}x, \text{ using the conditions of Eq. (RT4-13)} \\ &= \int [A_x]^2\mathrm{d}x \\ &\geq 0 \end{aligned}$$

(c) Using the conditions of Eq. (RT2-34C, D):

$$\int (b_s - b_1)A^2\,\mathrm{d}x \geq 0 \qquad \text{(iii)}$$

Hence, the conditions (a), (b), and (c) reduce Eq. (RT2-37) to the following inequality:

$$1/2(\mathrm{d}/\mathrm{d}t)||A||^2 + \int (a_s - a_2)AB\,\mathrm{d}x \leq 0 \qquad \text{(RT4-14A)}$$

Similarly, from Eq. (RT2-35B), multiplying by B, and integrating with respect to all possible values of x, one obtains

$$1/2(\mathrm{d}/\mathrm{d}t)||B||^2 + m\int (a_s - a_2)AB\,\mathrm{d}x \leq 0 \qquad \text{(RT4-14B)}$$

From Eq. (RT-38A):

$$1/2(\mathrm{d}/\mathrm{d}t)||A||^2 \leq -\int (a_s - a_2)AB\,\mathrm{d}x$$

or

$$\begin{aligned} (\mathrm{d}/\mathrm{d}t)||A||^2 &\leq -2\int (a_s - a_2)AB\,\mathrm{d}x \\ &\leq \left|2\int (a_s - a_2)AB\,\mathrm{d}x\right| \\ &\leq \left|2\int (a_s - a_2)\mathrm{d}x\right|\left|\int AB\,\mathrm{d}x\right|, * \\ &\leq 2\left|\int (a_s - a_2)\mathrm{d}x\right|\left|\left(\int A^2\,\mathrm{d}x\right)^{1/2}\right|\left|\left(\int B^2\,\mathrm{d}x\right)^{1/2}\right|, ** \\ &\leq 2||a_s - a_2||L^\infty(0,\infty)||A||\,||B|| \\ &\leq 2\alpha||A||\,||B||, \text{ where } \alpha = ||a_s - a_2||L^\infty(0,\infty) \\ &\leq \alpha(||A||^2 + ||B||^2), ** \end{aligned}$$

* Holder's inequality [1].

**Theorem of the arithmetic and geometric means [34].

Hence,

$$(\mathrm{d}/\mathrm{d}t)\|A\|^2 - \alpha(\|A\|^2 + \|B\|^2) \leq 0 \tag{RT4-15}$$

and similarly,

$$(\mathrm{d}/\mathrm{d}t)\|B\|^2 - \beta(\|A\|^2 + \|B\|^2) \leq 0 \tag{RT4-16}$$

where $\beta = m\alpha$

Now let

$$\lambda = \lambda(t) = \|A\|^2 + \|B\|^2$$

then

$$\lambda \geq 0 \tag{RT4-17}$$

And, on adding Eqs. (RT-39) and (RT-40), one obtains

$$(\mathrm{d}/\mathrm{d}t)\lambda - (\alpha + \beta)\lambda \leq 0$$

so that

$$\lambda \leq \lambda(0)\exp(\alpha + \beta)t$$
$$= 0$$

because $\lambda(0) = 0$, by the condition of homogeneity of the boundaries.

Thus,

$$\lambda \leq 0 \tag{RT4-18}$$

The conditions required by Eqs. (RT2-41) and (RT2-42) imply that

$$\lambda \equiv 0 \tag{RT4-19}$$

which means

$$A = 0 = B, \quad \text{for all } t \tag{RT4-20}$$

This completes the proof of the ***Uniqueness Theorem***.

3.9 Theory of Simultaneous Bimolecular Reactions and Mass Transfer in Two Dimensions: Further Cases of Practical Interests

3.9.1 Case of Stagnant Film of Finite Thickness – Second-Order Irreversible Reactions

Some basic simplifying assumptions of this model are [35]

It is assumed that

(1) *The chemical reaction steps are infinitely rapid, compared to the translational movements of the reactants and products. Thus, the reaction rates of the process*

are fundamentally determined by the diffusion rates of the reactants A, B, C, etc., in a moving reaction site.

(2) *At a given instance, the concentration profiles of A, B, and C in the liquid may be established separately and distinctly.*
(3) *The temperature is uniform.*
(4) *The liquid layer is flowing with a uniform speed parallel to the interface or is stationary.*
(5) *All the resistance to mass transfer is essentially in the liquid phase.*
(6) *The interfacial concentrations of unreacted, but dissolved, A and B are constant: given by A_i and B_i, respectively.*
(7) *The two reactions, of A and B with substance C, consume ν_A and ν_B moles of C, respectively, per mole of A and B.*
(8) *The liquid diffusivities are constant.*

For the second-order irreversible reaction in the liquid phase, it should be noted that it is within the liquid film that the reaction takes place. Thus, if a solute gas A is absorbed, from a mixture of gases by a solution containing a substance B which combines with A on the basis of the equation

$$A + B \rightarrow AB \tag{3.126}$$

When the solution contacts the gas mixture, A dissolves and reacts with B at the phase boundary, forming the product AB which will diffuse toward the main body of the liquid phase. Consequently, the liquid near the surface will be depleted of B which will start to diffuse from the main body of the liquid toward the interface. This movement of B from the space close to the surface leads the movement of A to diffuse through the liquid film to replace B diffusing in the opposite direction. Thus, the reaction zone of A and B will move away from the liquid–gas interface, to a position such that the diffusion rate of A from the gas phase is to be equal (mole for mole) to the diffusion rate of B from the main body of the liquid. Generally, this process occurs quickly, during which the rate of absorption falls quickly, and quickly reaching an equilibrium position for the system. An idealized model of gradients of the interface may be described in terms of a double film with irreversible instantaneous reaction following absorption:

Figure 3.11 shows a model representation of the concentration gradients of the interaction of the gas–liquid system under discussion: the ordinates represent the gaseous pressures or liquid concentrations, and the abscissa represents the physical position in the film.

Let a vertical plane PQ represents the interfacial plane of the liquid and gaseous phases.

Let VW and UT represent the outer boundaries of liquid films and gas phases.

Let SR represents the equilibrium position of the reaction zone, to which both A and B diffuse, and from which the reaction product AB diffuses toward the main body of the liquid.

Now, under the driving force $(p - p_i)$, A diffuses through the gas film and diffuses to the reaction zone – owing to the driving force c_i in the liquid phase.

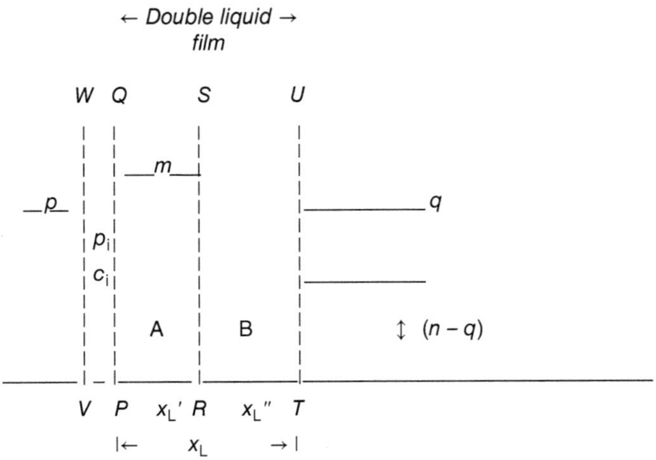

Figure 3.11 Gradients in double film with irreversible instantaneous reaction following absorption. Source: Sherwood and Pigford (1952).

Simultaneously:

- After the equilibrium reaction position is attained, for component A let p_i be the interfacial pressure at the reaction zone, and p be the partial pressure of a component outside of the reaction zone, then A will diffuse to the reaction zone under the influence of the partial pressure driving force $(p - p_i)$ in the gas film, ***and***, to the reaction zone owing to driving force c_i in the liquid phase.
- B will diffuse from the main body of the liquid to the reaction zone – under the influence of the driving force q, and the product of this reaction AB will diffuse back to the main body of the liquid by the potential $m - (n - q)$, where m = the concentration of the product *within* the reaction zone, and $(n - q)$ = the concentration of the product *outside of* the reaction zone.

Thus, the solution is n-normal in free plus combined B and is q-normal in B. The product of the reaction will diffuse from the reaction zone to the gas phase and toward the main body of the liquid.

Here, the product AB is assumed to be nonvolatile, and the concentration of AB is rapidly equalized within the reaction zone.

For the first section QS of the liquid film:

The diffusion equation for the gas film, as well as for both sections of the liquid film on either side of the reaction zone, may be expressed as:

$$N_A = k_G(p - p_i) \tag{3.127}$$

for the gas film is high compared to the concentrations of A, B, and AB, then Eq. (3.127) may be expressed as:

$$N_A = \left(D_A/x_L'\right)(c_i - 0) \tag{3.128}$$

where

D_A = The diffusivity of component A in the liquid phase
$x_L{}'$ = The width of the reaction zone outside of the reaction zone
c_i = The concentration of the component A at the interface between the bulk liquid and the reaction zone

For the second section SU of the liquid film:

$$-N_B = (D_B/X_L{}'')\,(q - 0) = +N_A \tag{3.129}$$

and

$$N_{AB} = (D_{AB}/X_L{}'')\,(m - n + q) = N_A \tag{3.130}$$

Again, assuming the liquid and gas phases at the actual interface QP is assumed to be in equilibrium, and as A does not react in the section of the film QS but dissolves physically, one may assume that the equilibrium may be approximately given by Henry's law, viz.,

$$p_i = Hc_i \tag{3.131}$$

where H is the Henry's law constant for the system.

As the molal diffusion rates of B and AB are equal, the potentials $[m - (n - q)]$ and q are in inverse proportion to the diffusivities of AB and B, as given by Eqs. (3.129) and (3.130), respectively. Thus, by combining these two equations, the result is

$$m - (n - q) = (D_B/D_{AB})q \tag{3.132}$$

Now, Eqs. (3.127), (3.128), (3.129), and (3.131) may now be combined to eliminate p_i, c_i, m, n, $x_L{}'$, and $x_L{}''$, resulting in

$$N_A = \{(p/H) + (D_B/D_A)q\}/\{(x_L/D_A) + (1/Hk_G)\} \tag{3.133}$$

which states that the absorption rate is proportional to the overall driving force, as expressed by the sum of the two terms in the numerator, and inversely proportional to the overall resistance, as expressed by the two terms in the denominator. The first term of the denominator represents the combined resistances of both sections of the liquid film, while the second term of Eq. (3.133) states that, for a constant gas composition and constant film conditions, the absorption rate should be a linear function of the residual concentration of B in the liquid.

By eliminating the gas-phase concentrations, Eq. (3.133) may be written in an equivalent and useful form by eliminating the gas-phase concentrations to give

$$N_A = \{c_{Ai} + (D_B/D_A)q\}/(x_L/D_A) \tag{3.134}$$

Now, if the liquid film coefficient k_L is defined on the basis of the liquid film driving force $(c_{Ai} - 0)$, one finds that

$$k_L = (D_A/x_L)\{1 + (D_B/D_A)(q/c_{Ai})\} \tag{3.135}$$

Remarks:

(1) When there is no chemical reaction, according to Eq. (3.135), the coefficient is given simply by (D_A/x_L).

(2) The factor in parenthesis {} provides the effect of q on the apparent coefficient.

(3) Since p_i is nonnegative, Eq. (3.127) does not apply where p_i is negative. Solving Eqs. (3.127) and (3.133) for p_i, the result is

$$p_i = \{k_G p - (D_s q/x_L)\}/\{k_G + (D_A/Hx_L)\} > 0 \tag{3.136}$$

Hence

$$q < x_L k_G p/D_B \tag{3.137}$$

constituting a ***necessary*** condition for the application of Eq. (3.133).

(4) When

$$q > x_L k_G p/D_B \tag{3.138}$$

p_i approaches 0, and absorption occurs at a constant rate of

$$N_A = k_G p \tag{3.139}$$

(5) The condition given by Eq. (3.137) expresses the ***critical*** value of q. If absorption has begun using a solution in which the concentration of B is greater than this ***critical*** value, the rate of absorption will be constant as q decreases to the critical value. The rate will then decrease linearly with q according to Eq. (3.133), providing p and other conditions are maintained unchanged.

3.9.2 Case of Unsteady-State Absorption in the Stagnant Liquid – Slow First-Order Reaction (S&P 325, 328)

When the reaction is ***slow*** compared with the rate of diffusion, the reaction zone will not be concentrated in a thin region, as shown in Figure 3.11, but will be spread ***throughout*** the liquid film. **Sherwood and Pigford [36]** reported the following approach to describe the process of simultaneous gas absorption and chemical reaction for slow first-order reactions for the absorption and simultaneous slow irreversible reaction of a solute A, with a large excess of the solvent.

The solute first diffuses through the gas film under the influence of the driving force $(p - p_i)$. Upon entry into the liquid phase, it starts to diffuse toward the main body of the liquid and immediately begins to react with the dissolved substance B. As the diffusing current of solute is being depleted as it diffuses into the liquid, the gradient will be concave upward (Figure 3.12).

The total rate of diffusion of reacted and unreacted forms of A is constant as there is no accumulation of A in the film.

If the rate equation for the reaction is known, then one may express mathematically the conditions of the process: thus, for the first-order reaction, viz., the rate of elimination of A is proportional to the concentration of A at any point, viz.,

$$-\mathrm{d}A/\mathrm{d}\theta = k_c cV \tag{3.140}$$

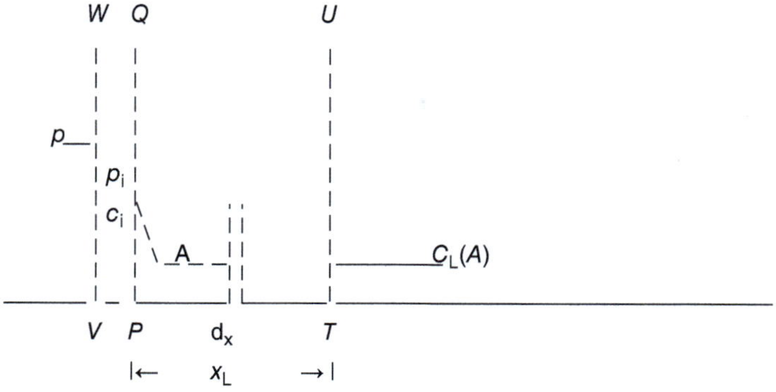

Figure 3.12 Sketch of film conditions for absorption followed by slow first-order reaction in the liquid film.

where

k_c is the specific reaction rate constant,
c is the variable concentration of A, and
V is the liquid volume considered.

Now, consider a differential element, if unit area and of thickness dx, of the liquid film of thickness x_L. Then, at $x = x$, the rate of diffusion *into this element* is

$$N_A = -D_A(dc/dx) \tag{3.141}$$

And the rate of diffusion *out of this element, at* $x = x + dx$, *will be*

$$N_A = -D_A(dc/dx + d^2c/dx^2 \, dx) \tag{3.142}$$

And the disappearance of A within this element, at *owing to reaction with* B, is

$$dA/d\theta = -k_c c \, dx \tag{3.143}$$

since the volume of the element is dx.

The difference in diffusion rates in and out of the element must equal the rate of elimination of A by reaction with B. Hence,

$$D_A \, d^2c/dx^2 = k_c c \tag{3.144}$$

If one assumes k_c remains constant throughout the film, then the solution to Eq. (3.144) may be given as:

$$c = A_1 \exp(a_0 x) + A_2 \exp(-a_0 x) \tag{3.145}$$

where a_0 e.

The constants A_1 and A_2 in Eq. (3.145) may be obtained by substituting the limits: $c = c_i$, at $x = 0$, and $c = c_L$, at $x = x_L$, and the solution becomes

$$c = [c_L \sinh(a_0 x) + c_i \sinh\{a_0(x_L - x)\}]/\sinh(a_0 x_L) \tag{3.146}$$

which reduces to

$$c = ([\sinh\{a_0(x_L - x)\}]/\sinh(a_0 x_L))c_i \tag{3.147}$$

By differentiating Eq. (3.146) with respect to x, the slope of the concentration curve may be obtained:

$$dc/dx = [a_0 c_L \cosh(a_0 x) + a_0 c_i \cosh\{a_0(x_L - x)\}]/\sinh(a_0 x_L) \tag{3.148}$$

The rate of diffusion into the liquid may be obtained by multiplying the slope at $x = 0$ by the coefficient of diffusivity D_A:

$$N_A = -D_A(dc/dx)_{x=0} = D_A a_0\{c_i \cosh(a_0 x_L) - c_L\}/\sinh(a_0 x_L) \tag{3.149}$$

and the rate of diffusion of A into the main body of liquid may be similarly obtained by substituting $x = x_L$:

$$N_A{}' = -D_A(dc/dx)_{x=x_L} = D_A a_0\{c_i - c_L \cosh(a_0 x_L)\}/\sinh(a_0 x_L) \tag{3.150}$$

From the solute A entering the liquid phase, the fraction F reaching the main body of the liquid ***without*** reacting is given by:

$$F = N_A{}'/N_A = \{c_L \cosh(a_0 x_L) - c_i\}/\{c_L - c_i \cosh(a_0 x_L)\} \tag{3.151}$$

If the concentration A in the main body of liquid is low, Eq. (3.149) may be simplified to:

$$N_A = bD_A(c_i - c_L)/x_L \tag{3.152}$$

where

$$b = a_0 x_L/\tanh(a_0 x_L) \tag{3.153}$$

Remark: ***It may be easily shown that, if the chemical reaction is slow, the absorption process becomes the same as for pure physical absorption, and the reaction takes place in the main body of the liquid!***

3.9.3 Simultaneous Absorption of Two Gases in a Liquid in Which Each Then Reacts With a *Third* Component in the Liquid

Roper, Hatch, and Pigford [37] reported the following approach to describe the process of simultaneous gas absorption and chemical reaction for two gases in a liquid which also contains a third component capable of reacting chemically with both of the dissolved gases.

Roper, Hatch, and Pigford in this seminal paper presented the theory of diffusion and simultaneous chemical combination of two gases dissolving in a liquid and reacting near the interface was presented mathematically. The nonlinear partial differential equations representing mass balances and corresponding to the penetration theory are solved numerically. The chemical reaction rate expression is second order, corresponding to a bimolecular mechanism. The results show that chemical reaction between the two dissolved substances increases the rate of absorption of each by an amount that depends on the ratio of solubilities, the reaction rate constant, and the diffusion coefficients. Mathematical solutions are also presented for the simultaneous absorption of two gases in a liquid which contains a third component that is capable of reacting chemically with both of the dissolved gases.

Moreover, Hatch and Pigford [37] also studied experimentally the process of gas absorption accompanied by chemical reaction of two gases and a liquid, using the system CO_2–NH_3–water. The two gases (CO_2 and NH_3) were first mixed, then contacted with a smooth laminar jet of water (H_2O): after dissolution, they reacted with each other by a bimolecular reaction. The rates of absorption were determined by chemical analysis of the liquid leaving the jet absorption device. The rates of absorption of CO_2, significantly influenced by the reaction with an excess of NH_3 in the liquid, was in agreement with the prediction based on the penetration theory for the absorption of two gases followed by the second-order, irreversible reaction. The high solubility of NH_3 in water, resulting in a large gas-side resistance to its absorption, made it impossible to determine the effect of the rate of reaction on its rate of absorption because of the difficulty in knowing the NH_3 concentration at the liquid–gas interface. The confirmation of the diffusion-reaction theory was therefore limited to the observation on CO_2 absorption, and its validity depends on the accuracy of the NH_3 gas-phase resistance measurements.

3.9.3.1 Mathematical Modeling

The chemical reaction has the stoichiometry represented by the chemical equation:

$$\mathrm{A} + \nu\mathrm{B} \rightarrow \text{Products} \tag{3.154}$$

and is irreversible. It is also assumed that, kinetically, it is second order with respect to the two concentrations.

From a mass balance on a differential element, with dimension $dx \times dz$, in the direction of flow parallel to the interface, and perpendicular to the interface, results in the following pair of semi-linear partial differential equations:

$$D_\mathrm{A}\partial^2 A/\partial x^2 = u\partial A/\partial z + kAB \tag{3.155}$$

$$D_\mathrm{B}\partial^2 B/\partial x^2 = u\partial B/\partial z + \nu kAB \tag{3.156}$$

In both Eqs. (3.155) and (3.156):

- the terms on the LHS represent the amounts of materials, for species A and B, respectively, consumed by diffusion into the reaction zone,
- the first terms on the RHS represent the amounts of materials, for species A and B, respectively, transported by the convective flow of materials, and
- the second terms on the RHS represent the amounts of materials consumed by chemical reaction of the two species. They represent the rates of homogeneous chemical reaction between species A and B in the solution per unit volume of solution.

In comparing the rates of absorption ***with*** and ***without*** the chemical reaction, it has been mentioned in Section 3.10.1 that, in the study of the enhancement of rates of mass transport with reactions *over* the rate of pure physical mass transport, a typical measure of these absorption rates may be expressed in terms of the ***transport coefficients***, defined as the rates of transport of dissolved reactants *per unit of interfacial area*.

In other words, the results are expressed by the ratio of rate of absorption of the gas, ***with the chemical reaction***, to the rate that would occur ***without the chemical reaction***, but with all other conditions the same.

It may be considered as the enhancement of the absorption rate. This enhancement ratio would depend on

- the reaction rate constants,
- the time of exposure of the liquid surface to the gas,
- the diffusion coefficients in the liquid,
- certain stoichiometric ratios.

Remarks:

(1) *Theoretical calculations of such phenomena may be based on the assumption that diffusion and chemical reaction may be based on the assumption that takes place in a laminar flow regime near the interface, and that the chemical reaction per unit volume being so fast that only a small space in the liquid phase is sufficient to maintain the diffusion of reactants to the reaction zone.*

(2) *With these assumptions, the mathematical analysis is that of solving the time-dependent semi-linear diffusion equations obtainable by mass balances, with the reaction term included.*

(3) *Clearly, there can be as many cases as there are orders of reaction, as well as ways of bringing various numbers of reactants together!*

(4) *For the simplest problems linear partial differential equations are obtained – for first-order reactions of a single diffusing species when it dissolves in the liquid at the interface. The computed results for both reversible and irreversible first-order reactions are known.*

(5) *Computations have been undertaken for reactions for two dissolved substances, A and B, in which A is gas while B is a nonvolatile reagent initially present in the liquid. Exact solutions for this class of reactions have been obtained for fast reactions which are limited by diffusion of A and B to a reaction boundary that moves into the liquid in due course. Numerical solutions are often used for such cases.*

(6) *For the physically interesting situation in which A and B are both gases which are brought into contact with a laminar liquid layer in which both A and B are soluble and within which they react with one another by a second-order chemical mechanism. A typical example for such a system is the simultaneous absorption of NH_3 and CO_2 into water. In the vapor phase, these gases do not react with each other – but in the dissolved state they do react with each other, and with water to form ammonium carbamate.*

For this system, the chemical equation is of the type:

$$\mathrm{A} + \mathrm{B} \rightarrow \tag{3.157}$$

(7) *Another similar example is the washing of a gas mixture of H_2S and CO_2, which are washed with an alkaline liquid in which both gases dissolve, and react – competing for the same alkaline reagent!*

For this system, the chemical equation is of the type:

$$\mathrm{A} + \mathrm{C} \rightarrow \tag{3.158A}$$

and, simultaneously,

$$B + C \rightarrow \tag{3.158B}$$

Analyses of these two cases will follow.

3.9.3.2 Analysis of the Model: A + B →

In Section 3.11.3.1, the model for this system was described, showing that the chemical reaction has the stoichiometry represented by the chemical equation:

$$\mathrm{A} + \nu\mathrm{B} \rightarrow \text{Products} \tag{3.159}$$

and is irreversible. It is also assumed that, kinetically, it is second order with respect to the two concentrations.

From a mass balance on a differential element, with dimension $dx \times dz$, in the direction of flow parallel to the interface, and perpendicular to the interface, results in the following pair of semi-linear partial differential equations:

$$D_{\mathrm{A}}\partial^2 A/\partial x^2 = u\partial A/\partial z + kAB \tag{3.160}$$

$$D_{\mathrm{B}}\partial^2 B/\partial x^2 = u\partial B/\partial z + \nu kAB \tag{3.161}$$

in which the last terms on the right-hand side of these two equations, viz., kAB and νkAB, represent the rates of homogeneous reaction between the substances A and B in solution per unit volume of solution. The reaction is kinetically assumed to be second order with respect to the concentrations of the two species A and B.

Consider the following situation for this system: at certain fixed concentrations A_{i} and B_{i} at the interface, as indicated by the partial pressures of A and B in the corresponding gas phase and solubilities of these substances. Moreover, the liquid layer may be considered to be infinitely deep, far from the interface, and assume that initially prior to the liquid coming into contact with the gas phase that concentrations of dissolved substances are zero. These conditions are equivalent to the following boundary conditions on $A(x,z)$ and $B(x,z)$:

$$A(0,z) = A_{\mathrm{i}}, \quad B(0,z) = B_{\mathrm{i}}$$

$$A(\infty,z) = 0, \quad B(\infty,z) = 0$$

$$A(x,0) = 0, \quad B(x,0) = 0 \tag{3.162}$$

3.9.3.3 Discussions

Liquid diffusion coefficients is the penetration depth of the dissolved but unreacted gas into the liquid so small that the fluid velocity is everywhere near the velocity of the interface. For these circumstances, Eqs. (3.160) and (3.161) may be reduced by substituting the following new variables:

$$a = A/A_{\mathrm{i}}, \quad b = B/B_{\mathrm{i}} \tag{3.163}$$

$$y = (kB_{\mathrm{i}}/D_{\mathrm{A}})^{1/2}x, \quad \theta = (kB_{\mathrm{i}}/u)x \tag{3.164}$$

to the condensed form:

$$\partial^2 a/\partial y^2 = \partial a/\partial\theta + ab \tag{3.165}$$

$$R\partial^2 b/\partial x^2 = \partial b/\partial \theta + mab \tag{3.166}$$

in which the dimensionless parameters are defined as:

$$R = (D_A/D_B), \quad \text{and} \quad m = \nu A_i/B_i \tag{3.167}$$

with the boundary conditions on $a = a(y, \theta)$ and $b = b(y, \theta)$

$$\begin{aligned} a(0, \theta) &= 1 = b(0, \theta) \\ a(\infty, \theta) &= 0 = b(\infty, \theta) \\ a(y, 0) &= 0 = b(y, 0) \end{aligned} \tag{3.168}$$

In this analysis, the parameters of interest are the mass transfer coefficients, defined to provide the rate of absorption of the dissolving gas per unit of interfacial area. Generally, the mass transfer coefficient, k_{LA} and k_{LB}, will be different for A and for B, each being influenced by the speed of the reaction between the substances:

$$k_{LA} = k_{LA}(\theta) = -(D_A/A_i)(\partial A/\partial x)_{x=0} = -\sqrt{(kB_iD_A)}(\partial a/\partial y)_{y=0} \tag{3.169}$$

$$k_{LA} = k_{LB}(\theta) = -(D_B/B_i)(\partial B/\partial x)_{x=0} = -\sqrt{(kB_iD_A)}R(\partial b/\partial y)_{y=0} \tag{3.170}$$

Equations (3.165) and (3.166) may then be solved, subject to the boundary conditions given by Eq. (3.168), to determine the mass transfer coefficients, k_{LA} and k_{LB}, as given by Eqs. (3.169) and (3.170).

The mass transfer coefficients are usually expressed as the products of the standard mass transfer coefficient, $\sqrt{(kB_iD_A)}$, and a nondimensional factor which depend on the parameters R, m, and θ.

For the whole surface, the mean mass transfer coefficients for the whole surface are

$$\underline{k_{LA}} = \underline{k_{LA}(\theta)} = -\sqrt{(kB_iD_A)}\left[\left\{\int_0^\theta (\partial a/\partial y)_{y=0}\, d\theta\right\}\Big/\theta\right] \tag{3.171}$$

$$\underline{k_{LB}} = \underline{k_{LB}(\theta)} = -\sqrt{(kB_iD_A)}\left[\left\{R\int_0^\theta (\partial b/\partial y)_{y=0}\, d\theta\right\}\Big/\theta\right] \tag{3.172}$$

Equations (3.165)–(3.172) are nonlinear, and numerical procedures may be used to obtain numerical solutions.

3.9.3.4 Further Theoretical Analysis

Three additional aspects of theoretical analysis of the system may be considered:

1. Steady-State Solutions,
2. Solutions by Method of Moments
3. Perturbation Solution for Small θ

3.9.3.4.1 Steady-State Solution

The steady-state solution may be obtained by considering the concentration distributions that the system approaches asymptotically as θ (viz., time) tends toward infinity. Such special solutions can provide guidance in making certain arbitrary assumptions that are pertinent in the numerical solutions of the system equations.

Moreover, these special solutions may also provide important guidance in some assumptions in the numerical analysis, and they may also lead to some arbitrarily useful assumptions in the analysis. For example, these particular solutions can give a useful solution when the pseudo-first-order reaction rate constant kB_t is sufficiently large for the early transients in the concentration distribution to be completed in a fraction of the total time θ.

Consider the following two examples.

Example 3.6 ***Pseudo-First-Order Reactions***

Such conditions are satisfied when the concentration of B is substantially constant within the entire liquid phase, viz.,

$$B(x, z) = B_{\mathrm{i}} \tag{3.173}$$

This condition obtains whenever the concentration of component B far exceeds that of A or when the diffusion coefficient of B is much greater than that of A, i.e. when $m \to 0$, or $R \to \infty$.

The problem then becomes a linear one for which the solution leads to:

$$k_{\mathrm{LA}} = \sqrt{(kB_{\mathrm{i}}D_{\mathrm{A}})} \tag{3.174}$$

Example 3.7 ***Second-Order Conditions – Substances Identical***

For $m = 1$, $R = 1$, and $\partial/\partial\theta = 0$, while recognizing that the boundary conditions are equivalent for a and b, viz., $a = b$, may then applied in the system equations:

$$\partial^2 a/\partial y^2 = \partial a/\partial\theta + ab \tag{3.175}$$

$$R\partial^2 b/\partial x^2 = \partial b/\partial\theta + mab \tag{3.176}$$

so that ***both*** Eqs. (3.175) ***and*** (3.176) become the steady-state configuration:

$$\mathrm{d}^2 a/\mathrm{d}y^2 = a^2 \tag{3.177}$$

for which a closed-form solution is

$$a = 1/\{1 + (y/\sqrt{6})\}^2 \tag{3.178}$$

Checking:

$$\mathrm{d}a/\mathrm{d}y = (\mathrm{d}/\mathrm{d}y)[a]$$

$$= (\mathrm{d}/\mathrm{d}y)[1/\{1 + (y/\sqrt{6})\}^2]$$

$$= (\mathrm{d}/\mathrm{d}y)\{1 + (y/\sqrt{6})\}^{-2}$$

$$= \mathrm{d}\{1 + (y/\sqrt{6})\}^{-2}/\mathrm{d}\{1 + (y/\sqrt{6})\} \times \{\mathrm{d}\{1 + (y/\sqrt{6})\}/\mathrm{d}y\}$$

$$= -2\{1 + (y/\sqrt{6})\}^{-3} \times (1/\sqrt{6})$$

and

$$\begin{aligned}
d^2a/dy^2 &= (d/dy)\{da/dy\} \\
&= (d/dy)[-2\{1 + (y/\sqrt{6})\}^{-3} \times (1/\sqrt{6})] \\
&= [+6\{1 + (y/\sqrt{6})\}^{-4} \times (1/\sqrt{6})](1/\sqrt{6}) \\
&= \{1 + (y/\sqrt{6})\}^{-4} \\
&= [\{1 + (y/\sqrt{6})\}^{-2}]^2 \\
&= [1/\{1 + (y/\sqrt{6})\}^2]^2 \\
&= a^2
\end{aligned}$$

as required, and for which

$$k_{LA} = k_{LB} = \sqrt{\{(2/3)kB_iD_A\}} \tag{3.179}$$

It appears that, under these conditions, the correction factor by which the standard mass transfer coefficient is $\sqrt{(2/3)} = 0.816\,50$, *not* 1.000 00!

3.9.3.4.2 Solution by the Method of Moments

Yet another approach is the "Method of Moments," commonly used in the Theory of the Laminar Boundary Layer in Fluid Mechanics. Its approach is an assumption of an approximate solution that:

(i) Satisfies the boundary conditions ***exactly***, and
(ii) Satisfies the system of differential equations ***only on the average***.

This approach can be useful, particularly in boundary layer problems where the assumed function represents the velocity distribution in the section perpendicular to the boundary plane: $y = 0$, and where the contour of the distribution function at one value of z is similar to that at another value of z. Under such conditions, it is usually assumed that the concentration profiles may be represented as follows:

$$\begin{aligned}
a(y,\theta) = A(\xi) &= A_0 + A_1\xi + A_2\xi^2 + A_3\xi^3 + \cdots, \qquad && 0 \le \xi \le 1 \\
&= 0 && 1 \le \xi < \infty
\end{aligned} \tag{3.180}$$

and

$$\begin{aligned}
b(y,\theta) = B(\eta) &= B_0 + B_1\eta + B_2\eta^2 + B_3\eta^3 + \cdots, \qquad && 0 \le \eta \le 1 \\
&= 0 && 1 \le \eta < \infty
\end{aligned} \tag{3.181}$$

where the As and Bs are constants, and where $\xi = y/\Delta_a$ and $\eta = y/\Delta_b$.

Remarks:

(1) The two Δs are the thicknesses of the boundary layer – in this case, concentration boundary layers, viz., for each equation, the infinite boundary in the real problem is being replaced by a finite boundary where the concentrations of A and B

reach zero, respectively. The boundary layer thicknesses are functions of time:

$$\Delta_a = \Delta_a(\theta) \tag{3.182}$$

$$\Delta_b = \Delta_b(\theta) \tag{3.183}$$

which will be derived from the differential equations.

(2) The two *approximating functions*, $A(\xi)$ and $B(\eta)$, are required to satisfy the boundary conditions and to be solutions of the system equations – in the sense that they are required to render the ***moments*** of the residuals in the differential equation to be zero, that is, using Eqs. (3.165) and (3.166), and writing the conditions corresponding to the zeroth and the first moments:

$$\int_0^1 \{(\partial^2 a/\partial y^2) - (\partial a/\partial\theta) - ab\}\mathrm{d}\xi = 0 \tag{3.184}$$

$$\int_0^1 \xi\{(\partial^2 a/\partial y^2) - (\partial a/\partial\theta) - ab\}\mathrm{d}\xi = 0 \tag{3.185}$$

$$\int_0^1 \{(R\partial^2 b/\partial y^2) - (\partial b/\partial\theta) - mab\}\mathrm{d}\eta = 0 \tag{3.186}$$

$$\int_0^1 \eta\{(R\partial^2 b/\partial y^2) - (\partial b/\partial\theta) - mab\}\mathrm{d}\eta = 0 \tag{3.187}$$

Presumably, in this way, it is possible to obtain sufficient conditions to fix all the required constants in the trial solutions and then to find the true solution. Note that the derivatives with respect to θ and y are obtainable from:

$$\partial a/\partial\theta = -\xi F'(\xi){\Delta_a}^{-1}(\mathrm{d}\Delta_a/\mathrm{d}\theta); \quad \partial^2 a/\partial y^2 = {\Delta_a}^{-2}F''(\xi) \tag{3.188}$$

$$\partial b/\partial\theta = -\eta G'(\xi){\Delta_b}^{-1}(\mathrm{d}\Delta_b/\mathrm{d}\theta); \quad \partial^2 b/\partial y^2 = {\Delta_b}^{-2}G''(\xi) \tag{3.189}$$

For the boundary thicknesses, the four ***partial*** differential equations (3.184)–(3.187), are transformed into the following ***ordinary*** differential equations for the boundary layer thicknesses:

$$\left[-1/2\int_0^1 \xi F'(\xi)\mathrm{d}\xi\right](\mathrm{d}{\Delta_a}^2/\mathrm{d}\theta) + \left[\int_0^1 F(\xi)G(\eta)\mathrm{d}\xi\right]{\Delta_a}^2 = \left[\int_0^1 F''(\xi)\mathrm{d}\xi\right] \tag{3.190}$$

$$\left[-1/2\int_0^1 \xi^2 F'(\xi)\mathrm{d}\xi\right](\mathrm{d}{\Delta_a}^2/\mathrm{d}\theta) + \left[\int_0^1 \xi F(\xi)G(\eta)\mathrm{d}\xi\right]{\Delta_a}^2 = \left[\int_0^1 \xi F''(\xi)\mathrm{d}\xi\right] \tag{3.191}$$

$$\left[-1/2\int_0^1 G'(\eta)\mathrm{d}\eta\right](\mathrm{d}{\Delta_b}^2/\mathrm{d}\theta) + m\left[\int_0^1 F(\xi)G(\eta)\mathrm{d}\eta\right]{\Delta_b}^2 = \left[\int_0^1 G''(\eta)\mathrm{d}\eta\right]R \tag{3.192}$$

$$\left[-1/2\int_0^1 \eta G'(\eta)\mathrm{d}\eta\right]\left(\mathrm{d}\Delta_b{}^2/\mathrm{d}\theta\right)+m\left[\int_0^1 \eta F(\xi)G(\eta)\mathrm{d}\eta\right]\Delta_b{}^2$$
$$=\left[\int_0^1 \eta G''(\eta)\mathrm{d}\eta\right]R \tag{3.193}$$

Of the foregoing four equations, Eqs. (3.190)–(3.193), it is possible to satisfy only two. However, the coefficients A_n and B_n in Eqs. (3.180) and (3.181) may be selected so that both equations in a pair will give the same result. It remains to choose the coefficients before obtaining a numerical solution to Eqs. (3.190)–(3.193). In addition to the conditions equivalent to Eq. (3.168):

$$F(0)=1;\quad G(0)=1;\quad F(1)=0;\quad G(1)=0 \tag{3.194}$$

one may also set

$$F'(1)=0=G'(1) \tag{3.195}$$

so that the approximating functions F and G may fit smoothly into the constant (zero) values of a and b at the inner edges of the two concentration boundary layers. And if the series in Eqs. (3.180) and (3.181) are terminated at the *third* powers of their arguments, viz.,

$$F(\xi)=f(1-\xi)^2+(1-f)(1-\xi)^3 \tag{3.196}$$
$$G(\eta)=g(1-\eta)^2+(1-g)(1-\eta)^3 \tag{3.197}$$

where f and g are constants.

The values of the integral coefficients in the four equations (3.190)–(3.193), corresponding to Eqs. (3.196) and (3.197), are shown in Table 3.1.

3.9.3.4.3 Perturbation Solution for Small θ

By the method of perturbations, it is possible to obtain useful solutions of this problem, when the effect of the chemical reaction is small, but not wholly negligible.

Starting with equations:

$$\partial^2 a/\partial y^2=\partial a/\partial\theta+ab \tag{3.198}$$

$$R\partial^2 b/\partial x^2=\partial b/\partial\theta+mab \tag{3.199}$$

one may introduce two new independent variables, τ and z, defined by:

$$\tau=\theta \tag{3.200}$$

$$z=y/(2\sqrt{\theta}) \tag{3.201}$$

to obtain the corresponding pair of partial differential equations:

$$a_{zz}+2za_z=4\theta(a_\theta+ab) \tag{3.202}$$

$$Rb_{zz}+2zb_z=4\theta(b_\theta+mab) \tag{3.203}$$

in which the subscripts indicate partial differentials.

Table 3.1 Coefficients in moment equations.

$-(1/2)\int_0^1 \xi F'(\xi)\mathrm{d}\xi = 1/8 + (f/24)$	$\int_0^1 \xi F''(\xi)\mathrm{d}\xi = -F'(0) = 3 - f$
$-(1/2)\int_0^1 \xi^2 F'(\xi)\mathrm{d}\xi = 1/20 + (f/30)$	$\int_0^1 \xi F''(\xi)\mathrm{d}\xi = F(0) = 1$
$-(1/2)\int_0^1 \eta G'(\eta)\mathrm{d}\eta = 1/8 + (g/24)$	$\int_0^1 G''(\eta)\mathrm{d}\eta = -G'(0) = 3 - g$
$-(1/2)\int_0^1 \eta^2 G'(\eta)\mathrm{d}\eta = 1/20 + (g/30)$	$\int_0^1 \eta G''(\eta)\mathrm{d}\eta = G'(0) = 1$
$M_0 = \int_0^1 F(\xi)G(\eta)\mathrm{d}\xi = \int_0^1 F(\xi)G(\rho\xi)\mathrm{d}\xi$ $+fg(1/3 - \rho/6 + \rho^2/30)$ $+f(1-g)(1/3 - \rho/4 + \rho^2/10 - \rho^3/60)$ $+(1-f)(1-g)(1/4 - 3\rho/20 + \rho^2/20 - \rho^3/140)$	$M_1 = \int_0^1 \xi F(\xi)GB(\eta)\mathrm{d}\xi$ $= \int_0^1 \xi F(\xi)G(\rho\xi)\mathrm{d}\xi$ $+fg(1/12 - \rho/15 + \rho^2/60)$ $+f(1-g)(1/12 - \rho/10 + \rho^2/20 - \rho^3/105)$ $+(1-f)(1-g)(1/20 - \rho/30 + \rho^2/140)$ $+(1-f)(1-g)(1/20 - \rho/20$ $+3\rho^2/140 - \rho^3/280)$

Remarks:

(a) If the reaction did not take place, the solution of the pair of Eqs. (3.202) and (3.203), consistent with boundary equations in Eq. (3.204):

$$\begin{aligned} a(0,\theta) &= 1 = b(0,\theta) \\ a(\infty,\theta) &= 0 = b(\infty,\theta) \\ a(y,0) &= 0 = b(y,0) \end{aligned} \tag{3.204}$$

could be expressed in terms of z alone.

(b) The said reaction causes θ, the second variable, to affect the solution so that $\partial a/\partial\theta$ is not zero.

(c) In Eqs. (3.202) and (3.203), the magnitude of the terms expressing the effect of the reaction is directly proportional to θ and lead to the following perturbation series to be tried as solutions:

$$a(z,\theta) = a_0(z) + \theta a_1(z) + \theta^2 a_2(z) + \cdots \tag{3.205}$$

$$b(z,\theta) = b_0(z) + \theta b_1(z) + \theta^2 b_2(z) + \cdots \tag{3.206}$$

Substituting Eqs. (3.205) and (3.206) into Eqs. (3.202) and (3.203) and *equating the coefficient of each power of θ to zero*, one obtains the following infinite series of linear ordinary differential equations given by Eqs. (3.207) and (3.208):

$$a_0'' + 2za_0' = 0$$

$$a_0(0) = 1$$
$$a_0(\infty) = 0$$
$$a_1'' + 2za_1' - 4a_1 = 4a_0b_0$$
$$a_1(0) = a_1(\infty) = 0 \tag{3.207}$$

$$Rb_0'' + 2zb_0' = 0$$
$$b_0(0) = 1$$
$$b_0(\infty) = 0$$
$$Rb_1'' + 2zb_1'' - 4b_1'' = 4ma_0b_0$$
$$b_1(0) = b_1(\infty) = 0 \tag{3.208}$$

(d) The boundary conditions on the perturbation functions selected for satisfying the original condition in Eq. (3.204) for all dissolved (but unreacted) A and B are constant and equal to A_i and B_i, respectively,
(e) The two reactions with substance C used up ν_A and ν_B moles of C, respectively, per mole of A and B in the reaction.
(f) The diffusivities are assumed constant.
(g) Since the two reactions are assumed to be infinitely rapid, the co-existence of A and C, or of B and C, at any point is impossible: both reactions occur at a single moving plane: the "reaction zone," which is located at the interface when $\theta = 0$, and moves into the liquid as θ increases.
(h) The differential equations representing diffusion on the left of the reaction zone, $x < X$, and on right: $x > X$, are

$$x < X: \quad D_A(\partial^2 A/\partial x^2) = \partial A/\partial t \tag{3.209}$$

$$x < X: \quad D_B(\partial^2 B/\partial x^2) = \partial B/\partial t \tag{3.210}$$

$$x > X: \quad D_C(\partial^2 C/\partial x^2) = \partial C/\partial t \tag{3.211}$$

with the following boundary conditions:

$$A(0, t) = A_i \tag{3.212A}$$

$$B(0, t) = B_i \tag{3.212B}$$

$$A[X(t), t] = B[X(t), t] = C[X(t), t] = 0 \tag{3.212C}$$

$$C(x, 0) = C_0 = \text{constant} \tag{3.212D}$$

$$C(\infty, t) = C_0 \tag{3.212E}$$

3.9.3.4.4 Solving the Differential Equations for the Perturbation Functions for Small θ

Equations (3.209)–(3.211) may now be solved in turn, with the higher perturbations depending on the ones of lower perturbations. Thus, the zeroth-order perturbations are simply the well-known solutions of the simple diffusion problem:

$$a_0(z) = \mathrm{erfc}(z) \tag{3.213}$$

$$b_0(z) = \mathrm{erfc}(z/R^{1/2}) \tag{3.214}$$

Introducing these results on the right-hand side of the equation for $a_1(z)$, viz., Eq. (3.215):

$$a_1'' + 2za_1' - 4a_1 = 4a_0b_0 \tag{3.215}$$

one obtains

$$a_1'' + 2za_1' - 4a_1 = 4\,\mathrm{erfc}(z)\mathrm{erfc}(z/\sqrt{R}) \tag{3.216}$$

It is known that for the homogeneous form of the second-order equation (3.216), the method of variation of parameters may be applied, and by differentiating the resulting equation to extract the gradient of the concentration at the interface, one may obtain the following result:

$$-a_1'(0) = 4\int_0^{\infty} \left[(1 + 2s^2)\exp(s^2)\mathrm{erfc}(s) - \{2/(\sqrt{\pi})\}s\right] \mathrm{erfc}(s)\mathrm{erfc}(s/\sqrt{R})\,\mathrm{d}s \tag{3.217}$$

The definite integral given by Eq. (3.217) has been evaluated numerically, and the results are as follows.

And the rate of absorption of component A into the liquid may be evaluated from Eq. (3.218):

$$a(z,\theta) = a_0(z) + \theta a_1(z) + \theta^2 a_2(z) + \cdots \tag{3.218}$$

by differentiating with respect to z and setting $z = 0$. Upon integrating with respect to time t, leading to the equation for the ratio of mass transfer coefficients:

$$\underline{k}_L(\theta)/\underline{k}_L{}^{\circ}(\theta) = 1 + (\sqrt{\pi}/6)[-a_{1z}(0) \times \theta^{3/2} + \cdots] \tag{3.219}$$

where $a_{1z}(0)$ may be obtained from Table 3.2.

It appears that the results obtained with the method of moments which covered a greater range of θ than that given Eq. (3.219) is valid, confirming the validity of the previous computations. This result supported the adjustment of the θ-axis which was used to force the agreement between the boundary layer results and the known behavior of the solution at infinite θ.

Table 3.2 Evaluating the factor $-a_{1s}(0)$.

$R = D_B/D_A$	$-a_{1s}(0)$
0	0
0.2	0.5812
0.5	0.7235
1.0	0.8196
2.0	0.8995
5.0	0.9786

3.9.4 Simultaneous Absorption of Two Gases in a Liquid in Which Each Then Reacts with a *Third* Component in the Liquid

3.9.4.1 The Mathematical Model

This model analysis assumes that the chemical steps are ***infinitely*** rapid, and that the rate of the process may be determined by the rates of diffusion of the reactants A, B, and C to a moving reaction region. Additional assumptions for this model include the following:

(a) All the resistance to mass transfers are in the liquid state.
(b) All the interfacial concentrations of the unreacted, but dissolved, A and B are constant and are given by A_{i} and B_{i}, respectively.
(c) The two reactions of A and B, with the third substance C, use up ν_{A} and ν_{B} of C, respectively, per mole of A and B reacting.
(d) The diffusivities remain constant,
(e) The temperature is uniform,
(f) The liquid layer is either stationary or is flowing with a uniform speed parallel to the interface and should be infinitely rapid: there can be no co-existence of A and C, or of B and C, at any point, and both reactions occur at a single moving plane, the "Reaction Zone," located at the interface when $\theta = 0$ and moves into the liquid as θ increases.

Thus, the system differential equations representing diffusion on the left of the reaction zone (where $x < X$) and on the right (where $x > X$) are

$$x < X: \quad D_{\mathrm{A}}(\partial^2 A/\partial x^2) = \partial A/\partial t \tag{3.220}$$

$$x < X: \quad D_{\mathrm{B}}(\partial^2 B/\partial x^2) = \partial B/\partial t \tag{3.221}$$

$$x > X: \quad D_{\mathrm{C}}(\partial^2 C/\partial x^2) = \partial C/\partial t \tag{3.222}$$

with the following boundary conditions:

$$A(0, t) = A_{\mathrm{i}} \quad B(0, t) = B_{\mathrm{i}} \tag{3.223}$$

$$A(x, 0) = 0 = B(x, 0) \tag{3.224}$$

$$A[X(t), t] = B[X(t), t] = C[X(t), t] = 0 \tag{3.225}$$

$$C(x, 0) = C_0 = \text{constant} \tag{3.226}$$

$$C(\infty, t) = C_0 \tag{3.227}$$

3.9.4.2 Analysis of the Model

One approach to solve this problem is to ascertain the position of the reaction zone as a function of time, viz., the function $X(t)$. The velocity of this reaction zone is determined by the velocity of diffusion of reactants to this zone from the left and from the right, and hence it can be related to concentration gradients as described by the following set of equations:

$$(\partial A/\partial t)\mathrm{d}t = (\partial A/\partial x)(-\mathrm{d}X) \tag{3.228}$$

$$(\partial B/\partial t)\mathrm{d}t = (\partial B/\partial x)(-\mathrm{d}X) \tag{3.229}$$

$$(\partial C/\partial t)(-\mathrm{d}t) = (\partial C/\partial x)(\mathrm{d}X) \tag{3.230}$$

and it follows that:

at $x = X$

$$-(\mathrm{d}X/\mathrm{d}t) = (\partial A/\partial t)/(\partial A/\partial x) = (\partial B/\partial t)/(\partial B/\partial x) = (\partial C/\partial t)/(\partial C/\partial x) \tag{3.231}$$

Additionally, another system requirement is that, at the reaction zone, the following stoichiometric condition should be satisfied:

at $x = X$

$$\nu_\mathrm{A} D_\mathrm{A}(\partial A/\partial x) + \nu_\mathrm{B} D_\mathrm{B}(\partial B/\partial x) + D_\mathrm{C}(\partial C/\partial x) \tag{3.232}$$

As an initial trial function for A, B, and C to fit the boundary conditions (to test if it will fit the boundary conditions), it has been suggested that the Error Function should be tried to obtain satisfactory particular solutions for these partial differential equations, viz., let

$$A(x,t) = c_1 + c_2\,\mathrm{erf}\{(x/2)\sqrt{(D_\mathrm{A} t)}\}; \quad x < X \tag{3.233}$$

$$B(x,t) = c_2 + c_4\,\mathrm{erf}\{(x/2)\sqrt{(D_\mathrm{B} t)}\}; \quad x < X \tag{3.234}$$

$$C(x,t) = c_5 + c_6\,\mathrm{erf}\{(x/2)\sqrt{(D_\mathrm{C} t)}\}; \quad x > X \tag{3.235}$$

Using Eq. (3.231), it may be shown that the three solutions are consistent with the movement of a single reaction boundary if

$$\mathrm{d}X/\mathrm{d}t = X/2t \tag{3.236}$$

viz.,

$$X = 2(\alpha t)^{1/2} \tag{3.237}$$

Checking: If $X = 2(\alpha t)^{1/2}$, then

$$\begin{aligned}
\mathrm{d}X/\mathrm{d}t = (\mathrm{d}/\mathrm{d}t)(X) = (\mathrm{d}/\mathrm{d}t)\{2(\alpha t)^{1/2}\} &= 2(\mathrm{d}/\mathrm{d}t)\{(\alpha t)^{1/2}\}\\
&= 2\alpha^{1/2}\{(\mathrm{d}/\mathrm{d}t)(t)^{1/2}\}\\
&= 2\alpha^{1/2}\{1/2(t)^{-1/2}\}\\
&= \alpha^{1/2}(t)^{-1/2}\\
&= \alpha^{1/2}/t^{1/2}\\
&= (X/2t^{1/2})/t^{1/2}\\
&= X/2t
\end{aligned}$$

viz.,

$$\mathrm{d}X/\mathrm{d}t = X/2t$$

which is Eq. (3.236).

Finally, one may determine the constant K, in Eq. (3.124) from the other boundary conditions.

3.9.4.3 Boundary Conditions

The constant a, in Eq. (3.237), may now be determined from the other boundary conditions. The results are as follows:

$$c_1 = A_i \tag{3.238A}$$

$$c_3 = B_i \tag{3.238B}$$

$$c_5 + c_6 = C_0 \tag{3.238C}$$

$$c_1 + c_2 \,\mathrm{erf}(\alpha/D_A)^{1/2} = 0 \tag{3.238D}$$

$$c_3 + c_4 \,\mathrm{erf}(\alpha/D_B)^{1/2} = 0 \tag{3.238E}$$

$$c_5 + c_6 \,\mathrm{erf}(\alpha/D_C)^{1/2} = 0 \tag{3.238F}$$

$$c_2 \nu_A D_A \exp(-\alpha/D_A) + c_4 \nu_B D_B \exp(-\alpha/D_B) + c_6 \nu_A D_C \exp(-\alpha/D_C) = 0 \tag{3.238G}$$

Thus, $\alpha = \beta D_A$ *may be given by:*

$$1/\left\{1 - \mathrm{erf}(\beta R_A)^{1/2}\right\} = [\sqrt{(R_A)} T_A \exp\{\beta(R_A - 1)\}]/\{\mathrm{erf}(\beta)^{1/2}\} + [\sqrt{(R_B)} T_B \exp\{\beta(R_A - (R_A/R_B)\}]/\left[\mathrm{erf}(\beta R_A/R_B)^{1/2}\right] \tag{3.239}$$

in which the dimensionless parameters are

$$R_A = D_A/D_C, \quad R_B = D_B/D_C, \quad T_A = \nu_A A_i/C_0, \quad T_B = \nu_B B_i/C_0 \tag{3.240}$$

3.9.4.4 Mass Transfer Coefficients

The mass transfer coefficients for gases A and B may be obtained from the solutions obtained here which are derived by calculating the concentration gradients at the interface. These coefficients vary with time, in the same way as if the reaction did not occur. Thus, one way to represent the results is by the ratio of the coefficients with and without reaction. Thus, for species A, one has

$$k_{LA}/k*_{LA} = \underline{k}_{LA}/\underline{k}*_{LA} = 1/\mathrm{erf}(\beta)^{1/2} \tag{3.241}$$

$$k_{LB}/k*_{LB} = \underline{k}_{LB}/\underline{k}*_{LB} = 1/\mathrm{erf}(\beta R_A/R_B)^{1/2} \tag{3.242}$$

Simplifications of these results may be obtained by writing series approximations for the error functions and the exponential functions:

Let $w = \beta^{1/2}$.

For a large excess of the reagent C, such as when T_A and T_B are small, the reaction zone moves slowly, and w will be small. Then the usual convergent series gives

$$\mathrm{erf}(w) \approx (2/\sqrt{\pi})w \tag{3.243}$$

and

$$\mathrm{erf}(w^2) \approx 1 + w^2 \tag{3.244}$$

etc. and Eq. (3.239) becomes, approximately

$$2w/\sqrt{\pi} \approx 1/\{1 + (R_A T_A + R_B T_B)^{-1} \tag{3.245}$$

Upon making the same approximations in Eqs. (3.241) and (3.242), the ratios of mass transfer coefficients then become

$$k_{LA}/k^{o}{}_{LA} = \underline{k}_{LA}/\underline{k}^{o}{}_{LA} = \sqrt{R_A} + \{\sqrt{R_A}/(R_A T_A + R_B T_B)\} \quad (3.246)$$

$$k_{LB}/k^{o}{}_{LB} = \underline{k}_{LB}/\underline{k}^{o}{}_{LB} = \sqrt{R_B} + \{\sqrt{R_B}/(R_A T_A + R_B T_B)\} \quad (3.247)$$

Remarks:

(a) The approximations are *not* valid when $D_c = 0$, and similar approximations result in:

$$k_{LA}/k^{o}{}_{LA} = \underline{k}_{LA}/\underline{k}^{o}{}_{LA} = \sqrt{(\pi/[2D_A\{(T_A/D_A) + (T_B/D_B) + \cdots\}])} \quad (3.248)$$

for T_A and T_B sufficiently small.

(b) The mathematical approximations are exact if $R_A = 1 = R_B$.

(c) A generalization of these results for any number, n, of gaseous reactants competing for a single reagent in the liquid phase may be expressed by extending the summation over all the reactant quantities:

$$k_{LA}/k^{o}{}_{LA} = \underline{k}_{LA}/\underline{k}^{o}{}_{LA} = \sqrt{R_A} + \sqrt{R_A}\Big/\left\{\sum(R_i T_i)\right\} \quad (3.249)$$

3.9.5 Cases of Slow First-Order Reactions

Sherwood and Pigford [25] discussed the theory of slow first-order unsteady-state absorption of gases in a stagnant liquid – for a slow first-order reaction.

3.9.5.1 Case of Unsteady-State Absorption in the Stagnant Liquid

When the reaction is not very rapid (compared with the rate of diffusion), the reaction zone may be considered to be concentrated in a thin region and will be spread throughout the liquid film. Here the solute diffuses through the gas film under the driving force $(p - p_i)$, where

p = vapor pressure in the gas phase, and
p_i = vapor pressure at the interphase

Upon entering the liquid phase, it begins to diffuse towards the main body of the liquid ***and*** begins to react with the dissolved substance A. As the diffusing current of solute is being used up as it diffuse into the liquid, the concentration will diminish accordingly. Over the time period of initial reaction of the liquid to the gas, the solute may accumulate in the liquid. Nevertheless, the total rate of diffusion of reacted and unreacted forms of A remains constant, since there is no net accumulation of A in the film.

The conditions of the process may be described mathematically if the rate equation for the reaction is known. Thus, for a first-order reaction, the rate of reaction, and hence elimination, of A is proportional to the concentration of A at any point, viz.,

$$-dA/dt = kcV \quad (3.250)$$

where

k is the specific reaction rate constant,
c is the variable concentration of A, and
V is the liquid volume considered.

At $x = x$, for a differential element, of *unit* area, of thickness dx of the liquid film, and of the liquid film of thickness x_L, the rate of diffusion ***into*** this element is

$$N_A = -D_A(dc/dx) \tag{3.251}$$

Also, the rate of diffusion out of this element, at $x = x + dx$, will be

$$N_A = -D_A\{dc/dx + (d^2c/dx^2)dx\} \tag{3.252}$$

The consummation of A within this element, owing to the reaction with B, is

$$dA/d\theta = -kc\, dx \tag{3.253A}$$

since the volume of the element is $1 \cdot dx = dx$. The difference in the diffusion rates in and out of the same element is equal to the rate of elimination of A by the reaction with B, so that

$$D_A\, d^2c/dx^2 = k_e c \tag{3.253B}$$

Now, the rate of elimination of the species A (by reaction with B) must be equal to the difference in the diffusion rates (in and out of the element) of consummation of A, by reaction with B, viz.,

$$k = D_A\, d^2c/dx^2 \tag{3.254}$$

For a fixed value of k_c throughout the film. The solution of Eq. (3.254) takes the form:

$$c = K_1 e^{a_0 x} + K_2 e^{-a_0 x} \tag{3.255}$$

in which $a_0 = \sqrt{(k/D_A)}$.

Upon substituting the limits: $c = c_i$ at $x = 0$, and $c = c_L$ at $x = x_L$, the constants K_1 and K_2 may be obtained, and the solution takes the form:

$$c = [c_L \sinh(a_0 x) + c_i \sinh\{a_0(x_L - x)\}]/\sinh(a_0 x_L) \tag{3.256}$$

Remarks:

(a) For $c = c_L = 0$, one has

$$c = [\sinh\{a_0(x_L - x)\}/\sinh(a_0 x_L)]c_i \tag{3.257}$$

(b) By differentiating Eq. (3.256), one obtains the slope of the concentration curve:

$$dc/dx = [a_0 c_L\ \cosh(a_0 x) + a_0 c_i\ \cosh\{a_0(x_L - x)\}]/\sinh(a_0 x_L) \tag{3.258}$$

(c) By multiplying the diffusivity of A, D_A, by the slope at $x = 0$, one obtains the rate of diffusion of A into the liquid:

$$N_A = -D_A(dc/dx)_{x=0} = [D_A a_0\{c_i \cosh(a_0 x_L) - c_L\}]/\sinh(a_0 x_L) \tag{3.259}$$

(d) By substituting $x = x_L$ into Eq. (3.258), one obtains the rate of diffusion of A into the main body of the liquid:

$$N_A' = -D_A(\mathrm{d}c/\mathrm{d}x)_{x=x_L} = [D_A a_0\{c_i - c_L \cosh(a_0 x_L)\}]/\sinh(a_0 x_L) \tag{3.260}$$

(e) The fraction F of the solute A, entering the liquid phase *and* reaching the main body of the liquid without reacting, from Eqs. (3.259) and (3.260), is

$$F = N_A'/N_A = \{c_i - c_L \cosh(a_0 x_L)\}/\{c_i \cosh(a_0 x_L) - c_L\}$$

or,

$$F = \{c_L \cosh(a_0 x_L) - c_i\}/\{c_L - c_i \cosh(a_0 x_L)\} \tag{3.261}$$

(f) If, in the main body of the liquid, the concentration of A is low, viz., c_L is small, *then* N_A, in Eq. (3.259), may be simplified to

$$N_A = bD_A(c_i - c_L)/x_L \tag{3.262}$$

where

$$b = a_0 x_L/\tanh(a_0 x_L) \tag{3.263}$$

(g) In interphasal mass transfer analysis, the rate of transfer is proportional to the difference in the bulk composition of the phases, expressed in the same units, provided that there exists a linear relationship between the compositions of the phases at equilibrium. For example:

$$\begin{aligned} N_A a\,\mathrm{d}V &= K_G a(p - p_e)\mathrm{d}V \\ &= k_G a(p - p_i)\mathrm{d}V \\ &= k_a aP(y - y_i)\mathrm{d}V \end{aligned} \tag{3.264A}$$

or:

$$\begin{aligned} N_A a\,\mathrm{d}V &= K_L a(c_e - c)\mathrm{d}V \\ &= k_L a(c_i - c)\mathrm{d}V \\ &= k_L a\rho_M(x_i - x)\mathrm{d}V \end{aligned} \tag{3.264B}$$

where p_e is the equilibrium vapor pressure over a solution having a concentration c_i, and where p_i and p are similarly related to c_i and c_e, and x and y are mole fractions of solute in liquid and gas, respectively, and ρ_M is the molal density of the liquid, viz., ρ_G divided by the average molecular weight of the liquid stream.

(**Note**: The equalities involving c_i and p_i hold *only if* a plot of p_i vs. c_i is *linear* over the concentration range under consideration. Under such conditions:

$$1/K_G a = 1/k_G a + H/k_L a \tag{3.265A}$$

$$1/K_L a = 1/k_L a + 1/Hk_G a \tag{3.265B}$$

where the constant $H = (p_i - p_e)/(c_i - c_e)$, in which Henry's law applies: viz., H is the Henry's law constant, defined by $p_i = Hc_i$.)

Equation (3.262) is similar to Eqs. (3.265A) and (3.265B), with k_L replaced by bD_A/x_L.

For overall resistance, Eq. (3.265B) becomes

$$1/K_L = x_L/bD_A + 1/Hk_G \tag{3.266}$$

so that

$$N_A = (c_e - c_L)/\{(x_L/bD_A) + (1/Hk_G)\tag{3.267}$$

which is the same as the equation for absorption through combined liquid and gas film resistance, with b introduced into the term representing liquid film resistance.

3.9.5.2 Case of Unsteady-State Absorption in the Stagnant Liquid – Slow First-Order Reactions

Within the period of initial exposure of the liquid to the gas, during which the solute may accumulate in the liquid, an additional event may be considered by including as additional term in Eq. (3.268):

$$D_A \, d^2c/dx^2 = k_e c \tag{3.268}$$

For a body of liquid adjacent to the interface, with a vertical side of unit cross-sectional area parallel to the interface and a thickness dx. The diffusion rate in and out of this interfacial element are given by, as before:

$$N_A = -D_A(dc/dx) \tag{3.269}$$

and

$$N_A = -D_A\{dc/dx + (d^2c/dx^2)dx\} \tag{3.270}$$

And the net rate of consumption per unit volume, owing to reaction, is given by:

$$k_e c_A - k_c{}' c_B \tag{3.271}$$

Now, at equilibrium, the opposing rates are equal, viz., $k_e{}' = k_e/K$, where K is the chemical equilibrium constant. The rate of depletion of A from the interfacial prism is $-(\partial c_A/\partial \theta)dx$ in the unsteady state. Equating the rate of input of A by diffusion to the sum of the rates of output by diffusion, elimination of reaction, and depletion, one obtains

$$D_A \partial^2 c_A/\partial x^2 = k_c c_A - (k_e/K)c_B + \partial c_A/\partial \theta \tag{3.272}$$

For component B, a similar mass balance yields

$$D_B \partial^2 c_B/\partial x^2 = (k_c/K)c_B - k_e c_A + \partial c_B/\partial \theta \tag{3.273}$$

Eliminating the concentration of B and assuming the diffusivities to be equal, one obtains the following fourth-order linear partial differential equation:

$$D\partial^4 c_A/\partial x^4 - 2\partial^3 c_A/\partial \theta\, \partial x^2 - k_c\{1 + (1/K)\}\, \partial^2 c_A/\partial x^2 (1/D)\partial^2 c_A/\partial \theta^2 + (k_c/D)\{1 + (1/K)\}(\partial c_A/\partial \theta) = 0 \tag{3.274}$$

In this model, it is assumed that:

(i) The product of the reaction is assumed to be nonvolatile.
(ii) The interfacial concentration of A is assumed to be equal to the equilibrium value with the gas where the partial pressure of A is constant.
(iii) The initial concentrations of A and B in the liquid before exposure are assumed to be uniformly equal to their equilibrium values.
(iv) The body of liquid is considered as infinite.

Hence, the following boundary conditions are appropriate:

(1) $c_A = c_i = p_A/H$ at $x = 0$
(2) $\partial c_B/\partial x = 0$ at $x = 0$
(3) $\underline{c_B} = Kc_A = c_0\{K/(1+K)\}$, a constant, at $\theta = 0$
(4) $\underline{c_B} = Kc_B = c_0\{K/(1+K)\}$, a constant, at $x = \infty$

where c_0 is the sum of the initial values of c_A and c_B, which are assumed to $b\sqrt{e}$ in equilibrium with one another.

Using the Laplace transform method, the solution of Eq. (3.274), subject to the foregoing boundary conditions, is

$$[\{(K+1)c_{av} - c_0\}/\{(K+1)c_i - c_0\}] = (3/\sqrt{\pi})\sqrt{(D_L\theta/B_{F^2})\,[f(K,k_c\theta)/(K+1)]} \tag{3.275}$$

where

$$\begin{aligned} f(K,k_c\theta) = {} & K + 1 + [(\pi K^5)/\{4(K-1)(k_c\theta)\}]^{1/2}\exp[(k_c\theta/\{K(K-1)\}] \\ & \times [\text{erf}\{Kk_c\theta)/(K-1)\}^{1/2} - \text{erf}\{(k_c\theta/K(K-1)\}] \\ & - [\pi K^3(K+1)/4k_e\theta]^{1/2}\text{erf}[\{(K+1)k_e\theta\}/K]^{1/2} \end{aligned} \tag{3.276}$$

and where $(K+1)c_{av}$ represents the ***average*** total concentration of solute in the liquid in both the reacted and unreacted forms, *cnv* represents the **average** concentration of A after absorption if A and B were in chemical equilibrium at every point, and $(K+1)c_i$ represents the ***total*** concentration of reacted and unreacted solute in the liquid if it were permitted to become physically and chemically in equilibrium with the gas. Using the following functional notations:

$$\exp(q) = e^q \tag{3.277A}$$

and

$$\text{erf}(q) = (2/\sqrt{\pi})\int^{q} \exp(-q^2)\text{d}p \tag{3.277B}$$

Remarks:

(a) Equation (3.277B) is the Error Function in mathematics.
(b) For **large** values of $k_e\theta$, the *approximate equivalent* of Eq. (3.275) is

$$\begin{aligned} f(K,k_e\theta) = (1+K)\Big[& 1 - (\pi^{1/2}/2)K\{K/(K+1)\}^{1/2}(\underline{k_e\theta})^{-1/2} \\ & + K^3/\{2(K+1)\}(k_e\theta)^{-1}\Big] \end{aligned} \tag{3.278}$$

(c) For **small** values of $k_e\theta$, the *approximate equivalent* of Eq. (3.275) is

$$f(K, k_c\theta) = 1 + [\{K^{1/2}(K+1)^{3/2}(K-1)^2 - K^4 + 3K^2 - 2K\}/\{3(K-1)^2\}]k_c\theta \tag{3.279}$$

(d) For an irreversible reaction, viz., $K = \infty$,

$$f(\infty, k_c\theta) = 1/2\ \exp(-k_e\theta) + \sqrt{(\pi/4)}\left[(1 + 2k_e\theta)/(k_e\theta)^{1/2}\right] \mathrm{erf}(k_c\theta)^{1/2} \tag{3.280}$$

(e) When these equations are applied to the absorption from a gas in a falling liquid film, the exposure time is that of the liquid surface, which may be estimated from the surface velocity:

$$\theta = Z/u_s = (2/3)(Z/u_{av}) \tag{3.281}$$

where Z is the height of the liquid surface.

(f) From a material balance on the liquid layer of thickness T_F, the liquid film absorption coefficient k_L may be obtained as follows:

$$u_{av}B_F\{(K+1)c_{av} - c_0\} = k_L Z\{(K+1)c_i - c_0\} \tag{3.282}$$

and

$$k_L Z/u_{av}B_F = \{(K+1)c_{av} - c_0\}/\{(K+1)c_i - c_0\} \tag{3.283A}$$

$$= (3/\sqrt{\pi})\left(D_L\theta/B_F{}^2\right)^{1/2}\{f(K, k_c\theta/(K+1))\} \tag{3.283B}$$

As before, in these equations, c_0 represents the concentration of the total dissolved solute (both A and B forms) before the exposure of the liquid surface, and $(K+1)c_i$ represents the concentration of total dissolved solute, both A and B forms, in a liquid in phase equilibrium and reaction equilibrium with the gas.

(g) For pure physical absorption, one has, from Sherwood and Pigford, p. 266, Equation (327):

$$k_L = \sqrt{(6/\pi)}\sqrt{(D_L u_{av}/Z)} = \sqrt{(6/\pi)}\sqrt{(D_L \Gamma/\rho_L B_F Z)} \tag{3.284}$$

developed for pure physical absorption, or

$$k_L Z/u_{av}B_F = (3/\sqrt{\pi})\sqrt{\left(D_L\theta/B_F{}^2\right)} \tag{3.285}$$

which shows that $f(K,\ k_c\theta)/(K+1)$ is a multiplying factor that provides the effect of the chemical reaction on the coefficient for no reaction. Inspection of Eq. (3.279) indicates that $f(K,\ k_c\theta)$ equals 1 for the condition of a very unfavorable equilibrium: $K = 0$. Under such a condition, the correction factor for the absorption coefficient equals unity.

3.10 Further Theoretical Analysis

These several theorems have been established for a two-dimensional system. The same conclusions may likely be extended to a three-dimensional system, as well as to systems of higher dimensions!

Further Reading

Backhurst, J.R., Harker, J.H., Coulson, J.M., and Richardson, F. (1994). *Chemical Engineering: Volume 4: Solutions to the Problems in Volume 1*. London, UK: Butterworth Heinemann.

Crocker, B.B., SM, PE, Consulting Chemical Engineer; Fellow, American Institute of Chemical Engineers; Member, Air Pollution Control Association. (Phase Separation).

Coulson, J.M. and Richardson, J.F. (1996). *Chemical Engineering: Volume 1: Fluid Flow, Heat Transfer and Mass Transfer*, 5e. London, UK: Butterworth Heinemann.

Danckwerts, P.V. and Kennedy, B.E. (1954). Kinetics of liquid-film process in gas absorption. Part I: Models of the absorption process. *Trans. Inst. Chem. Eng.* 32: S49–S52.

Fair, J.R., PhD, PE, Professor of Chemical Engineering, University of Texas; Fellow, American Institute of Chemical Engineers; Member, American Chemical Society, American Society for Engineering Education, National Society of Professional Engineers. (Section Editor, Absorption, Gas-Liquid Contacting).

Foust, A.S., Wenzel, L.A., Clump, C.W., Maus, L., and Andersen, L.B. (1980). *Principles of Unit Operations*, 2e. Melbourne, FL: Krieger Publishing Company.

Franks, R.G.E. (1967). *Mathematical Modeling in Chemical Engineering*. New York: Wiley. pp. 4–6.

Green, D.W. and Perry, R.H. (eds.) (2008). *Perry's Chemical Engineers' Handbook*. 1/e to 8/e. Section 14: Absorption and Gas–Liquid System Design. New York: McGraw-Hill Professional. 2008 1997 1984 1973 1963 1950 1941 1934 ftp://ftp.feq.ufu.br/Claudio/Perry/DOCS/Chap14.pdf.

Green, D.W. and Southard, M.Z. (eds.) (17 August 2018). *Perry's Chemical Engineers' Handbook*. 9/e. New York: McGraw-Hill Professional.

Higbie, R. (1935). The rate of absorption of a pure gas into a still liquid during short period of exposure. *Trans. Am. Inst. Chem. Eng.* 31: 365–388.

Kutepov, A.M., Bondareva, T.I., and Berengarten, M.G. (1988). *Basic Chemical Engineering with Practical Applications*. Moscow, Russia: Mir Publishers. pp. 95–110.

Levinspiel, O. (1972). *Chemical Reaction Engineering*, 2e. New York: Wiley. pp. 210–213, 320–326.

Luyben, W.L. (1989). *Process Modeling, Simulation, and Control for Chemical Engineers*, 2e. Columbus, OH: McGraw-Hill Education. pp. 22–24.

Lynn, S., Straatemeier, J.R., and Kramers, H. (1955). Absorption studies in the light of the penetration theory I. Long wetted-wall columns. *Chem. Eng. Sci.* 4: 49–57.

Majumder, S.K. and Das, C. (2012). National Programme on Technology Enhanced Learning (NPTEL) – Mass Transfer Operation-I. Dept. of Chem. Eng., IIT-Guwahati, India. http://nptel.ac.in/courses/103103035/1.

Majumder, S.K., and Das, C., Dept. of Chemical Engineering, Indian Institute of Technology, Guwahati, Guwahati – 781039, Assam, India. National Programme on Technology Enhanced Learning (NPTEL) - Phase II, Course Name: Mass Transfer Operation- I. http://nptel.ac.in/courses/103103035/1.

Olutoye, M.A. and Mohammed, A. (2006). Modelling of a gas-absorption packed column for carbon dioxide–sodium hydroxide system. *AU J.T.* 10 (2): 132–140. https://www.researchgate.net/publication/254071709.

Olutoye, M. and Alhassan, M. (2006). Modelling of a gas-absorption packed column (PDF Download Available). https://www.researchgate.net/publication/254071709_Modelling_of_a_Gas-Absorption_Packed_Column_for_Carbon_Dioxide-Sodium_Hydroxide_System (accessed 23 April 2018).

Penetration Theory:

1 Danckwerts, P.V. (1950). *Trans. Faraday Soc.* 46: 300.
2 Danckwerts, P.V. (1951). *Trans. Faraday Soc.* 47: 300.
3 Danckwerts, P.V. (1970). *Gas–Liquid Reactions*. New York: McGraw-Hill.
4 Higbie, R. (1935). *J. Am. Inst. Chem. Eng.* 31: 365.

Penney, W.R., PhD, PE, Professor of Chemical Engineering, University of Arkansas; Member, American Institute of Chemical Engineers. (Gas-in-Liquid Dispersions).

Section 14 Gas Absorption and Gas-Liquid System Design: ftp://ftp.feq.ufu.br/AQ36 Claudio/Perry/DOCS/Chap14.pdf.

Steinmeyer, D.E., MA, MS, PE, Distinguished Fellow, Monsanto Company; Fellow, American Institute of Chemical Engineers; Member, American Chemical Society. (Liquid-in-Gas Dispersions).

Surface Renewal Theory:

1 Danckwerts, P.V. (1951). *Ind. Eng. Chem.* 43: 1460.

Two-film theory:

1 Whitman, W.G. (1923). *Chem. Met. Eng.* 29: 147.
2 Lewis, W.K. and Whitman, W.G. (1924). *Ind. Eng. Chem.* 16: 215.

Whitman, W.G. (1923). Preliminary experimental confirmation of the two-film theory of gas absorption. *Chem. Met. Eng.* 29: 146–148.

4

Numerical Worked Examples Using R for Simultaneous Mass Transfer and Chemical Reactions

4.1 Advection and Convection

Several numerical examples are chosen to illustrate and highlight the application of the R Program for numerical computations in Simultaneous Mass Transfer and Chemical Reactions in typical transport phenomena in Chemical Engineering Science.

Example A: Simultaneous Mass Transfer and Chemical Reactions Using R
* **1D, 2D, and 3D Rectangular Geometries**
* **Cylindrical Geometry**
* **Polar Coordinates**
* **1D, 2D, and 3D Volumetric Advective–Diffusive Transport in an Aquatic System**

Example B: Diffusive Models of Environmental Transport Using R
* **Equations for Diffusion in Environmental Transport**

Example C: Numerical Worked Examples of Advection Using R for:
* **1D, 2D, and 3D Rectangular Geometries**
* **Cylindrical Geometry**
* **Polar Coordinates**

4.1.1 Advection

In the field of science and engineering, ***advection*** is the transport of a substance by *bulk motion*. The properties of that substance are carried with it. Generally, the majority of the advected substance is a fluid (gas or liquid). The properties that are carried with the advected substance are conserved properties such as energy. An example of advection is the transport of pollutants or silt in a river by bulk water

Simultaneous Mass Transfer and Chemical Reactions in Engineering Science, First Edition. Bertram K. C. Chan.

flowing downstream. Another commonly advected quantity is energy or enthalpy. Here, the fluid may be any material that contains thermal energy, such as air or water. In general, any substance or conserved, extensive quantity may be advected by a fluid that can hold or contain the quantity or substance.

During advection, a fluid transports some conserved quantity or material via bulk motion. The fluid motion may be described mathematically as a vector field, and the transported material is described by a scalar field showing its distribution over space. Advection requires currents in the fluid, and so it cannot happen in rigid solids. It does ***not*** include transport of substances by molecular diffusion. Advection is not to be equated with the process of ***convection*** which is the ***combination*** of advective transport and diffusive transport.

In meteorology and physical oceanography, advection often refers to the transport of some property of the atmosphere or ocean, such as heat, humidity, or salinity. (Advection is important for the formation of orographic clouds and the precipitation of water from clouds, as part of the hydrological cycle.)

4.1.2 Advection vs. Convection

The term *advection* is sometimes considered as a synonym for *convection*, but technically, *convection* covers the sum of transport both by diffusion and by advection. Advective transport describes the movement of some quantity via the bulk flow of a fluid (as in a river or pipeline).

4.1.2.1 Meteorology

In meteorology and physical oceanography, advection often refers to the horizontal transport of some property of the atmosphere or ocean, such as heat, humidity, or salinity, and convection generally refers to vertical transport (vertical advection). Advection is important for the formation of orographic clouds (terrain-forced convection) and the precipitation of water from clouds, as part of the hydrological cycle. Advection also applies if the quantity being advected is represented by a probability density function at each point.

4.1.2.2 The Mathematics of Advection

The **advection equation** is the partial differential equation that governs the motion of a conserved scalar field as it is advected by a known velocity vector field. It is derived from the scalar field's conservation law, together with Gauss's theorem and taking the infinitesimal limit.

A simple visualized example of advection is the transport of ink dumped into a river. As the river flows, ink will move downstream in a "pulse" via advection, as the movement of bulk water itself transports the ink. Note that as it moves downstream, the "pulse" of ink will also spread via diffusion. The *sum* of these processes is called **convection**.

Note: If the ink is added to a lake without significant bulk water flow, the ink would simply disperse outwards from its source via diffusion alone, which is ***not*** advection.

4.1.2.3 The Advection Equation

In Cartesian coordinates the advection *operator* is

$$\mathbf{u} \cdot \nabla = u_x \frac{\partial}{\partial_x} + u_y \frac{\partial}{\partial_y} + u_z \frac{\partial}{\partial_z} \tag{4.1}$$

where $\mathbf{u} = (u_x, u_y, u_z)$ is the *velocity field*, and $\nabla =$ is the del operator.

Note that Cartesian coordinates are used here.

The advection equation for a conserved quantity described by a scalar field ψ is expressed mathematically by a *continuity equation*:

$$\frac{\partial \psi}{\partial t} + \nabla \cdot (\psi \mathbf{u}) = 0 \tag{4.2}$$

where $\nabla \cdot$ is the divergence operator, and $\mathbf{u}$ is the velocity vector field.

Frequently, it is assumed that the flow is incompressible, that is, the velocity field satisfies the equation

$$\nabla \cdot \mathbf{u} = 0 \tag{4.3}$$

and $\mathbf{u}$ is said to be ***solenoidal***. Under such conditions, the above equation may be written as

$$\frac{\partial \psi}{\partial t} + \mathbf{u} \cdot \nabla \psi = 0 \tag{4.4}$$

In particular, if the flow is steady, then $\partial/\partial t = 0$, and

$$\mathbf{u} \cdot \nabla \psi = 0 \tag{4.5}$$

which shows that ψ is constant along a *streamline*. Hence,

$$\partial \psi / \partial t = 0 \tag{4.6}$$

so ψ does not vary in time.

If a vector quantity $\mathbf{a}$ (such as a magnetic field) is being advected by the solenoidal velocity field $\mathbf{u}$, the advection equation above becomes

$$\frac{\partial a}{\partial t} + (\mathbf{u} \cdot \nabla)\,\mathbf{a} = 0 \tag{4.7}$$

4.1.2.3.1 Solving the Advection Equation (4.7)

The advection equation is not simple to solve numerically: the system is a hyperbolic partial differential equation, and interest typically centers on discontinuous "shock" solutions, which are difficult for numerical schemes to solve. Even with one space dimension and a constant velocity field, the system remains difficult to simulate. The equation becomes

$$\frac{\partial \psi}{\partial t} + u_x \frac{\partial \psi}{\partial x} = 0 \tag{4.8}$$

where $\psi = \psi(x, t)$ is the scalar field being advected, and u_x is the x component of the vector $\mathbf{u} = (u_x, 0, 0)$ (Figure 4.1).

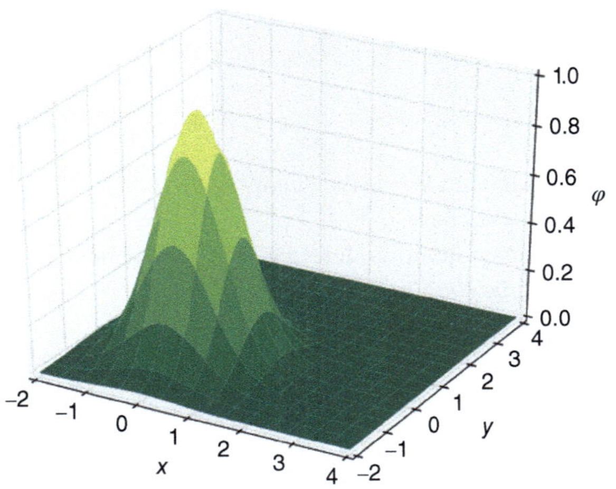

Figure 4.1 A simulation of the advection equation where $\mathbf{u} = (\sin t, \cos t)$ is solenoidal. Source: Shiyu Ji/Wikimedia Commons/CC BY-SA 4.0.

4.1.2.4 The Advection Operator in the Incompressible Navier–Stokes Equations

Numerical simulation can be aided by considering the *skew symmetric* form for the advection operator:

$$\frac{1}{2}\mathbf{u}\cdot\nabla\mathbf{u} + \frac{1}{2}\nabla(\mathbf{uu}) \tag{4.9}$$

where

$$\nabla(\mathbf{uu}) = \left[\nabla\left(\mathbf{u}u_x\right), \nabla\left(\mathbf{u}u_y\right), \nabla\left(\mathbf{u}u_z\right)\right] \tag{4.10}$$

and $\mathbf{u}$ is the same as above.

Since skew symmetry implies only imaginary eigenvalues, this form reduces the "blow up" and "spectral blocking" often experienced in numerical solutions with sharp discontinuities.

Using vector calculus identities, these operators can also be expressed in other ways, available in more software packages for more coordinate systems.

$$\mathbf{u}\cdot\nabla\mathbf{u} = \nabla\left(\frac{\|\mathbf{u}\|^2}{2}\right) + (\nabla\times\mathbf{u})\times\mathbf{u} \tag{4.11}$$

$$\frac{1}{2}\mathbf{u}\cdot\nabla\mathbf{u} + \frac{1}{2}\nabla(\mathbf{uu}) = \nabla\left(\frac{\|\mathbf{u}\|^2}{2}\right) + (\nabla\times\mathbf{u})\times\mathbf{u} + \frac{1}{2}\mathbf{u}(\nabla\cdot\mathbf{u}) \tag{4.12}$$

This form also makes visible that the ***skew symmetric*** operator introduces error when the velocity field diverges. Solving the advection equation by numerical methods is challenging and there is a large scientific literature on this subject!

4.2 Worked Examples

Worked Example 4.1 ***Solving Reactive Transport Equations Using*** R

Package 'ReacTran' **August 16, 2017**

Version 1.4.3.1

Title Reactive Transport Modelling in 1D, 2D, and 3D

Authors Karline Soetaert <karline.soetaert@nioz.nl>, Filip Meysman <filip.meysman@nioz.nl>

Maintainer Karline Soetaert <karline.soetaert@nioz.nl>

Depends R (>= 2.10), rootSolve, deSolve, shape

Imports stats, graphics

Description Routines for developing models that describe reaction and advective–diffusive transport in one, two, or three dimensions. Includes transport routines in porous media, in estuaries, and in bodies with variable shape

License GPL (>=3)

LazyData Yes

Repository CRAN

Repository/R-Forge/Project reactran

Repository/R-Forge/Revision 100

Repository/R-Forge/DateTimeStamp 2017-08-15 06:57:39

Date/Publication 2017-08-15 22:13:03 UTC

Needs Compilation Yes

R topics documented:

1. ReacTran-package
2. advection.1D
3. fiadeiro
4. g.sphere
5. p.exp
6. setup.compaction.1D
7. setup.grid.1D
8. setup.grid.2D
9. setup.prop.1D
10. setup.prop.2D
11. tran.1D
12. tran.2D
13. tran.3D
14. tran.cylindrical
15. tran.polar
16. tran.volume.1D

ReacTran-Package Reactive Transport Modeling in 1D, 2D, and 3D

Description

R-package ReacTran contains routines that enable the development of reactive transport models in aquatic systems (rivers, lakes, oceans), porous media (floc aggregates, sediments, etc.), and even idealized organisms (spherical cells, cylindrical worms, etc.).

The geometry of the model domain is either one-dimensional, two-dimensional, or three-dimensional.

The package contains:

- Functions to setup a finite-difference grid (1D or 2D)
- Functions to attach parameters and properties to this grid (1D or 2D)
- Functions to calculate the advective–diffusive transport term over the grid (1D, 2D, or 3D)
- Utility functions.

Details

Package	ReacTran
Type	Package
Version	1.4.3
Date	2017-08-14
License	GNU Public License 2 or above

Authors

Karline Soetaert (Maintainer)
Filip Meysman

See Also

Functions ode.1D, ode.2D, ode.3D from package deSolve to integrate the reactive-transport model.

Functions steady.1D, steady.2D, steady.3D from the package rootSolve to find the steady-state solution of the reactive-transport model.

tran.1D, tran.2D, tran.3D for a discretization of the general transport equations.

tran.volume.1D for discretization of the 1D transport equations using finite volumes.

tran.cylindrical, tran.spherical for a discretization of 3D transport equations in cylindrical and spherical coordinates.

tran.polar, for a discretization of 2D transport equations in polar coordinates.

setup.grid.1D, setup.grid.2D for the creation of grids in 1D and in 2D.

setup.prop.1D, setup.prop.2D for defining properties on these grids.

Examples

```
## show examples (see respective help pages for details)
## 1-dimensional transport
example(tran.1D)
example(tran.volume.1D)

## 2-dimensional transport
example(tran.2D)
example(tran.polar)

## 3-dimensional transport
example(tran.3D)
example(tran.cylindrical)
example(tran.spherical)

## open the directory with documents
browseURL(paste(system.file(package="ReacTran"), "/doc", sep=""))

## open the directory with fortran codes of the transport functions
browseURL(paste(system.file(package="ReacTran"), "/doc/fortran", sep=""))

## show package vignette with how to use ReacTran and how to solve PDEs
## + source code of the vignettes
vignette("ReacTran")
vignette("PDE")
edit(vignette("ReacTran"))

## a directory with fortran implementations of the transport
browseURL(paste(system.file(package="ReacTran"), "/doc/fortran", sep=""))

## End
```

Worked Example 4.2 trans.1D ***General One-Dimensional Advective–Diffusive Transport***

Description

This program estimates the transport term (i.e. the rate of change of a concentration owing to diffusion and advection) in a one-dimensional model of a liquid (volume fraction constant and equal to 1) or in a porous medium (volume fraction variable and lower than 1).

The interfaces between grid cells can have a variable cross-sectional area, e.g. when modeling spherical or cylindrical geometries (see example).

Usage

```
tran.1D(C, C.up = C[1], C.down = C[length(C)],
        flux.up = NULL, flux.down = NULL,
```

```
a.bl.up = NULL, a.bl.down = NULL,
D = 0, v = 0, AFDW = 1, VF = 1, A = 1, dx,
full.check = FALSE, full.output = FALSE)
```

Arguments

C	Concentration, expressed per unit of phase volume
C.up	Concentration at upstream boundary. One value $[M/L^3]$
C.down	Concentration at downstream boundary. One value $[M/L^3]$
flux.up	Flux across the upstream boundary, positive = INTO model domain. One value, expressed per unit of total surface $[M/L^2/T]$. If NULL, the boundary is prescribed as a concentration or a convective transfer boundary
flux.down	Flux across the downstream boundary, positive = OUT of model domain. One value, expressed per unit of total surface $[M/L^2/T]$. If NULL, the boundary is prescribed as a concentration or a convective transfer boundary
a.bl.up	Convective transfer coefficient across the upstream boundary layer
Flux = a.bl.up*(C.up-C0)	One value [L/T]
a.bl.down	Convective transfer coefficient across the downstream boundary layer (L). Flux = a.bl.down*(CL-C.down). One value [L/T]
D	Diffusion coefficient, defined on grid cell interfaces. One value, a vector of length N+1 $[L^2/T]$, or a 1D property list; the list contains at least the element int (see setup.prop.1D) $[L^2/T]$
v	Advective velocity, defined on the grid cell interfaces. Can be positive (downstream flow) or negative (upstream flow). One value, a vector of length N+1 [L/T], or a 1D property list; the list contains at least the element int (see setup.prop.1D) [L/T]
AFDW	Weight used in the finite difference scheme for advection, defined on grid cell interfaces; backward = 1, centered = 0.5, forward = 0; default is backward. One value, a vector of length N+1, or a 1D property list; the list contains at least the element int (see setup.prop.1D) [–]
VF	Volume fraction defined at the grid cell interfaces. One value, a vector of Length N+1, or a 1D property list; the list contains at least the elements int and mid (see setup.prop.1D) [–]

A	Interface area defined at the grid cell interfaces. One value, a vector of length N+1, or a 1D grid property list; the list contains at least the elements int and mid (see setup.prop.1D) [L^2]
dx	Distance between adjacent cell interfaces (thickness of grid cells). One value, a vector of length N, or a 1D grid list containing at least the elements dx and dx.aux (see setup.grid.1D) [L]
full.check	Logical flag enabling a full check of the consistency of the arguments (default = FALSE; TRUE slows down execution by 50%)
full.output	Logical flag enabling a full return of the output (default = FALSE; TRUE slows down execution by 20%)

Details

The **boundary conditions may take one of the following values:**

(1) zero gradient
(2) fixed concentration
(3) convective boundary layer
(4) fixed flux.

The above order also shows the priority. The default condition is the zero gradient. The fixed concentration condition overrules the zero gradient. The convective boundary layer condition overrules the fixed concentration and zero gradient. The fixed flux overrules all other specifications.

To ensure that the boundary conditions are well defined: for instance, it does not make sense to specify an influx in a boundary cell with the advection velocity pointing outward.

Transport Properties

The *diffusion coefficient* (D), the *advective velocity* (v), the *volume fraction* (VF), the *interface surface* (A), and the *advective finite difference weight* (AFDW) can either be specified as one value, a vector or a 1D property list as generated by setup.prop.1D.

When a vector, this vector must be of length $N+1$, defined at all grid cell interfaces, including the upper and lower boundary.

The **finite difference grid** (dx) is specified either as one value, a vector or a 1D grid list, as generated by setup.grid.1D.

Value

dC	The rate of change of the concentration **C** due to transport, defined in the center of each grid cell. The rate of change is expressed per unit of phase volume [$M/L^3/T$]

`C.up`	Concentration at the upstream interface. One value [M/L^3] only when (`full.output = TRUE`)
`C.down`	Concentration at the downstream interface. One value [M/L^3] only when (`full.output = TRUE`)
`dif.flux`	Diffusive flux across at the interface of each grid cell. A vector of length N+1 [M/L^2/T] only when (`full.output = TRUE`)
`adv.flux`	Advective flux across at the interface of each grid cell. A vector of length N+1 [M/L^2/T] only when (`full.output = TRUE`)
`flux`	Total flux across at the interface of each grid cell. A vector of length N+1 [M/L^2/T] only when (`full.output = TRUE`)
`flux.up`	Flux across the upstream boundary, positive = `INTO` model domain. One value [M/L^2/T]
`flux.down`	Flux across the downstream boundary, positive = `OUT` of model domain. One value [M/L^2/T]

Note

The advective equation is not checked for mass conservation. Sometimes, this is not an issue, for instance when `v` represents a sinking velocity of particles or a swimming velocity of organisms. In other cases, however, mass conservation needs to be accounted for. To ensure mass conservation, the advective velocity must obey certain continuity constraints: in essence the product of the volume fraction (VF), interface surface area (A), and advective velocity (v) should be constant. In sediments, one can use `setup.compaction.1D` to ensure that the advective velocities for the pore water and solid phase meet these constraints.

In terms of the units of concentrations and fluxes, one may follow the convention in the geosciences. The concentration `C`, `C.up`, `C.down` as well at the rate of change of the concentration `dC` are always expressed per unit of phase volume (i.e. per unit volume of solid or liquid).

Total concentrations (e.g. per unit volume of bulk sediment) may be obtained by multiplication with the appropriate volume fraction. In contrast, fluxes are always expressed per unit of total interface area (so here the volume fraction is accounted for).

Authors

Filip Meysman <filip.meysman@nioz.nl>,
Karline Soetaert <karline.soetaert@nioz.nl>.

See Also

`tran.volume.1D` for a discretization the transport equation using finite volumes.
`tran.2D, tran.3D`
`advection.1D,` for more sophisticated advection schemes.

Examples

```
# ==================================================#
# EXAMPLE 1: O2 and OC consumption in sediments      #
# ==================================================#

# This example uses only the volume fractions
# in the reactive transport term

#==============#
# Model formulation    #
#==============#

# Monod consumption of oxygen (O2)

O2.model <- function (t = 0, O2, pars = NULL) {
   tran <- tran.1D(C = O2, C.up = C.ow.O2, D = D.grid,
                  v = v.grid, VF = por.grid, dx = grid)$dC
   reac <- - R.O2*(O2/(Ks+O2))
   return(list(dCdt = tran + reac))
}

# First order consumption of organic carbon (OC)


OC.model <- function (t = 0, OC, pars = NULL) {

      tran <- tran.1D(C = OC, flux.up = F.OC, D = Db.grid,
                     v = v.grid, VF = svf.grid, dx = grid)$dC
      reac <- - k*OC
      return(list(dCdt = tran + reac))
}

#============== #
# Parameter definition  #
#==============#

# Parameter values

F.OC    <- 25 # input flux organic carbon [micromol cm-2 yr-1]
C.ow.O2 <- 0.25 # concentration O2 in overlying water [micromol cm-3]
por     <- 0.8 # porosity
D       <- 400 # diffusion coefficient O2 [cm2 yr-1]
Db      <- 10 # mixing coefficient sediment [cm2 yr-1]
v       <- 1 # advective velocity [cm yr-1]
k       <- 1 # decay constant organic carbon [yr-1]
R.O2    <- 10 # O2 consumption rate [micromol cm-3 yr-1]
Ks      <- 0.005 # O2 consumption saturation constant
```

```
# Grid definition

L     <- 10 # depth of sediment domain [cm]
N     <- 100 # number of grid layers
grid <- setup.grid.1D(x.up = 0, L = L, N = N)

# Volume fractions

por.grid <- setup.prop.1D(value = por, grid = grid)
svf.grid <- setup.prop.1D(value = (1-por), grid = grid)
D.grid   <- setup.prop.1D(value = D, grid = grid)
Db.grid <- setup.prop.1D(value = Db, grid = grid)
v.grid   <- setup.prop.1D(value = v, grid = grid)

#===========#
# Model solution   #
#===========#
=
# Initial conditions + simulation O2

yini <- rep(0, length.out = N)
O2 <- steady.1D(y = yini, func = O2.model, nspec = 1)

# Initial conditions + simulation OC

yini <- rep(0, length.out = N)
OC <- steady.1D(y = yini, func = OC.model, nspec = 1)

# Plotting output, using S3 plot method of package rootSolve"

plot(O2, grid = grid$x.mid, xyswap = TRUE, main = "O2 concentration",
      ylab = "depth [cm]", xlab = "", mfrow = c(1,2), type = "p", pch = 16)

plot(OC, grid = grid$x.mid, xyswap = TRUE, main = "C concentration",
      ylab = "depth [cm]", xlab = "", mfrow = NULL)

## ======================================
## EXAMPLE 2: O2 in a cylindrical and spherical organism
## ======================================

# This example uses only the surface areas
# in the reactive transport term
```

```
#============#
# Model formulation #
#============#

# The numerical model - rate of change = transport-consumption
Cylinder.Model <- function(time, O2, pars)
return (list(
  tran.1D(C = O2, C.down = BW, D = Da, A = A.cyl, dx = dx)$dC - Q
  ))

Sphere.Model <- function(time, O2, pars)
  return (list(
    tran.1D(C = O2, C.down = BW, D = Da, A = A.sphere, dx = dx)$dC - Q
     ))

#==============#
# Parameter definition  #
#==============#

# Parameter values

BW <- 2 # mmol/m3, oxygen conc in surrounding water
Da <- 0.5 # cm2/d effective diffusion coeff in organism
R  <- 0.0025 # cm radius of organism
Q  <- 250000 # nM/cm3/d oxygen consumption rate/volume/day
L   <- 0.05 # cm length of organism (if a cylinder)

# the numerical model

N <- 40                                     # layers in the body
dx <- R/N                                   # thickness of each layer
x.mid <- seq(dx/2, by=dx, length.out=N)  # distance of center to mid-layer
x.int  <- seq(0, by=dx, length.out = N+1) # distance to layer interface


# Cylindrical surfaces
A.cyl <- 2*pi*x.int*L                       # surface at mid-layer depth

# Spherical surfaces
A.sphere <- 4*pi*x.int^2               # surface of sphere, at each mid-layer

#=========== #
# Model solution     #
#=========== #
```

```
# the analytical solution of cylindrical and spherical model

cylinder <- function(Da, Q, BW, R, r)  BW + Q/(4*Da)*(r^2-R^2)

sphere  <- function(Da, Q, BW, R, r) BW + Q/(6*Da)*(r^2-R^2)

# solve the model numerically for a cylinder
O2.cyl <- steady.1D (y = runif(N), name = "O2",
           func = Cylinder.Model, nspec = 1, atol = 1e-10)

# solve the model numerically for a sphere
O2.sphere <- steady.1D (y = runif(N), name = "O2",
              func = Sphere.Model, nspec = 1, atol = 1e-10)

#===========#
# Plotting output  #
#===========#

# Analytical solution - "observations"
Ana.cyl <- cbind(x.mid, O2 = cylinder(Da, Q, BW, R, x.mid))
Ana.spher <- cbind(x.mid, O2 = sphere(Da, Q, BW, R, x.mid))

plot(O2.cyl, O2.sphere, grid = x.mid, lwd = 2, lty = 1, col = 1:2,
     xlab = "distance from center cm",
     ylab = "mmol/m3", main = "tran.1D",
     sub = "diffusion-reaction in a cylinder and sphere",
     obs = list(Ana.cyl, Ana.spher), obspar = list(pch = 16, col =1:2))

legend ("topleft", lty = c(1, NA), pch = c(NA, 18),
        c("numerical approximation", "analytical solution"))

legend ("bottomright", pch = 16, lty = 1, col = 1:2,
        c("cylinder", "sphere"))
#=========================================== #
# EXAMPLE 3: O2 consumption in a spherical aggregate                #
#=========================================== #

# This example uses both the surface areas and the volume fractions
# in the reactive transport term

#=============#
# Model formulation  #
#=============#

Aggregate.Model <- function(time, O2, pars) {
```

```
  tran <- tran.1D(C = O2, C.down = C.ow.O2,
                  D = D.grid, A = A.grid,
                  VF = por.grid, dx = grid)$dC

reac <- - R.O2*(O2/(Ks+O2))*(O2>0)
return(list(dCdt = tran + reac, consumption = -reac))
}

#============= #
# Parameter definition #
#============= #

# Parameters

C.ow.O2 <- 0.25      # concentration O2 water [micromol cm-3]
por       <- 0.8     # porosity
D         <- 400     # diffusion coefficient O2 [cm2 yr-1]
v         <- 0       # advective velocity [cm yr-1]
R.O2 <- 1000000      # O2 consumption rate [micromol cm-3 yr-1]
Ks    <- 0.005          # O2 saturation constant [micromol cm-3]

# Grid definition
R <- 0.025              # radius of the aggregate [cm]
N <- 100                #  number of grid layers
grid <- setup.grid.1D(x.up = 0, L = R, N = N)

# Volume fractions
por.grid <- setup.prop.1D(value = por, grid = grid)
D.grid   <- setup.prop.1D(value = D, grid = grid)

# Surfaces

A.mid <- 4*pi*grid$x.mid^2 # surface of sphere at middle of grid cells
A.int  <- 4*pi*grid$x.int^2 # surface of sphere at interface
A.grid <- list(int = A.int, mid = A.mid)

#===========#
# Model solution   #
#===========#

# Numerical solution: steady state

O2.agg <- steady.1D (runif(N), func = Aggregate.Model, nspec = 1,
                     atol = 1e-10, names = "O2")

#===========#
# Plotting output  #
#===========#
```

```
par(mfrow = c(1,1))

plot(grid$x.mid, O2.agg$y, xlab = "distance from center, cm",
     ylab  = "mmol/m3",
     main = "Diffusion-reaction of O2 in a spherical aggregate")

legend ("bottomright", pch = c(1, 18), lty = 1, col = "black",
        c("O2 concentration"))


# Similar, using S3 plot method of package rootSolve"
plot(O2.agg, grid = grid$x.mid, which = c("O2", "consumption"),
     xlab = "distance from center, cm", ylab = c("mmol/m3","mmol/m3/d"))
```

In the R domain:

```
>
> install.packages("ReacTran")
Installing package into 'C:/Users/Bert/Documents/R/win-library/3.3'
(as 'lib' is unspecified)
--- Please select a CRAN mirror for use in this session ---
A CRAN mirror is selected.
also installing the dependencies 'rootSolve', 'deSolve'
trying URL 'https://mirrors.tuna.tsinghua.edu.cn/CRAN/bin/windows/contrib/3.3/
rootSolve_1.7.zip'
Content type 'application/zip' length 683839 bytes (667 KB)
downloaded 667 KB
trying URL 'https://mirrors.tuna.tsinghua.edu.cn/CRAN/bin/windows/contrib/3.3/
deSolve_1.20.zip'
Content type 'application/zip' length 2667028 bytes (2.5 MB)
downloaded 2.5 MB
trying URL 'https://mirrors.tuna.tsinghua.edu.cn/CRAN/bin/windows/contrib/3.3/
ReacTran_1.4.3.1.zip'
Content type 'application/zip' length 915996 bytes (894 KB)
downloaded 894 KB

package 'rootSolve' successfully unpacked and MD5 sums checked
package 'deSolve' successfully unpacked and MD5 sums checked
package 'ReacTran' successfully unpacked and MD5 sums checked
The downloaded binary packages are in

C:\Users\Bert\AppData\Local\Temp\RtmpmEkAb2\downloaded_packages
> require(ReacTran)
Loading required package: ReacTran
Loading required package: rootSolve
Loading required package: deSolve
```

```
Loading required package: shape
> library(ReacTran)
> ls("package:ReacTran") # Listing all the programs in this package:
 [1] "advection.1D"       "advection.volume.1D"   "fiadeiro"
 [4] "g.cylinder"         "g.sphere"              "g.spheroid"
 [7] "p.exp"              "p.lin"                 "p.sig"
[10] "polar2cart"         "setup.compaction.1D"   "setup.grid.1D"
[13] "setup.grid.2D"      "setup.prop.1D"         "setup.prop.2D"
[16] "tran.1D"            "tran.2D"               "tran.3D"
[19] "tran.cylindrical"   "tran.polar"            "tran.spherical"
[22] "tran.volume.1D"     "tran.volume.2D"        "tran.volume.3D"
> tran.1D # Outputting the program listing of "tran.1D"
function (C, C.up = C[1], C.down = C[length(C)], flux.up = NULL,
   flux.down = NULL, a.bl.up = NULL, a.bl.down = NULL, D = 0,
   v = 0, AFDW = 1, VF = 1, A = 1, dx, full.check = FALSE, full.output = FALSE)
{
   if (is.null(dx))
      stop("error: dx should be inputted ")
   N <- length(C)
 if (N == 0)
      stop("C should be a vector with numeric values")
   if (is.null(C.up))
      C.up <- C[1]
   if (is.null(C.down))
      C.down <- C[N]
   if (!is.list(AFDW))
      AFDW <- list(int = rep_len(AFDW, N + 1))
   if (!is.list(D))
      D <- list(int = rep_len(D, N + 1))
   if (!is.list(v))
      v <- list(int = rep_len(v, N + 1))
   if (!is.list(VF))
      VF <- list(int = rep_len(VF, N + 1), mid = 0.5 * (rep_len(VF,
         N + 1)[1:N] + rep_len(VF, N + 1)[2:(N + 1)]))
   if (!is.list(A))
      A <- list(int = rep_len(A, N + 1), mid = 0.5 * (rep_len(A,
         N + 1)[1:N] + rep_len(A, N + 1)[2:(N + 1)]))
   if (is.list(dx))
      grid <- dx
   if (!is.list(dx))
      grid <- list(dx = rep_len(dx, N), dx.aux = 0.5 * (c(0,
         rep_len(dx, N)) + c(rep_len(dx, N), 0)))

  if (full.check) {
      gn <- names(AFDW)
      if (!"int" %in% gn)
```

```
        stop("error: AFDW should be a list that contains 'int', the AFDW
values at the interface of the grid cell ")
    if (is.null(AFDW$int))
        stop("error: AFDW is NULL, should contain (numeric) values")
    if (!is.null(AFDW$int)) {
        if (!((length(AFDW$int) == 1) || (length(AFDW$int) ==
            (N + 1))))
            stop("error: AFDW should be a vector of length 1 or N+1")
    }
    if (any(AFDW$int < 0) || any(AFDW$int > 1))
        stop("error: the AFDW should always range between 0 and 1")
    gn <- names(D)
    if (!"int" %in% gn)
        stop("error: D should be a list that contains 'int', the D values at the
interface of the grid cell ")
    if (is.null(D$int))
        stop("error: D is NULL, should contain (numeric) values")
    if (!is.null(D$int)) {
        if (!((length(D$int) == 1) || (length(D$int) == (N +
            1))))
            stop("error: D should be a vector of length 1 or N+1")
    }

 if (any(D$int < 0))
        stop("error: the diffusion coefficient should always be positive")
    gn <- names(v)
    if (!"int" %in% gn)
        stop("error: v should be a list that contains 'int', the v values at the
interface of the grid cell ")
    if (is.null(v$int))
        stop("error: the advective velocity v is NULL, should contain (numeric)
values")
    if (!is.null(v$int)) {
        if (!((length(v$int) == 1) || (length(v$int) == (N +
            1))))
            stop("error: v should be a vector of length 1 or N+1")
    }
    gn <- names(VF)
    if (!"int" %in% gn)
        stop("error: VF should be a list that contains 'int', the area values at the
interface of the grid cell ")
    if (!"mid" %in% gn)
        stop("error: VF should be a list that contains 'mid', the area at the
middle of the grid cells")
    if (is.null(VF$int) || is.null(VF$mid))
        stop("error: the volume fraction VF should contain (numeric) values")
```

```
        if (!is.null(VF$int)) {
           if (!((length(VF$int) == 1) || (length(VF$int) ==
              (N + 1))))

         stop("error: VF$int should be a vector of length 1 or N+1")
        }
        if (!is.null(VF$mid)) {
           if (!((length(VF$mid) == 1) || (length(VF$mid) ==
              (N))))
              stop("error: VF$mid should be a vector of length 1 or N")
        }
        if (any(VF$int < 0) || any(VF$mid < 0) || any(VF$int >
           1) || any(VF$mid > 1))
           stop("error: the volume fraction should range between 0 and 1")
        gn <- names(A)
        if (!"int" %in% gn)
           stop("error: A should be a list that contains 'int', the area values at the
interface of the grid cell ")
        if (!"mid" %in% gn)
           stop("error: A should be a list that contains 'mid', the area at the
middle of the grid cells")
        if (is.null(A$int) || is.null(A$mid))
           stop("error: the surface area A is NULL, should contain (numeric) values")
        if (!is.null(A$int)) {
           if (!((length(A$int) == 1) || (length(A$int) == (N +
              1))))
              stop("error: A$int should be a vector of length 1 or N+1")
        }

if (!is.null(A$mid)) {
           if (!((length(A$mid) == 1) || (length(A$mid) == (N))))
              stop("error: A$mid should be a vector of length 1 or N")
        }
        if (any(A$int < 0) || any(A$mid < 0))
           stop("error: the area A should always be positive")
        gn <- names(grid)
        if (!"dx" %in% gn)
           stop("error: grid should be a list that contains 'dx' ")
        if (!"dx.aux" %in% gn)
           stop("error: grid should be a list that contains 'dx.aux' ")
        if (is.null(grid$dx) || is.null(grid$dx.aux))
           stop("error: the grid should be a list with (numeric) values for 'dx' and
'dx.aux' ")
        if (!is.null(grid$dx)) {
           if (!((length(grid$dx) == 1) || (length(grid$dx) ==
              N)))
```

```
            stop("error: dx should be a vector of length 1 or N")
        }
        if (!is.null(grid$dx.aux)) {
          if (!((length(grid$dx.aux) == 1) || (length(grid$dx.aux) ==
            (N + 1))))
            stop("error: dx.aux should be a vector of length 1 or N+1")
        }

    if (any(grid$dx <= 0) || any(grid$dx.aux <= 0))
          stop("error: the grid distances dx and dx.aux should always be positive")
        if (!is.null(a.bl.up) & !is.null(C.up)) {
          if (a.bl.up < 0)
            stop("error: the boundary layer transfer coefficient should be positive")
        }
        if (!is.null(a.bl.down) & !is.null(C.down)) {
          if (a.bl.down < 0)
            stop("error: the boundary layer transfer coefficient should be positive")
        }
      }
      if (full.output) {
        if (!is.null(flux.up)) {
          if (v$int[1] >= 0) {
            nom <- flux.up + VF$int[1] * (D$int[1]/grid$dx.aux[1] +
              (1 - AFDW$int[1]) * v$int[1]) * C[1]
            denom <- VF$int[1] * (D$int[1]/grid$dx.aux[1] +
              AFDW$int[1] * v$int[1])
          }
          else {
            nom <- flux.up + VF$int[1] * (D$int[1]/grid$dx.aux[1] +
              AFDW$int[1] * v$int[1]) * C[1]


            denom <- VF$int[1] * (D$int[1]/grid$dx.aux[1] +
              (1 - AFDW$int[1]) * v$int[1])
          }
          C.up <- nom/denom
        }
        if (!is.null(flux.down)) {
          if (v$int[N + 1] >= 0) {
            nom <- flux.down - VF$int[N + 1] * (D$int[N +
              1]/grid$dx.aux[N + 1] + AFDW$int[N + 1] * v$int[N +
              1]) * C[N]
            denom <- -VF$int[N + 1] * (D$int[N + 1]/grid$dx.aux[N +
              1] + (1 - AFDW$int[N + 1]) * v$int[N + 1])
          }
          else {
```

```
      nom <- flux.down - VF$int[N + 1] * (D$int[N +
        1]/grid$dx.aux[N + 1] + (1 - AFDW$int[N + 1]) *
        v$int[N + 1]) * C[N]
      denom <- -VF$int[N + 1] * (D$int[N + 1]/grid$dx.aux[N +
        1] + AFDW$int[N + 1] * v$int[N + 1])
    }
    C.down <- nom/denom
}
if (!is.null(a.bl.up) & !is.null(C.up)) {
    if (v$int[1] >= 0) {
      nom <- a.bl.up * C.up + VF$int[1] * (D$int[1]/grid$dx.aux[1] +
        (1 - AFDW$int[1]) * v$int[1]) * C[1]
      denom <- a.bl.up + VF$int[1] * (D$int[1]/grid$dx.aux[1] +
        AFDW$int[1] * v$int[1])
    }
    else {
      nom <- a.bl.up * C.up + VF$int[1] * (D$int[1]/grid$dx.aux[1] +
        AFDW$int[1] * v$int[1]) * C[1]
      denom <- a.bl.up + VF$int[1] * (D$int[1]/grid$dx.aux[1] +
        (1 - AFDW$int[1]) * v$int[1])
    }
    C.up <- nom/denom
}
if (!is.null(a.bl.down) & !is.null(C.down)) {
    if (v$int[N + 1] >= 0) {
      nom <- a.bl.down * C.down + VF$int[N + 1] * (D$int[N +
        1]/grid$dx.aux[N + 1] + (1 - AFDW$int[N + 1]) *
        v$int[N + 1]) * C[N]
      denom <- a.bl.down + VF$int[N + 1] * (D$int[N +
        1]/grid$dx.aux[N + 1] + AFDW$int[N + 1] * v$int[N +
        1])
    }
    else {
      nom <- a.bl.down * C.down + VF$int[N + 1] * (D$int[N +

        1]/grid$dx.aux[N + 1] + AFDW$int[N + 1] * v$int[N +
        1]) * C[N]
      denom <- a.bl.down + VF$int[N + 1] * (D$int[N +
        1]/grid$dx.aux[N + 1] + (1 - AFDW$int[N + 1]) *
        v$int[N + 1])
    }
    C.down <- nom/denom
}
dif.flux <- as.vector(-VF$int * D$int * diff(c(C.up,
    C, C.down))/grid$dx.aux)
adv.flux <- rep(0, length.out = length(dif.flux))
```

```
    if (any(v$int > 0)) {
      vv <- v$int
      vv[v$int < 0] <- 0
      conc <- AFDW$int * c(C.up, C)
      if (any(AFDW$int < 1))
        conc <- conc + (1 - AFDW$int) * c(C, C.down)
      adv.flux <- adv.flux + as.vector(VF$int * vv * conc)
    }
    if (any(v$int < 0)) {
      vv <- v$int
      vv[v$int > 0] <- 0
      conc <- AFDW$int * c(C, C.down)
      if (any(AFDW$int < 1))
        conc <- conc + (1 - AFDW$int) * c(C.up, C)
      adv.flux <- adv.flux + as.vector(VF$int * vv * conc)
    }
    flux <- dif.flux + adv.flux
  }
  else {
    if (!is.null(a.bl.up) & !is.null(C.up)) {
      if (v$int[1] >= 0) {
        nom <- a.bl.up * C.up + VF$int[1] * (D$int[1]/grid$dx.aux[1] +
          (1 - AFDW$int[1]) * v$int[1]) * C[1]
        denom <- a.bl.up + VF$int[1] * (D$int[1]/grid$dx.aux[1] +
          AFDW$int[1] * v$int[1])
      }
      else {
        nom <- a.bl.up * C.up + VF$int[1] * (D$int[1]/grid$dx.aux[1] +
          AFDW$int[1] * v$int[1]) * C[1]
        denom <- a.bl.up + VF$int[1] * (D$int[1]/grid$dx.aux[1] +
          (1 - AFDW$int[1]) * v$int[1])
      }
      C.up <- nom/denom
    }
    if (!is.null(a.bl.down) & !is.null(C.down)) {
      if (v$int[N + 1] >= 0) {
       nom <- a.bl.down * C.down + VF$int[N + 1] * (D$int[N +
          1]/grid$dx.aux[N + 1] + (1 - AFDW$int[N + 1]) *
          v$int[N + 1]) * C[N]
      denom <- a.bl.down + VF$int[N + 1] * (D$int[N +
          1]/grid$dx.aux[N + 1] + AFDW$int[N + 1] * v$int[N +
          1])
      }
      else {
        nom <- a.bl.down * C.down + VF$int[N + 1] * (D$int[N +
          1]/grid$dx.aux[N + 1] + AFDW$int[N + 1] * v$int[N +
```

```
                1]) * C[N]
            denom <- a.bl.down + VF$int[N + 1] * (D$int[N +
              1]/grid$dx.aux[N + 1] + (1 - AFDW$int[N + 1]) *
              v$int[N + 1])
          }
          C.down <- nom/denom
        }
        flux <- as.vector(-(VF$int) * D$int * diff(c(C.up, C,
          C.down))/grid$dx.aux)
        if (any(v$int > 0)) {
          vv <- v$int
          vv[v$int < 0] <- 0
          conc <- AFDW$int * c(C.up, C)
          if (any(AFDW$int < 1))
            conc <- conc + (1 - AFDW$int) * c(C, C.down)
          flux <- flux + as.vector(VF$int * vv * conc)
        }
        if (any(v$int < 0)) {
          vv <- v$int
          vv[v$int > 0] <- 0
          conc <- AFDW$int * c(C, C.down)
          if (any(AFDW$int < 1))
            conc <- conc + (1 - AFDW$int) * c(C.up, C)
          flux <- flux + as.vector(VF$int * vv * conc)
        }
      }
      if (!is.null(flux.up))
        flux[1] <- flux.up
      if (!is.null(flux.down))
        flux[N + 1] <- flux.down
      dC <- -diff(A$int * flux)/A$mid/VF$mid/grid$dx
      if (!full.output) {
        return(list(dC = dC, flux.up = flux[1], flux.down = flux[length(flux)]))
      }
      else {
        return(list(dC = dC, C.up = C.up, C.down = C.down, dif.flux = dif.flux,
          adv.flux = adv.flux, flux = flux, flux.up = flux[1],
          flux.down = flux[length(flux)]))
      }
    }
    <environment: namespace:ReacTran>
    >
    > install.packages("ReacTran")
    Installing package into 'C:/Users/Bert/Documents/R/win-library/3.3'
    (as 'lib' is unspecified)
    --- Please select a CRAN mirror for use in this session ---
```

```
A CRAN mirror is selected.
trying URL 'https://mirrors.tuna.tsinghua.edu.cn/CRAN/bin/windows/contrib/3.3/
ReacTran_1.4.3.1.zip'
Content type 'application/zip' length 915996 bytes (894 KB)
downloaded 894 KB

package 'ReacTran' successfully unpacked and MD5 sums checked

The downloaded binary packages are in

C:\Users\Bert\AppData\Local\Temp\RtmpCiBQxG\downloaded_packages
> require(ReacTran)
Loading required package: ReacTran
Loading required package: rootSolve
Loading required package: deSolve
Loading required package: shape
> library(ReacTran)
> ls("package:ReacTran")
 [1] "advection.1D"       "advection.volume.1D" "fiadeiro"
 [4] "g.cylinder"         "g.sphere"            "g.spheroid"
[7]  "p.exp"              "p.lin"               "p.sig"
[10] "polar2cart"         "setup.compaction.1D" "setup.grid.1D"
[13] "setup.grid.2D"      "setup.prop.1D"       "setup.prop.2D"
[16] "tran.1D"            "tran.2D"             "tran.3D"
[19] "tran.cylindrical"   "tran.polar"          "tran.spherical"
[22] "tran.volume.1D"     "tran.volume.2D"      "tran.volume.3D"
>
> tran.1D #Listing the program "tran.1D"
function (C, C.up = C[1], C.down = C[length(C)], flux.up = NULL,
   flux.down = NULL, a.bl.up = NULL, a.bl.down = NULL, D = 0,
   v = 0, AFDW = 1, VF = 1, A = 1, dx, full.check = FALSE, full.output = FALSE)
{
   if (is.null(dx))
      stop("error: dx should be inputted ")
   N <- length(C)
   if (N == 0)
      stop("C should be a vector with numeric values")
   if (is.null(C.up))
      C.up <- C[1]
   if (is.null(C.down))
      C.down <- C[N]
   if (!is.list(AFDW))
      AFDW <- list(int = rep_len(AFDW, N + 1))
if (!is.list(D))
      D <- list(int = rep_len(D, N + 1))
   if (!is.list(v))
```

```
    v <- list(int = rep_len(v, N + 1))
  if (!is.list(VF))
    VF <- list(int = rep_len(VF, N + 1), mid = 0.5 * (rep_len(VF,
      N + 1)[1:N] + rep_len(VF, N + 1)[2:(N + 1)]))
  if (!is.list(A))
    A <- list(int = rep_len(A, N + 1), mid = 0.5 * (rep_len(A,
      N + 1)[1:N] + rep_len(A, N + 1)[2:(N + 1)]))
  if (is.list(dx))
    grid <- dx
  if (!is.list(dx))
    grid <- list(dx = rep_len(dx, N), dx.aux = 0.5 * (c(0,
      rep_len(dx, N)) + c(rep_len(dx, N), 0)))
  if (full.check) {
    gn <- names(AFDW)
    if (!"int" %in% gn)
      stop("error: AFDW should be a list that contains 'int', the AFDW
values at the interface of the grid cell ")
    if (is.null(AFDW$int))
      stop("error: AFDW is NULL, should contain (numeric) values")
    if (!is.null(AFDW$int)) {
      if (!((length(AFDW$int) == 1) || (length(AFDW$int) ==
        (N + 1))))
        stop("error: AFDW should be a vector of length 1 or N+1")

    }
    if (any(AFDW$int < 0) || any(AFDW$int > 1))
      stop("error: the AFDW should always range between 0 and 1")
    gn <- names(D)
    if (!"int" %in% gn)
      stop("error: D should be a list that contains 'int', the D values at the
interface of the grid cell ")
    if (is.null(D$int))
      stop("error: D is NULL, should contain (numeric) values")
    if (!is.null(D$int)) {
      if (!((length(D$int) == 1) || (length(D$int) == (N +
        1))))
        stop("error: D should be a vector of length 1 or N+1")
    }
    if (any(D$int < 0))
      stop("error: the diffusion coefficient should always be positive")
    gn <- names(v)
    if (!"int" %in% gn)
      stop("error: v should be a list that contains 'int', the v values at the
interface of the grid cell ")
    if (is.null(v$int))
```

```
        stop("error: the advective velocity v is NULL, should contain (numeric)
values")
      if (!is.null(v$int)) {
        if (!((length(v$int) == 1) || (length(v$int) == (N +
          1))))
          stop("error: v should be a vector of length 1 or N+1")
}
      gn <- names(VF)
      if (!"int" %in% gn)
        stop("error: VF should be a list that contains 'int', the area values at the
interface of the grid cell ")
      if (!"mid" %in% gn)
        stop("error: VF should be a list that contains 'mid', the area at the middle
of the grid cells")
      if (is.null(VF$int) || is.null(VF$mid))
        stop("error: the volume fraction VF should contain (numeric) values")
      if (!is.null(VF$int)) {
        if (!((length(VF$int) == 1) || (length(VF$int) ==
          (N + 1))))
          stop("error: VF$int should be a vector of length 1 or N+1")
      }
      if (!is.null(VF$mid)) {
        if (!((length(VF$mid) == 1) || (length(VF$mid) ==
          (N))))
          stop("error: VF$mid should be a vector of length 1 or N")
      }
      if (any(VF$int < 0) || any(VF$mid < 0) || any(VF$int >
        1) || any(VF$mid > 1))
        stop("error: the volume fraction should range between 0 and 1")
      gn <- names(A)
      if (!"int" %in% gn)
        stop("error: A should be a list that contains 'int', the area values at the
interface of the grid cell ")
      if (!"mid" %in% gn)
        stop("error: A should be a list that contains 'mid', the area at the middle
of the grid cells")
      if (is.null(A$int) || is.null(A$mid))
        stop("error: the surface area A is NULL, should contain (numeric) values")
      if (!is.null(A$int)) {
        if (!((length(A$int) == 1) || (length(A$int) == (N +
          1))))
          stop("error: A$int should be a vector of length 1 or N+1")
      }
      if (!is.null(A$mid)) {
        if (!((length(A$mid) == 1) || (length(A$mid) == (N))))
          stop("error: A$mid should be a vector of length 1 or N")
```

```
    }
    if (any(A$int < 0) || any(A$mid < 0))
      stop("error: the area A should always be positive")
    gn <- names(grid)
    if (!"dx" %in% gn)
      stop("error: grid should be a list that contains 'dx' ")
    if (!"dx.aux" %in% gn)
      stop("error: grid should be a list that contains 'dx.aux' ")
    if (is.null(grid$dx) || is.null(grid$dx.aux))
      stop("error: the grid should be a list with (numeric) values for 'dx' and
'dx.aux' ")
  if (!is.null(grid$dx)) {
      if (!((length(grid$dx) == 1) || (length(grid$dx) ==
        N)))
        stop("error: dx should be a vector of length 1 or N")
    }
    if (!is.null(grid$dx.aux)) {
      if (!((length(grid$dx.aux) == 1) || (length(grid$dx.aux) ==
        (N + 1))))
        stop("error: dx.aux should be a vector of length 1 or N+1")
    }
    if (any(grid$dx <= 0) || any(grid$dx.aux <= 0))
      stop("error: the grid distances dx and dx.aux should always be positive")
    if (!is.null(a.bl.up) & !is.null(C.up)) {
      if (a.bl.up < 0)
        stop("error: the boundary layer transfer coefficient should be positive")
    }
    if (!is.null(a.bl.down) & !is.null(C.down)) {
      if (a.bl.down < 0)
        stop("error: the boundary layer transfer coefficient should be positive")
    }
  }
  if (full.output) {
    if (!is.null(flux.up)) {
      if (v$int[1] >= 0) {
        nom <- flux.up + VF$int[1] * (D$int[1]/grid$dx.aux[1] +
          (1 - AFDW$int[1]) * v$int[1]) * C[1]
        denom <- VF$int[1] * (D$int[1]/grid$dx.aux[1] +
          AFDW$int[1] * v$int[1])
      }
      else {
        nom <- flux.up + VF$int[1] * (D$int[1]/grid$dx.aux[1] +
          AFDW$int[1] * v$int[1]) * C[1]
        denom <- VF$int[1] * (D$int[1]/grid$dx.aux[1] +
          (1 - AFDW$int[1]) * v$int[1])
      }
```

```
    C.up <- nom/denom
  }
  if (!is.null(flux.down)) {
    if (v$int[N + 1] >= 0) {
      nom <- flux.down - VF$int[N + 1] * (D$int[N +
       1]/grid$dx.aux[N + 1] + AFDW$int[N + 1] * v$int[N +
       1]) * C[N]
      denom <- -VF$int[N + 1] * (D$int[N + 1]/grid$dx.aux[N +
       1] + (1 - AFDW$int[N + 1]) * v$int[N + 1])
    }
    else {
      nom <- flux.down - VF$int[N + 1] * (D$int[N +
       1]/grid$dx.aux[N + 1] + (1 - AFDW$int[N + 1]) *
       v$int[N + 1]) * C[N]

     denom <- -VF$int[N + 1] * (D$int[N + 1]/grid$dx.aux[N +
       1] + AFDW$int[N + 1] * v$int[N + 1])
    }
    C.down <- nom/denom
  }
  if (!is.null(a.bl.up) & !is.null(C.up)) {
    if (v$int[1] >= 0) {
      nom <- a.bl.up * C.up + VF$int[1] * (D$int[1]/grid$dx.aux[1] +
       (1 - AFDW$int[1]) * v$int[1]) * C[1]
      denom <- a.bl.up + VF$int[1] * (D$int[1]/grid$dx.aux[1] +
       AFDW$int[1] * v$int[1])
    }
    else {
      nom <- a.bl.up * C.up + VF$int[1] * (D$int[1]/grid$dx.aux[1] +
       AFDW$int[1] * v$int[1]) * C[1]
      denom <- a.bl.up + VF$int[1] * (D$int[1]/grid$dx.aux[1] +
       (1 - AFDW$int[1]) * v$int[1])
    }
    C.up <- nom/denom
  }
  if (!is.null(a.bl.down) & !is.null(C.down)) {
    if (v$int[N + 1] >= 0) {
      nom <- a.bl.down * C.down + VF$int[N + 1] * (D$int[N +
       1]/grid$dx.aux[N + 1] + (1 - AFDW$int[N + 1]) *
       v$int[N + 1]) * C[N]
     denom <- a.bl.down + VF$int[N + 1] * (D$int[N +
       1]/grid$dx.aux[N + 1] + AFDW$int[N + 1] * v$int[N +
       1])
    }
    else {
      nom <- a.bl.down * C.down + VF$int[N + 1] * (D$int[N +
```

```
          1]/grid$dx.aux[N + 1] + AFDW$int[N + 1] * v$int[N +
          1]) * C[N]
        denom <- a.bl.down + VF$int[N + 1] * (D$int[N +
          1]/grid$dx.aux[N + 1] + (1 - AFDW$int[N + 1]) *
          v$int[N + 1])
      }
      C.down <- nom/denom
    }
    dif.flux <- as.vector(-VF$int * D$int * diff(c(C.up,
      C, C.down))/grid$dx.aux)
    adv.flux <- rep(0, length.out = length(dif.flux))
    if (any(v$int > 0)) {
      vv <- v$int
      vv[v$int < 0] <- 0
      conc <- AFDW$int * c(C.up, C)
      if (any(AFDW$int < 1))
        conc <- conc + (1 - AFDW$int) * c(C, C.down)
      adv.flux <- adv.flux + as.vector(VF$int * vv * conc)
}
    if (any(v$int < 0)) {
      vv <- v$int
      vv[v$int > 0] <- 0
      conc <- AFDW$int * c(C, C.down)
      if (any(AFDW$int < 1))
        conc <- conc + (1 - AFDW$int) * c(C.up, C)
      adv.flux <- adv.flux + as.vector(VF$int * vv * conc)
    }
    flux <- dif.flux + adv.flux
  }
  else {
    if (!is.null(a.bl.up) & !is.null(C.up)) {
      if (v$int[1] >= 0) {
        nom <- a.bl.up * C.up + VF$int[1] * (D$int[1]/grid$dx.aux[1] +
          (1 - AFDW$int[1]) * v$int[1]) * C[1]
        denom <- a.bl.up + VF$int[1] * (D$int[1]/grid$dx.aux[1] +
          AFDW$int[1] * v$int[1])
      }
      else {
        nom <- a.bl.up * C.up + VF$int[1] * (D$int[1]/grid$dx.aux[1] +
          AFDW$int[1] * v$int[1]) * C[1]
        denom <- a.bl.up + VF$int[1] * (D$int[1]/grid$dx.aux[1] +
          (1 - AFDW$int[1]) * v$int[1])
      }
      C.up <- nom/denom
    }
    if (!is.null(a.bl.down) & !is.null(C.down)) {
```

```
      if (v$int[N + 1] >= 0) {
         nom <- a.bl.down * C.down + VF$int[N + 1] * (D$int[N +
           1]/grid$dx.aux[N + 1] + (1 - AFDW$int[N + 1]) *
           v$int[N + 1]) * C[N]
         denom <- a.bl.down + VF$int[N + 1] * (D$int[N +
           1]/grid$dx.aux[N + 1] + AFDW$int[N + 1] * v$int[N +
           1])
      }
      else {
         nom <- a.bl.down * C.down + VF$int[N + 1] * (D$int[N +
           1]/grid$dx.aux[N + 1] + AFDW$int[N + 1] * v$int[N +
           1]) * C[N]
         denom <- a.bl.down + VF$int[N + 1] * (D$int[N +
           1]/grid$dx.aux[N + 1] + (1 - AFDW$int[N + 1]) *
           v$int[N + 1])
      }
      C.down <- nom/denom
   }
   flux <- as.vector(-(VF$int) * D$int * diff(c(C.up, C,
      C.down))/grid$dx.aux)
   if (any(v$int > 0)) {
      vv <- v$int
    vv[v$int < 0] <- 0
      conc <- AFDW$int * c(C.up, C)
      if (any(AFDW$int < 1))
         conc <- conc + (1 - AFDW$int) * c(C, C.down)
      flux <- flux + as.vector(VF$int * vv * conc)
   }
   if (any(v$int < 0)) {
      vv <- v$int
      vv[v$int > 0] <- 0
      conc <- AFDW$int * c(C, C.down)
      if (any(AFDW$int < 1))
         conc <- conc + (1 - AFDW$int) * c(C.up, C)
      flux <- flux + as.vector(VF$int * vv * conc)
   }
}
if (!is.null(flux.up))
   flux[1] <- flux.up
if (!is.null(flux.down))
   flux[N + 1] <- flux.down
dC <- -diff(A$int * flux)/A$mid/VF$mid/grid$dx
if (!full.output) {
   return(list(dC = dC, flux.up = flux[1], flux.down = flux[length(flux)]))
}
else {
```

```
      return(list(dC = dC, C.up = C.up, C.down = C.down, dif.flux = dif.flux,
        adv.flux = adv.flux, flux = flux, flux.up = flux[1],
          flux.down = flux[length(flux)]))
   }
}
<environment: namespace:ReacTran
```

Computing EXAMPLE 1 in the R ***environment:***

```
> ## ===============================================
> ## EXAMPLE 1: O2 and OC (Organic Carbon) consumption in sediments =
> ## ===============================================
> # This example uses only the volume fractions
> # in the reactive transport term
> #============#
> # Model formulation #
> #============#
> # Monod consumption of oxygen (O2)
> O2.model <- function (t = 0, O2, pars = NULL) {
+ tran <- tran.1D(C = O2, C.up = C.ow.O2, D = D.grid,
+ v = v.grid, VF = por.grid, dx = grid)$dC
+
reac <- - R.O2*(O2/(Ks+O2))
+ return(list(dCdt = tran + reac))
+ }
> # First order consumption of organic carbon (OC)
> OC.model <- function (t = 0, OC, pars = NULL) {
+ tran <- tran.1D(C = OC, flux.up = F.OC, D = Db.grid,
+ v = v.grid, VF = svf.grid, dx = grid)$dC
+ reac <- - k*OC
+ return(list(dCdt = tran + reac))
+ }
> #==============#
> # Parameter definition  #
> #==============#
> # Parameter values
> F.OC <- 25 # input flux organic carbon [micromol cm-2 yr-1]
> C.ow.O2 <- 0.25 # concentration O2 in overlying water [micromol cm-3]
> por <- 0.8 # porosity
> D <- 400 # diffusion coefficient O2 [cm2 yr-1]
> Db <- 10 # mixing coefficient sediment [cm2 yr-1]
> v <- 1 # advective velocity [cm yr-1]
> k <- 1 # decay constant organic carbon [yr-1]
> R.O2 <- 10 # O2 consumption rate [micromol cm-3 yr-1]
> Ks <- 0.005 # O2 consumption saturation constant
> # Grid definition
```

```
> L <- 10 # depth of sediment domain [cm]
> N <- 100 # number of grid layers
> grid <- setup.grid.1D(x.up = 0, L = L, N = N)
> # Volume fractions
> por.grid <- setup.prop.1D(value = por, grid = grid)
> svf.grid <- setup.prop.1D(value = (1-por), grid = grid)
> D.grid <- setup.prop.1D(value = D, grid = grid)
> Db.grid <- setup.prop.1D(value = Db, grid = grid)
> v.grid <- setup.prop.1D(value = v, grid = grid)
> #==========#
> # Model solution #
> #==========#
> # Initial conditions + simulation O2
> yini <- rep(0, length.out = N)
> O2 <- steady.1D(y = yini, func = O2.model, nspec = 1)
> # Initial conditions + simulation OC
> yini <- rep(0, length.out = N)
> OC <- steady.1D(y = yini, func = OC.model, nspec = 1)
> # Plotting output, using S3 plot method of package rootSolve"
> plot(O2, grid = grid$x.mid, xyswap = TRUE, main = "O2 concentration",
+ ylab = "depth [cm]", xlab = "", mfrow = c(1,2), type = "p", pch = 16)
> #
> # Outputting: Figure A1-1: tran.1D-1
> #
> plot(OC, grid = grid$x.mid, xyswap = TRUE, main = "C concentration",
+ ylab = "depth [cm]", xlab = "", mfrow = NULL)
> #
> # Outputting: Figure A1-2: tran.1D-2
> #
> ## =======================================
> ## EXAMPLE 2: O2 in a cylindrical and spherical organism   =
> ## =======================================
> # This example uses only the surface areas
> # in the reactive transport term
> #=============#
> # Model formulation   #
> #=============#
> # The numerical model - rate of change = transport-consumption
> Cylinder.Model <- function(time, O2, pars)
+ return (list(
+ tran.1D(C = O2, C.down = BW, D = Da, A = A.cyl, dx = dx)$dC - Q
+ ))
> Sphere.Model <- function(time, O2, pars)
+ return (list(
+ tran.1D(C = O2, C.down = BW, D = Da, A = A.sphere, dx = dx)$dC - Q
+ ))
```

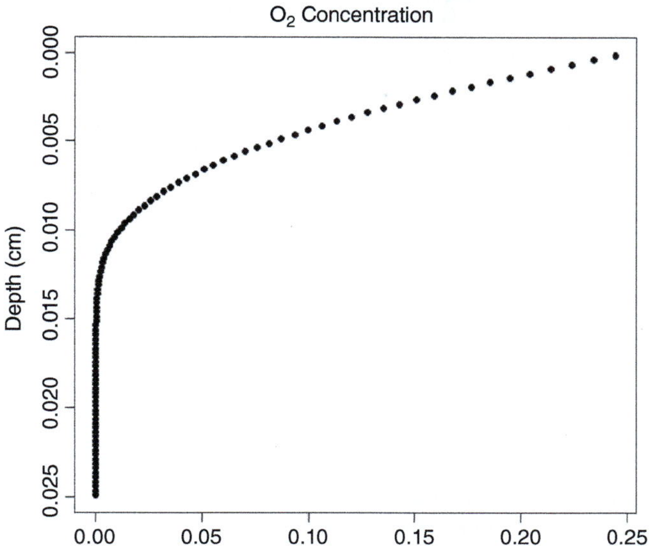

Figure A1-1 trans.1D **1** Reactive transport modeling in 1D.

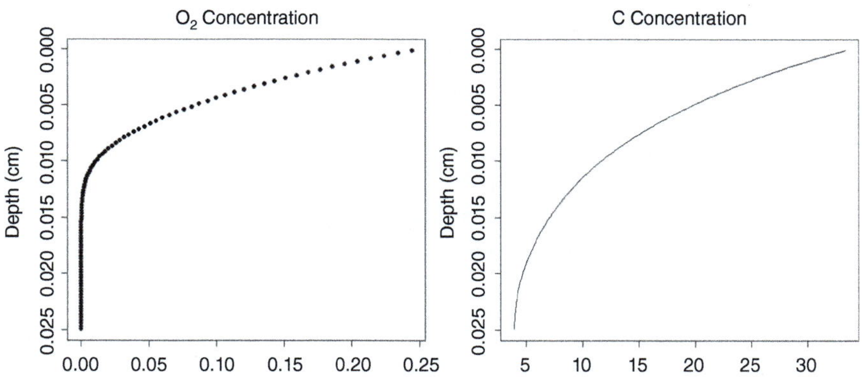

Figure A1-2 trans.1D **2** Reactive transport modeling in 1D.

```
> #==============#
> # Parameter definition  #
> #==============#
> # Parameter values
> BW <- 2 # mmol/m3, oxygen conc in surrounding water
> Da <- 0.5 # cm2/d effective diffusion coeff in organism
> R <- 0.0025 # cm radius of organism
> Q <- 250000 # nM/cm3/d oxygen consumption rate/volume/day
> L <- 0.05 # cm length of organism (if a cylinder)
> # the numerical model
> N <- 40 # layers in the body
> dx <- R/N # thickness of each layer
> x.mid <- seq(dx/2, by = dx, length.out = N) # distance of center to mid-layer
```

```
> x.int <- seq(0, by = dx, length.out = N+1) # distance to layer interface
> # Cylindrical surfaces
> A.cyl <- 2*pi*x.int*L # surface at mid-layer depth
> # Spherical surfaces
> A.sphere <- 4*pi*x.int^2 # surface of sphere, at each mid-layer
> #===========#
> # Model solution   #
> #===========#
> # The analytical solution of cylindrical and spherical model
> cylinder <- function(Da, Q, BW, R, r) BW + Q/(4*Da)*(r^2-R^2)
> sphere <- function(Da, Q, BW, R, r) BW + Q/(6*Da)*(r^2-R^2)
> # solve the model numerically for a cylinder
> O2.cyl <- steady.1D (y = runif(N), name = "O2",
+ func = Cylinder.Model, nspec = 1, atol = 1e-10)
> # solve the model numerically for a sphere
> O2.sphere <- steady.1D (y = runif(N), name = "O2",
+ func = Sphere.Model, nspec = 1, atol = 1e-10)
> #==========#
> # Plotting output #
> #==========#
> # Analytical solution - "observations"
> Ana.cyl <- cbind(x.mid, O2 = cylinder(Da, Q, BW, R, x.mid))
> Ana.spher <- cbind(x.mid, O2 = sphere(Da, Q, BW, R, x.mid))
> plot(O2.cyl, O2.sphere, grid = x.mid, lwd = 2, lty = 1, col = 1:2,
+ xlab = "distance from center, cm",
+ ylab = "mmol/m3", main = "tran.1D",
+ sub = "diffusion-reaction in a cylinder and sphere",
+ obs = list(Ana.cyl, Ana.spher), obspar = list(pch = 16, col =1:2))
> #
> # Outputting: Figure A1-3: tran.1D-3
> #
> legend ("topleft", lty = c(1, NA), pch = c(NA, 18),
+ c("numerical approximation", "analytical solution"))
> #
> # Outputting: Figure A1-4: tran.1D-4
> #
> legend ("bottomright", pch = 16, lty = 1, col = 1:2,
+ c("cylinder", "sphere"))
> #
> # Outputting: Figure A1-5: tran.1D-5
> #
>====================================
> ## EXAMPLE 3: O2 consumption in a spherical aggregate =
> ## ================================ =
> # This example uses both the surface areas and the volume fraction
> # in the reactive transport term
```

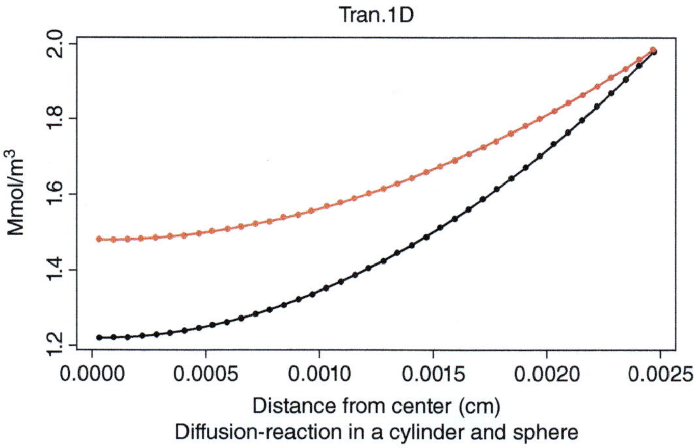

Figure A1-3 trans.1D **3** Reactive transport modeling in 1D.

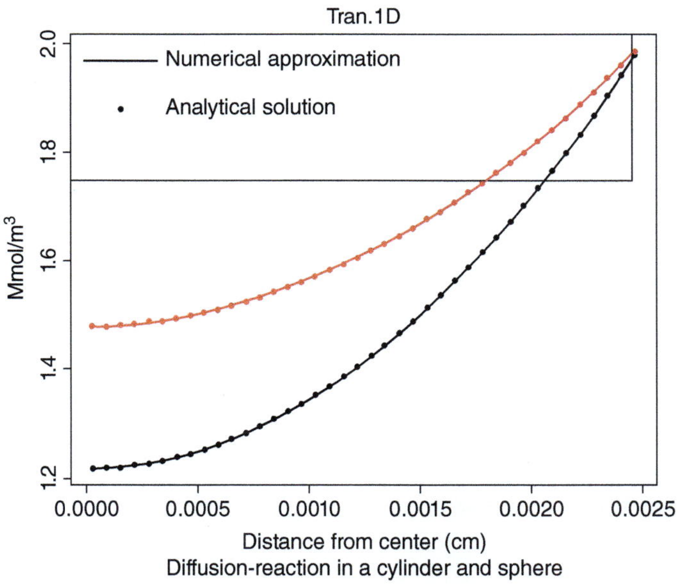

Figure A1-4 trans.1D **4** Reactive transport modeling in 1D.

```
> #==============#
> # Model formulation    #
> #==============#
> Aggregate.Model <- function(time, O2, pars) {
+ tran <- tran.1D(C = O2, C.down = C.ow.O2,
+ D = D.grid, A = A.grid,
+ VF = por.grid, dx = grid )$dC
+ reac <- - R.O2*(O2/(Ks+O2))*(O2>0)
+ return(list(dCdt = tran + reac, consumption = -reac))
+ }
```

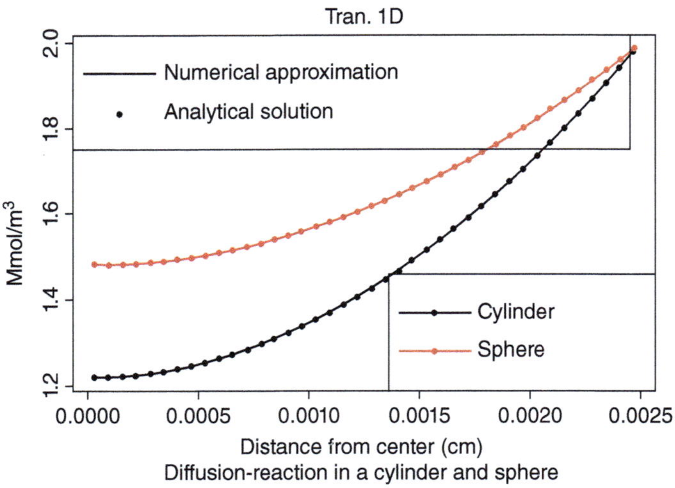

Figure A1-5 trans.1D **5** Reactive transport modeling in 1D.

```
> #==============#
> # Parameter definition #
> #==============#
>
> # Parameters
> C.ow.O2 <- 0.25 # concentration O2 water [micromol cm-3]
> por <- 0.8 # porosity
> D <- 400 # diffusion coefficient O2 [cm2 yr-1]
> v <- 0 # advective velocity [cm yr-1]
> R.O2 <- 1000000 # O2 consumption rate [micromol cm-3 yr-1]
> Ks <- 0.005 # O2 saturation constant [micromol cm-3]
> # Grid definition
> R <- 0.025 # radius of the aggregate [cm]
> N <- 100 # number of grid layers
> grid <- setup.grid.1D(x.up = 0, L = R, N = N)
> # Volume fractions
> por.grid <- setup.prop.1D(value = por, grid = grid)
> D.grid <- setup.prop.1D(value = D, grid = grid)
> # Surfaces
> A.mid <- 4*pi*grid$x.mid^2 # surface of sphere at middle of grid cells
> A.int <- 4*pi*grid$x.int^2 # surface of sphere at interface
> A.grid <- list(int = A.int, mid = A.mid)
> #=========#
> # Model solution #
> #=========#
> # Numerical solution: steady state
> O2.agg <- steady.1D (runif(N), func = Aggregate.Model, nspec = 1,
+ atol = 1e-10, names = "O2")
```

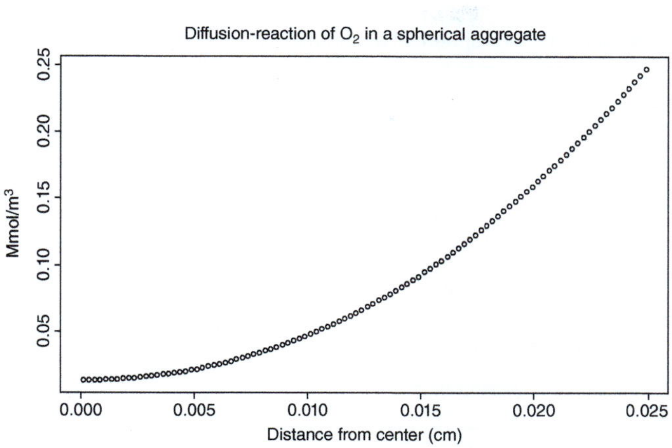

Figure A1-6 trans.1D **6** Reactive transport modeling in 1D.

```
> #============#
> # Plotting output  #
> #============#
> par(mfrow = c(1,1))
> plot(grid$x.mid, O2.agg$y, xlab = "distance from center, cm",
+ ylab = "mmol/m3",
+ main = "Diffusion-reaction of O2 in a spherical aggregate")
> #
> # Outputting: Figure A1-6: tran.1D-6
> #
> legend ("bottomright", pch = c(1, 18), lty = 1, col = "black",
+ c("O2 concentration"))
> #
> # Outputting: Figure A1-7: tran.1D-7
> #
> # Similar, using S3 plot method of package rootSolve"
> plot(O2.agg, grid = grid$x.mid, which = c("O2", "consumption"),
+ xlab = "distance from center, cm", ylab = c("mmol/m3","mmol/m3/d"))
> #
> # Outputting: Figure A1-8: tran.1D-8
> #
```

Worked Example 4.3 tran.2D ***General Two-Dimensional Advective–Diffusive Transport***

Description

This program estimates the transport term (i.e. the rate of change of a concentration owing to diffusion and advection) in a two-dimensional model domain.

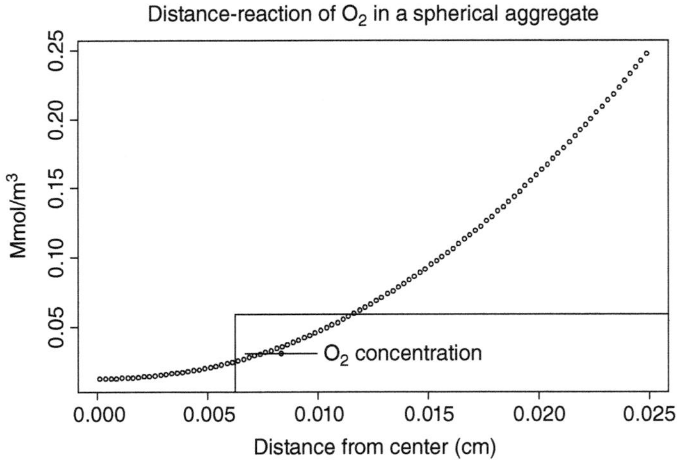

Figure A1-7 trans.1D **7** Reactive transport modeling in 1D.

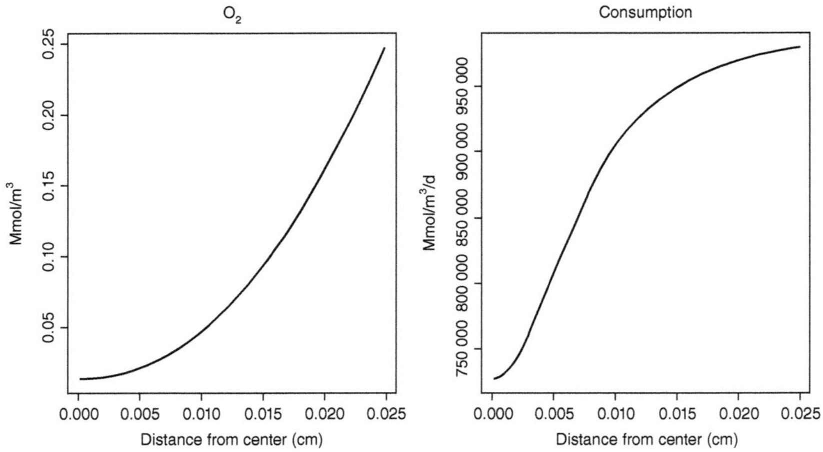

Figure A1-8 trans.1D **8** Reactive transport modeling in 1D.

Usage

```
tran.2D (C, C.x.up = C[1,], C.x.down = C[nrow(C),],
         C.y.up = C[,1], C.y.down = C[ ,ncol(C)],
         flux.x.up = NULL, flux.x.down = NULL,
         flux.y.up = NULL, flux.y.down = NULL,
         a.bl.x.up = NULL, a.bl.x.down = NULL,
         a.bl.y.up = NULL, a.bl.y.down = NULL,
         D.grid = NULL, D.x = NULL, D.y = D.x,
         v.grid = NULL, v.x = 0, v.y = 0,
        AFDW.grid = NULL, AFDW.x = 1, AFDW.y = AFDW.x,
        VF.grid = NULL, VF.x = 1, VF.y = VF.x,
        A.grid = NULL, A.x = 1, A.y = 1,
```

```
grid = NULL, dx = NULL, dy = NULL,
full.check = FALSE, full.output = FALSE)
```

Arguments

C	Concentration, expressed per unit volume, defined at the center of each grid cell; Nx*Ny matrix [M/L^3]
C.x.up	Concentration at upstream boundary in x-direction; vector of length Ny [M/L^3]
C.x.down	Concentration at downstream boundary in x-direction; vector of length Ny [M/L^3]
C.y.up	Concentration at upstream boundary in y-direction; vector of length Nx [M/L^3]
C.y.down	Concentration at downstream boundary in y-direction; vector of length Nx [M/L^3]
flux.x.up	Flux across the upstream boundary in x-direction, positive = INTO model domain; vector of length Ny [M/L^2/T]
flux.x.down	Flux across the downstream boundary in x-direction, positive = OUT of model domain; vector of length Ny [M/L^2/T]
flux.y.up	Flux across the upstream boundary in y-direction, positive = INTO model domain; vector of length Nx [M/L^2/T]
flux.y.down	Flux across the downstream boundary in y-direction, positive = OUT of model domain; vector of length Nx [M/L^2/T]
a.bl.x.up	Transfer coefficient across the upstream boundary layer in x-direction; Flux=a.bl.x.up*(C.x.up-C[1,]). One value [L/T]
a.bl.x.down	Transfer coefficient across the downstream boundary layer in x-direction; Flux=a.bl.x.down*(C[Nx,]-C.x.down). One value [L/T]
a.bl.y.up	Transfer coefficient across the upstream boundary layer in y-direction; Flux=a.bl.y.up*(C.y.up-C[,1]). One value [L/T]
a.bl.y.down	Transfer coefficient across the downstream boundary layer in y-direction; Flux=a.bl.y.down*(C[,Ny]-C.y.down). One value [L/T]
D.grid	Diffusion coefficient defined on all grid cell interfaces. A prop.2D list created by setup.prop.2D [L^2/T]. See last example for creating spatially varying diffusion coefficients
D.x	Diffusion coefficient in x-direction, defined on grid cell interfaces. One value, a vector of length (Nx+1), a prop.1D list created by setup.prop.1D, or a (Nx+1)*Ny matrix [L^2/T]
D.y	Diffusion coefficient in y-direction, defined on grid cell interfaces. One value, a vector of length (Ny+1), a prop.1D list created by setup.prop.1D, or a Nx*(Ny+1) matrix [L^2/T]
v.grid	Advective velocity defined on all grid cell interfaces. Can be positive (downstream flow) or negative (upstream flow). A prop.2D list created by setup.prop.2D [L/T]

`v.x`	Advective velocity in the *x*-direction, defined on grid cell interfaces. Can be positive (downstream flow) or negative (upstream flow). One value, a vector of length (Nx+1), a prop.1D list created by `setup.prop.1D`, or a (Nx+1)*Ny matrix [L/T]
`v.y`	Advective velocity in the *y*-direction, defined on grid cell interfaces. Can be positive (downstream flow) or negative (upstream flow). One value, a vector of length (Ny+1), a prop.1D list created by `setup.prop.1D`, or a Nx*(Ny+1) matrix [L/T]
`AFDW.grid`	Weight used in the finite difference scheme for advection in the *x*- and *y*-direction, defined on grid cell interfaces; backward = 1, centered = 0.5, forward = 0; default is backward. A `prop.2D` list created by `setup.prop.2D [-]`
`AFDW.x`	Weight used in the finite difference scheme for advection in the *x*-direction, defined on grid cell interfaces; backward = 1, centered = 0.5, forward = 0; default is backward. One value, a vector of length (Nx+1), a prop.1D list created by `setup.prop.1D`, or a (Nx+1)*Ny matrix [–]
`AFDW.y`	Weight used in the finite difference scheme for advection in the *y*-direction, defined on grid cell interfaces; backward = 1, centered = 0.5, forward = 0; default is backward. One value, a vector of length (Ny+1), a prop.1D list created by `setup.prop.1D`, or a Nx*(Ny+1) matrix [–]
`VF.grid`	Volume fraction. A prop.2D list created by `setup.prop.2D [-]`
`VF.x`	Volume fraction at the grid cell interfaces in the *x*-direction. One value, a vector of length (Nx+1), a prop.1D list created by `setup.prop.1D`, or a (Nx+1)*Ny matrix [–]
`VF.y`	Volume fraction at the grid cell interfaces in the *y*-direction. One value, a vector of length (Ny+1), a prop.1D list created by `setup.prop.1D`, or a Nx*(Ny+1) matrix [–]
`A.grid`	Interface area. A prop.2D list created by `setup.prop.2D` $[L^2]$
`A.x`	Interface area defined at the grid cell interfaces in the *x*-direction. One value, a vector of length (Nx+1), a prop.1D list created by `setup.prop.1D`, or a (Nx+1)*Ny matrix $[L^2]$
`A.y`	Interface area defined at the grid cell interfaces in the *y*-direction. One value, a vector of length (Ny+1), a prop.1D list created by `setup.prop.1D`, or a Nx*(Ny+1) matrix $[L^2]$
`dx`	Distance between adjacent cell interfaces in the *x*-direction (thickness of grid cells). One value or vector of length Nx [L]
`dy`	Distance between adjacent cell interfaces in the *y*-direction (thickness of grid cells). One value or vector of length Ny [L]
`grid`	Discretization grid, a list containing at least elements dx, dx.aux, dy, dy.aux (see `setup.grid.2D`) [L]

full.check	Logical flag enabling a full check of the consistency of the arguments (default = FALSE; TRUE slows down execution by 50%)
full.output	Logical flag enabling a full return of the output (default = FALSE; TRUE slows down execution by 20%)

Details

The boundary conditions are either

(1) zero-gradient
(2) fixed concentration
(3) convective boundary layer
(4) fixed flux.

This is also the order of priority. The zero gradient is the default, the fixed flux overrules all other.

Value

A list containing

dC	The rate of change of the concentration *C* due to transport, defined in the center of each grid cell, a Nx*Ny matrix. [M/L^3/T]
C.x.up	Concentration at the upstream interface in *x*-direction. A vector of length Ny [M/L^3]. Only when full.output = TRUE
C.x.down	Concentration at the downstream interface in *x*-direction. A vector of length Ny [M/L^3]. Only when full.output = TRUE
C.y.up	Concentration at the upstream interface in *y*-direction. A vector of length Nx [M/L^3]. Only when full.output = TRUE
C.y.down	Concentration at the downstream interface in *y*-direction. A vector of length Nx [M/L^3]. Only when full.output = TRUE
x.flux	Flux across the interfaces in *x*-direction of the grid cells. A (Nx+1)*Ny matrix [M/L^2/T]. Only when full.output = TRUE
y.flux	Flux across the interfaces in *y*-direction of the grid cells. A Nx*(Ny+1) matrix [M/L^2/T]. Only when full.output = TRUE
flux.x.up	Flux across the upstream boundary in *x*-direction, positive = INTO model domain. A vector of length Ny [M/L^2/T]
flux.x.down	Flux across the downstream boundary in *x*-direction, positive = OUT of model domain. A vector of length Ny [M/L^2/T]
flux.y.up	Flux across the upstream boundary in *y*-direction, positive = INTO model domain. A vector of length Nx [M/L^2/T]
flux.y.down	Flux across the downstream boundary in *y*-direction, positive = OUT of model domain. A vector of length Nx [M/L^2/T]

Note

It is much more efficient to use the *grid* input rather than vectors or single numbers.

Thus: to optimize the code, use `setup.grid.2D` to create the grid, and use `setup.prop.2D` to create `D.grid`, `v.grid`, `AFDW.grid`, `VF.grid`, and `A.grid`, even if the values are 1 or remain constant.

There is no provision (yet) to deal with *cross-diffusion*. Set `D.x` and `D.y` different only if cross-diffusion effects are unimportant.

Authors

Filip Meysman <filip.meysman@nioz.nl>,
Karline Soetaert <karline.soetaert@nioz.nl>.

See Also

`tran.polar` for a discretization of 2D transport equations in polar coordinates.
`tran.1D`, `tran.3D`

Examples

```
## =============================================
## Testing the functions
## =============================================
# Parameters
F       <- 100          # input flux [micromol cm-2 yr-1]
por     <- 0.8          # constant porosity
D       <- 400          # mixing coefficient [cm2 yr-1]
v       <- 1            # advective velocity [cm yr-1]

# Grid definition
x.N <- 4     # number of cells in x-direction
y.N <- 6     # number of cells in y-direction
x.L <- 8     # domain size x-direction [cm]
y.L <- 24    # domain size y-direction [cm]
dx <- x.L/x.N                # cell size x-direction [cm]
dy <- y.L/y.N                # cell size y-direction [cm]

# Initial conditions
C <- matrix(nrow = x.N, ncol = y.N, data = 0, byrow = FALSE)

# Boundary conditions: fixed concentration
C.x.up     <- rep(1, times = y.N)
```

```
C.x.down  <- rep(0, times = y.N)
C.y.up    <- rep(1, times = x.N)
C.y.down  <- rep(0, times = x.N)

# Only diffusion
tran.2D(C = C, D.x = D, D.y = D, v.x = 0, v.y = 0,
  VF.x = por, VF.y = por, dx = dx, dy = dy,
  C.x.up = C.x.up, C.x.down = C.x.down,
  C.y.up = C.y.up, C.y.down = C.y.down, full.output = TRUE)

# Strong advection, backward (default), central and forward
#finite difference schemes
tran.2D(C = C, D.x = D, v.x = 100*v,
  VF.x = por, dx = dx, dy = dy,
  C.x.up = C.x.up, C.x.down = C.x.down,
  C.y.up = C.y.up, C.y.down = C.y.down)

tran.2D(AFDW.x = 0.5, C = C, D.x = D, v.x = 100*v,
  VF.x = por, dx = dx, dy = dy,
  C.x.up = C.x.up, C.x.down = C.x.down,
  C.y.up = C.y.up, C.y.down = C.y.down)

tran.2D(AFDW.x = 0, C = C, D.x = D, v.x = 100*v,
  VF.x = por, dx = dx, dy = dy,
  C.x.up = C.x.up, C.x.down = C.x.down,
  C.y.up = C.y.up, C.y.down = C.y.down)

# Boundary conditions: fixed fluxes

flux.x.up <- rep(200, times = y.N)
flux.x.down <- rep(-200, times = y.N)
flux.y.up <- rep(200, times = x.N)
flux.y.down <- rep(-200, times = x.N)
tran.2D(C = C, D.x = D, v.x = 0,
  VF.x = por, dx = dx, dy = dy,
  flux.x.up = flux.x.up, flux.x.down = flux.x.down,
   flux.y.up = flux.y.up, flux.y.down = flux.y.down)

# Boundary conditions: convective boundary layer on all sides

a.bl <- 800 # transfer coefficient
C.x.up <- rep(1, times = (y.N)) # fixed conc at boundary layer
C.y.up <- rep(1, times = (x.N)) # fixed conc at boundary layer
tran.2D(full.output = TRUE, C = C, D.x = D, v.x = 0,
```

```
  VF.x = por, dx = dx, dy = dy,
  C.x.up = C.x.up, a.bl.x.up = a.bl,
  C.x.down = C.x.up, a.bl.x.down = a.bl,
  C.y.up = C.y.up, a.bl.y.up = a.bl,
  C.y.down = C.y.up, a.bl.y.down = a.bl)

# Runtime test with and without argument checking

n.iterate <-500

test1 <- function() {
  for (i in 1:n.iterate )
    ST <- tran.2D(full.check = TRUE, C = C, D.x = D,
       v.x = 0, VF.x = por, dx = dx, dy = dy,
       C.x.up = C.x.up, a.bl.x.up = a.bl, C.x.down = C.x.down)
}
system.time(test1())
test2 <- function() {
  for (i in 1:n.iterate )
    ST <- tran.2D(full.output = TRUE, C = C, D.x = D,
        v.x = 0, VF.x = por, dx = dx, dy = dy,
        C.x.up = C.x.up, a.bl.x.up = a.bl, C.x.down = C.x.down)
}
system.time(test2())

test3 <- function() {
  for (i in 1:n.iterate )
    ST <- tran.2D(full.output = TRUE, full.check = TRUE, C = C,
       D.x = D, v.x = 0, VF.x = por, dx = dx, dy = dy,
       C.x.up = C.x.up, a.bl.x.up = a.bl, C.x.down = C.x.down)
}
system.time(test3())

## ===============================================
## 8A 2D model with diffusion in x- and y-direction and first-order
##  consumption - inefficient implementation
## ===============================================

N  <- 51                  # number of grid cells
XX <- 10                 # total size
dy <- dx <- XX/N         # grid size
Dy <- Dx <- 0.1          # diffusion coeff, X- and Y-direction
r  <- 0.005               # consumption rate
ini <- 1                 # initial value at x=0

N2 <- ceiling(N/2)
X  <- seq (dx, by = dx, len = (N2-1))
```

```
X  <- c(-rev(X), 0, X)

# The model equations

Diff2D <- function (t, y, parms) {

CONC <- matrix(nrow = N, ncol = N, y)
dCONC <- tran.2D(CONC, D.x = Dx, D.y=Dy, dx=dx, dy=dy)$dC + r*CONC
return (list(dCONC))

}
# initial condition: 0 everywhere, except in central point

y <- matrix(nrow = N, ncol = N, data = 0)
y[N2, N2] <- ini # initial concentration in the central point...

# solve for 10 time units
times <- 0:10
out <- ode.2D (y = y, func = Diff2D, t = times, parms = NULL,
               dim = c(N,N), lrw = 160000)

pm <- par (mfrow = c(2, 2))

# Compare solution with analytical solution...
for (i in seq(2, 11, by = 3)) {
  tt <- times[i]
  mat <- matrix(nrow = N, ncol = N,
                data = subset(out, time == tt))
  plot(X, mat[N2,], type = "l", main = paste("time=", times[i]),
      ylab = "Conc", col = "red")
  ana <- ini*dx^2/(4*pi*Dx*tt)*exp(r*tt-X^2/(4*Dx*tt))
  points(X, ana, pch = "+")
}

legend ("bottom", col = c("red","black"), lty = c(1, NA),
  pch = c(NA, "+"), c("tran.2D", "exact"))
par("mfrow" = pm )

## ===========================================================
## A 2D model with diffusion in x- and y-direction and first-order
## consumption - more efficient implementation, specifying ALL 2D grids
## ===========================================================

N    <- 51              # number of grid cells
Dy   <- Dx <- 0.1       # diffusion coeff, X- and Y-direction
r    <- 0.005           # consumption rate
```

```
ini     <- 1                    # initial value at x=0

x.grid      <- setup.grid.1D(x.up = -5, x.down = 5, N = N)
y.grid      <- setup.grid.1D(x.up = -5, x.down = 5, N = N)
grid2D      <- setup.grid.2D(x.grid, y.grid)

D.grid      <- setup.prop.2D(value = Dx, y.value = Dy, grid = grid2D)
v.grid      <- setup.prop.2D(value = 0, grid = grid2D)
A.grid      <- setup.prop.2D(value = 1, grid = grid2D)
AFDW.grid   <- setup.prop.2D(value = 1, grid = grid2D)
VF.grid     <- setup.prop.2D(value = 1, grid = grid2D)
# The model equations - using the grids

Diff2Db <- function (t, y, parms) {

    CONC <- matrix(nrow = N, ncol = N, data = y)

    dCONC <- tran.2D(CONC, grid = grid2D, D.grid = D.grid,
          A.grid = A.grid, VF.grid = VF.grid, AFDW.grid = AFDW.grid,
          v.grid = v.grid)$dC + r * CONC

   return (list(dCONC))
}

# initial condition: 0 everywhere, except in central point
y <- matrix(nrow = N, ncol = N, data = 0)
y[N2,N2] <- ini # initial concentration in the central point...

# solve for 8 time units
Times <- 0:8

outb <- ode.2D (y = y, func = Diff2Db, t = times, parms = NULL,
                  dim = c(N, N), lrw = 160000)

outtimes <- c(1, 3, 5, 7)
image(outc, ask = FALSE, mfrow = c(2, 2), main = paste("time", outtimes),
       legend = TRUE, add.contour = TRUE, subset = time %in% outtimes)
```

```
# In the R domain:
> install.packages("ReacTran")
Installing package into 'C:/Users/Bert/Documents/R/win-library/3.3'
(as 'lib' is unspecified)
--- Please select a CRAN mirror for use in this session ---
```

A CRAN mirror is selected.

```
trying URL 'https://mirrors.tuna.tsinghua.edu.cn/CRAN/bin/windows/contrib/3.3/
ReacTran_1.4.3.1.zip'
Content type 'application/zip' length 915996 bytes (894 KB)
downloaded 894 KB

package 'ReacTran' successfully unpacked and MD5 sums checked

The downloaded binary packages are in
C:\Users\Bert\AppData\Local\Temp\RtmpEvJcoa\downloaded_packages
> require(ReacTran)
Loading required package: ReacTran
Loading required package: rootSolve


Loading required package: deSolve
Loading required package: shape
> library(ReacTran)
> ls("package:ReacTran")
 [1] "advection.1D"       "advection.volume.1D"  "fiadeiro"
 [4] "g.cylinder"         "g.sphere"             "g.spheroid"
 [7] "p.exp"              "p.lin"                "p.sig"
[10] "polar2cart"         "setup.compaction.1D"  "setup.grid.1D"
[13] "setup.grid.2D"      "setup.prop.1D"        "setup.prop.2D"
[16] "tran.1D"            "tran.2D"              "tran.3D"
[19] "tran.cylindrical"   "tran.polar"           "tran.spherical"
[22] "tran.volume.1D"     "tran.volume.2D"       "tran.volume.3D"

> tran.2D # Outputting the program listing of tran.2D
function (C, C.x.up = C[1, ], C.x.down = C[nrow(C), ], C.y.up = C[,
   1], C.y.down = C[, ncol(C)], flux.x.up = NULL, flux.x.down = NULL,
   flux.y.up = NULL, flux.y.down = NULL, a.bl.x.up = NULL, a.bl.x.down =
   NULL,
   a.bl.y.up = NULL, a.bl.y.down = NULL, D.grid = NULL, D.x = NULL,
   D.y = D.x, v.grid = NULL, v.x = 0, v.y = 0, AFDW.grid = NULL,
   AFDW.x = 1, AFDW.y = AFDW.x, VF.grid = NULL, VF.x = 1, VF.y = VF.x,
   A.grid = NULL, A.x = 1, A.y = 1, grid = NULL, dx = NULL,
   dy = NULL, full.check = FALSE, full.output = FALSE)
{

   if (is.null(grid))
      if (is.null(dx) | is.null(dy))
         stop("error: either grid or dx and dy should be specified ")
   Nx <- nrow(C)
   Ny <- ncol(C)
   if (is.null(grid)) {
      DX <- if (is.list(dx))
```

```
        dx$dx
      else rep(dx, length.out = Nx)
      DXaux <- if (is.list(dx))
        dx$dx.aux
      else 0.5 * (c(0, rep(dx, length.out = Nx)) + c(rep(dx,
        length.out = Nx), 0))
      DY <- if (is.list(dy))
        dy$dx
      else rep(dy, length.out = Ny)
      DYaux <- if (is.list(dy))
        dy$dx.aux
      else 0.5 * (c(0, rep(dy, length.out = Ny)) + c(rep(dy,
        length.out = Ny), 0))
      grid <- list(dx = matrix(nrow = Nx, ncol = Ny, data = DX),
        dx.aux = matrix(nrow = Nx + 1, ncol = Ny, data = DXaux),
        dy = matrix(nrow = Nx, ncol = Ny, data = DY, byrow = TRUE),
        dy.aux = matrix(nrow = Nx, ncol = Ny + 1, data = DYaux,
          byrow = TRUE))
}
else if (class(grid) != "grid.2D") {
      if (length(grid$dx) != Nx * Ny)
        grid$dx <- matrix(nrow = Nx, ncol = Ny, grid$dx)
      if (length(grid$dx.aux) != (Nx + 1) * Ny)
        grid$dx.aux <- matrix(nrow = Nx + 1, ncol = Ny, grid$dx.aux)
      if (length(grid$dy) != Nx * Ny)
        grid$dy <- matrix(nrow = Nx, ncol = Ny, data = grid$dy,
          byrow = TRUE)
      if (length(grid$dy.aux) != Nx * (Ny + 1))
        grid$dy.aux <- matrix(nrow = Nx, ncol = Ny + 1, grid$dy.aux,
          byrow = TRUE)
  }
  gridFill <- function(G.x, G.y, Name) {
      if (is.null(G.x) | is.null(G.y))
        stop((paste("error: ", Name, "and (", Name, ".x and",
          Name, ".y) cannot be NULL at the same time",
          del = "")))
      G.grid <- list()
      if (is.matrix(G.x)) {
        if (sum(abs(dim(G.x) - c(Nx + 1, Ny))) != 0)
          stop(paste("error: ", Name, ".x matrix not of correct (Nx+1) Ny dimen-
sions",
           del = ""))
        G.grid$x.int <- G.x
        G.grid$x.mid <- 0.5 * (G.x[1:Nx, ] + G.x[2:(Nx + 1), ])
        }
         else if (class(G.x) == "prop.1D") {
```

```
    G.grid$x.int <- matrix(data = G.x$int, nrow = (Nx +
      1), ncol = Ny)
    G.grid$x.mid <- matrix(data = G.x$mid, nrow = Nx,
      ncol = Ny)
}
else if (length(G.x) == 1) {
    G.grid$x.int <- matrix(data = G.x, nrow = (Nx + 1),
      ncol = Ny)
    G.grid$x.mid <- matrix(data = G.x, nrow = Nx, ncol = Ny)
}
else if (length(G.x) != Nx + 1) {
    stop(paste("error: ", Name, ".x should be a vector of length 1 or Nx+1",
      del = ""))
}
else {
    G.grid$x.int <- matrix(data = G.x, nrow = (Nx + 1),
      ncol = Ny)
    G.grid$x.mid <- matrix(data = 0.5 * (G.x[1:Nx] +
      G.x[2:(Nx + 1)]), nrow = Nx, ncol = Ny)
}
if (is.matrix(G.y)) {
    if (sum(abs(dim(G.y) - c(Nx, Ny + 1))) != 0)
   stop(paste("error: ", Name, ".y matrix not of correct Nx(Ny+1)dimensions",
      del = ""))
    G.grid$y.int <- G.y
    G.grid$y.mid <- 0.5 * (G.y[, 1:Ny] + G.y[, 2:(Ny +
      1)])
}
else if (class(G.y) == "prop.1D") {
    G.grid$y.int <- matrix(data = G.y$int, nrow = Nx,
      ncol = (Ny + 1))
    G.grid$y.mid <- matrix(data = G.y$mid, nrow = Nx,
      ncol = Ny)
}
else if (length(G.y) == 1) {
    G.grid$y.int <- matrix(data = G.y, nrow = Nx, ncol = (Ny +
      1))
    G.grid$y.mid <- matrix(data = G.y, nrow = Nx, ncol = Ny)
}
else if (length(G.y) != Ny + 1) {
    stop(paste("error: ", Name, ".y should be a vector of length 1 or Ny+1",
      del = ""))
}
else {
    G.grid$y.int <- matrix(data = G.y, nrow = Nx, ncol = (Ny +
      1))
```

```
        G.grid$y.mid <- matrix(data = 0.5 * (G.y[1:Nx] +
          G.y[2:(Nx + 1)]), nrow = Nx, ncol = Ny)
   }
  G.grid
   }
   if (is.null(VF.grid))
      VF.grid <- gridFill(VF.x, VF.y, "VF")
   if (is.null(A.grid))
      A.grid <- gridFill(A.x, A.y, "A")
   gridInt <- function(G.x, G.y, Name) {
      if (is.null(G.x) | is.null(G.y))
        stop((paste("error: ", Name, "and (", Name, ".x and",
          Name, ".y) cannot be NULL at the same time",
          del = "")))
      G.grid <- list()
      if (is.matrix(G.x)) {
        if (sum(abs(dim(G.x) - c(Nx + 1, Ny))) != 0)
          stop(paste("error: ", Name, ".x matrix not of correct (Nx+1) Ny
dimensions",
           del = ""))
        G.grid$x.int <- G.x
      }
      else if (class(G.x) == "prop.1D") {
        G.grid$x.int <- matrix(data = G.x$int, nrow = (Nx +
          1), ncol = Ny)
      }
      else if (length(G.x) == 1) {
        G.grid$x.int <- matrix(data = G.x, nrow = (Nx + 1),
       ncol = Ny)
      }
      else if (length(G.x) != Nx + 1) {
        stop(paste("error: ", Name, ".x should be a vector of length 1 or Nx+1",
          del = ""))
      }
      else {
        G.grid$x.int <- matrix(data = G.x, nrow = (Nx + 1),
          ncol = Ny)
      }
      if (is.matrix(G.y)) {
        if (sum(abs(dim(G.y) - c(Nx, Ny + 1))) != 0)
       stop(paste("error: ", Name, ".y matrix not of correct Nx(Ny+1)dimensions",
           del = ""))
        G.grid$y.int <- G.y
      }
      else if (class(G.y) == "prop.1D") {
        G.grid$y.int <- t(matrix(data = G.y$int, ncol = Nx,
```

```
        nrow = (Ny + 1)))
    }
    else if (length(G.y) == 1) {
      G.grid$y.int <- matrix(data = G.y, nrow = Nx, ncol = (Ny +
        1))
    }
    else if (length(G.y) != Ny + 1) {
      stop(paste("error: ", Name, ".y should be a vector of length 1 or Ny+1",
        del = ""))
    }
    else {
      G.grid$y.int <- matrix(data = G.y, nrow = Nx, ncol = (Ny +
        1))
    }
    G.grid
  }
  if (is.null(AFDW.grid))
    AFDW.grid <- gridInt(AFDW.x, AFDW.y, "AFDW")
  if (is.null(D.grid))
    D.grid <- gridInt(D.x, D.y, "D")
  if (is.null(v.grid))
    v.grid <- gridInt(v.x, v.y, "v")
  if (full.check) {
    if (!is.null(C.x.up)) {
      if (!((length(C.x.up) == 1) || (length(C.x.up) ==
        (Ny))))
        stop("error: C.x.up should be a vector of length 1 or ncol(C)")
    }
    if (!is.null(C.x.down)) {
      if (!((length(C.x.down) == 1) || (length(C.x.down) ==
    (Ny))))
        stop("error: C.x.down should be a vector of length 1 or ncol(C)")
    }
    if (!is.null(C.y.up)) {
      if (!((length(C.y.up) == 1) || (length(C.y.up) ==
        (Nx))))
        stop("error: C.y.up should be a vector of length 1 or nrow(C)")
    }
    if (!is.null(C.y.down)) {
      if (!((length(C.y.down) == 1) || (length(C.y.down) ==
        (Nx))))
        stop("error: C.y.down should be a vector of length 1 or nrow(C)")
    }
    if (!is.null(flux.x.up)) {
      if (!((length(flux.x.up) == 1) || (length(flux.x.up) ==
        (Ny))))
```

```
            stop("error: flux.x.up should be a vector of length 1 or ncol(C)")
      }
      if (!is.null(flux.x.down)) {
         if (!((length(flux.x.down) == 1) || (length(flux.x.down) ==
            (Ny))))
            stop("error: flux.x.down should be a vector of length 1 or ncol(C)")
      }
      if (!is.null(flux.y.up)) {
         if (!((length(flux.y.up) == 1) || (length(flux.y.up) ==

           (Nx)))
         stop("error: flux.y.up should be a vector of length 1 or nrow(C)")
      }
      if (!is.null(flux.y.down)) {
         if (!((length(flux.y.down) == 1) || (length(flux.y.down) ==
            (Nx))))
            stop("error: flux.y.down should be a vector of length 1 or nrow(C)")
      }
      if (is.null(dx) && is.null(dy) && is.null(grid))
         stop("error: dx, dy, and grid cannot be NULL at the same time")
      gn <- names(grid)
      if (!"dx" %in% gn)
         stop("error: grid should be a list that contains 'dx' ")
      if (!"dx.aux" %in% gn)
         stop("error: grid should be a list that contains 'dx.aux' ")
      if (!"dy" %in% gn)
         stop("error: grid should be a list that contains 'dy' ")
      if (!"dy.aux" %in% gn)
         stop("error: grid should be a list that contains 'dy.aux' ")
      if (is.null(grid$dx) || is.null(grid$dx.aux))
         stop("error: the grid should be a list with (numeric)
values for 'dx' and 'dx.aux' ")
      if (is.null(grid$dy) || is.null(grid$dy.aux))
         stop("error: the grid should be a list with (numeric)
values for 'dy' and 'dy.aux' ")
      if (any(grid$dx <= 0) || any(grid$dx.aux <= 0))

         stop("error: the grid distances dx and dx.aux should always be positive")
      if (any(grid$dy <= 0) || any(grid$dy.aux <= 0))
         stop("error: the grid distances dy and dy.aux should always be positive")
      gn <- names(AFDW.grid)
      if (!"x.int" %in% gn)
         stop("error: AFDW.grid should be a list that contains 'x.int', the AFDW
values at the interfaces of the grid cells in x-direction")
      if (!"y.int" %in% gn)
```

```
        stop("error: AFDW.grid should be a list that contains 'y.int', the AFDW
values at the interfaces of the grid cells in y-direction")
    if (is.null(AFDW.grid$x.int))
        stop("error: AFDW.grid$x.int should be a list with (numeric) values")
    if (is.null(AFDW.grid$y.int))
        stop("error: AFDW.grid$y.int should be a list with (numeric) values")
    if (any(AFDW.grid$x.int < 0) || any(AFDW.grid$x.int >
        1))
        stop("error: the AFDW should range between 0 and 1")
    if (any(AFDW.grid$y.int < 0) || any(AFDW.grid$y.int >
        1))
        stop("error: the AFDW should range between 0 and 1")
    gn <- names(D.grid)
    if (!"x.int" %in% gn)
        stop("error: D.grid should be a list that contains 'x.int', the D values
at the interfaces of the grid cells in x-direction")
  if (!"y.int" %in% gn)
        stop("error: D.grid should be a list that contains 'y.int', the D values
at the interfaces of the grid cells in y-direction")
    if (is.null(D.grid$x.int))
        stop("error: D.grid$x.int should be a list with (numeric) values")
    if (is.null(D.grid$y.int))
        stop("error: D.grid$y.int should be a list with (numeric) values")
    if (any(D.grid$x.int < 0) || any(D.grid$y.int < 0))
        stop("error: the diffusion coefficient should always be positive")
    gn <- names(v.grid)
    if (!"x.int" %in% gn)
        stop("error: v.grid should be a list that contains 'x.int', the velocity
            values at the interfaces of the grid cells in x-direction")
    if (!"y.int" %in% gn)
        stop("error: v.grid should be a list that contains 'y.int', the velocity values
at the interfaces of the grid cells in y-direction")
    if (is.null(v.grid$x.int))
    stop("error: the advective velocity v.grid$x.int should be a list with (numeric)
values")
    if (is.null(v.grid$y.int))
    stop("error: the advective velocity v.grid$y.int should be a list with (numeric)
values")
    gn <- names(VF.grid)
    if (!"x.int" %in% gn)
        stop("error: VF.grid should be a list that contains 'x.int'")
    if (!"y.int" %in% gn)
        stop("error: VF.grid should be a list that contains 'y.int'")
    if (!"x.mid" %in% gn)
        stop("error: VF.grid should be a list that contains 'x.mid'")
    if (!"y.mid" %in% gn)
```

```
      stop("error: VF.grid should be a list that contains 'y.mid'")
    if (is.null(VF.grid$x.int) || is.null(VF.grid$y.int) ||
      is.null(VF.grid$x.mid) || is.null(VF.grid$y.mid))
      stop("error: VF should contain (numeric) values")
    if (any(VF.grid$x.int < 0) || any(VF.grid$y.int < 0) ||
      any(VF.grid$x.mid < 0) || any(VF.grid$y.mid < 0))
      stop("error: the VF values should always be positive")
    gn <- names(A.grid)
    if (!"x.int" %in% gn)
      stop("error: A.grid should be a list that contains 'x.int'")
    if (!"y.int" %in% gn)
      stop("error: A.grid should be a list that contains 'y.int'")
    if (!"x.mid" %in% gn)
      stop("error: A.grid should be a list that contains 'x.mid'")
    if (!"y.mid" %in% gn)
      stop("error: A.grid should be a list that contains 'y.mid'")
    if (is.null(A.grid$x.int) || is.null(A.grid$y.int) ||
      is.null(A.grid$x.mid) || is.null(A.grid$y.mid))
      stop("error: the VF should contain (numeric) values")
    if (any(A.grid$x.int < 0) || any(A.grid$y.int < 0) ||
      any(A.grid$x.mid < 0) || any(A.grid$y.mid < 0))
   stop("error: the A values should always be positive")
  }
  if (!is.null(flux.x.up[1])) {
    nom <- flux.x.up + VF.grid$x.int[1, ] * (D.grid$x.int[1,
      ]/grid$dx.aux[1, ] + (1 - AFDW.grid$x.int[1, ]) *
      v.grid$x.int[1, ]) * C[1, ]
    denom <- VF.grid$x.int[1, ] * (D.grid$x.int[1, ]/grid$dx.aux[1,
      ] + AFDW.grid$x.int[1, ] * v.grid$x.int[1, ])
    C.x.up <- nom/denom
  }
  if (!is.null(flux.x.down[1])) {
    nom <- flux.x.down - VF.grid$x.int[(Nx + 1), ] * (D.grid$x.int[(Nx +
      1), ]/grid$dx.aux[Nx + 1, ] + AFDW.grid$x.int[(Nx +
      1), ] * v.grid$x.int[(Nx + 1), ]) * C[Nx, ]
    denom <- -VF.grid$x.int[(Nx + 1), ] * (D.grid$x.int[(Nx +
      1), ]/grid$dx.aux[Nx + 1, ] + (1 - AFDW.grid$x.int[(Nx +
      1), ]) * v.grid$x.int[(Nx + 1), ])
    C.x.down <- nom/denom
  }
  if (!is.null(flux.y.up[1])) {
    nom <- flux.y.up + VF.grid$y.int[, 1] * (D.grid$y.int[,
      1]/grid$dy.aux[, 1] + (1 - AFDW.grid$y.int[, 1]) *
      v.grid$y.int[, 1]) * C[, 1]
    denom <- VF.grid$y.int[, 1] * (D.grid$y.int[, 1]/grid$dy.aux[,
      1] + AFDW.grid$y.int[, 1] * v.grid$y.int[, 1])
```

```
  C.y.up <- nom/denom
}
if (!is.null(flux.y.down[1])) {
   nom <- flux.y.down - VF.grid$y.int[, (Ny + 1)] * (D.grid$y.int[,
      (Ny + 1)]/grid$dy.aux[, (Ny + 1)] + AFDW.grid$y.int[,
      (Ny + 1)] * v.grid$y.int[, (Ny + 1)]) * C[, Ny]
   denom <- -VF.grid$y.int[, (Ny + 1)] * (D.grid$y.int[,
      (Ny + 1)]/grid$dy.aux[, Ny + 1] + (1 - AFDW.grid$y.int[,
      (Ny + 1)]) * v.grid$y.int[, (Ny + 1)])
   C.y.down <- nom/denom
}
if (!is.null(a.bl.x.up) & !is.null(C.x.up[1])) {
   nom <- a.bl.x.up * C.x.up + VF.grid$x.int[1, ] * (D.grid$x.int[1,
      ]/grid$dx.aux[1, ] + (1 - AFDW.grid$x.int[1, ]) *
      v.grid$x.int[1, ]) * C[1, ]
   denom <- a.bl.x.up + VF.grid$x.int[1, ] * (D.grid$x.int[1,
      ]/grid$dx.aux[1, ] + AFDW.grid$x.int[1, ] * v.grid$x.int[1,
      ])
   C.x.up <- nom/denom
}
if (!is.null(a.bl.x.down) & !is.null(C.x.down[1])) {
   nom <- a.bl.x.down * C.x.down + VF.grid$x.int[(Nx + 1),
      ] * (D.grid$x.int[(Nx + 1), ]/grid$dx.aux[(Nx + 1),
      ] + (1 - AFDW.grid$x.int[(Nx + 1), ]) * v.grid$x.int[(Nx +
      1), ]) * C[Nx, ]
   denom <- a.bl.x.down + VF.grid$x.int[(Nx + 1), ] * (D.grid$x.int[(Nx
      + 1), ]/grid$dx.aux[(Nx + 1), ] + AFDW.grid$x.int[(Nx +
      1), ] * v.grid$x.int[(Nx + 1), ])
   C.x.down <- nom/denom
}
if (!is.null(a.bl.y.up) & !is.null(C.y.up[1])) {
   nom <- a.bl.y.up * C.y.up + VF.grid$y.int[, 1] * (D.grid$y.int[,
      1]/grid$dy.aux[, 1] + (1 - AFDW.grid$y.int[, 1]) *
      v.grid$y.int[, 1]) * C[, 1]
   denom <- a.bl.y.up + VF.grid$y.int[, 1] * (D.grid$y.int[,
      1]/grid$dy.aux[, 1] + AFDW.grid$y.int[, 1] * v.grid$y.int[,
      1])
   C.y.up <- nom/denom
}
if (!is.null(a.bl.y.down) & !is.null(C.y.down[1])) {
   nom <- a.bl.y.down * C.y.down + VF.grid$y.int[, (Ny +
      1)] * (D.grid$y.int[, (Ny + 1)]/grid$dy.aux[, (Ny +
      1)] + (1 - AFDW.grid$y.int[, (Ny + 1)]) * v.grid$y.int[,
      (Ny + 1)]) * C[, Ny]
   denom <- a.bl.y.down + VF.grid$y.int[, (Ny + 1)] * (D.grid$y.int[,
      (Ny + 1)]/grid$dy.aux[, (Ny + 1)] + AFDW.grid$y.int[,
```

```
            (Ny + 1)] * v.grid$y.int[, (Ny + 1)])
        C.y.down <- nom/denom
    }
    x.Dif.flux <- as.matrix(-VF.grid$x.int * D.grid$x.int * diff(rbind(C.x.up,
        C, C.x.down, deparse.level = 0))/grid$dx.aux)
y.Dif.flux <- as.matrix(-VF.grid$y.int * D.grid$y.int * t(diff(t(cbind(C.y.up,
        C, C.y.down, deparse.level = 0))))/grid$dy.aux)
    x.Adv.flux <- 0
    if (any(v.grid$x.int > 0)) {
        vv <- v.grid$x.int
        vv[vv < 0] <- 0
        x.Adv.flux <- x.Adv.flux + as.matrix(VF.grid$x.int *
            vv * (AFDW.grid$x.int * rbind(C.x.up, C, deparse.level = 0) +
            (1 - AFDW.grid$x.int) * rbind(C, C.x.down, deparse.level = 0)))
    }
    if (any(v.grid$x.int < 0)) {
        vv <- v.grid$x.int
        vv[vv > 0] <- 0
        x.Adv.flux <- x.Adv.flux + as.matrix(VF.grid$x.int *
            vv * ((1 - AFDW.grid$x.int) * rbind(C.x.up, C, deparse.level = 0) +
            AFDW.grid$x.int * rbind(C, C.x.down, deparse.level = 0)))
    }
    y.Adv.flux <- 0
    if (any(v.grid$y.int > 0)) {
        vv <- v.grid$y.int
        vv[vv < 0] <- 0
        y.Adv.flux <- y.Adv.flux + as.matrix(VF.grid$y.int *
            vv * (AFDW.grid$y.int * cbind(C.y.up, C, deparse.level = 0) +
            (1 - AFDW.grid$y.int) * cbind(C, C.y.down, deparse.level = 0)))
    }
    if (any(v.grid$y.int < 0)) {
        vv <- v.grid$y.int
        vv[vv > 0] <- 0
        y.Adv.flux <- y.Adv.flux + as.matrix(VF.grid$y.int *
            vv * ((1 - AFDW.grid$y.int) * cbind(C.y.up, C, deparse.level = 0) +
            AFDW.grid$y.int * cbind(C, C.y.down, deparse.level = 0)))
    }
    x.flux <- x.Dif.flux + x.Adv.flux
    y.flux <- y.Dif.flux + y.Adv.flux
    if (!is.null(flux.x.up[1]))
        x.flux[1, ] <- flux.x.up
    if (!is.null(flux.x.down[1]))
        x.flux[nrow(x.flux), ] <- flux.x.down
    if (!is.null(flux.y.up[1]))
        y.flux[, 1] <- flux.y.up
```

```
  if (!is.null(flux.y.down[1]))
    y.flux[, ncol(y.flux)] <- flux.y.down
  dFdx <- -(diff(A.grid$x.int * x.flux)/A.grid$x.mid)/grid$dx/VF.grid$x.mid
  dFdy <- -t(diff(t(A.grid$y.int * y.flux))/t(A.grid$y.mid))/grid$dy/VF.grid$y.mid
  if (!full.output) {
    return(list(dC = dFdx + dFdy, flux.x.up = x.flux[1, ],
      flux.x.down = x.flux[nrow(x.flux), ], flux.y.up = y.flux[,
        1], flux.y.down = y.flux[, ncol(y.flux)]))
  }
  else {
    return(list(dC = dFdx + dFdy, C.x.up = C.x.up, C.x.down = C.x.down,
      C.y.up = C.y.up, C.y.down = C.y.down, x.flux = x.flux,
      y.flux = y.flux, flux.x.up = x.flux[1, ], flux.x.down = x.flux[nrow(x.flux),
        ], flux.y.up = y.flux[, 1], flux.y.down = y.flux[,
        ncol(y.flux)]))
  }
}
<environment: namespace:ReacTran>
> ## ==========================================
> ## Testing the functions
> ## ==========================================
> # Parameters
> F <- 100 # input flux [micromol cm-2 yr-1]
> por <- 0.8 # constant porosity
> D <- 400 # mixing coefficient [cm2 yr-1]
> v <- 1 # advective velocity [cm yr-1]
> # Grid definition
> x.N <- 4 # number of cells in x-direction
> y.N <- 6 # number of cells in y-direction
> x.L <- 8 # domain size x-direction [cm]
> y.L <- 24 # domain size y-direction [cm]
> dx <- x.L/x.N # cell size x-direction [cm]
> dy <- y.L/y.N # cell size y-direction [cm]
> # Initial conditions
> C <- matrix(nrow = x.N, ncol = y.N, data = 0, byrow = FALSE)
> # Boundary conditions: fixed concentration
> C.x.up <- rep(1, times = y.N)
> C.x.down <- rep(0, times = y.N)
> C.y.up <- rep(1, times = x.N)
> C.y.down <- rep(0, times = x.N)
> # Only diffusion
> tran.2D(C = C, D.x = D, D.y = D, v.x = 0, v.y = 0,
+ VF.x = por, VF.y = por, dx = dx, dy = dy,
+ C.x.up = C.x.up, C.x.down = C.x.down,
+ C.y.up = C.y.up, C.y.down = C.y.down, full.output = TRUE)
```

 # Outputting:

```
 $dC
     [,1] [,2] [,3] [,4] [,5] [,6]
[1,]  250  200  200  200  200  200
[2,]   50    0    0    0    0    0
[3,]   50    0    0    0    0    0
[4,]   50    0    0    0    0    0

$C.x.up
[1] 1 1 1 1 1 1

$C.x.down
[1] 0 0 0 0 0 0

$C.y.up
[1] 1 1 1 1

$C.y.down
[1] 0 0 0 0

$x.flux
     [,1] [,2] [,3] [,4] [,5] [,6]
[1,]  320  320  320  320  320  320
[2,]    0    0    0    0    0    0
[3,]    0    0    0    0    0    0
[4,]    0    0    0    0    0    0
[5,]    0    0    0    0    0    0

$y.flux
     [,1] [,2] [,3] [,4] [,5] [,6] [,7]
[1,]  160    0    0    0    0    0    0
[2,]  160    0    0    0    0    0    0
[3,]  160    0    0    0    0    0    0
[4,]  160    0    0    0    0    0    0

$flux.x.up
[1] 320 320 320 320 320 320

$flux.x.down
[1] 0 0 0 0 0 0

$flux.y.up
[1] 160 160 160 160

$flux.y.down
[1] 0 0 0 0
```

```
> # Strong advection, backward (default), central and forward
> #finite difference schemes
> tran.2D(C = C, D.x = D, v.x = 100*v,
+ VF.x = por, dx = dx, dy = dy,
+ C.x.up = C.x.up, C.x.down = C.x.down,
+ C.y.up = C.y.up, C.y.down = C.y.down)
# Outputting:
$dC
     [,1] [,2] [,3] [,4] [,5] [,6]
[1,]  300  250  250  250  250  250
[2,]   50    0    0    0    0    0
[3,]   50    0    0    0    0    0
[4,]   50    0    0    0    0    0
$flux.x.up
[1] 400 400 400 400 400 400

$flux.x.down
[1] 0 0 0 0 0 0

$flux.y.up
[1] 160 160 160 160

$flux.y.down
[1] 0 0 0 0

> tran.2D(AFDW.x = 0.5, C = C, D.x = D, v.x = 100*v,
+ VF.x = por, dx = dx, dy = dy,
+ C.x.up = C.x.up, C.x.down = C.x.down,
+ C.y.up = C.y.up, C.y.down = C.y.down)
# Outputting:
$dC
     [,1] [,2] [,3] [,4] [,5] [,6]
[1,]  275  225  225  225  225  225
[2,]   50    0    0    0    0    0
[3,]   50    0    0    0    0    0
[4,]   50    0    0    0    0    0
$flux.x.up
[1] 360 360 360 360 360 360

$flux.x.down
[1] 0 0 0 0 0 0
$flux.y.up
[1] 160 160 160 160
$flux.y.down
[1] 0 0 0 0
```

```
> tran.2D(AFDW.x = 0, C = C, D.x = D, v.x = 100*v,
+ VF.x = por, dx = dx, dy = dy,
+ C.x.up = C.x.up, C.x.down = C.x.down,
+ C.y.up = C.y.up, C.y.down = C.y.down)
# Outputting:
$dC
     [,1] [,2]  [,3] [,4] [,5] [,6]
[1,] 250  200  200  200  200  200
[2,]  50    0    0    0    0    0
[3,]  50    0    0    0    0    0
[4,]  50    0    0    0    0    0

$flux.x.up
[1] 320 320 320 320 320 320

$flux.x.down
[1] 0 0 0 0 0 0


$flux.y.up
[1] 160 160 160 160

$flux.y.down
[1] 0 0 0 0

> # Boundary conditions: fixed fluxes
> flux.x.up <- rep(200, times = y.N)
> flux.x.down <- rep(-200, times = y.N)
> flux.y.up <- rep(200, times = x.N)
> flux.y.down <- rep(-200, times = x.N)
> tran.2D(C = C, D.x = D, v.x = 0,
+ VF.x = por, dx = dx, dy = dy,
+ flux.x.up = flux.x.up, flux.x.down = flux.x.down,
+ flux.y.up = flux.y.up, flux.y.down = flux.y.down)

# Outputting:
$dC
      [,1]  [,2] [,3] [,4]  [,5]   [,6]
[1,] 187.5  125  125  125   125  187.5
[2,]  62.5    0    0    0     0   62.5
[3,]  62.5    0    0    0     0   62.5
[4,] 187.5  125  125  125   125  187.5
$flux.x.up
[1] 200 200 200 200 200 200

$flux.x.down
```

```
[1] -200 -200 -200 -200 -200 -200

$flux.y.up
[1] 200 200 200 200

$flux.y.down
[1] -200 -200 -200 -200

> # Boundary conditions: convective boundary layer on all sides
> a.bl <- 800 # transfer coefficient
> C.x.up <- rep(1, times = (y.N)) # fixed conc at boundary layer
> C.y.up <- rep(1, times = (x.N)) # fixed conc at boundary layer
> tran.2D(full.output = TRUE, C = C, D.x = D, v.x = 0,
+ VF.x = por, dx = dx, dy = dy,
+ C.x.up = C.x.up, a.bl.x.up = a.bl,
+ C.x.down = C.x.up, a.bl.x.down = a.bl,
+ C.y.up = C.y.up, a.bl.y.up = a.bl,
+ C.y.down = C.y.up, a.bl.y.down = a.bl)
```

Outputting:

```
$dC
          [,1]     [,2]     [,3]     [,4]     [,5]      [,6]
[1,] 184.52381   2.8571 142.8571 142.8571 142.8571 184.52381
[2,]  41.66667   0.0000   0.0000   0.0000   0.0000  41.66667
[3,]  41.66667   0.0000   0.0000   0.0000   0.0000  41.66667
[4,] 184.52381 142.8571 142.8571 142.8571 142.8571 184.52381
$C.x.up
[1]  0.7142857 0.7142857 0.7142857 0.7142857 0.7142857 0.7142857
$C.x.down
[1] 0.7142857 0.7142857 0.7142857 0.7142857 0.7142857 0.7142857
$C.y.up
[1] 0.8333333 0.8333333 0.8333333 0.8333333
$C.y.down
[1] 0.8333333 0.8333333 0.8333333 0.8333333

$x.flux
          [,1]      [,2]      [,3]      [,4]      [,5]      [,6]
[1,]  228.5714  228.5714  228.5714  228.5714  228.5714  228.5714
[2,]    0.0000    0.0000    0.0000    0.0000    0.0000    0.0000
[3,]    0.0000    0.0000    0.0000    0.0000    0.0000    0.0000
[4,]    0.0000    0.0000    0.0000    0.0000    0.0000    0.0000
[5,] -228.5714 -228.5714 -228.5714 -228.5714 -228.5714 -228.5714

$y.flux
         [,1] [,2] [,3] [,4] [,5] [,6]      [,7]
[1,] 133.3333    0    0    0    0    0 -133.3333
```

```
[2,] 133.3333    0    0    0    0    0 -133.3333
[3,] 133.3333    0    0    0    0    0 -133.3333
[4,] 133.3333    0    0    0    0    0 -133.3333

$flux.x.up
[1] 228.5714 228.5714 228.5714 228.5714 228.5714 228.5714
$flux.x.down
[1] -228.5714 -228.5714 -228.5714 -228.5714 -228.5714 -228.5714

$flux.y.up
[1] 133.3333 133.3333 133.3333 133.3333

$flux.y.down
[1] -133.3333 -133.3333 -133.3333 -133.3333
> # Runtime test with and without argument checking
> n.iterate <-500
> test1 <- function() {
+ for (i in 1:n.iterate )
+ ST <- tran.2D(full.check = TRUE, C = C, D.x = D,
+ v.x = 0, VF.x = por, dx = dx, dy = dy,
+ C.x.up = C.x.up, a.bl.x.up = a.bl, C.x.down = C.x.down)
+ }
> system.time(test1())
```

Outputting:

```
  user  system elapsed
  0.35    0.00    0.35
> test2 <- function() {
+ for (i in 1:n.iterate )
+ ST <- tran.2D(full.output = TRUE, C = C, D.x = D,
+ v.x = 0, VF.x = por, dx = dx, dy = dy,
+ C.x.up = C.x.up, a.bl.x.up = a.bl, C.x.down = C.x.down)
+ }
> system.time(test2())
```

Outputting:

```
  user  system elapsed
  0.25    0.00    0.24
> test3 <- function() {
+ for (i in 1:n.iterate )
+ ST <- tran.2D(full.output = TRUE, full.check = TRUE, C = C,
+ D.x = D, v.x = 0, VF.x = por, dx = dx, dy = dy,
+ C.x.up = C.x.up, a.bl.x.up = a.bl, C.x.down = C.x.down)
+ }
> system.time(test3())
```

```
# Outputting:
  user  system elapsed
  0.36    0.00    0.36
> ## =============================================
> ## A 2D model with diffusion in x- and y-direction and first-order =
> ## consumption - inefficient implementation                        =
> ## =============================================
> N <- 51 # number of grid cells
> XX <- 10 # total size
> dy <- dx <- XX/N # grid size
> Dy <- Dx <- 0.1 # diffusion coeff, X- and Y-direction
> r <- 0.005 # consumption rate
> ini <- 1 # initial value at x=0
> N2 <- ceiling(N/2)
> X <- seq (dx, by = dx, len = (N2-1))
> X <- c(-rev(X), 0, X)
> # The model equations
> Diff2D <- function (t, y, parms) {
+ CONC <- matrix(nrow = N, ncol = N, y)
+ dCONC <- tran.2D(CONC, D.x = Dx, D.y = Dy, dx = dx, dy = dy)$dC + r
+ * CONC
+ return (list(dCONC))
+ }
> # initial condition: 0 everywhere, except in central point
> y <- matrix(nrow = N, ncol = N, data = 0)
> y[N2, N2] <- ini # initial concentration in the central point...
> # solve for 10 time units
> times <- 0:10
> out <- ode.2D (y = y, func = Diff2D, t = times, parms = NULL,
+ dim = c(N,N), lrw = 160000)
> pm <- par (mfrow = c(2, 2))
> # Compare solution with analytical solution...
> for (i in seq(2, 11, by = 3)) {
+ tt <- times[i]
+ mat <- matrix(nrow = N, ncol = N,
+ data = subset(out, time == tt))
+ plot(X, mat[N2,], type = "l", main = paste("time=", times[i]),
+ ylab = "Conc", col = "red")
+ ana <- ini*dx^2/(4*pi*Dx*tt)*exp(r*tt-X^2/(4*Dx*tt))
+ points(X, ana, pch = "+")
+ }
> legend ("bottom", col = c("red","black"), lty = c(1, NA),
+ pch = c(NA, "+"), c("tran.2D", "exact"))
> par("mfrow" = pm )
```

```
> #=============================================================
> #  A 2D model with diffusion in x- and y-direction and first-order          =
> #  consumption - more efficient implementation, specifying ALL 2D grids . =
> # =============================================================
> N <- 51 # number of grid cells
> Dy <- Dx <- 0.1 # diffusion coeff, X- and Y-direction
> r <- 0.005 # consumption rate
> ini <- 1 # initial value at x=0
> x.grid <- setup.grid.1D(x.up = -5, x.down = 5, N = N)
> y.grid <- setup.grid.1D(x.up = -5, x.down = 5, N = N)
> grid2D <- setup.grid.2D(x.grid, y.grid)
> D.grid <- setup.prop.2D(value = Dx, y.value = Dy, grid = grid2D)
> v.grid <- setup.prop.2D(value = 0, grid = grid2D)
> A.grid <- setup.prop.2D(value = 1, grid = grid2D)
> AFDW.grid <- setup.prop.2D(value = 1, grid = grid2D)
> VF.grid <- setup.prop.2D(value = 1, grid = grid2D)
>
> # The model equations - using the grids
> Diff2Db <- function (t, y, parms) {
+ CONC <- matrix(nrow = N, ncol = N, data = y)
+ dCONC <- tran.2D(CONC, grid = grid2D, D.grid = D.grid,
+ A.grid = A.grid, VF.grid = VF.grid, AFDW.grid = AFDW.grid,
+ v.grid = v.grid)$dC + r * CONC
+ return (list(dCONC))
+ }
> # initial condition: 0 everywhere, except in central point
> y <- matrix(nrow = N, ncol = N, data = 0)
> y[N2,N2] <- ini # initial concentration in the central point...
> # solve for 8 time units
> times <- 0:8
> outb <- ode.2D (y = y, func = Diff2Db, t = times, parms = NULL,
+ dim = c(N, N), lrw = 160000)
> image(outb, ask = FALSE, mfrow = c(3, 3), main = paste("time", times))
>
> # =============================================================
> # Same 2D model, but now with spatially variable diffusion coefficients  =
> # =============================================================
> N <- 51 # number of grid cells
> r <- 0.005 # consumption rate
> ini <- 1 # initial value at x=0
> N2 <- ceiling(N/2)
> D.grid <- list()
> # Diffusion on x-interfaces
> D.grid$x.int <- matrix(nrow = N+1, ncol = N, data = runif(N*(N+1)))
> # Diffusion on y-interfaces
> D.grid$y.int <- matrix(nrow = N, ncol = N+1, data = runif(N*(N+1)))
```

```
> dx <- 10/N
> dy <- 10/N
> # The model equations
> Diff2Dc <- function (t, y, parms) {
+ CONC <- matrix(nrow = N, ncol = N, data = y)
+ dCONC <- tran.2D(CONC, dx = dx, dy = dy, D.grid = D.grid)$dC + r *
+ CONC
+ return (list(dCONC))
+ }
> # initial condition: 0 everywhere, except in central point
> y <- matrix(nrow = N, ncol = N, data = 0)
> y[N2, N2] <- ini # initial concentration in the central point...
> # solve for 8 time units
> times <- 0:8
> outc <- ode.2D (y = y, func = Diff2Dc, t = times, parms = NULL,
+ dim = c(N, N), lrw = 160000)
> outtimes <- c(1, 3, 5, 7)
> image(outc, ask = FALSE, mfrow = c(2, 2), main = paste("time", outtimes),
+ legend = TRUE, add.contour = TRUE, subset = time %in% outtimes)
>
```

```
> # In the R domain:
>
> install.packages("ReacTran")
Installing package into 'C:/Users/Bert/Documents/R/win-library/3.3'
(as 'lib' is unspecified)
--- Please select a CRAN mirror for use in this session ---
A CRAN mirror is selected.
also installing the dependencies 'rootSolve', 'deSolve'

trying URL 'https://mirrors.tuna.tsinghua.edu.cn/CRAN/bin/windows/contrib/3.3/
rootSolve_1.7.zip'
Content type 'application/zip' length 683839 bytes (667 KB)
downloaded 667 KB

trying URL 'https://mirrors.tuna.tsinghua.edu.cn/CRAN/bin/windows/contrib/3.3/
deSolve_1.20.zip'
Content type 'application/zip' length 2667028 bytes (2.5 MB)
downloaded 2.5 MB

trying URL 'https://mirrors.tuna.tsinghua.edu.cn/CRAN/bin/windows/contrib/3.3/
ReacTran_1.4.3.1.zip'
Content type 'application/zip' length 915996 bytes (894 KB)
downloaded 894 KB
```

```
package 'rootSolve' successfully unpacked and MD5 sums checked
package 'deSolve' successfully unpacked and MD5 sums checked
package 'ReacTran' successfully unpacked and MD5 sums checked

The downloaded binary packages are in

C:\Users\Bert\AppData\Local\Temp\RtmpmEkAb2\downloaded_packages
> require(ReacTran)
Loading required package: ReacTran
Loading required package: rootSolve
Loading required package: deSolve
Loading required package: shape
> library(ReacTran)
> ls("package:ReacTran") # Listing all the programs in this package:
 [1] "advection.1D"       "advection.volume.1D"  "fiadeiro"
 [4] "g.cylinder"         "g.sphere"             "g.spheroid"
 [7] "p.exp"              "p.lin"                "p.sig"
[10] "polar2cart"         "setup.compaction.1D"  "setup.grid.1D"
[13] "setup.grid.2D"      "setup.prop.1D"        "setup.prop.2D"
[16] "tran.1D"            "tran.2D"              "tran.3D"
[19] "tran.cylindrical"   "tran.polar"           "tran.spherical"
[22] "tran.volume.1D"     "tran.volume.2D"       "tran.volume.3D"
> tran.2D # Outputting the program listing of "tran.2D"
> tran.2D
function (C, C.x.up = C[1, ], C.x.down = C[nrow(C), ], C.y.up = C[,
   1], C.y.down = C[, ncol(C)], flux.x.up = NULL, flux.x.down = NULL,

   flux.y.up = NULL, flux.y.down = NULL, a.bl.x.up = NULL, a.bl.x.down = NULL,
   a.bl.y.up = NULL, a.bl.y.down = NULL, D.grid = NULL, D.x = NULL,
   D.y = D.x, v.grid = NULL, v.x = 0, v.y = 0, AFDW.grid = NULL,
   AFDW.x = 1, AFDW.y = AFDW.x, VF.grid = NULL, VF.x = 1, VF.y = VF.x,
   A.grid = NULL, A.x = 1, A.y = 1, grid = NULL, dx = NULL,
   dy = NULL, full.check = FALSE, full.output = FALSE)
{
   if (is.null(grid))
      if (is.null(dx) | is.null(dy))
         stop("error: either grid or dx and dy should be specified  ")
   Nx <- nrow(C)
   Ny <- ncol(C)
   if (is.null(grid)) {
      DX <- if (is.list(dx))
         dx$dx
      else rep(dx, length.out = Nx)
      DXaux <- if (is.list(dx))
         dx$dx.aux
      else 0.5 * (c(0, rep(dx, length.out = Nx)) + c(rep(dx,
```

```
        length.out = Nx), 0))
    DY <- if (is.list(dy))
        dy$dx
    else rep(dy, length.out = Ny)
    DYaux <- if (is.list(dy))
   dy$dx.aux
    else 0.5 * (c(0, rep(dy, length.out = Ny)) + c(rep(dy,
        length.out = Ny), 0))
    grid <- list(dx = matrix(nrow = Nx, ncol = Ny, data = DX),
        dx.aux = matrix(nrow = Nx + 1, ncol = Ny, data = DXaux),
        dy = matrix(nrow = Nx, ncol = Ny, data = DY, byrow = TRUE),
        dy.aux = matrix(nrow = Nx, ncol = Ny + 1, data = DYaux,
            byrow = TRUE))
  }
  else if (class(grid) != "grid.2D") {
    if (length(grid$dx) != Nx * Ny)
        grid$dx <- matrix(nrow = Nx, ncol = Ny, grid$dx)
    if (length(grid$dx.aux) != (Nx + 1) * Ny)
        grid$dx.aux <- matrix(nrow = Nx + 1, ncol = Ny, grid$dx.aux)
    if (length(grid$dy) != Nx * Ny)
        grid$dy <- matrix(nrow = Nx, ncol = Ny, data = grid$dy,
            byrow = TRUE)
    if (length(grid$dy.aux) != Nx * (Ny + 1))
        grid$dy.aux <- matrix(nrow = Nx, ncol = Ny + 1, grid$dy.aux,
            byrow = TRUE)
  }
  gridFill <- function(G.x, G.y, Name) {
    if (is.null(G.x) | is.null(G.y))
        stop((paste("error: ", Name, "and (", Name, ".x and",
            Name, ".y) cannot be NULL at the same time",
            del = "")))
 G.grid <- list()
    if (is.matrix(G.x)) {
        if (sum(abs(dim(G.x) - c(Nx + 1, Ny))) != 0)
            stop(paste("error: ", Name, ".x matrix not of correct (Nx+1) Ny
dimensions",
             del = ""))
        G.grid$x.int <- G.x
        G.grid$x.mid <- 0.5 * (G.x[1:Nx, ] + G.x[2:(Nx +
            1), ])
    }
    else if (class(G.x) == "prop.1D") {
        G.grid$x.int <- matrix(data = G.x$int, nrow = (Nx +
            1), ncol = Ny)
        G.grid$x.mid <- matrix(data = G.x$mid, nrow = Nx,
            ncol = Ny)
```

```
}
else if (length(G.x) == 1) {
   G.grid$x.int <- matrix(data = G.x, nrow = (Nx + 1),
      ncol = Ny)
   G.grid$x.mid <- matrix(data = G.x, nrow = Nx, ncol = Ny)
}
else if (length(G.x) != Nx + 1) {
   stop(paste("error: ", Name, ".x should be a vector of length 1 or Nx+1",
      del = ""))
}

else {
   G.grid$x.int <- matrix(data = G.x, nrow = (Nx + 1),
      ncol = Ny)
   G.grid$x.mid <- matrix(data = 0.5 * (G.x[1:Nx] +
      G.x[2:(Nx + 1)]), nrow = Nx, ncol = Ny)
}
if (is.matrix(G.y)) {
   if (sum(abs(dim(G.y) - c(Nx, Ny + 1))) != 0)
   stop(paste("error: ", Name, ".y matrix not of correct Nx(Ny+1)dimensions",
       del = ""))
   G.grid$y.int <- G.y
   G.grid$y.mid <- 0.5 * (G.y[, 1:Ny] + G.y[, 2:(Ny +
      1)])
}
else if (class(G.y) == "prop.1D") {
   G.grid$y.int <- matrix(data = G.y$int, nrow = Nx,
      ncol = (Ny + 1))
   G.grid$y.mid <- matrix(data = G.y$mid, nrow = Nx,
      ncol = Ny)
}
else if (length(G.y) == 1) {
   G.grid$y.int <- matrix(data = G.y, nrow = Nx, ncol = (Ny +
      1))
   G.grid$y.mid <- matrix(data = G.y, nrow = Nx, ncol = Ny)
}
   else if (length(G.y) != Ny + 1) {
   stop(paste("error: ", Name, ".y should be a vector of length 1 or Ny+1",
      del = ""))
}
else {
   G.grid$y.int <- matrix(data = G.y, nrow = Nx, ncol = (Ny +
      1))
   G.grid$y.mid <- matrix(data = 0.5 * (G.y[1:Nx] +
      G.y[2:(Nx + 1)]), nrow = Nx, ncol = Ny)
}
```

```
    G.grid
}
if (is.null(VF.grid))
    VF.grid <- gridFill(VF.x, VF.y, "VF")
if (is.null(A.grid))
    A.grid <- gridFill(A.x, A.y, "A")
gridInt <- function(G.x, G.y, Name) {
    if (is.null(G.x) | is.null(G.y))
        stop((paste("error: ", Name, "and (", Name, ".x and",
            Name, ".y) cannot be NULL at the same time",
            del = "")))
    G.grid <- list()
    if (is.matrix(G.x)) {
        if (sum(abs(dim(G.x) - c(Nx + 1, Ny))) != 0)
            stop(paste("error: ", Name, ".x matrix not of correct (Nx+1) Ny
      dimensions",
              del = ""))
        G.grid$x.int <- G.x
    }
    else if (class(G.x) == "prop.1D") {
        G.grid$x.int <- matrix(data = G.x$int, nrow = (Nx +
            1), ncol = Ny)
    }
    else if (length(G.x) == 1) {
        G.grid$x.int <- matrix(data = G.x, nrow = (Nx + 1),
            ncol = Ny)
    }
    else if (length(G.x) != Nx + 1) {
        stop(paste("error: ", Name, ".x should be a vector of length 1 or Nx+1",
            del = ""))
    }
    else {
        G.grid$x.int <- matrix(data = G.x, nrow = (Nx + 1),
            ncol = Ny)
    }
    if (is.matrix(G.y)) {
        if (sum(abs(dim(G.y) - c(Nx, Ny + 1))) != 0)
       stop(paste("error: ", Name, ".y matrix not of correct Nx(Ny+1)dimensions",
              del = ""))
      G.grid$y.int <- G.y
    }
    else if (class(G.y) == "prop.1D") {
        G.grid$y.int <- t(matrix(data = G.y$int, ncol = Nx,
            nrow = (Ny + 1)))
    }
    else if (length(G.y) == 1) {
```

```
        G.grid$y.int <- matrix(data = G.y, nrow = Nx, ncol = (Ny +
          1))
      }
      else if (length(G.y) != Ny + 1) {
        stop(paste("error: ", Name, ".y should be a vector of length 1 or Ny+1",
          del = ""))
      }
      else {
        G.grid$y.int <- matrix(data = G.y, nrow = Nx, ncol = (Ny +
          1))
      }
      G.grid
    }
    if (is.null(AFDW.grid))
      AFDW.grid <- gridInt(AFDW.x, AFDW.y, "AFDW")
    if (is.null(D.grid))
      D.grid <- gridInt(D.x, D.y, "D")
    if (is.null(v.grid))
      v.grid <- gridInt(v.x, v.y, "v")
    if (full.check) {
      if (!is.null(C.x.up)) {
        if (!((length(C.x.up) == 1) || (length(C.x.up) ==
          (Ny))))
          stop("error: C.x.up should be a vector of length 1 or ncol(C)")
      }
      if (!is.null(C.x.down)) {
        if (!((length(C.x.down) == 1) || (length(C.x.down) ==
          (Ny))))
          stop("error: C.x.down should be a vector of length 1 or ncol(C)")
      }
      if (!is.null(C.y.up)) {
        if (!((length(C.y.up) == 1) || (length(C.y.up) ==
          (Nx))))
          stop("error: C.y.up should be a vector of length 1 or nrow(C)")
      }
      if (!is.null(C.y.down)) {
        if (!((length(C.y.down) == 1) || (length(C.y.down) ==
          (Nx))))
          stop("error: C.y.down should be a vector of length 1 or nrow(C)")
      }
      if (!is.null(flux.x.up)) {
        if (!((length(flux.x.up) == 1) || (length(flux.x.up) ==
         (Ny))))
          stop("error: flux.x.up should be a vector of length 1 or ncol(C)")
      }
      if (!is.null(flux.x.down)) {
```

```
    if (!((length(flux.x.down) == 1) || (length(flux.x.down) ==
      (Ny))))
      stop("error: flux.x.down should be a vector of length 1 or ncol(C)")
  }
  if (!is.null(flux.y.up)) {
    if (!((length(flux.y.up) == 1) || (length(flux.y.up) ==
      (Nx))))
      stop("error: flux.y.up should be a vector of length 1 or nrow(C)")
  }
  if (!is.null(flux.y.down)) {
    if (!((length(flux.y.down) == 1) || (length(flux.y.down) ==
      (Nx))))
      stop("error: flux.y.down should be a vector of length 1 or nrow(C)")
  }
  if (is.null(dx) && is.null(dy) && is.null(grid))
    stop("error: dx, dy, and grid cannot be NULL at the same time")
  gn <- names(grid)
  if (!"dx" %in% gn)
    stop("error: grid should be a list that contains 'dx' ")
  if (!"dx.aux" %in% gn)
    stop("error: grid should be a list that contains 'dx.aux' ")
  if (!"dy" %in% gn)
    stop("error: grid should be a list that contains 'dy' ")
  if (!"dy.aux" %in% gn)
    stop("error: grid should be a list that contains 'dy.aux' ")
  if (is.null(grid$dx) || is.null(grid$dx.aux))
          stop("error: the grid should be a list with (numeric) val-
ues for 'dx' and 'dx.aux' ")
  if (is.null(grid$dy) || is.null(grid$dy.aux))
          stop("error: the grid should be a list with (numeric) val-
ues for 'dy' and 'dy.aux' ")
  if (any(grid$dx <= 0) || any(grid$dx.aux <= 0))
    stop("error: the grid distances dx and dx.aux should always be positive")
  if (any(grid$dy <= 0) || any(grid$dy.aux <= 0))
    stop("error: the grid distances dy and dy.aux should always be positive")
  gn <- names(AFDW.grid)
  if (!"x.int" %in% gn)
    stop("error: AFDW.grid should be a list that contains 'x.int', the AFDW val-
ues at the interfaces of the grid cells in x-direction")
  if (!"y.int" %in% gn)
    stop("error: AFDW.grid should be a list that contains 'y.int', the AFDW val-
ues at the interfaces of the grid cells in y-direction")
  if (is.null(AFDW.grid$x.int))
    stop("error: AFDW.grid$x.int should be a list with (numeric) values")
  if (is.null(AFDW.grid$y.int))
    stop("error: AFDW.grid$y.int should be a list with (numeric) values")
```

```
    if (any(AFDW.grid$x.int < 0) || any(AFDW.grid$x.int >
      1))
      stop("error: the AFDW should range between 0 and 1")
    if (any(AFDW.grid$y.int < 0) || any(AFDW.grid$y.int >
      1))
      stop("error: the AFDW should range between 0 and 1")
    gn <- names(D.grid)
    if (!"x.int" %in% gn)
      stop("error: D.grid should be a list that contains 'x.int', the D values at the
interfaces of the grid cells in x-direction")
    if (!"y.int" %in% gn)
      stop("error: D.grid should be a list that contains 'y.int', the D values at the
interfaces of the grid cells in y-direction")
    if (is.null(D.grid$x.int))
      stop("error: D.grid$x.int should be a list with (numeric) values")
    if (is.null(D.grid$y.int))
      stop("error: D.grid$y.int should be a list with (numeric) values")
    if (any(D.grid$x.int < 0) || any(D.grid$y.int < 0))
      stop("error: the diffusion coefficient should always be positive")
    gn <- names(v.grid)
    if (!"x.int" %in% gn)
      stop("error: v.grid should be a list that contains 'x.int', the velocity values at
the interfaces of the grid cells in x-direction")
    if (!"y.int" %in% gn)
      stop("error: v.grid should be a list that contains 'y.int', the velocity val-
ues at the interfaces of the grid cells in y-direction")
    if (is.null(v.grid$x.int))
    stop("error: the advective velocity v.grid$x.int should be a list with (numeric)
values")
    if (is.null(v.grid$y.int))
     stop("error: the advective velocity v.grid$y.int should be a list with (numeric)
values")
    gn <- names(VF.grid)
    if (!"x.int" %in% gn)
      stop("error: VF.grid should be a list that contains 'x.int'")
    if (!"y.int" %in% gn)
      stop("error: VF.grid should be a list that contains 'y.int'")
    if (!"x.mid" %in% gn)
      stop("error: VF.grid should be a list that contains 'x.mid'")
    if (!"y.mid" %in% gn)
      stop("error: VF.grid should be a list that contains 'y.mid'")
    if (is.null(VF.grid$x.int) || is.null(VF.grid$y.int) ||
      is.null(VF.grid$x.mid) || is.null(VF.grid$y.mid))
      stop("error: VF should contain (numeric) values")
    if (any(VF.grid$x.int < 0) || any(VF.grid$y.int < 0) ||
      any(VF.grid$x.mid < 0) || any(VF.grid$y.mid < 0))
```

```
      stop("error: the VF values should always be positive")
    gn <- names(A.grid)
    if (!"x.int" %in% gn)
      stop("error: A.grid should be a list that contains 'x.int'")
    if (!"y.int" %in% gn)
      stop("error: A.grid should be a list that contains 'y.int'")
    if (!"x.mid" %in% gn)
      stop("error: A.grid should be a list that contains 'x.mid'")
    if (!"y.mid" %in% gn)
      stop("error: A.grid should be a list that contains 'y.mid'")
    if (is.null(A.grid$x.int) || is.null(A.grid$y.int) ||
      is.null(A.grid$x.mid) || is.null(A.grid$y.mid))
      stop("error: the VF should contain (numeric) values")
    if (any(A.grid$x.int < 0) || any(A.grid$y.int < 0) ||
      any(A.grid$x.mid < 0) || any(A.grid$y.mid < 0))
      stop("error: the A values should always be positive")
  }
  if (!is.null(flux.x.up[1])) {
    nom <- flux.x.up + VF.grid$x.int[1, ] * (D.grid$x.int[1,
      ]/grid$dx.aux[1, ] + (1 - AFDW.grid$x.int[1, ]) *
      v.grid$x.int[1, ]) * C[1, ]
    denom <- VF.grid$x.int[1, ] * (D.grid$x.int[1, ]/grid$dx.aux[1,
      ] + AFDW.grid$x.int[1, ] * v.grid$x.int[1, ])
    C.x.up <- nom/denom
  }
  if (!is.null(flux.x.down[1])) {
    nom <- flux.x.down - VF.grid$x.int[(Nx + 1), ] * (D.grid$x.int[(Nx +
      1), ]/grid$dx.aux[Nx + 1, ] + AFDW.grid$x.int[(Nx +
      1), ] * v.grid$x.int[(Nx + 1), ]) * C[Nx, ]
    denom <- -VF.grid$x.int[(Nx + 1), ] * (D.grid$x.int[(Nx +
      1), ]/grid$dx.aux[Nx + 1, ] + (1 - AFDW.grid$x.int[(Nx +
      1), ]) * v.grid$x.int[(Nx + 1), ])
    C.x.down <- nom/denom
  }
   if (!is.null(flux.y.up[1])) {
    nom <- flux.y.up + VF.grid$y.int[, 1] * (D.grid$y.int[,
      1]/grid$dy.aux[, 1] + (1 - AFDW.grid$y.int[, 1]) *
      v.grid$y.int[, 1]) * C[, 1]
    denom <- VF.grid$y.int[, 1] * (D.grid$y.int[, 1]/grid$dy.aux[,
      1] + AFDW.grid$y.int[, 1] * v.grid$y.int[, 1])
    C.y.up <- nom/denom
  }
  if (!is.null(flux.y.down[1])) {
    nom <- flux.y.down - VF.grid$y.int[, (Ny + 1)] * (D.grid$y.int[,
      (Ny + 1)]/grid$dy.aux[, (Ny + 1)] + AFDW.grid$y.int[,
      (Ny + 1)] * v.grid$y.int[, (Ny + 1)]) * C[, Ny]
```

```
    denom <- -VF.grid$y.int[, (Ny + 1)] * (D.grid$y.int[,
      (Ny + 1)]/grid$dy.aux[, Ny + 1] + (1 - AFDW.grid$y.int[,
      (Ny + 1)]) * v.grid$y.int[, (Ny + 1)])
    C.y.down <- nom/denom
}
if (!is.null(a.bl.x.up) & !is.null(C.x.up[1])) {
    nom <- a.bl.x.up * C.x.up + VF.grid$x.int[1, ] * (D.grid$x.int[1,
      ]/grid$dx.aux[1, ] + (1 - AFDW.grid$x.int[1, ]) *
      v.grid$x.int[1, ]) * C[1, ]
    denom <- a.bl.x.up + VF.grid$x.int[1, ] * (D.grid$x.int[1,
      ]/grid$dx.aux[1, ] + AFDW.grid$x.int[1, ] * v.grid$x.int[1,
      ])
    C.x.up <- nom/denom
}
  if (!is.null(a.bl.x.down) & !is.null(C.x.down[1])) {
    nom <- a.bl.x.down * C.x.down + VF.grid$x.int[(Nx + 1),
      ] * (D.grid$x.int[(Nx + 1), ]/grid$dx.aux[(Nx + 1),
      ] + (1 - AFDW.grid$x.int[(Nx + 1), ]) * v.grid$x.int[(Nx +
      1), ]) * C[Nx, ]
    denom <- a.bl.x.down + VF.grid$x.int[(Nx + 1), ] * (D.grid$x.int[(Nx +
      1), ]/grid$dx.aux[(Nx + 1), ] + AFDW.grid$x.int[(Nx +
      1), ] * v.grid$x.int[(Nx + 1), ])
    C.x.down <- nom/denom
}
if (!is.null(a.bl.y.up) & !is.null(C.y.up[1])) {
    nom <- a.bl.y.up * C.y.up + VF.grid$y.int[, 1] * (D.grid$y.int[,
      1]/grid$dy.aux[, 1] + (1 - AFDW.grid$y.int[, 1]) *
      v.grid$y.int[, 1]) * C[, 1]
    denom <- a.bl.y.up + VF.grid$y.int[, 1] * (D.grid$y.int[,
      1]/grid$dy.aux[, 1] + AFDW.grid$y.int[, 1] * v.grid$y.int[,
      1])
    C.y.up <- nom/denom
}
if (!is.null(a.bl.y.down) & !is.null(C.y.down[1])) {
    nom <- a.bl.y.down * C.y.down + VF.grid$y.int[, (Ny +
      1)] * (D.grid$y.int[, (Ny + 1)]/grid$dy.aux[, (Ny +
      1)] + (1 - AFDW.grid$y.int[, (Ny + 1)]) * v.grid$y.int[,
      (Ny + 1)]) * C[, Ny]
    denom <- a.bl.y.down + VF.grid$y.int[, (Ny + 1)] * (D.grid$y.int[,
     (Ny + 1)]/grid$dy.aux[, (Ny + 1)] + AFDW.grid$y.int[,
     (Ny + 1)] * v.grid$y.int[, (Ny + 1)])
    C.y.down <- nom/denom
}
x.Dif.flux <- as.matrix(-VF.grid$x.int * D.grid$x.int * diff(rbind(C.x.up,
    C, C.x.down, deparse.level = 0))/grid$dx.aux)
y.Dif.flux <- as.matrix(-VF.grid$y.int * D.grid$y.int * t(diff(t(cbind(C.y.up,
```

```
    C, C.y.down, deparse.level = 0))))/grid$dy.aux)
  x.Adv.flux <- 0
  if (any(v.grid$x.int > 0)) {
    vv <- v.grid$x.int
    vv[vv < 0] <- 0
    x.Adv.flux <- x.Adv.flux + as.matrix(VF.grid$x.int *
      vv * (AFDW.grid$x.int * rbind(C.x.up, C, deparse.level = 0) +
      (1 - AFDW.grid$x.int) * rbind(C, C.x.down, deparse.level = 0)))
  }
  if (any(v.grid$x.int < 0)) {
    vv <- v.grid$x.int
    vv[vv > 0] <- 0
    x.Adv.flux <- x.Adv.flux + as.matrix(VF.grid$x.int *
      vv * ((1 - AFDW.grid$x.int) * rbind(C.x.up, C, deparse.level = 0) +
      AFDW.grid$x.int * rbind(C, C.x.down, deparse.level = 0)))
  }
  y.Adv.flux <- 0
  if (any(v.grid$y.int > 0)) {
    vv <- v.grid$y.int
    vv[vv < 0] <- 0
    y.Adv.flux <- y.Adv.flux + as.matrix(VF.grid$y.int *
      vv * (AFDW.grid$y.int * cbind(C.y.up, C, deparse.level = 0) +
      (1 - AFDW.grid$y.int) * cbind(C, C.y.down, deparse.level = 0)))
  }
  if (any(v.grid$y.int < 0)) {
    vv <- v.grid$y.int
    vv[vv > 0] <- 0
    y.Adv.flux <- y.Adv.flux + as.matrix(VF.grid$y.int *
      vv * ((1 - AFDW.grid$y.int) * cbind(C.y.up, C, deparse.level = 0) +
      AFDW.grid$y.int * cbind(C, C.y.down, deparse.level = 0)))
  }
  x.flux <- x.Dif.flux + x.Adv.flux
  y.flux <- y.Dif.flux + y.Adv.flux
  if (!is.null(flux.x.up[1]))
    x.flux[1, ] <- flux.x.up
  if (!is.null(flux.x.down[1]))
    x.flux[nrow(x.flux), ] <- flux.x.down
  if (!is.null(flux.y.up[1]))
    y.flux[, 1] <- flux.y.up
  if (!is.null(flux.y.down[1]))
    y.flux[, ncol(y.flux)] <- flux.y.down
  dFdx <- -(diff(A.grid$x.int * x.flux)/A.grid$x.mid)/grid$dx/VF.grid$x.mid
  dFdy <- -t(diff(t(A.grid$y.int * y.flux))/t(A.grid$y.mid))/grid$dy/VF.grid$y.mid
 if (!full.output) {
    return(list(dC = dFdx + dFdy, flux.x.up = x.flux[1, ],
      flux.x.down = x.flux[nrow(x.flux), ], flux.y.up = y.flux[,
```

```
            1], flux.y.down = y.flux[, ncol(y.flux)]))
    }
    else {
        return(list(dC = dFdx + dFdy, C.x.up = C.x.up, C.x.down = C.x.down,
            C.y.up = C.y.up, C.y.down = C.y.down, x.flux = x.flux,
            y.flux = y.flux, flux.x.up = x.flux[1, ], flux.x.down = x.flux[nrow(x.flux),
                ], flux.y.up = y.flux[, 1], flux.y.down = y.flux[,
                ncol(y.flux)]))
    }
}
<environment: namespace:ReacTran>
((((
<environment: namespace:ReacTran>
>##
> # Running the program tran.2D

> ##================
> ## Testing the functions  =
> ## ===============
> # Parameters
> F <- 100 # input flux [micromol cm-2 yr-1]
> por <- 0.8 # constant porosity
> D <- 400 # mixing coefficient [cm2 yr-1]
> v <- 1 # advective velocity [cm yr-1]
> # Grid definition
> x.N <- 4 # number of cells in x-direction
> y.N <- 6 # number of cells in y-direction
> x.L <- 8 # domain size x-direction [cm]
> y.L <- 24 # domain size y-direction [cm]
> dx <- x.L/x.N # cell size x-direction [cm]
> dy <- y.L/y.N # cell size y-direction [cm]
> # Initial conditions
> C <- matrix(nrow = x.N, ncol = y.N, data = 0, byrow = FALSE)
> # Boundary conditions: fixed concentration
> C.x.up <- rep(1, times = y.N)
> C.x.down <- rep(0, times = y.N)
> C.y.up <- rep(1, times = x.N)
> C.y.down <- rep(0, times = x.N)
> # Only diffusion
> tran.2D(C = C, D.x = D, D.y = D, v.x = 0, v.y = 0,
+ VF.x = por, VF.y = por, dx = dx, dy = dy,
+ C.x.up = C.x.up, C.x.down = C.x.down,
+ C.y.up = C.y.up, C.y.down = C.y.down, full.output = TRUE)
$dC
```

```
     [,1] [,2]  [,3] [,4]  [,5] [,6]
[1,]  250  200  200  200  200  200
[2,]   50    0    0    0    0    0
[3,]   50    0    0    0    0    0
[4,]   50    0    0    0    0    0

$C.x.up
[1] 1 1 1 1 1 1

$C.x.down
[1] 0 0 0 0 0 0

$C.y.up
[1] 1 1 1 1

$C.y.down
[1] 0 0 0 0

$x.flux
     [,1] [,2] [,3] [,4] [,5] [,6]
[1,]  320  320  320  320  320  320
[2,]    0    0    0    0    0    0
[3,]    0    0    0    0    0    0
[4,]    0    0    0    0    0    0
[5,]    0    0    0    0    0    0

$y.flux
     [,1] [,2] [,3] [,4] [,5] [,6] [,7]
[1,]  160    0    0    0    0    0    0
[2,]  160    0    0    0    0    0    0
[3,]  160    0    0    0    0    0    0
[4,]  160    0    0    0    0    0    0

$flux.x.up
[1] 320 320 320 320 320 320

$flux.x.down
[1] 0 0 0 0 0 0

$flux.y.up
[1] 160 160 160 160

$flux.y.down
[1] 0 0 0 0
```

```
> # Strong advection, backward (default), central and forward
> #finite difference schemes
> tran.2D(C = C, D.x = D, v.x = 100*v,
+ VF.x = por, dx = dx, dy = dy,
+ C.x.up = C.x.up, C.x.down = C.x.down,
+ C.y.up = C.y.up, C.y.down = C.y.down)
$dC

     [,1] [,2] [,3] [,4] [,5] [,6]
[1,]  300  250  250  250  250  250
[2,]   50    0    0    0    0    0
[3,]   50    0    0    0    0    0
[4,]   50    0    0    0    0    0

$flux.x.up
[1] 400 400 400 400 400 400

$flux.x.down
[1] 0 0 0 0 0 0

$flux.y.up
[1] 160 160 160 160

$flux.y.down
[1] 0 0 0 0

> tran.2D(AFDW.x = 0.5, C = C, D.x = D, v.x = 100*v,
+ VF.x = por, dx = dx, dy = dy,
+ C.x.up = C.x.up, C.x.down = C.x.down,
+ C.y.up = C.y.up, C.y.down = C.y.down)
$dC

     [,1] [,2] [,3] [,4] [,5] [,6]
[1,]  275  225  225  225  225  225
[2,]   50    0    0    0    0    0
[3,]   50    0    0    0    0    0
[4,]   50    0    0    0    0    0

$flux.x.up
[1] 360 360 360 360 360 360

$flux.x.down
[1] 0 0 0 0 0 0

$flux.y.up
[1] 160 160 160 160
```

```
$flux.y.down
[1] 0 0 0 0

> tran.2D(AFDW.x = 0, C = C, D.x = D, v.x = 100*v,
+ VF.x = por, dx = dx, dy = dy,
+ C.x.up = C.x.up, C.x.down = C.x.down,
+ C.y.up = C.y.up, C.y.down = C.y.down)
$dC

     [,1] [,2] [,3] [,4] [,5] [,6]
[1,]  250  200  200  200  200  200
[2,]   50    0    0    0    0    0
[3,]   50    0    0    0    0    0
[4,]   50    0    0    0    0    0

$flux.x.up
[1] 320 320 320 320 320 320

$flux.x.down
[1] 0 0 0 0 0 0

$flux.y.up
[1] 160 160 160 160

$flux.y.down
[1] 0 0 0 0

> # Boundary conditions: fixed fluxes
> flux.x.up <- rep(200, times = y.N)
> flux.x.down <- rep(-200, times = y.N)
> flux.y.up <- rep(200, times = x.N)
> flux.y.down <- rep(-200, times = x.N)
> tran.2D(C = C, D.x = D, v.x = 0,
+ VF.x = por, dx = dx, dy = dy,
+ flux.x.up = flux.x.up, flux.x.down = flux.x.down,
+ flux.y.up = flux.y.up, flux.y.down = flux.y.down)
$dC

      [,1] [,2] [,3] [,4] [,5]  [,6]
[1,] 187.5  125  125  125  125 187.5
[2,]  62.5    0    0    0    0  62.5
[3,]  62.5    0    0    0    0  62.5
[4,] 187.5  125  125  125  125 187.5

$flux.x.up
[1] 200 200 200 200 200 200
```

```
$flux.x.down
[1] -200 -200 -200 -200 -200 -200

$flux.y.up
[1] 200 200 200 200

$flux.y.down
[1] -200 -200 -200 -200

> # Boundary conditions: convective boundary layer on all sides
> a.bl <- 800 # transfer coefficient
> C.x.up <- rep(1, times = (y.N)) # fixed conc at boundary layer
> C.y.up <- rep(1, times = (x.N)) # fixed conc at boundary layer
> tran.2D(full.output = TRUE, C = C, D.x = D, v.x = 0,
+ VF.x = por, dx = dx, dy = dy,
+ C.x.up = C.x.up, a.bl.x.up = a.bl,
+ C.x.down = C.x.up, a.bl.x.down = a.bl,
+ C.y.up = C.y.up, a.bl.y.up = a.bl,
+ C.y.down = C.y.up, a.bl.y.down = a.bl)
$dC

          [,1]     [,2]     [,3]     [,4]     [,5]      [,6]
[1,] 184.52381 142.8571 142.8571 142.8571 142.8571 184.52381
[2,]  41.66667   0.0000   0.0000   0.0000   0.0000  41.66667
[3,]  41.66667   0.0000   0.0000   0.0000   0.0000  41.66667
[4,] 184.52381 142.8571 142.8571 142.8571 142.8571 184.52381

$C.x.up
[1] 0.7142857 0.7142857 0.7142857 0.7142857 0.7142857 0.7142857

$C.x.down
[1] 0.7142857 0.7142857 0.7142857 0.7142857 0.7142857 0.7142857

$C.y.up
[1] 0.8333333 0.8333333 0.8333333 0.8333333

$C.y.down
[1] 0.8333333 0.8333333 0.8333333 0.8333333

$x.flux

         [,1]     [,2]     [,3]     [,4]     [,5]     [,6]
[1,] 228.5714 228.5714 228.5714 228.5714 228.5714 228.5714
[2,]   0.0000   0.0000   0.0000   0.0000   0.0000   0.0000
[3,]   0.0000   0.0000   0.0000   0.0000   0.0000   0.0000
```

```
[4,]     0.0000    0.0000    0.0000    0.0000    0.0000    0.0000
[5,] -228.5714 -228.5714 -228.5714 -228.5714 -228.5714 -228.5714

$y.flux
         [,1] [,2] [,3] [,4] [,5] [,6]      [,7]
[1,] 133.3333    0    0    0    0    0 -133.3333
[2,] 133.3333    0    0    0    0    0 -133.3333
[3,] 133.3333    0    0    0    0    0 -133.3333
[4,] 133.3333    0    0    0    0    0 -133.3333

$flux.x.up
[1] 228.5714 228.5714 228.5714 228.5714 228.5714 228.5714

$flux.x.down
[1] -228.5714 -228.5714 -228.5714 -228.5714 -228.5714 -228.5714

$flux.y.up
[1] 133.3333 133.3333 133.3333 133.3333

$flux.y.down
[1] -133.3333 -133.3333 -133.3333 -133.3333


> # Runtime test with and without argument checking
> n.iterate <-500
> test1 <- function() {
+ for (i in 1:n.iterate )
+ ST <- tran.2D(full.check = TRUE, C = C, D.x = D,
+ v.x = 0, VF.x = por, dx = dx, dy = dy,
+ C.x.up = C.x.up, a.bl.x.up = a.bl, C.x.down = C.x.down)
+ }
> system.time(test1())
   user  system elapsed
   0.35    0.00    0.35
> test2 <- function() {
+ for (i in 1:n.iterate )
+ ST <- tran.2D(full.output = TRUE, C = C, D.x = D,
+ v.x = 0, VF.x = por, dx = dx, dy = dy,
+ C.x.up = C.x.up, a.bl.x.up = a.bl, C.x.down = C.x.down)
+ }
> system.time(test2())
   user  system elapsed
   0.25    0.00    0.24
> test3 <- function() {
+ for (i in 1:n.iterate )
+ ST <- tran.2D(full.output = TRUE, full.check = TRUE, C = C,
```

```
+ D.x = D, v.x = 0, VF.x = por, dx = dx, dy = dy,
+ C.x.up = C.x.up, a.bl.x.up = a.bl, C.x.down = C.x.down)
+ }
> system.time(test3())
  user  system elapsed
  0.36    0.00    0.36
> ## ===========================================
> ## A 2D model with diffusion in x- and y-direction and first-order =
> ## consumption - inefficient implementation                        =
> ## ===========================================
> N <- 51 # number of grid cells
> XX <- 10 # total size
> dy <- dx <- XX/N # grid size
> Dy <- Dx <- 0.1 # diffusion coeff, X- and Y-direction
> r <- 0.005 # consumption rate
> ini <- 1 # initial value at x=0
> N2 <- ceiling(N/2)
> X <- seq (dx, by = dx, len = (N2-1))
> X <- c(-rev(X), 0, X)
> # The model equations
> Diff2D <- function (t, y, parms) {
+ CONC <- matrix(nrow = N, ncol = N, y)
+ dCONC <- tran.2D(CONC, D.x = Dx, D.y = Dy, dx = dx, dy = dy)$dC + r * CONC
+ return (list(dCONC))
+ }
> # initial condition: 0 everywhere, except in central point
> y <- matrix(nrow = N, ncol = N, data = 0)
> y[N2, N2] <- ini # initial concentration in the central point...

> # solve for 10 time units
> times <- 0:10
> out <- ode.2D (y = y, func = Diff2D, t = times, parms = NULL,
+ dim = c(N,N), lrw = 160000)
> pm <- par (mfrow = c(2, 2))
> # Compare solution with analytical solution...
> for (i in seq(2, 11, by = 3)) {
+ tt <- times[i]
+ mat <- matrix(nrow = N, ncol = N,
+ data = subset(out, time == tt))
+ #
+ # Fig 1
+ #
+ plot(X, mat[N2,], type = "l", main = paste("time=", times[i]),
+ ylab = "Conc", col = "red")
+ ana <- ini*dx^2/(4*pi*Dx*tt)*exp(r*tt-X^2/(4*Dx*tt))
+ points(X, ana, pch = "+")
```

```
+ }
> #
> # Fig 2
> #
> legend ("bottom", col = c("red","black"), lty = c(1, NA),
+ pch = c(NA, "+"), c("tran.2D", "exact"))
> #
> # Fig 3
> #
> par("mfrow" = pm )
> #
> # Fig 4
> #
> ## ===========================================================
> ## A 2D model with diffusion in x- and y-direction and first-order        =
> ## consumption - more efficient implementation, specifying ALL 2D grids  =
> ## ===========================================================
> N <- 51 # number of grid cells
> Dy <- Dx <- 0.1 # diffusion coeff, X- and Y-direction
> r <- 0.005 # consumption rate
> ini <- 1 # initial value at x=0
> x.grid <- setup.grid.1D(x.up = -5, x.down = 5, N = N)
> y.grid <- setup.grid.1D(x.up = -5, x.down = 5, N = N)
> grid2D <- setup.grid.2D(x.grid, y.grid)
> D.grid <- setup.prop.2D(value = Dx, y.value = Dy, grid = grid2D)
> v.grid <- setup.prop.2D(value = 0, grid = grid2D)
> A.grid <- setup.prop.2D(value = 1, grid = grid2D)
> AFDW.grid <- setup.prop.2D(value = 1, grid = grid2D)
> VF.grid <- setup.prop.2D(value = 1, grid = grid2D)
>
> # The model equations - using the grids
> Diff2Db <- function (t, y, parms) {
+ CONC <- matrix(nrow = N, ncol = N, data = y)
+ dCONC <- tran.2D(CONC, grid = grid2D, D.grid = D.grid,
+ A.grid = A.grid, VF.grid = VF.grid, AFDW.grid = AFDW.grid,
+ v.grid = v.grid)$dC + r * CONC
+ return (list(dCONC))
+ }
> # initial condition: 0 everywhere, except in central point
> y <- matrix(nrow = N, ncol = N, data = 0)
> y[N2,N2] <- ini # initial concentration in the central point...
> # solve for 8 time units
> times <- 0:8
> outb <- ode.2D (y = y, func = Diff2Db, t = times, parms = NULL,
+ dim = c(N, N), lrw = 160000)
> image(outb, ask = FALSE, mfrow = c(3, 3), main = paste("time", times))
```

```
> #
> # Fig 5
> #
>
> ## ===================================================
> ## Same 2D model, but now with spatially variable diffusion coefficients  =
> ## ===================================================
> N <- 51 # number of grid cells
> r <- 0.005 # consumption rate
> ini <- 1 # initial value at x=0
> N2 <- ceiling(N/2)
> D.grid <- list()
> # Diffusion on x-interfaces
> D.grid$x.int <- matrix(nrow = N+1, ncol = N, data = runif(N*(N+1)))
> # Diffusion on y-interfaces
> D.grid$y.int <- matrix(nrow = N, ncol = N+1, data = runif(N*(N+1)))
> dx <- 10/N
> dy <- 10/N
> # The model equations
> Diff2Dc <- function (t, y, parms) {
+ CONC <- matrix(nrow = N, ncol = N, data = y)
+ dCONC <- tran.2D(CONC, dx = dx, dy = dy, D.grid = D.grid)$dC + r * CONC
+ return (list(dCONC))
+ }
> # initial condition: 0 everywhere, except in central point
> y <- matrix(nrow = N, ncol = N, data = 0)
> y[N2, N2] <- ini # initial concentration in the central point...
> # solve for 8 time units
> times <- 0:8
> outc <- ode.2D (y = y, func = Diff2Dc, t = times, parms = NULL,
+ dim = c(N, N), lrw = 160000)
> outtimes <- c(1, 3, 5, 7)
> image(outc, ask = FALSE, mfrow = c(2, 2), main = paste("time", outtimes),
+ legend = TRUE, add.contour = TRUE, subset = time %in% outtimes)

> #
> # Fig 6
> #
>

> # Outputting Figure A2-1: tran.2D 1
> #
> #
> legend ("bottom", col = c("red","black"), lty = c(1, NA),
+ pch = c(NA, "+"), c("tran.2D", "exact"))
>
```

Figure A2-1 tran.2D 1 Preparation of space for subsequent graphical outputs.

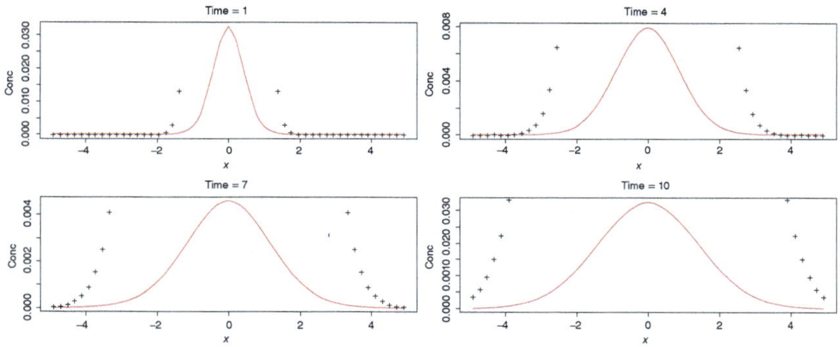

Figure A2-2 tran.2D 2 ini # initial concentration in the central point.

```
> # Outputting Figure A2-2: tran.2D 2
> #
> ##=====================================================
> ## A 2D model with diffusion in x- and y-direction and first-order      =
> ## consumption - more efficient implementation, specifying ALL 2D grids  =
> ## =====================================================
> N <- 51 # number of grid cells
> Dy <- Dx <- 0.1 # diffusion coeff, X- and Y-direction
> r <- 0.005 # consumption rate
> ini <- 1 # initial value at x=0
> x.grid <- setup.grid.1D(x.up = -5, x.down = 5, N = N)
> y.grid <- setup.grid.1D(x.up = -5, x.down = 5, N = N)
> grid2D <- setup.grid.2D(x.grid, y.grid)
> D.grid <- setup.prop.2D(value = Dx, y.value = Dy, grid = grid2D)
> v.grid <- setup.prop.2D(value = 0, grid = grid2D)
> A.grid <- setup.prop.2D(value = 1, grid = grid2D)
> AFDW.grid <- setup.prop.2D(value = 1, grid = grid2D)
> VF.grid <- setup.prop.2D(value = 1, grid = grid2D)
> # The model equations - using the grids
```

```
> Diff2Db <- function (t, y, parms) {
+ CONC <- matrix(nrow = N, ncol = N, data = y)
+ dCONC <- tran.2D(CONC, grid = grid2D, D.grid = D.grid,
+ A.grid = A.grid, VF.grid = VF.grid, AFDW.grid = AFDW.grid,
+ v.grid = v.grid)$dC + r * CONC
+ return (list(dCONC))
+ }
> # initial condition: 0 everywhere, except in central point
> y <- matrix(nrow = N, ncol = N, data = 0)
> y[N2,N2] <- ini # initial concentration in the central point...
> # solve for 8 time units
> times <- 0:8
> outb <- ode.2D (y = y, func = Diff2Db, t = times, parms = NULL,
+ dim = c(N, N), lrw = 160000)
> image(outb, ask = FALSE, mfrow = c(3, 3), main = paste("time", times))
> #
```

> # ***Outputting***: **Figure A2-3:** tran.2D 3

```
> #
> #
> ## ==================================================
> ## Same 2D model, but now with spatially variable diffusion coefficients  =
> ## ==================================================
> N <- 51 # number of grid cells
> r <- 0.005 # consumption rate
> ini <- 1 # initial value at x=0
> N2 <- ceiling(N/2)
> D.grid <- list()
> # Diffusion on x-interfaces
> D.grid$x.int <- matrix(nrow = N+1, ncol = N, data = runif(N*(N+1)))
```

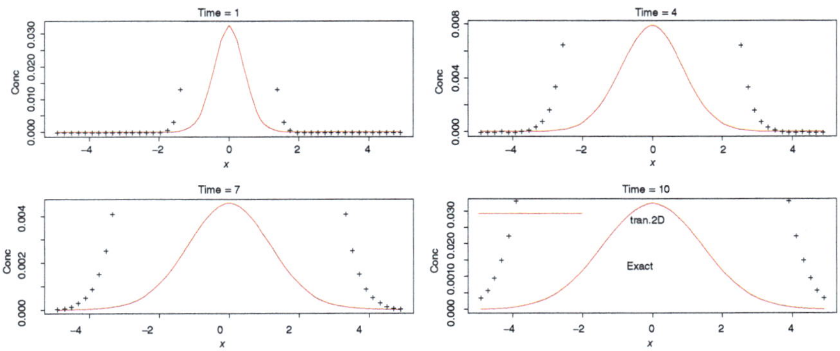

Figure A2-3 tran.2D 3 ini # initial concentration in the central point.

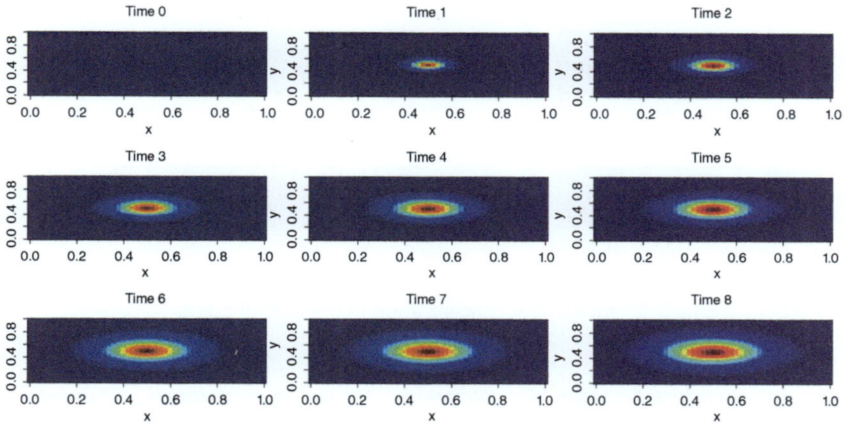

Figure A2-4 tran.2D 4 ini # initial concentration in the central point.

```
> # Diffusion on y-interfaces
> D.grid$y.int <- matrix(nrow = N, ncol = N+1, data = runif(N*(N+1)))
> dx <- 10/N
> dy <- 10/N
> # The model equations
> Diff2Dc <- function (t, y, parms) {
+ CONC <- matrix(nrow = N, ncol = N, data = y)
+ dCONC <- tran.2D(CONC, dx = dx, dy = dy, D.grid = D.grid)$dC + r * CONC
+ return (list(dCONC))
+ }
> # initial condition: 0 everywhere, except in central point
> y <- matrix(nrow = N, ncol = N, data = 0)
> y[N2, N2] <- ini # initial concentration in the central point...
> # solve for 8 time units
> times <- 0:8
> outc <- ode.2D (y = y, func = Diff2Dc, t = times, parms = NULL,
+ dim = c(N, N), lrw = 160000)
> outtimes <- c(1, 3, 5, 7)
> image(outc, ask = FALSE, mfrow = c(2, 2), main = paste("time", outtimes),
> #
```

> # ***Outputting* Figure A2-4:** tran.2D **4**

```
>
+ legend = TRUE, add.contour = TRUE, subset = time %in% outtimes)
> #
```

> # ***Outputting* Figure A2-5:** tran.2D **5**

```
> #
>
```

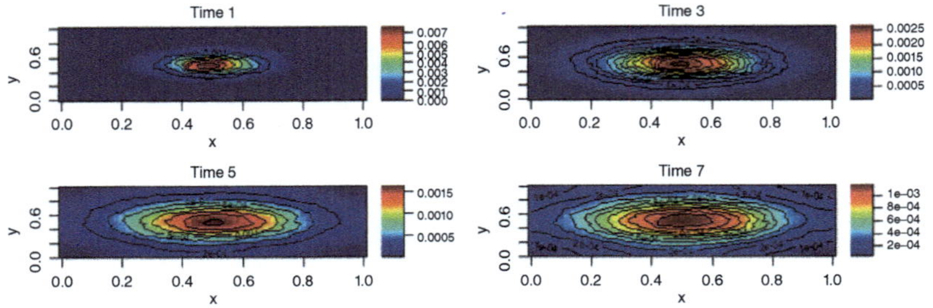

Figure A2-5 tran.2D 5 ini # initial concentration in the central point.

Worked Example 4.4 tran.3D *General Three-Dimensional Advective–Diffusive Transport*

Description

This program estimates the transport term (i.e. the rate of change of a concentration owing to diffusion and advection) in a three-dimensional rectangular model domain.

Usage

```
tran.3D (C, C.x.up = C[1,,], C.x.down = C[dim(C)[1],,],
          C.y.up = C[ ,1, ], C.y.down=C[ ,dim(C)[2], ],
          C.z.up = C[ , ,1], C.z.down=C[ , ,dim(C)[3]],
          flux.x.up = NULL, flux.x.down = NULL,
          flux.y.up = NULL, flux.y.down = NULL,
          flux.z.up = NULL, flux.z.down = NULL,
          a.bl.x.up = NULL, a.bl.x.down = NULL,
          a.bl.y.up = NULL, a.bl.y.down = NULL,
          a.bl.z.up = NULL, a.bl.z.down = NULL,
          D.grid = NULL, D.x = NULL, D.y = D.x, D.z = D.x,
          v.grid = NULL, v.x = 0, v.y = 0, v.z = 0,
          AFDW.grid = NULL, AFDW.x = 1, AFDW.y = AFDW.x,
          AFDW.z = AFDW.x,
          VF.grid = NULL, VF.x = 1, VF.y = VF.x, VF.z = VF.x,
          A.grid = NULL, A.x = 1, A.y = 1, A.z = 1,
          grid = NULL, dx = NULL, dy = NULL, dz = NULL,
          full.check = FALSE, full.output = FALSE)
```

Arguments

C	Concentration, expressed per unit volume, defined at the center of\each grid cell; Nx*Ny*Nz array [M/L^3]
C.x.up	Concentration at upstream boundary in x-direction; matrix of dimensions Ny*Nz [M/L^3]

C.x.down	Concentration at downstream boundary in *x*-direction; matrix of dimensions Ny*Nz [M/L^3]
C.y.up	Concentration at upstream boundary in *y*-direction; matrix of dimensions Nx*Nz [M/L^3]
C.y.down	Concentration at downstream boundary in *y*-direction; matrix of dimensions Nx*Nz [M/L^3]
C.z.up	Concentration at upstream boundary in *z*-direction; matrix of dimensions Nx*Ny [M/L^3]
C.z.down	Concentration at downstream boundary in *z*-direction; matrix of dimensions Nx*Ny [M/L^3]
flux.x.up	Flux across the upstream boundary in *x*-direction, positive = INTO model domain; matrix of dimensions Ny*Nz [M/L^2/T]
flux.x.down	Flux across the downstream boundary in *x*-direction, positive = OUT of model domain; matrix of dimensions Ny*Nz [M/L^2/T]
flux.y.up	Flux across the upstream boundary in *y*-direction, positive = INTO model domain; matrix of dimensions Nx*Nz [M/L^2/T]
flux.y.down	Flux across the downstream boundary in *y*-direction, positive = OUT of model domain; matrix of dimensions Nx*Nz [M/L^2/T]
flux.z.up	Flux across the upstream boundary in *z*-direction, positive = INTO model domain; matrix of dimensions Nx*Ny [M/L^2/T]
flux.z.down	Flux across the downstream boundary in *z*-direction, positive = OUT of model domain; matrix of dimensions Nx*Ny [M/L^2/T]
a.bl.x.up	Transfer coefficient across the upstream boundary layer in *x*-direction `Flux=a.bl.x.up*(C.x.up-C[1,,])`. One value [L/T]
a.bl.x.down	Transfer coefficient across the downstream boundary layer in *x*-direction; `Flux=a.bl.x.down*(C[Nx,,]-C.x.down)`. One value [L/T]
a.bl.y.up	Transfer coefficient across the upstream boundary layer in *y*-direction; `Flux=a.bl.y.up*(C.y.up-C[,1,])`. One value [L/T]
a.bl.y.down	Transfer coefficient across the downstream boundary layer in *y*-direction; `Flux=a.bl.y.down*(C[,Ny,]-C.y.down)`. One value [L/T]
a.bl.z.up	Transfer coefficient across the upstream boundary layer in *y*-direction; `Flux=a.bl.y.up*(C.y.up-C[,,1])`. One value [L/T]
a.bl.z.down	Transfer coefficient across the downstream boundary layer in *z*-direction; `Flux=a.bl.z.down*(C[,,Nz]-C.z.down)`. One value [L/T]

D.grid	Diffusion coefficient defined on all grid cell interfaces. Should contain elements x.int, y.int, z.int, arrays with the values on the interfaces in x-, y-, and z-directions, and with dimensions (Nx+1)*Ny*Nz, Nx*(Ny+1)*Nz, and Nx*Ny*(Nz+1), respectively [L^2/T]
D.x	Diffusion coefficient in x-direction, defined on grid cell interfaces. One value, a vector of length (Nx+1), or a (Nx+1)*Ny*Nz array [L^2/T]
D.y	Diffusion coefficient in y-direction, defined on grid cell interfaces. One value, a vector of length (Ny+1), or a Nx*(Ny+1)*Nz array [L^2/T]
D.z	Diffusion coefficient in z-direction, defined on grid cell interfaces. One value, a vector of length (Nz+1), or a Nx*Ny*(Nz+1) array [L^2/T]
v.grid	Advective velocity defined on all grid cell interfaces. Can be positive (downstream flow) or negative (upstream flow). Should contain elements x.int, y.int, z.int, arrays with the values on the interfaces in x-, y-, and z-directions, and with dimensions (Nx+1)*Ny*Nz, Nx*(Ny+1)*Nz, and Nx*Ny*(Nz+1), respectively [L/T]
v.x	Advective velocity in the x-direction, defined on grid cell interfaces. Can be positive (downstream flow) or negative (upstream flow). One value, a vector of length (Nx+1), or a (Nx+1)*Ny*Nz array [L/T]
v.y	Advective velocity in the y-direction, defined on grid cell interfaces. Can be positive (downstream flow) or negative (upstream flow). One value, a vector of length (Ny+1), or a Nx*(Ny+1)*Nz array [L/T]
v.z	Advective velocity in the z-direction, defined on grid cell interfaces. Can be positive (downstream flow) or negative (upstream flow). One value, a vector of length (Nz+1), or a Nx*Ny*(Nz+1) array [L/T]
AFDW.grid	Weight used in the finite difference scheme for advection in the x-direction, defined on grid cell interfaces; backward = 1, centered = 0.5, forward = 0; default is backward. Should contain elements x.int, y.int, z.int, arrays with the values on the interfaces in x-, y-, and z-directions, and with dimensions (Nx+1)*Ny*Nz, Nx*(Ny+1)*Nz, and Nx*Ny*(Nz+1), respectively [–]
AFDW.x	Weight used in the finite difference scheme for advection in the x-direction, defined on grid cell interfaces; backward = 1, centered = 0.5, forward = 0; default is backward. One value, a vector of length (Nx+1), a prop.1D list created by setup.prop.1D, or a (Nx+1)*Ny*Nz array [–]

AFDW.y	Weight used in the finite difference scheme for advection in the *y*-direction, defined on grid cell interfaces; backward = 1, centered = 0.5, forward = 0; default is backward. One value, a vector of length (Ny+1), a prop.1D list created by setup.prop.1D, or a Nx*(Ny+1)*Nz array [–]
AFDW.z	Weight used in the finite difference scheme for advection in the *z*-direction, defined on grid cell interfaces; backward = 1, centered = 0.5, forward = 0, default is backward. One value, a vector of length (Nz+1), a prop.1D list created by setup.prop.1D, or a Nx*Ny*(Nz+1) array [–]
VF.grid	Volume fraction. A list should contain elements x.int, y.int, z.int, arrays with the values on the interfaces in *x*-, *y*-, and *z*-directions, and with dimensions (Nx+1)*Ny*Nz, Nx*(Ny+1)*Nz, and Nx*Ny*(Nz+1), respectively [–]
VF.x	Volume fraction at the grid cell interfaces in the *x*-direction. One value, a vector of length (Nx+1), a prop.1D list created by setup.prop.1D, or a (Nx+1)*Ny*Nz array [–]
VF.y	Volume fraction at the grid cell interfaces in the *y*-direction. One value, a vector of length (Ny+1), a prop.1D list created by setup.prop.1D, or a Nx*(Ny+1)*Nz array [–]
VF.z	Volume fraction at the grid cell interfaces in the *z*-direction. One value, a vector of length (Nz+1), a prop.1D list created by setup.prop.1D, or a Nx*Ny*(Nz+1) array [–]
A.grid	Interface area, a list should contain elements x.int, y.int, z.int, arrays with the values on the interfaces in *x*-, *y*-, and *z*-directions, and with dimensions (Nx+1)*Ny*Nz, Nx*(Ny+1)*Nz, and Nx*Ny*(Nz+1), respectively [L^2]
A.x	Interface area defined at the grid cell interfaces in the *x*-direction. One value, a vector of length (Nx+1), a prop.1D list created by setup.prop.1D, or a (Nx+1)*Ny*Nz array [L^2]
A.y	Interface area defined at the grid cell interfaces in the *y*-direction. One value, a vector of length (Ny+1), a prop.1D list created by setup.prop.1D, or a Nx*(Ny+1)*Nz array [L^2]
A.z	Interface area defined at the grid cell interfaces in the *z*-direction. One value, a vector of length (Nz+1), a prop.1D list created by setup.prop.1D, or a Nx*Ny*(Nz+1) array [L^2]
dx	Distance between adjacent cell interfaces in the *x*-direction (thickness of grid cells). One value or vector of length Nx [L]
dy	Distance between adjacent cell interfaces in the *y*-direction (thickness of grid cells). One value or vector of length Ny [L]
dz	Distance between adjacent cell interfaces in the *z*-direction (thickness of grid cells). One value or vector of length Nz [L]
grid	Discretization grid, a list containing at least elements dx, dx.aux, dy, dy.aux, dz, dz.aux (see setup.grid.2D) [L]

full.check	Logical flag enabling a full check of the consistency of the arguments (default = FALSE; TRUE slows down execution by 50%)
full.output	Logical flag enabling a full return of the output (default = FALSE; TRUE slows down execution by 20%)

Details

Do not use this with too large grid.

The boundary conditions are either

(1) zero gradient
(2) fixed concentration
(3) convective boundary layer
(4) fixed flux.

This is also the order of priority. The zero gradient is the default, the fixed flux overrules all others.

Value

A list containing

dC	The rate of change of the concentration C owing to transport, defined in the center of each grid cell, an array with dimension Nx*Ny*Nz [M/L^3/T]
C.x.up	Concentration at the upstream interface in x-direction. A matrix of dimension Ny*Nz [M/L^3]. Only when full.output = TRUE
C.x.down	Concentration at the downstream interface in x-direction. A matrix of dimension Ny*Nz [M/L^3]. Only when full.output = TRUE
C.y.up	Concentration at the upstream interface in y-direction. A matrix of dimension Nx*Nz [M/L^3]. Only when full.output = TRUE
C.y.down	Concentration at the downstream interface in y-direction. A matrix of dimension Nx*Nz [M/L^3]. Only when full.output = TRUE
C.z.up	Concentration at the upstream interface in z-direction. A matrix of dimension Nx*Ny [M/L]. Only when full.output = TRUE
C.z.down	Concentration at the downstream interface in z-direction. A matrix of dimension Nx*Ny [M/L^3]. Only when full.output = TRUE

x.flux	Flux across the interfaces in x-direction of the grid cells. An (Nx+1)*Ny*Nz array [M/L^2/T]. Only when full.output = TRUE
y.flux	Flux across the interfaces in y-direction of the grid cells. An Nx*(Ny+1)*Nz array [M/L^2/T]. Only when full.output = TRUE
z.flux	Flux across the interfaces in z-direction of the grid cells. A Nx*Ny*(Nz+1) array [M/L^2/T]. Only when full.output=TRUE
flux.x.up	Flux across the upstream boundary in x-direction, positive = INTO model domain. A matrix of dimension Ny*Nz [M/L^2/T]
flux.x.down	Flux across the downstream boundary in x-direction, positive=OUT model domain. A matrix of dimension Ny*Nz [M/L^2/T]
flux.y.up	Flux across the upstream boundary in y-direction, positive = INTO model domain. A matrix of dimension Nx*Nz [M/L^2/T]
flux.y.down	Flux across the downstream boundary in y-direction, positive = OUT of model domain. A matrix of dimension Nx*Nz [M/L^2/T]
flux.z.up	Flux across the upstream boundary in z-direction, positive = INTO model domain. A matrix of dimension Nx*Ny [M/L^2/T]
flux.z.down	Flux across the downstream boundary in z-direction, positive = OUT of model domain. A matrix of dimension Nx*Ny [M/L^2/T]

Authors

Filip Meysman <filip.meysman@nioz.nl>,
Karline Soetaert <karline.soetaert@nioz.nl>.

See Also

tran.cylindrical, tran.spherical, for a discretization of 3D transport equations in cylindrical and spherical coordinates.
tran.1D, tran.2D

Examples

```
# =================================
# Diffusion in 3D; imposed boundary conditions =
# =================================
```

```
diffusion3D <- function(t, Y, par) {
    yy <- array(dim = c(n, n, n), data = Y) # vector to 3D array
    dY <- -r * yy                           # consumption
    BND <- matrix(nrow = n, ncol = n, 1)    # boundary concentration

    dY <- dY + tran.3D(C = yy,
            C.x.up = BND, C.y.up = BND, C.z.up = BND,
            C.x.down = BND, C.y.down = BND, C.z.down = BND,
            D.x = Dx, D.y = Dy, D.z = Dz,
            dx = dx, dy = dy, dz = dz, full.check = TRUE)$dC
    return(list(dY))
}

# parameters
dy <- dx <- dz <- 1  # grid size
Dy <- Dx <- Dz <- 1  # diffusion coeff, X- and Y-direction
r <- 0.025           # consumption rate

n <- 10
y <- array(dim = c(n, n, n), data = 10.)

print(system.time(
   ST3 <- steady.3D(y, func = diffusion3D, parms = NULL,
                    pos = TRUE, dimens = c(n, n, n),
                    lrw = 2000000, verbose = TRUE)
))

pm <- par(mfrow = c(1,1))
y <- array(dim = c(n, n, n), data = ST3$y)
filled.contour(y[ , ,n/2], color.palette = terrain.colors)

# a selection in the x-direction
image(ST3, mfrow = c(2, 2), add.contour = TRUE, legend = TRUE,
       dimselect = list(x = c(1, 4, 8, 10)))

par(mfrow = pm)
```

In the R domain:

```
>
> # In the R domain:
>
> install.packages("ReacTran")
Installing package into 'C:/Users/Bert/Documents/R/win-library/3.3'
(as 'lib' is unspecified)
```

```
--- Please select a CRAN mirror for use in this session ---
A CRAN mirror is selected.
also installing the dependencies 'rootSolve', 'deSolve'

trying URL 'https://mirrors.tuna.tsinghua.edu.cn/CRAN/bin/windows/contrib/3.3/
rootSolve_1.7.zip'
Content type 'application/zip' length 683839 bytes (667 KB)
downloaded 667 KB

trying URL 'https://mirrors.tuna.tsinghua.edu.cn/CRAN/bin/windows/contrib/3.3/
deSolve_1.20.zip'
Content type 'application/zip' length 2667028 bytes (2.5 MB)
downloaded 2.5 MB

trying URL 'https://mirrors.tuna.tsinghua.edu.cn/CRAN/bin/windows/contrib/3.3/
ReacTran_1.4.3.1.zip'
Content type 'application/zip' length 915996 bytes (894 KB)

downloaded 894 KB
package 'rootSolve' successfully unpacked and MD5 sums checked
package 'deSolve' successfully unpacked and MD5 sums checked
package 'ReacTran' successfully unpacked and MD5 sums checked

The downloaded binary packages are in
C:\Users\Bert\AppData\Local\Temp\RtmpmEkAb2\downloaded_packages
> require(ReacTran)
Loading required package: ReacTran
Loading required package: rootSolve
Loading required package: deSolve
Loading required package: shape
> library(ReacTran)
> ls("package:ReacTran") # Listing all the programs in this package:
 [1] "advection.1D"       "advection.volume.1D"    "fiadeiro"
 [4] "g.cylinder"         "g.sphere"               "g.spheroid"
 [7] "p.exp"              "p.lin"                  "p.sig"
[10] "polar2cart"         "setup.compaction.1D"    "setup.grid.1D"
[13] "setup.grid.2D"      "setup.prop.1D"          "setup.prop.2D"
[16] "tran.1D"            "tran.2D"                "tran.3D"
[19] "tran.cylindrical"   "tran.polar"             "tran.spherical"
[22] "tran.volume.1D"     "tran.volume.2D"         "tran.volume.3D"
> tran.3D # Outputting the program listing of "tran.3D"

> function (C, C.up = C[1], C.down = C[length(C)], flux.up = NULL,
   flux.down = NULL, a.bl.up = NULL, a.bl.down = NULL, D = 0,
   v = 0, AFDW = 1, VF = 1, A = 1, dx, full.check = FALSE, full.output = FALSE)
{
```

```
if (is.null(dx))
   stop("error: dx should be inputted ")
N <- length(C)
if (N == 0)
   stop("C should be a vector with numeric values")
if (is.null(C.up))
   C.up <- C[1]
if (is.null(C.down))
   C.down <- C[N]
if (!is.list(AFDW))
   AFDW <- list(int = rep_len(AFDW, N + 1))
if (!is.list(D))
   D <- list(int = rep_len(D, N + 1))
if (!is.list(v))
   v <- list(int = rep_len(v, N + 1))
if (!is.list(VF))
   VF <- list(int = rep_len(VF, N + 1), mid = 0.5 * (rep_len(VF,
      N + 1)[1:N] + rep_len(VF, N + 1)[2:(N + 1)]))
if (!is.list(A))
   A <- list(int = rep_len(A, N + 1), mid = 0.5 * (rep_len(A,
      N + 1)[1:N] + rep_len(A, N + 1)[2:(N + 1)]))
if (is.list(dx))
   grid <- dx
if (!is.list(dx))
   grid <- list(dx = rep_len(dx, N), dx.aux = 0.5 * (c(0,
      rep_len(dx, N)) + c(rep_len(dx, N), 0)))
if (full.check) {
   gn <- names(AFDW)
   if (!"int" %in% gn)
      stop("error: AFDW should be a list that contains 'int', the AFDW
values at the interface of the grid cell ")
   if (is.null(AFDW$int))
      stop("error: AFDW is NULL, should contain (numeric) values")
   if (!is.null(AFDW$int)) {

   if (!((length(AFDW$int) == 1) || (length(AFDW$int) ==
         (N + 1))))
   stop("error: AFDW should be a vector of length 1 or N+1"
   }
   if (any(AFDW$int < 0) || any(AFDW$int > 1))
      stop("error: the AFDW should always range between 0 and 1")
   gn <- names(D)
   if (!"int" %in% gn)
      stop("error: D should be a list that contains 'int', the D values at the
interface of the grid cell ")
   if (is.null(D$int))
```

```
        stop("error: D is NULL, should contain (numeric) values")
    if (!is.null(D$int)) {
        if (!((length(D$int) == 1) || (length(D$int) == (N +
            1))))
            stop("error: D should be a vector of length 1 or N+1")
    }
    if (any(D$int < 0))
        stop("error: the diffusion coefficient should always be positive")
    gn <- names(v)
    if (!"int" %in% gn)
        stop("error: v should be a list that contains 'int', the v values at the interface
of the grid cell ")
    if (is.null(v$int))
        stop("error: the advective velocity v is NULL, should contain (numeric)
values")
    if (!is.null(v$int)) {
        if (!((length(v$int) == 1) || (length(v$int) == (N +
            1))))
            stop("error: v should be a vector of length 1 or N+1")
    }
    gn <- names(VF)
    if (!"int" %in% gn)
        stop("error: VF should be a list that contains 'int', the area values at the
interface of the grid cell ")
    if (!"mid" %in% gn)
        stop("error: VF should be a list that contains 'mid', the area at the middle of
the grid cells")
    if (is.null(VF$int) || is.null(VF$mid))
        stop("error: the volume fraction VF should contain (numeric) values")
   if (!is.null(VF$int)) {
        if (!((length(VF$int) == 1) || (length(VF$int) ==
            (N + 1))))
            stop("error: VF$int should be a vector of length 1 or N+1")
            }
    if (!is.null(VF$mid)) {
        if (!((length(VF$mid) == 1) || (length(VF$mid) ==
            (N))))
            stop("error: VF$mid should be a vector of length 1 or N")
    }
    if (any(VF$int < 0) || any(VF$mid < 0) || any(VF$int >
        1) || any(VF$mid > 1))
        stop("error: the volume fraction should range between 0 and 1")
    gn <- names(A)
    if (!"int" %in% gn)
        stop("error: A should be a list that contains 'int', the area values at the
interface of the grid cell ")
```

```
        if (!"mid" %in% gn)
              stop("error: A should be a list that contains 'mid', the area at the mid-
dle of the grid cells")
        if (is.null(A$int) || is.null(A$mid))
           stop("error: the surface area A is NULL, should contain (numeric) values")
        if (!is.null(A$int)) {
           if (!((length(A$int) == 1) || (length(A$int) == (N +
              1))))
              stop("error: A$int should be a vector of length 1 or N+1")
        }
        if (!is.null(A$mid)) {
           if (!((length(A$mid) == 1) || (length(A$mid) == (N))))
              stop("error: A$mid should be a vector of length 1 or N")
        }
        if (any(A$int < 0) || any(A$mid < 0))
           stop("error: the area A should always be positive")
        gn <- names(grid)
        if (!"dx" %in% gn)
           stop("error: grid should be a list that contains 'dx' ")
        if (!"dx.aux" %in% gn)
           stop("error: grid should be a list that contains 'dx.aux' ")
        if (is.null(grid$dx) || is.null(grid$dx.aux))
                    stop("error: the grid should be a list with (numeric) val-
ues for 'dx' and 'dx.aux' ")
       if (!is.null(grid$dx)) {
       if (!((length(grid$dx) == 1) || (length(grid$dx) ==
              N)))
              stop("error: dx should be a vector of length 1 or N"
        }
        if (!is.null(grid$dx.aux)) {
           if (!((length(grid$dx.aux) == 1) || (length(grid$dx.aux) ==
              (N + 1))))
              stop("error: dx.aux should be a vector of length 1 or N+1")
        }
        if (any(grid$dx <= 0) || any(grid$dx.aux <= 0))
           stop("error: the grid distances dx and dx.aux should always be positive")
        if (!is.null(a.bl.up) & !is.null(C.up)) {
           if (a.bl.up < 0)
              stop("error: the boundary layer transfer coefficient should be positive")
        }
        if (!is.null(a.bl.down) & !is.null(C.down)) {
           if (a.bl.down < 0)
              stop("error: the boundary layer transfer coefficient should be positive")
        }
     }
     if (full.output) {
```

```
if (!is.null(flux.up)) {
   if (v$int[1] >= 0) {
      nom <- flux.up + VF$int[1] * (D$int[1]/grid$dx.aux[1] +
        (1 - AFDW$int[1]) * v$int[1]) * C[1]
      denom <- VF$int[1] * (D$int[1]/grid$dx.aux[1] +
        AFDW$int[1] * v$int[1])
   }
   else {
      nom <- flux.up + VF$int[1] * (D$int[1]/grid$dx.aux[1] +
        AFDW$int[1] * v$int[1]) * C[1]
      denom <- VF$int[1] * (D$int[1]/grid$dx.aux[1] +
        (1 - AFDW$int[1]) * v$int[1])
   }
   C.up <- nom/denom
}
 if (!is.null(flux.down)) {
   if (v$int[N + 1] >= 0) {
      nom <- flux.down - VF$int[N + 1] * (D$int[N +
        1]/grid$dx.aux[N + 1] + AFDW$int[N + 1] * v$int[N +
        1]) * C[N]
      denom <- -VF$int[N + 1] * (D$int[N + 1]/grid$dx.aux[N +
  1] + (1 - AFDW$int[N + 1]) * v$int[N + 1])
 }
   else {
      nom <- flux.down - VF$int[N + 1] * (D$int[N +
        1]/grid$dx.aux[N + 1] + (1 - AFDW$int[N + 1]) *
        v$int[N + 1]) * C[N]
      denom <- -VF$int[N + 1] * (D$int[N + 1]/grid$dx.aux[N +
        1] + AFDW$int[N + 1] * v$int[N + 1])
   }
   C.down <- nom/denom
}
if (!is.null(a.bl.up) & !is.null(C.up)) {
   if (v$int[1] >= 0) {
      nom <- a.bl.up * C.up + VF$int[1] * (D$int[1]/grid$dx.aux[1] +
        (1 - AFDW$int[1]) * v$int[1]) * C[1]
      denom <- a.bl.up + VF$int[1] * (D$int[1]/grid$dx.aux[1] +
        AFDW$int[1] * v$int[1])
   }
   else {
      nom <- a.bl.up * C.up + VF$int[1] * (D$int[1]/grid$dx.aux[1] +
        AFDW$int[1] * v$int[1]) * C[1]
      denom <- a.bl.up + VF$int[1] * (D$int[1]/grid$dx.aux[1] +
        (1 - AFDW$int[1]) * v$int[1])
   }
   C.up <- nom/denom
```

```
    }
    if (!is.null(a.bl.down) & !is.null(C.down)) {
      if (v$int[N + 1] >= 0) {
        nom <- a.bl.down * C.down + VF$int[N + 1] * (D$int[N +
          1]/grid$dx.aux[N + 1] + (1 - AFDW$int[N + 1]) *
          v$int[N + 1]) * C[N]
        denom <- a.bl.down + VF$int[N + 1] * (D$int[N +
          1]/grid$dx.aux[N + 1] + AFDW$int[N + 1] * v$int[N +
          1])
      }
      else {
          nom <- a.bl.down * C.down + VF$int[N + 1] * (D$int[N +
          1]/grid$dx.aux[N + 1] + AFDW$int[N + 1] * v$int[N +
          1]) * C[N]
        denom <- a.bl.down + VF$int[N + 1] * (D$int[N +
          1]/grid$dx.aux[N + 1] + (1 - AFDW$int[N + 1]) *
          v$int[N + 1])
    }
      C.down <- nom/denom
    }
    dif.flux <- as.vector(-VF$int * D$int * diff(c(C.up,
      C, C.down))/grid$dx.aux)
    adv.flux <- rep(0, length.out = length(dif.flux))
    if (any(v$int > 0)) {
      vv <- v$int
      vv[v$int < 0] <- 0
      conc <- AFDW$int * c(C.up, C)
      if (any(AFDW$int < 1))
        conc <- conc + (1 - AFDW$int) * c(C, C.down)
      adv.flux <- adv.flux + as.vector(VF$int * vv * conc)
    }
    if (any(v$int < 0)) {
      vv <- v$int
      vv[v$int > 0] <- 0
      conc <- AFDW$int * c(C, C.down)
      if (any(AFDW$int < 1))
        conc <- conc + (1 - AFDW$int) * c(C.up, C)
      adv.flux <- adv.flux + as.vector(VF$int * vv * conc)
    }
    flux <- dif.flux + adv.flux
  }
  else {
    if (!is.null(a.bl.up) & !is.null(C.up)) {
      if (v$int[1] >= 0) {
        nom <- a.bl.up * C.up + VF$int[1] * (D$int[1]/grid$dx.aux[1] +
          (1 - AFDW$int[1]) * v$int[1]) * C[1]
```

```
      denom <- a.bl.up + VF$int[1] * (D$int[1]/grid$dx.aux[1] +
        AFDW$int[1] * v$int[1])
    }
    else {
      nom <- a.bl.up * C.up + VF$int[1] * (D$int[1]/grid$dx.aux[1] +
        AFDW$int[1] * v$int[1]) * C[1]
      denom <- a.bl.up + VF$int[1] * (D$int[1]/grid$dx.aux[1] +
        (1 - AFDW$int[1]) * v$int[1])
}
    C.up <- nom/denom
}
if (!is.null(a.bl.down) & !is.null(C.down)) {
    if (v$int[N + 1] >= 0) {
     nom <- a.bl.down * C.down + VF$int[N + 1] * (D$int[N +
        1]/grid$dx.aux[N + 1] + (1 - AFDW$int[N + 1]) *
        v$int[N + 1]) * C[N]
      denom <- a.bl.down + VF$int[N + 1] * (D$int[N +
        1]/grid$dx.aux[N + 1] + AFDW$int[N + 1] * v$int[N +
        1])
    }
    else {
      nom <- a.bl.down * C.down + VF$int[N + 1] * (D$int[N +
        1]/grid$dx.aux[N + 1] + AFDW$int[N + 1] * v$int[N +
        1]) * C[N]
      denom <- a.bl.down + VF$int[N + 1] * (D$int[N +
        1]/grid$dx.aux[N + 1] + (1 - AFDW$int[N + 1]) *
        v$int[N + 1])
    }
    C.down <- nom/denom
}
flux <- as.vector(-(VF$int) * D$int * diff(c(C.up, C,
    C.down))/grid$dx.aux)
if (any(v$int > 0)) {
    vv <- v$int
    vv[v$int < 0] <- 0
    conc <- AFDW$int * c(C.up, C)
    if (any(AFDW$int < 1))
      conc <- conc + (1 - AFDW$int) * c(C, C.down)
    flux <- flux + as.vector(VF$int * vv * conc)
}
if (any(v$int < 0)) {
    vv <- v$int
    vv[v$int > 0] <- 0
    conc <- AFDW$int * c(C, C.down)
    if (any(AFDW$int < 1))
      conc <- conc + (1 - AFDW$int) * c(C.up, C)
```

```
          flux <- flux + as.vector(VF$int * vv * conc)
        }
      }


  if (!is.null(flux.up))
        flux[1] <- flux.up
    if (!is.null(flux.down))
        flux[N + 1] <- flux.down
    dC <- -diff(A$int * flux)/A$mid/VF$mid/grid$dx
    if (!full.output)
      return(list(dC = dC, flux.up = flux[1], flux.down = flux[length(flux)]))
    }
    else {
        return(list(dC = dC, C.up = C.up, C.down = C.down, dif.flux = dif.flux,
          adv.flux = adv.flux, flux = flux, flux.up = flux[1],
          flux.down = flux[length(flux)]))
    }
}
<environment: namespace:ReacTran>
>

> # Running the program tran.3D
>
> ## =====================================================
> ## Diffusion in 3D; imposed boundary conditions
> ## =====================================================
> diffusion3D <- function(t, Y, par) {
+ yy <- array(dim = c(n, n, n), data = Y) # vector to 3D array
+ dY <- -r * yy # consumption
+ BND <- matrix(nrow = n, ncol = n, 1) # boundary concentration
+ dY <- dY + tran.3D(C = yy,
+ C.x.up = BND, C.y.up = BND, C.z.up = BND,
+ C.x.down = BND, C.y.down = BND, C.z.down = BND,
+ D.x = Dx, D.y = Dy, D.z = Dz,
+ dx = dx, dy = dy, dz = dz, full.check = TRUE)$dC
+ return(list(dY))
+ }
> # parameters
> dy <- dx <- dz <- 1 # grid size
> Dy <- Dx <- Dz <- 1 # diffusion coeff, X- and Y-direction
> r <- 0.025 # consumption rate
> n <- 10
> y <- array(dim = c(n, n, n), data = 10.)
> print(system.time(
+ ST3 <- steady.3D(y, func = diffusion3D, parms = NULL,
```

```
+ pos = TRUE, dimens = c(n, n, n),
+ lrw = 2000000, verbose = TRUE)
+ ))
```

> *# Outputting:*

```
[1] "Steady-state settings"
 sparseType                              message
>
 sparseType                              message
1        3D sparse 3D jacobian, calculated internally
[1] "estimated number of nonzero elements:  6910"
[1] "estimated number of function calls:  1001"
[1] "number of species:  1"
[1] "dimensions:  10 10 10"
[1] "cyclic boundaries:  0 0 0"
mean residual derivative 4.73661e-07
[1] "precision at each steady state step"
[1] 1.105000e+01 4.736612e-07
[1] ""
[1] "-------------------------"
[1] " Memory requirements"
[1] "-------------------------"
  par                                        mess   val
1  nnz                    the number of nonzero elements:  6400
2  ngp  the number of independent groups of state variables:     12
3  nsp       the length of the work array actually required: 82366
  user  system elapsed
  0.05    0.03    0.11
>
> pm <- par(mfrow = c(1,1))
```

> *# Outputting:*

```
> #
> # Figure A3-1: tran-3D 1
> #
> y <- array(dim = c(n, n, n), data = ST3$y)
> filled.contour(y[ , ,n/2], color.palette = terrain.colors)
```

> *# Outputting:*

```
> #
> # Figure A3-2: tran-3D 2
> #

> # a selection in the x-direction
> image(ST3, mfrow = c(2, 2), add.contour = TRUE, legend = TRUE,
+ dimselect = list(x = c(1, 4, 8, 10)))
```

Figure A3-1 tran-3D 1 Preparation of space for subsequent graphical outputs.

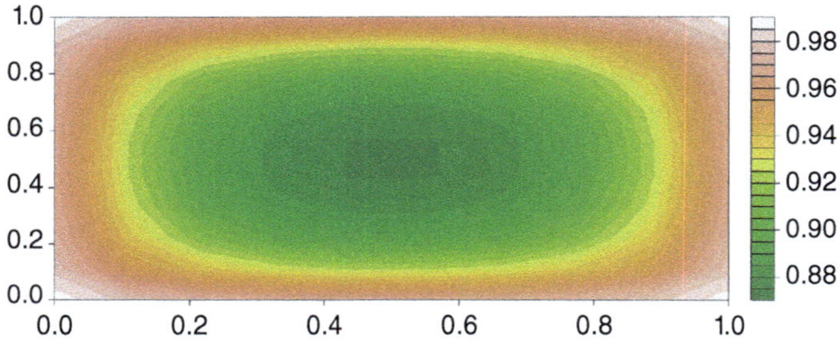

Figure A3-2

trans.3D **2**

```
> y <- array(dim = c(n, n, n), data = ST3$y)
                    > filled.contour(y[ , ,n/2], color.palette = terrain.colors)
```

Contours of flow patterns.

```
> #
> # Outputting:> #
> # Figure A3-3: tran-3D 3
> #
> par(mfrow = pm)
>
```

Worked Example 4.5 tran.cylindrical ***Diffusive Transport in Cylindrical (r, θ, z), and tran.spherical in Spherical (r, θ, ϕ) Coordinates***

Description

This program estimates the transport term (i.e. the rate of change of a concentration owing to diffusion) in a cylindrical (r, θ, z) or spherical (r, θ, ϕ) coordinate system.

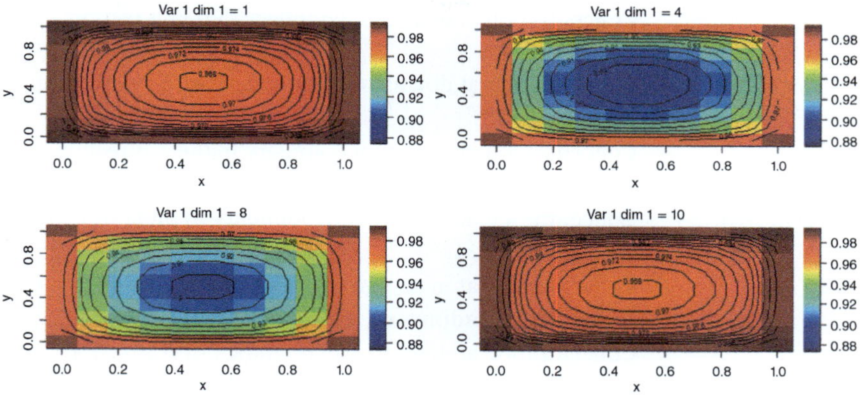

Figure A3-3

```
trans.3D 3
> image(ST3, mfrow = c(2, 2),
                               add.contour = TRUE,
                   + legend = TRUE, dimselect = list(x = c(1, 4, 8, 10)))
```

Contours of flow patterns.

Usage

```
tran.cylindrical (C, C.r.up = NULL, C.r.down = NULL,
                  C.theta.up = NULL, C.theta.down = NULL,
                  C.z.up = NULL, C.z.down = NULL,
                  flux.r.up = NULL, flux.r.down = NULL,
                  flux.theta.up = NULL, flux.theta.down = NULL,
                  flux.z.up = NULL, flux.z.down = NULL,
                  cyclicBnd = NULL,
                  D.r = NULL, D.theta = D.r, D.z = D.r,
                  r = NULL, theta = NULL, z = NULL)

tran.spherical  (C, C.r.up = NULL, C.r.down = NULL,
                  C.theta.up = NULL, C.theta.down = NULL,
                  C.phi.up = NULL, C.phi.down = NULL,
                  flux.r.up = NULL, flux.r.down = NULL,
                  flux.theta.up = NULL, flux.theta.down = NULL,
                  flux.phi.up = NULL, flux.phi.down = NULL,
                  cyclicBnd = NULL,
                  D.r = NULL, D.theta = D.r, D.phi = D.r,
                  r = NULL, theta = NULL, phi = NULL)
```

Arguments

C	Concentration, expressed per unit volume, defined at the center of each grid cell; Nr*Ntheta*Nz (cylindrical) or Nr*Ntheta*Nphi (spherical coordinates) array $[M/L^3]$

C.r.up	Concentration at upstream boundary in $r(x)$-direction; one value [M/L^3]
C.r.down	Concentration at downstream boundary in $r(x)$-direction; one value [M/L^3]
C.theta.up	Concentration at upstream boundary in theta-direction; one value [M/L^3]
C.theta.down	Concentration at downstream boundary in theta-direction; one value [M/L^3]
C.z.up	Concentration at upstream boundary in z-direction (cylindrical coordinates); one value [M/L^3]
C.z.down	Concentration at downstream boundary in z-direction (cylindrical coordinates); one value [M/L^3]
C.phi.up	Concentration at upstream boundary in phi-direction (spherical coordinates); one value [M/L^3]
C.phi.down	Concentration at downstream boundary in phi-direction (spherical coordinates); one value [M/L^3]
flux.r.up	Flux across the upstream boundary in r-direction, positive = INTO model domain; one value [M/L^2/T]
flux.r.down	Flux across the downstream boundary in r-direction, positive = OUT of model domain; one value [M/L^2/T]
flux.theta.up	Flux across the upstream boundary in theta-direction, positive = INTO model domain; one value [M/L^2/T]
flux.theta.down	Flux across the downstream boundary in theta-direction, positive = OUT of model domain; one value [M/L^2/T]
flux.z.up	Flux across the upstream boundary in z-direction (cylindrical coordinates); positive = INTO model domain; one value [M/L^2/T]
flux.z.down	Flux across the downstream boundary in z-direction, (cylindrical coordinates); positive = OUT of model domain; one value [M/L^2/T]
flux.phi.up	Flux across the upstream boundary in phi-direction (spherical coordinates); positive = INTO model domain; one value [M/L^2/T]
flux.phi.down	Flux across the downstream boundary in phi-direction, (spherical coordinates); positive = OUT of model domain; one value [M/L^2/T]
cyclicBnd	If not NULL, the direction in which a cyclic boundary is defined, i.e. cyclicBnd = 1 for the r-direction, cyclicBnd = 2 for the theta-direction and cyclicBnd = c(1,2) for both the r- and theta-directions
D.r	Diffusion coefficient in the r-direction, defined on grid cell interfaces. One value or a vector of length (Nr+1), [L^2/T]
D.theta	Diffusion coefficient in theta-direction, defined on grid cell interfaces. One value or a vector of length (Ntheta+1) [L^2/T]

D.z	Diffusion coefficient in z-direction, defined on grid cell interfaces for cylindrical coordinates. One value or a vector of length (Nz+1) [L^2/T]
D.phi	Diffusion coefficient in phi-direction, defined on grid cell interfaces for cylindrical coordinates. One value or a vector of length (Nphi+1) [L^2/T]
r	Position of adjacent cell interfaces in the r-direction. A vector of length Nr+1 [L]
theta	Position of adjacent cell interfaces in the theta-direction. A vector of length Ntheta+1 [L]. Theta should be within [0,2 pi]
z	Position of adjacent cell interfaces in the z-direction (cylindrical coordinates). A vector of length Nz+1 [L]
phi	Position of adjacent cell interfaces in the phi-direction (spherical coordinates). A vector of length Nphi+1 [L]. Phi should be within [0,2 pi]

Details

tran.cylindrical performs (diffusive) transport in cylindrical coordinates tran.spherical performs (diffusive) transport in spherical coordinates

The **boundary conditions** are either

(1) zero gradient
(2) fixed concentration
(3) fixed flux
(4) cyclic boundary.

This is also the order of priority. The cyclic boundary overrules the other. If fixed concentration, fixed flux, and cyclicBnd are NULL then the boundary is zero gradient.

A cyclic boundary condition has concentration and flux at upstream and downstream boundary the same. It is useful mainly for the theta- and phi-directions.

---Do not expect too much of this equation: do not try to use it with many boxes---

Value

A list containing

dC	The rate of change of the concentration C due to transport, defined in the center of each grid cell, a Nr*Ntheta*Nz (cylindrical) or Nr*Ntheta*Nphi (spherical coordinates) array [M/L^3/T]
flux.r.up	Flux across the upstream boundary in r-direction, positive = INTO model domain. A matrix of dimension Ntheta*Nz (cylindrical) or Ntheta*Nphi (spherical) [M/L^2/T]

flux.r.down	Flux across the downstream boundary in *r*-direction, positive = OUT of model domain. A matrix of dimension Ntheta*Nz (cylindrical) or Ntheta*Nphi (spherical) $[M/L^2/T]$
flux.theta.up	Flux across the upstream boundary in theta-direction, positive = INTO model domain. A matrix of dimension Nr*Nz (cylindrical) or Nr*Nphi (spherical) $[M/L^2/T]$
flux.theta.down	Flux across the downstream boundary in theta-direction, positive = OUT of model domain. A matrix of dimension Nr*Nz (cylindrical) or Nr*Nphi (spherical) $[M/L^2/T]$
flux.z.up	Flux across the upstream boundary in *z*-direction, for cylindrical coordinates; positive = OUT of model domain. A matrix of dimension Nr*Ntheta $[M/L^2/T]$
flux.z.down	Flux across the downstream boundary in *z*-direction for cylindrical coordinates; positive = OUT of model domain. A matrix of dimension Nr*Ntheta $[M/L^2/T]$
flux.phi.up	Flux across the upstream boundary in phi-direction, for spherical coordinates; positive = OUT of model domain. A matrix of dimension Nr*Ntheta $[M/L^2/T]$
flux.phi.down	Flux across the downstream boundary in phi-direction, for spherical coordinates; positive = OUT of model domain. A matrix of dimension Nr*Ntheta $[M/L^2/T]$

See Also

`tran.polar` for a discretization of 2D transport equations in polar coordinates.
`tran.1D`, `tran.2D`, `tran.3D`

Examples

```
# ================
# Testing the functions =
# ================
#
#Grid definition
r.N     <- 4      # number of cells in r-direction
theta.N <- 6      # number of cells in theta-direction
z.N     <- 3      # number of cells in z-direction

D       <- 100   # diffusion coefficient

r      <- seq(0, 8, len = r.N+1)          # cell size r-direction [cm]
theta <- seq(0,2*pi, len = theta.N+1) # theta-direction - theta: from 0, 2pi
phi    <- seq(0,2*pi, len = z.N+1)       # phi-direction (0,2pi)
z      <- seq(0,5, len = z.N+1)           # cell size z-direction [cm]

# Initial conditions
```

```
C <- array(dim = c(r.N, theta.N, z.N), data = 0)
# Concentration boundary conditions
tran.cylindrical (C = C, D.r = D, D.theta = D,
C.r.up = 1, C.r.down = 1,
C.theta.up = 1, C.theta.down = 1,
C.z.up = 1, C.z.down = 1,
r = r, theta = theta, z = z )

tran.spherical (C = C, D.r = D, D.theta = D,
  C.r.up = 1, C.r.down = 1, C.theta.up = 1, C.theta.down = 1,
  C.phi.up = 1, C.phi.down = 1,
  r = r, theta = theta, phi = phi)

# Flux boundary conditions
tran.cylindrical(C = C, D.r = D, r = r, theta = theta, z = z,
  flux.r.up = 10, flux.r.down = 10,
  flux.theta.up = 10, flux.theta.down = 10,
  flux.z.up = 10, flux.z.down = 10)

tran.spherical(C = C, D.r = D, r = r, theta = theta, phi = phi,
  flux.r.up = 10, flux.r.down = 10,
  flux.theta.up = 10, flux.theta.down = 10,
  flux.phi.up = 10, flux.phi.down = 10)

# cyclic boundary conditions
tran.cylindrical(C = C, D.r = D, r = r, theta = theta, z = z,
  cyclicBnd = 1:3)
tran.spherical(C = C, D.r = D, r = r, theta = theta, phi = phi,
  cyclicBnd = 1:3)

# zero-gradient boundary conditions
tran.cylindrical(C = C, D.r = D, r = r, theta = theta, z = z)
tran.spherical(C = C, D.r = D, r = r, theta = theta, phi = phi)

# ==========================================
# A model with diffusion and first-order consumption  =
# ==========================================

N <- 10                # number of grid cells
rr <- 0.005            # consumption rate
D <- 400

r          <- seq (2, 4, len = N+1)
theta      <- seq (0, 2*pi, len = N+1)
z          <- seq (0, 3, len = N+1)
phi        <- seq (0, 2*pi, len = N+1)
```

```
# The model equations
Diffcylin <- function (t, y, parms) {
  CONC <- array(dim = c(N, N, N), data = y)
  tran    <- tran.cylindrical(CONC,
            D.r = D, D.theta = D, D.z = D,
            r = r, theta = theta, z = z,
            C.r.up = 0, C.r.down = 1,
            cyclicBnd = 2)
 dCONC <- tran$dC - rr * CONC
 return (list(dCONC))
}

Diffspher <- function (t, y, parms) {
   CONC <- array(dim = c(N, N, N), data = y)
      tran <- tran.spherical (CONC,
            D.r = D, D.theta = D, D.phi = D,
            r = r, theta = theta, phi = phi,
            C.r.up = 0, C.r.down = 1,
            cyclicBnd = 2:3)
  dCONC <- tran$dC - rr * CONC
  return (list(dCONC))
}

# initial condition: 0 everywhere, except in central point
y    <- array(dim = c(N, N, N), data = 0)
N2   <- ceiling(N/2)

y[N2, N2, N2] <- 100    # initial concentration in the central point...

# solve to steady-state; cyclicBnd = 2,
outcyl <- steady.3D (y = y, func = Diffcylin, parms = NULL,
dim = c(N, N, N), lrw = 1e6, cyclicBnd = 2)

STDcyl <- array(dim = c(N, N, N), data = outcyl$y)
image(STDcyl[,,1])

# For spherical coordinates, cyclic Bnd = 2, 3
outspher <- steady.3D (y = y, func = Diffspher, parms = NULL, pos=TRUE,
                    dim = c(N, N, N), lrw = 1e6, cyclicBnd = 2:3)

#STDspher <- array(dim = c(N, N, N), data = outspher$y)
#image(STDspher[,,1])

## Not run:
    image(outspher)

## End(Not run)
```

In the R domain:

```
>
> # In the R domain:
>
> install.packages("ReacTran")
Installing package into 'C:/Users/Bert/Documents/R/win-library/3.3'
(as 'lib' is unspecified)
--- Please select a CRAN mirror for use in this session ---
A CRAN mirror is selected.
also installing the dependencies 'rootSolve', 'deSolve'

trying URL 'https://mirrors.tuna.tsinghua.edu.cn/CRAN/bin/windows/contrib/3.3/
rootSolve_1.7.zip'
Content type 'application/zip' length 683839 bytes (667 KB)
downloaded 667 KB

trying URL 'https://mirrors.tuna.tsinghua.edu.cn/CRAN/bin/windows/contrib/3.3/
deSolve_1.20.zip'
Content type 'application/zip' length 2667028 bytes (2.5 MB)
downloaded 2.5 MB

trying URL 'https://mirrors.tuna.tsinghua.edu.cn/CRAN/bin/windows/contrib/3.3/
ReacTran_1.4.3.1.zip'

Content type 'application/zip' length 915996 bytes (894 KB)
downloaded 894 KB

package 'rootSolve' successfully unpacked and MD5 sums checked
package 'deSolve' successfully unpacked and MD5 sums checked
package 'ReacTran' successfully unpacked and MD5 sums checked

The downloaded binary packages are in
C:\Users\Bert\AppData\Local\Temp\RtmpmEkAb2\downloaded_packages
> require(ReacTran)
Loading required package: ReacTran
Loading required package: rootSolve
Loading required package: deSolve
Loading required package: shape
> library(ReacTran)
> ls("package:ReacTran") # Listing all the programs in this package:
 [1] "advection.1D"      "advection.volume.1D"    "fiadeiro"
 [4] "g.cylinder"        "g.sphere"               "g.spheroid"
 [7] "p.exp"             "p.lin"                  "p.sig"
[10] "polar2cart"        "setup.compaction.1D"    "setup.grid.1D"
[13] "setup.grid.2D"     "setup.prop.1D"          "setup.prop.2D"
[16] "tran.1D"           "tran.2D"                "tran.3D"
```

```
[19] "tran.cylindrical"   "tran.polar"            "tran.spherical"
[22] "tran.volume.1D"   "tran.volume.2D"        "tran.volume.3D"
> tran.cylindrical # Outputting the program listing of "tran.cylindrical "
>
> ## ==============================================================
> ## Testing the functions
> ## ==============================================================
> # Grid definition
> r.N <- 4 # number of cells in r-direction
> theta.N <- 6 # number of cells in theta-direction
> z.N <- 3 # number of cells in z-direction
> D <- 100 # diffusion coefficient
> r <- seq(0, 8, len = r.N+1) # cell size r-direction [cm]
> theta <- seq(0,2*pi, len = theta.N+1) # theta-direction - theta: from 0, 2pi
> phi <- seq(0,2*pi, len = z.N+1) # phi-direction (0,2pi)
> z <- seq(0,5, len = z.N+1) # cell size z-direction [cm]
> # Initial conditions
> C <- array(dim = c(r.N, theta.N, z.N), data = 0)
> # Concentration boundary conditions
> tran.cylindrical (C = C, D.r = D, D.theta = D,
+ C.r.up = 1, C.r.down = 1,
+ C.theta.up = 1, C.theta.down = 1,
+ C.z.up = 1, C.z.down = 1,
+ r = r, theta = theta, z = z )
$dC
, , 1
            [,1 ]      [,2]      [,3]      [,4]      [,5]       [,6]
[1,] 426107.503  72.0000  72.0000  72.0000  72.0000 426107.503
[2,]  47409.278  72.0000  72.0000  72.0000  72.0000  47409.278
[3,]  17113.420  72.0000  72.0000  72.0000  72.0000  17113.420
[4,]   8823.745 129.1429 129.1429 129.1429 129.1429   8823.745
, , 2
          [,1]        [,2]      [,3]      [,4]      [,5]       [,6]
[1,] 426035.503  0.00000  0.00000  0.00000  0.00000 426035.503
[2,]  47337.278  0.00000  0.00000  0.00000  0.00000  47337.278
[3,]  17041.420  0.00000  0.00000  0.00000  0.00000  17041.420
[4,]   8751.745 57.14286 57.14286 57.14286 57.14286   8751.745
, , 3
            [,1]      [,2]      [,3]      [,4]      [,5]       [,6]
[1,] 426107.503  72.0000  72.0000  72.0000  72.0000 426107.503
[2,]  47409.278  72.0000  72.0000  72.0000  72.0000  47409.278
[3,]  17113.420  72.0000  72.0000  72.0000  72.0000  17113.420
[4,]   8823.745 129.1429 129.1429 129.1429 129.1429   8823.745

$flux.r.up
     [,1] [,2] [,3]
[1,] 100  100  100
```

```
[2,] 100 100 100
[3,] 100 100 100
[4,] 100 100 100
[5,] 100 100 100
[6,] 100 100 100

$flux.r.down
     [,1] [,2] [,3]
[1,] -100 -100 -100
[2,] -100 -100 -100
[3,] -100 -100 -100
[4,] -100 -100 -100
[5,] -100 -100 -100
[6,] -100 -100 -100

$flux.theta.up
         [,1]     [,2]     [,3]
[1,] 9230.769 9230.769 9230.769
[2,] 3076.923 3076.923 3076.923
[3,] 1846.154 1846.154 1846.154
[4,] 1318.681 1318.681 1318.681

$flux.theta.down
          [,1]      [,2]      [,3]
[1,] -9230.769 -9230.769 -9230.769
[2,] -3076.923 -3076.923 -3076.923
[3,] -1846.154 -1846.154 -1846.154
[4,] -1318.681 -1318.681 -1318.681

$flux.z.up
     [,1] [,2] [,3] [,4] [,5] [,6]
[1,]  120  120  120  120  120  120
[2,]  120  120  120  120  120  120
[3,]  120  120  120  120  120  120
[4,]  120  120  120  120  120  120

$flux.z.down
     [,1] [,2] [,3] [,4] [,5] [,6]
[1,] -120 -120 -120 -120 -120 -120
[2,] -120 -120 -120 -120 -120 -120
[3,] -120 -120 -120 -120 -120 -120
[4,] -120 -120 -120 -120 -120 -120

> tran.spherical (C = C, D.r = D, D.theta = D,
+ C.r.up = 1, C.r.down = 1, C.theta.up = 1, C.theta.down = 1,
+ C.phi.up = 1, C.phi.down = 1,
+ r = r, theta = theta, phi = phi)
```

```
$dC

, , 1

         [,1]      [,2]       [,3]       [,4]       [,5]      [,6]
[1,]  0.00000 151293676 45412970.6 21643848.5 12640411.2 8750598.0
[2,]  0.00000  16810408  5045885.6  2404872.1  1404490.1  972288.7
[3,]  0.00000   6051747  1816518.8   865753.9   505616.4  350023.9
[4,] 65.30612   3087691   926860.6   441776.5   258032.9  178648.9

, , 2

         [,1]     [,2]     [,3]     [,4]     [,5]       [,6]
[1,]  0.00000  0.00000  0.00000  0.00000  0.00000 464556.742
[2,]  0.00000  0.00000  0.00000  0.00000  0.00000  51617.416
[3,]  0.00000  0.00000  0.00000  0.00000  0.00000  18582.270
[4,] 65.30612 65.30612 65.30612 65.30612 65.30612   9546.056

, , 3

         [,1]      [,2]       [,3]        ,4]       [,5]      [,6]
[1,]  0.00000 151293676 45412970.6 21643848.5 12640411.2 8750598.0
[2,]  0.00000  16810408  5045885.6  2404872.1  1404490.1  972288.7
[3,]  0.00000   6051747  1816518.8   865753.9   505616.4  350023.9
[4,] 65.30612   3087691   926860.6   441776.5   258032.9  178648.9

$flux.r.up
     [,1] [,2] [,3]
[1,]  100  100  100
[2,]  100  100  100
[3,]  100  100  100
[4,]  100  100  100
[5,]  100  100  100
[6,]  100  100  100

$flux.r.down
     [,1] [,2] [,3]
[1,] -100 -100 -100
[2,] -100 -100 -100
[3,] -100 -100 -100
[4,] -100 -100 -100
[5,] -100 -100 -100
[6,] -100 -100 -100

$flux.theta.up
         [,1]     [,2]     [,3]
[1,] 9230.769 9230.769 9230.769
```

```
[2,] 3076.923 3076.923 3076.923
[3,] 1846.154 1846.154 1846.154
[4,] 1318.681 1318.681 1318.681

$flux.theta.down
          [,1]      [,2]      [,3]
[1,] -9230.769 -9230.769 -9230.769
[2,] -3076.923 -3076.923 -3076.923
[3,] -1846.154 -1846.154 -1846.154
[4,] -1318.681 -1318.681 -1318.681

$flux.phi.up
     [,1]      [,2]      [,3]     [,4]      [,5]      [,6]
[1,]    0 213034.42 106542.22 71055.94 53321.163 42686.998
[2,]    0  71011.47  35514.07 23685.31 17773.721 14228.999
[3,]    0  42606.88  21308.44 14211.19 10664.233  8537.400
[4,]    0  30433.49  15220.32 10150.85  7617.309  6098.143

$flux.phi.down
     [,1]       [,2]       [,3]      [,4]       [,5]       [,6]
[1,]    0 -213034.42 -106542.22 -71055.94 -53321.163 -42686.998
[2,]    0  -71011.47  -35514.07 -23685.31 -17773.721 -14228.999
[3,]    0  -42606.88  -21308.44 -14211.19 -10664.233  -8537.400
[4,]    0  -30433.49  -15220.32 -10150.85  -7617.309  -6098.143

> # Flux boundary conditions
> tran.cylindrical(C = C, D.r = D, r = r, theta = theta, z = z,
+ flux.r.up = 10, flux.r.down = 10,
+ flux.theta.up = 10, flux.theta.down = 10,
+ flux.z.up = 10, flux.z.down = 10)

$dC
, , 1

          [,1]      [,2]      [,3]      [,4]      [,5]       [,6]
[1,] 467.53846 6.0000000 6.0000000 6.0000000 6.0000000 -455.53846
[2,] 159.84615 6.0000000 6.0000000 6.0000000 6.0000000 -147.84615
[3,]  98.30769 6.0000000 6.0000000 6.0000000 6.0000000  -86.30769
[4,]  66.21978 0.2857143 0.2857143 0.2857143 0.2857143  -65.64835

, , 2

          [,1]      [,2]      [,3]      [,4]      [,5]       [,6]
[1,] 461.53846  0.000000  0.000000  0.000000  0.000000 -461.53846
[2,] 153.84615  0.000000  0.000000  0.000000  0.000000 -153.84615
[3,]  92.30769  0.000000  0.000000  0.000000  0.000000  -92.30769
[4,]  60.21978 -5.714286 -5.714286 -5.714286 -5.714286  -71.64835
```

```
, , 3

           [,1]      [,2]      [,3]      [,4]      [,5]       [,6]
[1,] 455.53846  -6.00000  -6.00000  -6.00000  -6.00000 -467.53846
[2,] 147.84615  -6.00000  -6.00000  -6.00000  -6.00000 -159.84615
[3,]  86.30769  -6.00000  -6.00000  -6.00000  -6.00000  -98.30769
[4,]  54.21978 -11.71429 -11.71429 -11.71429 -11.71429  -77.64835

$flux.r.up
     [,1] [,2] [,3]
[1,]   10   10   10
[2,]   10   10   10
[3,]   10   10   10
[4,]   10   10   10
[5,]   10   10   10
[6,]   10   10   10

$flux.r.down
     [,1] [,2] [,3]
[1,]   10   10   10
[2,]   10   10   10
[3,]   10   10   10
[4,]   10   10   10
[5,]   10   10   10
[6,]   10   10   10

$flux.theta.up
     [,1] [,2] [,3]
[1,]   10   10   10
[2,]   10   10   10
[3,]   10   10   10
[4,]   10   10   10

$flux.theta.down
     [,1] [,2] [,3]
[1,]   10   10   10
[2,]   10   10   10
[3,]   10   10   10
[4,]   10   10   10

$flux.z.up
     [,1] [,2] [,3] [,4] [,5] [,6]
[1,]   10   10   10   10   10   10
[2,]   10   10   10   10   10   10
```

```
[3,]  10  10  10  10  10  10
[4,]  10  10  10  10  10  10

$flux.z.down
     [,1] [,2] [,3] [,4] [,5] [,6]
[1,]  10  10  10  10  10  10
[2,]  10  10  10  10  10  10
[3,]  10  10  10  10  10  10
[4,]  10  10  10  10  10  10

> tran.spherical(C = C, D.r = D, r = r, theta = theta, phi = phi,
+ flux.r.up = 10, flux.r.down = 10,
+ flux.theta.up = 10, flux.theta.down = 10,
+ flux.phi.up = 10, flux.phi.down = 10)
$dC

, , 1
           [,1]      [,2]      [,3]      [,4]      [,5]      [,6]
[1,] 21302.192 7101.842 4262.4391 3046.0294 2370.6181 1437.8461
[2,]  7100.731 2367.281 1420.8130 1015.3431  790.2060  479.2820
[3,]  4260.438 1420.368  852.4878  609.2059  474.1236  287.5692
[4,]  3036.640 1008.018  602.3893  428.6164  332.1291  198.8760


, , 2
           [,1]      [,2]      [,3]      [,4]      [,5]      [,6]
[1,] 0.000000 0.000000 0.000000 0.000000 0.000000 -503.2698
[2,] 0.000000 0.000000 0.000000 0.000000 0.000000 -167.7566
[3,] 0.000000 0.000000 0.000000 0.000000 0.000000 -100.6540
[4,] -6.530612 -6.530612 -6.530612 -6.530612 -6.530612  -78.4263

, , 3

           [,1]      [,2]      [,3]      [,4]      [,5]      [,6]
[1,] -21302.192 -7101.842 -4262.4391 -3046.0294 -2370.6181 -2444.3858
[2,]  -7100.731 -2367.281 -1420.8130 -1015.3431  -790.2060  -814.7953
[3,]  -4260.438 -1420.368  -852.4878   -609.2059  -474.1236  -488.8772
[4,]  -3049.701 -1021.079  -615.4505   -441.6777   -345.1903  -355.7286

$flux.r.up
     [,1] [,2] [,3]
[1,]  10  10  10
[2,]  10  10  10
[3,]  10  10  10
```

```
[4,]  10  10  10
[5,]  10  10  10
[6,]  10  10  10

$flux.r.down
     [,1] [,2] [,3]
[1,]  10  10  10
[2,]  10  10  10
[3,]  10  10  10
[4,]  10  10  10
[5,]  10  10  10
[6,]  10  10  10

$flux.theta.up
     [,1] [,2] [,3]
[1,]  10  10  10
[2,]  10  10  10
[3,]  10  10  10
[4,]  10  10  10

$flux.theta.down
     [,1] [,2] [,3]
[1,]  10  10  10
[2,]  10  10  10
[3,]  10  10  10
[4,]  10  10  10

$flux.phi.up
     [,1] [,2] [,3] [,4] [,5] [,6]
[1,]  10  10  10  10  10  10
[2,]  10  10  10  10  10  10
[3,]  10  10  10  10  10  10
[4,]  10  10  10  10  10  10

$flux.phi.down
     [,1] [,2] [,3] [,4] [,5] [,6]
[1,]  10  10  10  10  10  10
[2,]  10  10  10  10  10  10
[3,]  10  10  10  10  10  10
[4,]  10  10  10  10  10  10

> # cyclic boundary conditions
> tran.cylindrical(C = C, D.r = D, r = r, theta = theta, z = z,
+ cyclicBnd = 1:3)
```

```
$dC
, , 1

     [,1] [,2] [,3] [,4] [,5] [,6]
[1,]    0    0    0    0    0    0
[2,]    0    0    0    0    0    0
[3,]    0    0    0    0    0    0
[4,]    0    0    0    0    0    0

, , 2

     [,1] [,2] [,3] [,4] [,5] [,6]
[1,]    0    0    0    0    0    0
[2,]    0    0    0    0    0    0
[3,]    0    0    0    0    0    0
[4,]    0    0    0    0    0    0

, , 3

     [,1] [,2] [,3] [,4] [,5] [,6]
[1,]    0    0    0    0    0    0
[2,]    0    0    0    0    0    0
[3,]    0    0    0    0    0    0
[4,]    0    0    0    0    0    0

$flux.r.up
     [,1] [,2] [,3]
[1,]    0    0    0
[2,]    0    0    0
[3,]    0    0    0
[4,]    0    0    0
[5,]    0    0    0
[6,]    0    0    0

$flux.r.down
     [,1] [,2] [,3]
[1,]    0    0    0
[2,]    0    0    0
[3,]    0    0    0
[4,]    0    0    0
[5,]    0    0    0
[6,]    0    0    0

$flux.theta.up
     [,1] [,2] [,3]
[1,]    0    0    0
```

```
[2,]    0    0    0
[3,]    0    0    0
[4,]    0    0    0

$flux.theta.down
     [,1] [,2] [,3]
[1,]    0    0    0
[2,]    0    0    0
[3,]    0    0    0
[4,]    0    0    0

$flux.z.up
     [,1] [,2] [,3] [,4] [,5] [,6]
[1,]    0    0    0    0    0    0
[2,]    0    0    0    0    0    0
[3,]    0    0    0    0    0    0
[4,]    0    0    0    0    0    0


$flux.z.down
     [,1] [,2] [,3] [,4] [,5] [,6]
[1,]    0    0    0    0    0    0
[2,]    0    0    0    0    0    0
[3,]    0    0    0    0    0    0
[4,]    0    0    0    0    0    0

> tran.spherical(C = C, D.r = D, r = r, theta = theta, phi = phi,
+ cyclicBnd = 1:3)
$dC

, , 1

     [,1] [,2] [,3] [,4] [,5] [,6]
[1,]    0    0    0    0    0    0
[2,]    0    0    0    0    0    0
[3,]    0    0    0    0    0    0
[4,]    0    0    0    0    0    0

, , 2

     [,1] [,2] [,3] [,4] [,5] [,6]
[1,]    0    0    0    0    0    0
[2,]    0    0    0    0    0    0
[3,]    0    0    0    0    0    0
[4,]    0    0    0    0    0    0
```

```
, , 3

     [,1] [,2] [,3] [,4] [,5] [,6]
[1,]    0    0    0    0    0    0
[2,]    0    0    0    0    0    0
[3,]    0    0    0    0    0    0
[4,]    0    0    0    0    0    0

$flux.r.up
     [,1] [,2] [,3]
[1,]    0    0    0
[2,]    0    0    0
[3,]    0    0    0
[4,]    0    0    0
[5,]    0    0    0
[6,]    0    0    0

$flux.r.down
     [,1] [,2] [,3]
[1,]    0    0    0
[2,]    0    0    0
[3,]    0    0    0
[4,]    0    0    0
[5,]    0    0    0
[6,]    0    0    0

$flux.theta.up
     [,1] [,2] [,3]
[1,]    0    0    0
[2,]    0    0    0
[3,]    0    0    0
[4,]    0    0    0

$flux.theta.down
     [,1] [,2] [,3]
[1,]    0    0    0
[2,]    0    0    0
[3,]    0    0    0
[4,]    0    0    0

$flux.phi.up
     [,1] [,2] [,3] [,4] [,5] [,6]
[1,]    0    0    0    0    0    0
[2,]    0    0    0    0    0    0
[3,]    0    0    0    0    0    0
[4,]    0    0    0    0    0    0
```

```
$flux.phi.down
     [,1] [,2] [,3] [,4] [,5] [,6]
[1,]    0    0    0    0    0    0
[2,]    0    0    0    0    0    0
[3,]    0    0    0    0    0    0
[4,]    0    0    0    0    0    0

> # zero-gradient boundary conditions
> tran.cylindrical(C = C, D.r = D, r = r, theta = theta, z = z)
$dC
, , 1

     [,1] [,2] [,3] [,4] [,5] [,6]
[1,]    0    0    0    0    0    0
[2,]    0    0    0    0    0    0
[3,]    0    0    0    0    0    0
[4,]    0    0    0    0    0    0

, , 2

     [,1] [,2] [,3] [,4] [,5] [,6]
[1,]    0    0    0    0    0    0
[2,]    0    0    0    0    0    0
[3,]    0    0    0    0    0    0
[4,]    0    0    0    0    0    0

, , 3

     [,1] [,2] [,3] [,4] [,5] [,6]
[1,]    0    0    0    0    0    0
[2,]    0    0    0    0    0    0
[3,]    0    0    0    0    0    0
[4,]    0    0    0    0    0    0

$flux.r.up
     [,1] [,2] [,3]
[1,]    0    0    0
[2,]    0    0    0
[3,]    0    0    0
[4,]    0    0    0
[5,]    0    0    0
[6,]    0    0    0

$flux.r.down
     [,1] [,2] [,3]
[1,]    0    0    0
[2,]    0    0    0
```

```
[3,]  0  0  0
[4,]  0  0  0
[5,]  0  0  0
[6,]  0  0  0

$flux.theta.up
     [,1] [,2] [,3]
[1,]  0  0  0
[2,]  0  0  0
[3,]  0  0  0
[4,]  0  0  0

$flux.theta.down
     [,1] [,2] [,3]
[1,]  0  0  0
[2,]  0  0  0
[3,]  0  0  0
[4,]  0  0  0

$flux.z.up
   [,1] [,2] [,3] [,4] [,5] [,6]
[1,]  0  0  0  0  0  0
[2,]  0  0  0  0  0  0
[3,]  0  0  0  0  0  0
[4,]  0  0  0  0  0  0

$flux.z.down
   [,1] [,2] [,3] [,4] [,5] [,6]
[1,]  0  0  0  0  0  0
[2,]  0  0  0  0  0  0
[3,]  0  0  0  0  0  0
[4,]  0  0  0  0  0  0

> tran.spherical(C = C, D.r = D, r = r, theta = theta, phi = phi)
$dC
, , 1

    [,1] [,2] [,3] [,4] [,5] [,6]
[1,]  0  0  0  0   0   0
[2,]  0  0  0  0   0   0
[3,]  0  0  0  0   0   0
[4,]  0  0  0  0   0   0

, , 2

     [,1] [,2] [,3] [,4] [,5] [,6]
[1,]  0  0   0   0  0  0
[2,]  0  0   0   0  0  0
```

```
[3,]  0  0  0  0  0  0
[4,]  0  0  0  0  0  0

, , 3

     [,1] [,2] [,3] [,4] [,5] [,6]
[1,]    0    0    0    0    0    0
[2,]    0    0    0    0    0    0
[3,]    0    0    0    0    0    0
[4,]    0    0    0    0    0    0

$flux.r.up
     [,1] [,2] [,3]
[1,]    0    0    0
[2,]    0    0    0
[3,]    0    0    0
[4,]    0    0    0
[5,]    0    0    0
[6,]    0    0    0

$flux.r.down
     [,1] [,2] [,3]
[1,]    0    0    0
[2,]    0    0    0
[3,]    0    0    0
[4,]    0    0    0
[5,]    0    0    0
[6,]    0    0    0

$flux.theta.up
     [,1] [,2] [,3]
[1,]    0    0    0
[2,]    0    0    0
[3,]    0    0    0
[4,]    0    0    0

$flux.theta.down
     [,1] [,2] [,3]
[1,]    0    0    0
[2,]    0    0    0
[3,]    0    0    0
[4,]    0    0    0

$flux.phi.up
     [,1] [,2] [,3] [,4] [,5] [,6]
[1,]    0    0    0    0    0    0
```

```
[2,]  0  0  0   0  0  0
[3,]  0  0  0   0  0  0
[4,]  0  0  0   0  0  0

$flux.phi.down
     [,1] [,2] [,3] [,4] [,5] [,6]
[1,]  0  0   0   0   0  0
[2,]  0  0   0   0   0  0
[3,]  0  0   0   0   0  0
[4,]  0  0   0   0   0  0

> # =====================================
> # A model with diffusion and first-order consumption  =
> # =====================================
> N <- 10 # number of grid cells
> rr <- 0.005 # consumption rate
> D <- 400
> r <- seq (2, 4, len = N+1)
> theta <- seq (0, 2*pi, len = N+1)
> z <- seq (0, 3, len = N+1)
> phi <- seq (0, 2*pi, len = N+1)
> # The model equations
> Diffcylin <- function (t, y, parms) {
+ CONC <- array(dim = c(N, N, N), data = y)
+ tran <- tran.cylindrical(CONC,
+ D.r = D, D.theta = D, D.z = D,
+ r = r, theta = theta, z = z,
+ C.r.up = 0, C.r.down = 1,
+ cyclicBnd = 2)
+ dCONC <- tran$dC - rr * CONC
+ return (list(dCONC))
+ }
> Diffspher <- function (t, y, parms) {
+ CONC <- array(dim = c(N, N, N), data = y)
+ tran <- tran.spherical (CONC,
+ D.r = D, D.theta = D, D.phi = D,
+ r = r, theta = theta, phi = phi,
+ C.r.up = 0, C.r.down = 1,
+ cyclicBnd = 2:3)
+ dCONC <- tran$dC - rr * CONC
+ return (list(dCONC))
+ }
> # initial condition: 0 everywhere, except in central point
> y <- array(dim = c(N, N, N), data = 0)
> N2 <- ceiling(N/2)
> y[N2, N2, N2] <- 100 # initial concentration in the central point...
```

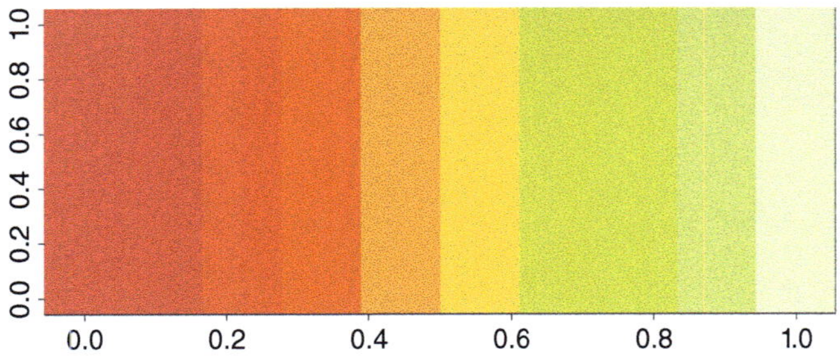

Figure A4-1 Cylindrical A model with diffusion and first-order consumption diffusive transport in cylindrical (r, θ, z), and spherical (r, θ, ϕ) coordinates.

```
> # solve to steady-state; cyclicBnd = 2,
> outcyl <- steady.3D (y = y, func = Diffcylin, parms = NULL,
+ dim = c(N, N, N), lrw = 1e6, cyclicBnd = 2)
> STDcyl <- array(dim = c(N, N, N), data = outcyl$y)
> image(STDcyl[,,1])
> #
>
> # Outputting: Figure A4-1: Cylindrical
> #
> # For spherical coordinates, cyclic Bnd = 2, 3
> outspher <- steady.3D (y = y, func = Diffspher, parms = NULL, pos=TRUE,
+ dim = c(N, N, N), lrw = 1e6, cyclicBnd = 2:3)

> #STDspher <- array(dim = c(N, N, N), data = outspher$y)
> #image(STDspher[,,1])
> ## Not run:
> image(outspher)
> #
>
> # Outputting: Figure A4-2: Spherical

>
```

Worked Example 4.6 tran.polar ***Diffusive Transport in Polar (r, θ) Coordinates***

Description

This program, in R, estimates the transport term (i.e. the rate of change of a concentration owing to diffusion) in a polar (r, θ) coordinate system.

Usage

```
tran.cylindrical (C, C.r.up = NULL, C.r.down = NULL,
                  C.theta.up = NULL, C.theta.down = NULL,
```

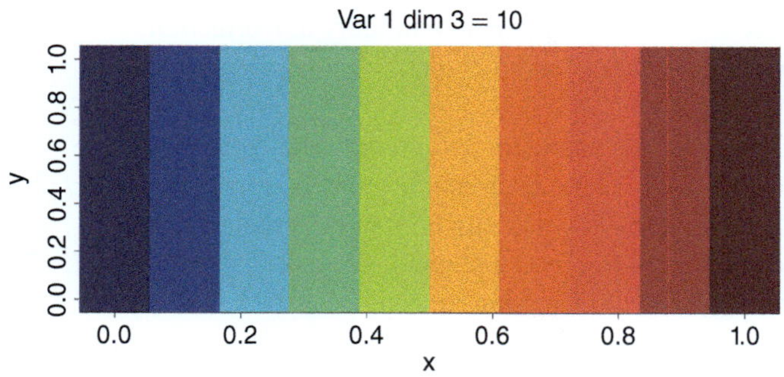

Figure A4-2 Spherical A model with diffusion and first-order consumption diffusive transport in cylindrical (r, θ, z), and spherical (r, θ, ϕ) coordinates.

```
                C.z.up = NULL, C.z.down = NULL,
                flux.r.up = NULL, flux.r.down = NULL,
                flux.theta.up = NULL, flux.theta.down = NULL,
                flux.z.up = NULL, flux.z.down = NULL,
                cyclicBnd = NULL,
                D.r = NULL, D.theta = D.r, D.z = D.r,
                r = NULL, theta = NULL, z = NULL)

tran.spherical (C, C.r.up = NULL, C.r.down = NULL,
                C.theta.up = NULL, C.theta.down = NULL,
                C.phi.up = NULL, C.phi.down = NULL,
                flux.r.up = NULL, flux.r.down = NULL,
                flux.theta.up = NULL, flux.theta.down = NULL,
                flux.phi.up = NULL, flux.phi.down = NULL,
                cyclicBnd = NULL,
                D.r = NULL, D.theta = D.r, D.phi = D.r,
                r = NULL, theta = NULL, phi = NULL)
```

Arguments

C	Concentration, expressed per unit volume, defined at the center of each grid cell; Nr*Ntheta matrix [M/L^3]
C.r.up	Concentration at upstream boundary in $r(x)$-direction; vector of length Ntheta [M/L^3]
C.r.down	Concentration at downstream boundary in $r(x)$-direction; vector of length Ntheta [M/L^3]
C.theta.up	Concentration at upstream boundary in theta-direction; vector of length Nr [M/L^3]
C.theta.down	Concentration at downstream boundary in theta-direction; vector of length Nr [M/L^3]
flux.r.up	Flux across the upstream boundary in r-direction, positive = INTO model domain; vector of length Ntheta [M/L^2/T]

flux.r.down	Flux across the downstream boundary in *r*-direction, positive = OUT of model domain; vector of length Ntheta [M/L²/T]
flux.theta.up	Flux across the upstream boundary in theta-direction, positive = INTO model domain; vector of length Nr [M/L²/T]
flux.theta.down	Flux across the downstream boundary in theta-direction, positive = OUT of model domain; vector of length Nr [M/L²/T]
cyclicBnd	If not NULL, the direction in which a cyclic boundary is defined, i.e. cyclicBnd = 1 for the r-direction, cyclicBnd = 2 for the theta-direction and cyclicBnd = c(1,2) for both the r- and theta-directions
D.r	Diffusion coefficient in *r*-direction, defined on grid cell interfaces. One value, a vector of length (Nr+1), a prop.1D list created by setup.prop.1D, or a (Nr+1)*Ntheta matrix [L²/T]
D.theta	Diffusion coefficient in theta-direction, defined on grid cell interfaces. One value, a vector of length (Ntheta+1), a prop.1D list created by setup.prop.1D, or a Nr*(Ntheta+1) matrix [L²/T]
r	Position of adjacent cell interfaces in the *r*-direction. A vector of length Nr+1 [L]
theta	Position of adjacent cell interfaces in the theta-direction. A vector of length Ntheta+1 [L]. Theta should be within [0, 2 pi]
full.output	Logical flag enabling a full return of the output (default = FALSE; TRUE slows down execution by 20%)
out	Output as returned by tran.polar, and which is to be mapped from polar to Cartesian coordinates
x	The Cartesian *x*-coordinates to which the polar coordinates are to be mapped
y	The Cartesian *y*-coordinates to which the polar coordinates are to be mapped

Details

tran.polar performs (simplified) transport in polar coordinates

The boundary conditions are either

(1) zero gradient
(2) fixed concentration
(3) fixed flux
(4) cyclic boundary.

This is also the order of priority. The cyclic boundary overrules the other. If fixed concentration, fixed flux, and cyclicBnd are NULL then the boundary is zero gradient.

A cyclic boundary condition has concentration and flux at upstream and downstream boundary the same.

polar2cart maps the polar coordinates to Cartesian coordinates.

If x and y is not provided, then it will create an (x,y) grid based on r: seq(-maxr, maxr, length.out=Nr), where maxr is the maximum value of r, and Nr is the number of elements in r.

Value

A list containing

dC	The rate of change of the concentration C due to transport, defined in the center of each grid cell, a Nr*Ntheta matrix [$M/L^3/T$]
C.r.up	Concentration at the upstream interface in r-direction. A vector of length Ntheta [M/L^3]. Only when full.output = TRUE
C.r.down	Concentration at the downstream interface in r-direction. A vector of length Ntheta [M/L^3]. Only when full.output = TRUE
C.theta.up	Concentration at the upstream interface in theta-direction. A vector of length Nr [M/L^3]. Only when full.output = TRUE
C.theta.down	Concentration at the downstream interface in theta-direction. A vector of length Nr [M/L^3]. Only when full.output = TRUE
r.flux	Flux across the interfaces in x-direction of the grid cells. A (Nr+1)*Ntheta matrix [$M/L^2/T$]. Only when full.output = TRUE
theta.flux	Flux across the interfaces in y-direction of the grid cells. A Nr*(Ntheta+1) matrix [$M/L^2/T$]. Only when full.output = TRUE
flux.r.up	Flux across the upstream boundary in r-direction, positive = INTO model domain. A vector of length Ntheta [$M/L^2/T$]
flux.r.down	Flux across the downstream boundary in r-direction, positive = OUT of model domain. A vector of length Ntheta [$M/L^2/T$]
flux.theta.up	Flux across the upstream boundary in theta-direction, positive = INTO model domain. A vector of length Nr [$M/L^2/T$]
flux.theta.down	Flux across the downstream boundary in theta-direction, positive = OUT of model domain. A vector of length Nr [$M/L^2/T$]

See Also

tran.cylindrical, tran.spherical for a discretization of 3D transport equations in cylindrical and spherical coordinates.

Examples

```
# ===============
# Testing the functions  =
# ===============
# Parameters
F          <- 100                # input flux [micromol cm-2 yr-1]
D          <- 400                # mixing coefficient [cm2 yr-1]

# Grid definition
r.N <- 4 # number of cells in r-direction
theta.N <- 6 # number of cells in theta-direction
r.L <- 8 # domain size r-direction [cm]
r <- seq(0, r.L, len = r.N+1) # cell size r-direction [cm]
theta <- seq(0, 2*pi,len = theta.N+1) # theta-direction - theta: from 0, 2pi

# Initial conditions
C <- matrix(nrow = r.N, ncol = theta.N, data = 0)

# Boundary conditions: fixed concentration
C.r.up            <- rep(1, times = theta.N)
C.r.down          <- rep(0, times = theta.N)
C.theta.up        <- rep(1, times = r.N)
C.theta.down      <- rep(0, times = r.N)

# Concentration boundary conditions
tran.polar(C = C, D.r = D, D.theta = D,
  r = r, theta = theta,
  C.r.up = C.r.up, C.r.down = C.r.down,
  C.theta.up = C.theta.up, C.theta.down = C.theta.down)

# Flux boundary conditions
flux.r.up <- rep(200, times = theta.N)
flux.r.down <- rep(-200, times = theta.N)
flux.theta.up <- rep(200, times = r.N)
flux.theta.down <- rep(-200, times = r.N)

tran.polar(C = C, D.r = D, r = r, theta = theta,
  flux.r.up = flux.r.up, flux.r.down = flux.r.down,
  flux.theta.up = flux.theta.up, flux.theta.down = flux.theta.down,
  full.output = TRUE)

## ====================================
## A model with diffusion and first-order consumption   =
## ====================================
```

```
N       <- 50          # number of grid cells
XX      <- 4           # total size
rr      <- 0.005       # consumption rate
ini     <- 1           # initial value at x=0
D       <- 400

r         <- seq (2, 4, len = N+1)
theta     <- seq(0, 2*pi, len = N+1)
theta.m <- 0.5*(theta[-1]+theta[-(N+1)])

# The model equations

 Diffpolar <- function (t, y, parms) {
   CONC <- matrix(nrow = N, ncol = N, data = y)
   tran   <- tran.polar(CONC, D.r = D, D.theta = D, r = r, theta = theta,
          C.r.up = 0, C.r.down = 1*sin(5*theta.m),
          cyclicBnd = 2, full.output=TRUE )
  dCONC <- tran$dC - rr * CONC
  return (list(dCONC))
}

# solve to steady-state; cyclicBnd = 2, because of C.theta.up, C.theta.down
out <- steady.2D (y = rep(0, N*N), func = Diffpolar, parms = NULL,
                    dim = c(N, N), lrw = 1e6, cyclicBnd = 2)

image(out)

cart <- polar2cart(out, r = r, theta = theta,
                          x = seq(-4, 4, len = 100),
                          y = seq(-4, 4, len = 100))
image(cart)
```

In the R domain:

```
> # In the R domain:
> install.packages("ReacTran")
Installing package into 'C:/Users/Bert/Documents/R/win-library/3.3'
(as 'lib' is unspecified)
--- Please select a CRAN mirror for use in this session ---
A CRAN mirror is selected.
also installing the dependencies 'rootSolve', 'deSolve'
trying URL 'https://mirrors.tuna.tsinghua.edu.cn/CRAN/bin/windows/contrib/3.3/
rootSolve_1.7.zip'
Content type 'application/zip' length 683839 bytes (667 KB)
```

```
downloaded 667 KB

trying URL 'https://mirrors.tuna.tsinghua.edu.cn/CRAN/bin/windows/contrib/3.3/
deSolve_1.20.zip'
Content type 'application/zip' length 2667028 bytes (2.5 MB)
downloaded 2.5 MB
trying URL 'https://mirrors.tuna.tsinghua.edu.cn/CRAN/bin/windows/contrib/3.3/
ReacTran_1.4.3.1.zip'
Content type 'application/zip' length 915996 bytes (894 KB)
downloaded 894 KB

package 'rootSolve' successfully unpacked and MD5 sums checked
package 'deSolve' successfully unpacked and MD5 sums checked
package 'ReacTran' successfully unpacked and MD5 sums checked

The downloaded binary packages are in
C:\Users\Bert\AppData\Local\Temp\RtmpmEkAb2\downloaded_packages
> require(ReacTran)
Loading required package: ReacTran
Loading required package: rootSolve
Loading required package: deSolve
Loading required package: shape
> library(ReacTran)
> ls("package:ReacTran") # Listing all the programs in this package:
 [1] "advection.1D"      "advection.volume.1D"   "fiadeiro"
 [4] "g.cylinder"        "g.sphere"              "g.spheroid"
 [7] "p.exp"             "p.lin"                 "p.sig"
[10] "polar2cart"        "setup.compaction.1D"   "setup.grid.1D"
[13] "setup.grid.2D"     "setup.prop.1D"         "setup.prop.2D"
[16] "tran.1D"           "tran.2D"               "tran.3D"
[19] "tran.cylindrical"  "tran.polar"            "tran.spherical"
[22] "tran.volume.1D"    "tran.volume.2D"        "tran.volume.3D"
```

In the R domain:

```
>
> install.packages("ReacTran")
Installing package into 'C:/Users/Bert/Documents/R/win-library/3.3'
(as 'lib' is unspecified)
--- Please select a CRAN mirror for use in this session ---
A CRAN mirror is selected.
also installing the dependencies 'rootSolve', 'deSolve'

trying URL 'https://mirrors.tuna.tsinghua.edu.cn/CRAN/bin/windows/contrib/3.3/
rootSolve_1.7.zip'
Content type 'application/zip' length 683839 bytes (667 KB)
downloaded 667 KB
```

```
trying URL 'https://mirrors.tuna.tsinghua.edu.cn/CRAN/bin/windows/contrib/3.3/
deSolve_1.20.zip'
Content type 'application/zip' length 2667028 bytes (2.5 MB)
downloaded 2.5 MB

trying URL 'https://mirrors.tuna.tsinghua.edu.cn/CRAN/bin/windows/contrib/3.3/
ReacTran_1.4.3.1.zip'
Content type 'application/zip' length 915996 bytes (894 KB)
downloaded 894 KB
package 'rootSolve' successfully unpacked and MD5 sums checked

package 'deSolve' successfully unpacked and MD5 sums checked
package 'ReacTran' successfully unpacked and MD5 sums checked

The downloaded binary packages are in
C:\Users\Bert\AppData\Local\Temp\RtmpmEkAb2\downloaded_packages
> require(ReacTran)
Loading required package: ReacTran
Loading required package: rootSolve
Loading required package: deSolve
Loading required package: shape
> library(ReacTran)
> ls("package:ReacTran") # Listing all the programs in this package:
 [1] "advection.1D"       "advection.volume.1D"   "fiadeiro"
 [4] "g.cylinder"         "g.sphere"              "g.spheroid"
 [7] "p.exp"              "p.lin"                 "p.sig"
[10] "polar2cart"         "setup.compaction.1D"   "setup.grid.1D"
[13] "setup.grid.2D"      "setup.prop.1D"         "setup.prop.2D"
[16] "tran.1D"            "tran.2D"               "tran.3D"
[19] "tran.cylindrical"   "tran.polar"            "tran.spherical"
[22] "tran.volume.1D"     "tran.volume.2D"        "tran.volume.3D"

> tran.polar # Outputting the program listing of " tran.polar"
> function (C, C.r.up = NULL, C.r.down = NULL, C.theta.up = NULL,
  C.theta.down = NULL, flux.r.up = NULL, flux.r.down = NULL,
  flux.theta.up = NULL, flux.theta.down = NULL, cyclicBnd = NULL,
  D.r = 1, D.theta = D.r, r = NULL, theta = NULL, full.output = FALSE)
{
  checkDim <- function(dr, dtheta, nx = "dr", ntheta = "dtheta") {
    if (length(dr) != 1)
      stop(nx, ", should have length 1")
    if (length(dtheta) != 1)
      stop(ntheta, ", should have length 1")
  }
  checkDim2 <- function(dr, dtheta, nx = "dr", ntheta = "dtheta") {
    if (length(dr) != dimC[1] + 1)
```

```
            stop(nx, ", should have length equal to dim(C)[1]+1 = ",
              dimC[1] + 1, " it is ", length(dr))
        if (length(dtheta) != dimC[2] + 1)
            stop(ntheta, ", should have length equal to dim(C)[2]+1 = ",
              dimC[2] + 1, " it is ", length(dtheta))
    }
    dimC <- dim(C)
    Type <- 2
    if (max(theta) > 2 * pi)
        stop("theta should be < 2pi")
    if (min(theta) < 0)
        stop("theta should be > 0 ")
    checkDim2(r, theta, "r, the grid in x-direction", "theta, the grid in theta-direction")
    Nr <- dimC[1]
    r.c <- 0.5 * (r[-1] + r[-(Nr + 1)])
    dr <- diff(r)
    drint <- c(r.c[1] - r[1], diff(r.c), r[Nr + 1] - r.c[Nr])
    divr <- 1/r
    divr[is.na(divr)] <- 0
    Ntheta <- dimC[2]
    theta.c <- 0.5 * (theta[-1] + theta[-(Ntheta + 1)])
    dtheta <- diff(theta)
    dthetaint <- c(theta.c[1] - theta[1], diff(theta.c), theta[Ntheta +
        1] - theta.c[Ntheta])
    checkDim(D.r, D.theta, "D.r, the diffusion in x-direction ",
        "D.theta, the diffusion in theta-direction ")
    if (!is.null(cyclicBnd)) {
    }
    if (!is.null(C.r.up)) {
        if (length(C.r.up) != 1 && length(C.r.up) != Ntheta)
            stop("'C.r.up' should have length 1 or equal to ",
              Ntheta)
    }
    else if (!is.null(flux.r.up)) {
        if (length(flux.r.up) != 1 && length(flux.r.up) != Ntheta)
            stop("'flux.r.up' should have length 1 or equal to ",
              Ntheta)
    }
    else if (!1 %in% cyclicBnd)
        C.r.up = C[1, ]
    if (!is.null(C.r.down)) {
        if (length(C.r.down) != 1 && length(C.r.down) != Ntheta)
            stop("'C.r.down' should have length 1 or equal to ",
              Ntheta)
    }
    else if (!is.null(flux.r.down)) {
```

```
  if (length(flux.r.down) != 1 && length(flux.r.down) !=
    Ntheta)
    stop("'flux.r.down' should have length 1 or equal to ",
      Ntheta)
}
else if (!1 %in% cyclicBnd)
  C.r.down = C[Nr, ]
if (!is.null(C.theta.up)) {
  if (length(C.theta.up) != 1 && length(C.theta.up) !=
    Nr)
    stop("'C.theta.up' should have length 1 or equal to ",
      Nr)
}
else if (!is.null(flux.theta.up)) {
  if (length(flux.theta.up) != 1 && length(flux.theta.up) !=
    Nr)
    stop("'flux.theta.up' should have length 1 or equal to ",
      Nr)
}
else if (!2 %in% cyclicBnd)
  stop("'flux.theta.up' OR 'C.theta.up' should be specified")
if (!is.null(C.theta.down)) {
  if (length(C.theta.down) != 1 && length(C.theta.down) !=
    Nr)
    stop("'C.theta.down' should have length 1 or equal to ",
      Nr)
}
else if (!is.null(flux.theta.down)) {
  if (length(flux.theta.down) != 1 && length(flux.theta.down) !=
    Nr)
    stop("'flux.theta.down' should have length 1 or equal to ",
      Nr)
}
else if (!2 %in% cyclicBnd)
  stop("'flux.theta.down' OR 'C.theta.down' should be specified")
Flux.r <- 0
Flux.theta <- 0
if (1 %in% cyclicBnd) {
  C.r.up <- (C[1, ] * drint[Nr + 1] + C[Nr, ] * drint[1])/(drint[1] +
    drint[Nr + 1])
  C.r.down <- C.r.up
}
if (is.null(C.r.up))
  C.r.up <- flux.r.up/D.r * drint[1] + C[1, ]
if (is.null(C.r.down))
  C.r.down <- -flux.r.down/D.r * drint[Nr + 1] + C[Nr,
```

```
        ]
    Flux.r <- D.r * (rbind(C.r.up, C, deparse.level = 0) - rbind(C,
        C.r.down, deparse.level = 0))/drint
    if (2 %in% cyclicBnd) {
        C.theta.up <- (C[, 1] * dthetaint[Ntheta + 1] + C[, Ntheta] *
            dthetaint[1])/(dthetaint[1] + dthetaint[Ntheta +
            1])
        C.theta.down <- C.theta.up
    }
    if (is.null(C.theta.up))
        C.theta.up <- flux.theta.up/D.theta * dthetaint[1] *
            r.c + C[, 1]
    if (is.null(C.theta.down))
        C.theta.down <- -flux.theta.down/D.theta * dthetaint[Ntheta +
            1] * r.c + C[, Ntheta]
    Flux.theta <- D.theta * (cbind(C.theta.up, C, deparse.level = 0) -
        cbind(C, C.theta.down, deparse.level = 0))/matrix(data = dthetaint,
        nrow = Nr, ncol = (Ntheta + 1), byrow = TRUE)/r.c
    dFdtheta <- 0
    dFdr <- -diff(Flux.r * r)/dr/r.c
    dFdtheta <- -t(diff(t(Flux.theta))/dtheta)/r.c
    if (full.output)
        return(list(dC = dFdr + dFdtheta, C.r.up = C.r.up, C.r.down =
     C.r.down,
          C.theta.up = C.theta.up, C.theta.down = C.theta.down,
            r.flux = Flux.r, theta.flux = Flux.theta, flux.r.up = Flux.r[1,
                ], flux.r.down = Flux.r[Nr + 1, ], flux.theta.up = Flux.theta[,
                1], flux.theta.down = Flux.theta[, Ntheta + 1]))
    else return(list(dC = dFdr + dFdtheta, flux.r.up = Flux.r[1,
        ], flux.r.down = Flux.r[Nr + 1, ], flux.theta.up = Flux.theta[,
        1], flux.theta.down = Flux.theta[, Ntheta + 1]))
}
<environment: namespace:ReacTran>
>
> # Running the program "tran.polar":
>
> #===============
> # Testing the functions =
> #===============
> # Parameters
> F <- 100 # input flux [micromol cm-2 yr-1]
> D <- 400 # mixing coefficient [cm2 yr-1]
> # Grid definition
> r.N <- 4 # number of cells in r-direction
> theta.N <- 6 # number of cells in theta-direction
> r.L <- 8 # domain size r-direction [cm]
> r <- seq(0, r.L, len = r.N+1) # cell size r-direction [cm]
```

```
> theta <- seq(0, 2*pi,len = theta.N+1) # theta-direction - theta: from 0,
+ # to 2pi
> # Initial conditions
> C <- matrix(nrow = r.N, ncol = theta.N, data = 0)
> # Boundary conditions: fixed concentration
> C.r.up <- rep(1, times = theta.N)
> C.r.down <- rep(0, times = theta.N)
> C.theta.up <- rep(1, times = r.N)
> C.theta.down <- rep(0, times = r.N)
> # Concentration boundary conditions
> tran.polar(C = C, D.r = D, D.theta = D,
+ r = r, theta = theta,
+ C.r.up = C.r.up, C.r.down = C.r.down,
+ C.theta.up = C.theta.up, C.theta.down = C.theta.down)
>
> # Outputting:

$dC
           [,1]  [,2] [,3] [,4] [,5] [,6]
[1,] 1704142.01    0    0    0    0    0
[2,]  189349.11    0    0    0    0    0
[3,]   68165.68    0    0    0    0    0
[4,]   34778.41    0    0    0    0    0

$flux.r.up
[1] 400 400 400 400 400 400

$flux.r.down
[1] 0 0 0 0 0 0

$flux.theta.up
[1] 36923.077 12307.692  7384.615  5274.725

$flux.theta.down
[1] 0 0 0 0

> # Flux boundary conditions
> flux.r.up <- rep(200, times = theta.N)
> flux.r.down <- rep(-200, times = theta.N)
> flux.theta.up <- rep(200, times = r.N)
> flux.theta.down <- rep(-200, times = r.N)
> tran.polar(C = C, D.r = D, r = r, theta = theta,
+ flux.r.up = flux.r.up, flux.r.down = flux.r.down,
+ flux.theta.up = flux.theta.up, flux.theta.down = flux.theta.down,
+ full.output = TRUE)

>
```

> *# Outputting:*

```
$dC
         [,1]     [,2]     [,3]     [,4]     [,5]     [,6]
[1,] 9230.769   0.0000   0.0000   0.0000   0.0000 9230.769
[2,] 3076.923   0.0000   0.0000   0.0000   0.0000 3076.923
[3,] 1846.154   0.0000   0.0000   0.0000   0.0000 1846.154
[4,] 1432.967 114.2857 114.2857 114.2857 114.2857 1432.967

$C.r.up
[1] 0.5 0.5 0.5 0.5 0.5 0.5

$C.r.down
[1] 0.5 0.5 0.5 0.5 0.5 0.5

$C.theta.up
[1] 0.005416667 0.016250000 0.027083333 0.037916667

$C.theta.down
[1] 0.005416667 0.016250000 0.027083333 0.037916667

$r.flux
     [,1] [,2] [,3] [,4] [,5] [,6]
[1,]  200  200  200  200  200  200
[2,]    0    0    0    0    0    0
[3,]    0    0    0    0    0    0
[4,]    0    0    0    0    0    0
[5,] -200 -200 -200 -200 -200 -200

$theta.flux
     [,1] [,2] [,3] [,4] [,5] [,6] [,7]
[1,]  200    0    0    0    0    0 -200
[2,]  200    0    0    0    0    0 -200
[3,]  200    0    0    0    0    0 -200
[4,]  200    0    0    0    0    0 -200

$flux.r.up
[1] 200 200 200 200 200 200

$flux.r.down
[1] -200 -200 -200 -200 -200 -200

$flux.theta.up
[1] 200 200 200 200

$flux.theta.down
[1] -200 -200 -200 -200
```

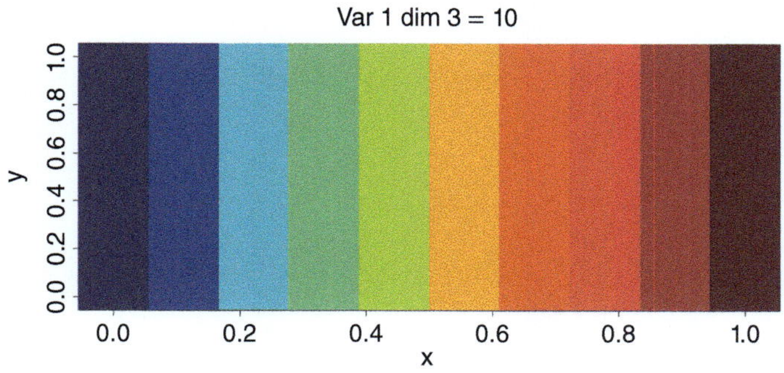

Figure A5-1 tran.polar 1 Diffusive transport in polar (r, θ) coordinates.

```
>
> # Outputting:
> #
> # Figure A5-1: tran.polar 1
> #
>
> ## ================================================
> ## A model with diffusion and first-order consumption
> ## ================================================
> N <- 50 # number of grid cells
> XX <- 4 # total size
> rr <- 0.005 # consumption rate
> ini <- 1 # initial value at x=0
> D <- 400
> r <- seq (2, 4, len = N+1)
> theta <- seq(0, 2*pi, len = N+1)
> theta.m <- 0.5*(theta[-1]+theta[-(N+1)])
> # The model equations
> Diffpolar <- function (t, y, parms) {
+ CONC <- matrix(nrow = N, ncol = N, data = y)
+ tran <- tran.polar(CONC, D.r = D, D.theta = D, r = r, theta = theta,
+ C.r.up = 0, C.r.down = 1*sin(5*theta.m),
+ cyclicBnd = 2, full.output=TRUE )
+ dCONC <- tran$dC - rr * CONC
+ return (list(dCONC))
+ }
>
> # solve to steady-state; cyclicBnd = 2, because of C.theta.up,
+ # C.theta.down
> out <- steady.2D (y = rep(0, N*N), func = Diffpolar, parms = NULL,
+ dim = c(N, N), lrw = 1e6, cyclicBnd = 2)
> image(out)
>
```

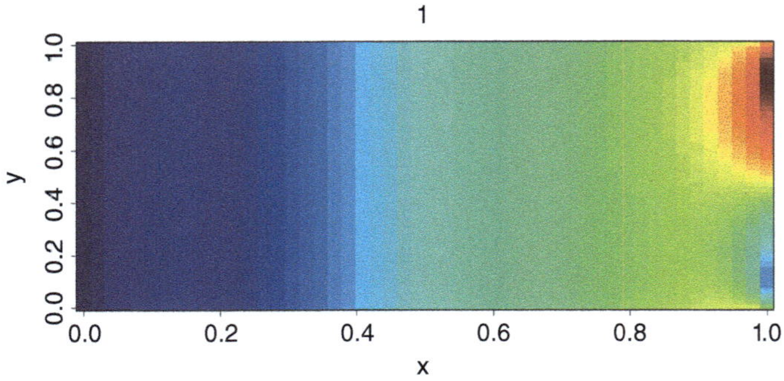

Figure A5-2 tran.polar 2 Diffusive transport in polar (r, θ) coordinates.

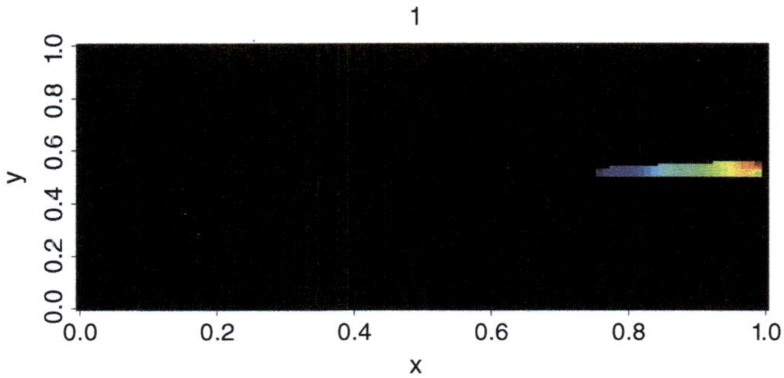

Figure A5-3 tran.polar 3 Diffusive transport in polar (r, θ) coordinates.

```
> # Figure A5-2: tran.polar 2
>
> cart <- polar2cart(out, r = r, theta = theta,
+ x = seq(-4, 4, len = 100),
+ y = seq(-4, 4, len = 100))
>
> image(cart)
>
> # Figure A5-3: tran.polar 3
>
```

Worked Example 4.7 tran.volume.1D, tran.volume.2D, tran.volume.3D, ***1D, 2D, and 3D Volumetric Advective–Diffusive Transport in an Aquatic System***

Description

This program estimates the volumetric transport term (i.e. the rate of change of the concentration owing to advection and diffusion) in a 1D, 2D, or 3D model of an aquatic system (e.g. river, estuary, and open sea).

Volumetric transport implies the use of flows (mass per unit of time) rather than fluxes (mass per unit of area per unit of time) as is done in tran.1D, tran.2D, or tran.3D.

The tran.volume.xD programs are suited for modeling channels (like rivers and estuaries) where the cross-sectional area changes, but where this area change needs not to be explicitly modeled as such.

Another difference with tran.1D is that the tran.volume.1D routine also allows lateral water or lateral mass input (such as input from side rivers or diffusive lateral ground water inflow).

The tran.volume.2D program can check for water balance and assume an influx or efflux in case the net flows in and out of a box are not = 0.

Usage

```
tran.volume.1D(C, C.up = C[1], C.down = C[length(C)],
               C.lat = C, F.up = NULL, F.down = NULL, F.lat = NULL,
               Disp,flow = 0, flow.lat = NULL, AFDW = 1,
               V = NULL, full.check = FALSE, full.output = FALSE)
tran.volume.2D(C, C.x.up = C[1, ], C.x.down = C[nrow(C), ],
               C.y.up = C[, 1], C.y.down = C[, ncol(C)],
               C.z = C, masscons = TRUE,
               F.x.up = NULL, F.x.down = NULL,
               F.y.up = NULL, F.y.down = NULL,
               Disp.grid = NULL, Disp.x = NULL, Disp.y = Disp.x,
               flow.grid = NULL, flow.x = NULL, flow.y = NULL,
               AFDW.grid = NULL, AFDW.x = 1, AFDW.y = AFDW.x,
               V = NULL, full.check = FALSE, full.output = FALSE)

tran.volume.3D(C, C.x.up = C[1, , ], C.x.down = C[dim(C)[1], , ],
               C.y.up = C[, 1, ], C.y.down = C[, dim(C)[2], ],
               C.z.up = C[, , 1], C.z.down = C[, , dim(C)[3]],
               F.x.up = NULL, F.x.down = NULL,
               F.y.up = NULL, F.y.down = NULL,
               F.z.up = NULL, F.z.down = NULL,
               Disp.grid = NULL,
               Disp.x = NULL, Disp.y = Disp.x, Disp.z = Disp.x,
               flow.grid = NULL, flow.x = 0, flow.y = 0, flow.z = 0,
               AFDW.grid = NULL, AFDW.x = 1, AFDW.y = AFDW.x,
               AFDW.z = AFDW.x,
               V = NULL, full.check = FALSE, full.output = FALSE)
```

Arguments

C — Tracer concentration, defined at the center of the grid cells. A vector of length N $[M/L^3]$ (tran.volume.1D), a matrix of dimension Nr*Nc (tran.volume.2D) or an Nx*Ny*Nz array (tran.volume.3D) $[M/L^3]$

`C.up`	Tracer concentration at the upstream interface. One value [M/L^3]
`C.down`	Tracer concentration at downstream interface. One value [M/L^3]
`C.lat`	Tracer concentration in the lateral input, defined at grid cell centers. One value, a vector of length *N*, or a list as defined by `setup.prop.1D` [M/L^3]. The default is `C.lat` = `C` (a zero-gradient condition). Setting `C.lat=0`, together with a positive `F.lat` will lead to dilution of the tracer concentration in the grid cells
`C.x.up`	Concentration at upstream boundary in *x*-direction; vector of length Ny (2D) or matrix of dimensions Ny*Nz (3D) [M/L^3]. `C.x.down` concentration at downstream boundary in *x*-direction; vector of length Ny (2D) or matrix of dimensions Ny*Nz (3D) [M/L^3]
`C.y.up`	Concentration at upstream boundary in *y*-direction; vector of length Nx (2D) or matrix of dimensions Nx*Nz (3D) [M/L^3]
`C.y.down`	Concentration at downstream boundary in *y*-direction; vector of length Nx (2D) or matrix of dimensions Nx*Nz (3D) [M/L^3]
`C.z.up`	Concentration at upstream boundary in *z*-direction; matrix of dimensions Nx*Ny [M/L^3]
`C.z.down`	Concentration at downstream boundary in *z*-direction; matrix of dimensions Nx*Ny [M/L^3]
`C.z`	Concentration at boundary in *z*-direction for 2D models where `masscons =TRUE`. Matrix of dimensions Nx*Ny [M/L^3]
`masscons`	When `TRUE`, will check flow balance in 2D model. The flow in the third direction will then be estimated
`F.up`	Total tracer input at the upstream interface. One value [M/T]
`F.down`	Total tracer input at downstream interface. One value [M/T]
`F.lat`	Total lateral tracer input, defined at grid cell centers. One value, a vector of length *N*, or a 1D list property as defined by `setup.prop.1D` [M/T]
`F.x.up`	Total tracer input at the upstream interface in *x*-direction. positive = INTO model domain. A vector of length Ny (2D) or a matrix of dimensions Ny*Nz (3D) [M/T]
`F.x.down`	Total tracer input at downstream interface in *x*-direction. Positive = `INTO` model domain. A vector of length Ny (2D) or a matrix of dimensions Ny*Nz (3D) [M/T]
`F.y.up`	Total tracer input at the upstream interface in *y*-direction. Positive = `INTO` model domain. A vector of length Nx (2D) or a matrix of dimensions Nx*Nz (3D) [M/T]
`F.y.down`	Total tracer input at the downstream interface in *y*-direction. Positive = `INTO` model domain. A vector of length Nx (2D) or a matrix of dimensions Nx*Nz (3D) [M/T]

F.z.up	Total tracer input at the upstream interface in *z*-direction. Positive = INTO model domain. A matrix of dimensions Nx*Ny [M/T]
F.z.down	Total tracer input at the downstream interface in *z*-direction. Positive = INTO model domain. A matrix of dimensions Nx*Ny [M/T]
Disp.grid BULK	Dispersion coefficients defined on all grid cell interfaces. For tran.volume.2D, should contain two matrices, x.int (dimension (Nx+1)*Ny) and y.int (dimension Nx*(Ny+1)). For tran.volume.3D should contain three arrays x.int (dim = (Nx+1)*Ny*Nz), y.int (dim = Nx*(Ny+1)*Nz), and z.int (dim = Nx*Ny*(Nz+1))
Disp BULK	Dispersion coefficient, defined on grid cell interfaces. One value, a vector of length N+1, or a 1D list property as defined by `setup.prop.1D` $[L^3/T]$
Disp.x BULK	Dispersion coefficient in *x*-direction, defined on grid cell interfaces. One value, a vector of length (Nx+1), a `prop.1D` list created by `setup.prop.1D`, a (Nx+1)*Ny matrix (2D) or a Nx*(Ny+1)*Nz array (3D) $[L^3/T]$
Disp.y BULK	Dispersion coefficient in *y*-direction, defined on grid cell interfaces. One value, a vector of length (Ny+1), a `prop.1D` list created by `setup.prop.1D`, or a Nx*(Ny+1) matrix (2D) or a Nx*(Ny+1)*Nz array (3D) $[L^3/T]$
Disp.z BULK	Dispersion coefficient in *z*-direction, defined on grid cell interfaces. One value, a vector of length (Nz+1), or a Nx*Ny*(Nz+1) array $[L^3/T]$
flow	Water flow rate, defined on grid cell interfaces. One value, a vector of length N+1, or a list as defined by `setup.prop.1D` $[L^3/T]$. If `flow`, `lat` is not `NULL` the flow should be one value containing the flow rate at the upstream boundary. If `flow.lat` is `NULL` then flow can be either one value, a vector or a list
flow.lat	Lateral water flow rate $[L^3/T]$ into each volume box, defined at grid cell centers. One value, a vector of length *N*, or a list as defined by `setup.prop.1D`. If `flow.lat` has a value, then flow should be the flow rate at the upstream interface (one value). For each grid cell, the `flow` at the downstream side of a grid cell is then estimated by water balance (adding `flow.lat` in the cell to flow rate at the upstream side of the grid cell). If `flow.lat` is `NULL`, then it is determined by water balance from `flow`

`flow.grid`	Flow rates defined on all grid cell interfaces. Can be positive (downstream flow) or negative (upstream flow). Should contain elements `x.int`, `y.int`, `z.int` (3D), arrays with the values on the interfaces in x-, y-, and z-directions [L^3/T]
`flow.x`	Flow rates in the x-direction, defined on grid cell interfaces. Can be positive (downstream flow) or negative (upstream flow). One value, a vector of length `(Nx+1)`, a `prop.1D` list created by `setup.prop.1D` (2D), a `(Nx+1)*Ny` matrix (2D) or a `(Nx+1)*Ny*Nz` array (3D) [L^3/T]
`flow.y`	Flow rates in the y-direction, defined on grid cell interfaces. Can be positive (downstream flow) or negative (upstream flow). One value, a vector of length (Ny+1), a `prop.1D` list created by `setup.prop.1D` (2D), a `Nx*(Ny+1)` matrix (2D) or a `Nx*(Ny+1)*Nz` array [L^3/T]
`flow.z`	Flow rates in the z-direction, defined on grid cell interfaces. Can be positive (downstream flow) or negative (upstream flow). One value, a vector of length `(Nz+1)`, or a `Nx*Ny*(Nz+1)` array [L^3/T]
`AFDW`	Weight used in the finite difference scheme for advection, defined on grid cell interfaces; backward = 1, centered = 0.5, forward = 0; default is backward. One value, a vector of length N+1, or a list as defined by `setup.prop.1D` `[-]`
`AFDW.grid`	Weight used in the finite difference scheme for advection in the x-direction, defined on grid cell interfaces; backward = 1, centered = 0.5, forward = 0; default is backward. For `tran.volume.3D` should contain elements `x.int`, `y.int`, `z.int (3D)`, for `tran.volume.2D` should contain elements `x.int` and `y.int` `[-]`
`AFDW.x`	Weight used in the finite difference scheme for advection in the x-direction, defined on grid cell interfaces; backward = 1, centered = 0.5, forward = 0; default is backward. One value, a vector of length (Nx+1), a `prop.1D` list created by `setup.prop.1D`, a (Nx+1)*Ny matrix (2D) or a (Nx+1)*Ny*Nz array (3D) [–]
`AFDW.y`	Weight used in the finite difference scheme for advection in the y-direction, defined on grid cell interfaces; backward = 1, centered = 0.5, forward = 0; default is backward. One value, a vector of length (Ny+1), a prop.1D list created by `setup.prop.1D`, a Nx*(Ny+1) matrix (2D) or a Nx*(Ny+1)*Nz array [–]
`AFDW.z`	Weight used in the finite difference scheme for advection in the z-direction, defined on grid cell interfaces; backward = 1, centered = 0.5, forward = 0; default is backward. One value, a vector of length (Nz+1), a prop.1D list created by `setup.prop.1D`, or a Nx*Ny*(Nz+1) array [–]

V	Grid cell volume, defined at grid cell centers [L^3]. One value, a vector of length N, or a list as defined by `setup.prop.1D`
full.check	Logical flag enabling a full check of the consistency of the arguments (default = FALSE; TRUE slows down execution by 50%)
full.output	Logical flag enabling a full return of the output (default = FALSE; TRUE slows down execution by 20%)

Details

The boundary conditions are of type

(1) zero gradient (default)
(2) fixed concentration
(3) fixed input.

The *bulk dispersion coefficient* (Disp) and the *flow rate* (flow) can be either one value or a vector of length N+1, defined at all grid cell interfaces, including upstream and downstream boundary.

The spatial discretization is given by the volume of each box (V), which can be one value or a vector of length N+1, defined at the center of each grid cell.

The water flow is mass conservative. Over each volume box, the routine calculates internally the downstream outflow of water in terms of the upstream inflow and the lateral inflow.

Value

dC	The rate of change of the concentration C owing to transport, defined in the center of each grid cell [$M/L^3/T$]
F.up	Mass flow across the upstream boundary, positive = INTO model domain. One value [M/T]
F.down	Mass flow across the downstream boundary, positive = OUT of model domain. One value [M/T]
F.lat	Lateral mass input per volume box, positive = INTO model domain. A vector of length N [M/T]
flow	Water flow across the interface of each grid cell. A vector of length N+1 [L^3/T]. Only provided when (full.output = TRUE)
flow.up	Water flow across the upstream (external) boundary, positive = INTO model domain. One value [L^3/T]. Only provided when (full.output = TRUE)
flow.down	Water flow across the downstream (external) boundary, positive = OUT of model domain. One value [L^3/T]. Only provided when (full.output = TRUE)
flow.lat	Lateral water input on each volume box, positive = INTO model domain. A vector of length N [L^3/T]. Only provided when (full.output = TRUE)

F — Mass flow across at the interface of each grid cell. A vector of length N+1 [M/T]. Only provided when (full.output = TRUE)

Authors

Filip Meysman <filip.meysman@nioz.nl>,
Karline Soetaert <karline.soetaert@nioz.nl>.

See Also

tran.1D
advection.volume.1D, for more sophisticated advection schemes.

Examples

```
## ===============================================
## EXAMPLE : organic carbon (OC) decay in a widening estuary
## ===============================================

# Two scenarios are simulated: the baseline includes only input
# of organic matter upstream. The second scenario simulates the
# input of an important side river half way the estuary.

#===============#
# Model formulation      #
#===============#
river.model <- function (t = 0, OC, pars = NULL) {
tran <- tran.volume.1D(C = OC, F.up = F.OC, F.lat = F.lat,
        Disp = Disp, flow = flow.up, flow.lat = flow.lat,
        V = Volume, full.output = TRUE)
reac <- - k*OC
return(list(dCdt = tran$dC + reac, Flow = tran$flow))
}

#==============#
# Parameter definition  #
#==============#

# Initializing morphology estuary:

nbox <- 500 # number of grid cells
lengthEstuary <- 100000 # length of estuary [m]
BoxLength <- lengthEstuary/nbox # [m]
```

```
Distance <- seq(BoxLength/2, by = BoxLength, len =nbox) # [m]
Int.Distance <- seq(0, by = BoxLength, len = (nbox+1)) # [m]

# Cross-sectional area: sigmoid function of estuarine distance [m2]
CrossArea <- 4000 + 72000 * Distance^5 /(Distance^5+50000^5)

# Volume of boxes (m3)
Volume <- CrossArea*BoxLength

# Transport coefficients
Disp <- 1000 # m3/s, bulk dispersion coefficient
flow.up <- 180 # m3/s, main river upstream inflow
flow.lat.0 <- 180 # m3/s, side river inflow

F.OC <- 180          # input organic carbon [mol s-1]
F.lat.0 <- 180       # lateral input organic carbon [mol s-1]

k <- 10/(365*24*3600) # decay constant organic carbon [s-1]

#============#
# Model solution  #
#============#

#scenario 1: without lateral input
F.lat <- rep(0, length.out = nbox)
flow.lat <- rep(0, length.out = nbox)

Conc1 <- steady.1D(runif(nbox), fun = river.model, nspec = 1, name = "OC")

#scenario 2: with lateral input
F.lat <- F.lat.0 * dnorm(x =Distance/lengthEstuary,
                   mean = Distance[nbox/2]/lengthEstuary,
                   sd = 1/20, log = FALSE)/nbox

flow.lat <- flow.lat.0 * dnorm(x = Distance/lengthEstuary,
                         mean = Distance[nbox/2]/lengthEstuary,
                         sd = 1/20, log = FALSE)/nbox

Conc2 <- steady.1D(runif(nbox), fun = river.model, nspec = 1, name = "OC")

#============#
# Plotting output   #
#============#
# use S3 plot method
```

```
plot(Conc1, Conc2, grid = Distance/1000, which = "OC",
      mfrow = c(2, 1), lwd = 2, xlab = "distance [km]",
     main = "Organic carbon decay in the estuary",
     ylab = "OC Concentration [mM]")

plot(Conc1, Conc2, grid = Int.Distance/1000, which = "Flow",
      mfrow = NULL, lwd = 2, xlab = "distance [km]",
      main = "Longitudinal change in the water flow rate",
      ylab = "Flow rate [m3 s-1]")

legend ("topright", lty = 1:2, col = 1:2, lwd = 2,
         c("baseline", "+ side river input"))
```

In the R domain:

```
> # In the R domain:
> install.packages("ReacTran")
Installing package into 'C:/Users/Bert/Documents/R/win-library/3.3'
(as 'lib' is unspecified)
--- Please select a CRAN mirror for use in this session ---
A CRAN mirror is selected.
also installing the dependencies 'rootSolve', 'deSolve'
trying URL 'https://mirrors.tuna.tsinghua.edu.cn/CRAN/bin/windows/contrib/3.3/
rootSolve_1.7.zip'
Content type 'application/zip' length 683839 bytes (667 KB)
downloaded 667 KB
trying URL 'https://mirrors.tuna.tsinghua.edu.cn/CRAN/bin/windows/contrib/3.3/
deSolve_1.20.zip'
Content type 'application/zip' length 2667028 bytes (2.5 MB)
downloaded 2.5 MB
trying URL 'https://mirrors.tuna.tsinghua.edu.cn/CRAN/bin/windows/contrib/3.3/
ReacTran_1.4.3.1.zip'
Content type 'application/zip' length 915996 bytes (894 KB)
downloaded 894 KB

package 'rootSolve' successfully unpacked and MD5 sums checked
package 'deSolve' successfully unpacked and MD5 sums checked
package 'ReacTran' successfully unpacked and MD5 sums checked


The downloaded binary packages are in
C:\Users\Bert\AppData\Local\Temp\RtmpmEkAb2\downloaded_packages
> require(ReacTran)
Loading required package: ReacTran
Loading required package: rootSolve
Loading required package: deSolve
```

```
Loading required package: shape
> library(ReacTran)
> ls("package:ReacTran") # Listing all the programs in this package:
 [1] "advection.1D"        "advection.volume.1D"  "fiadeiro"
 [4] "g.cylinder"          "g.sphere"             "g.spheroid"
 [7] "p.exp"               "p.lin"                "p.sig"
[10] "polar2cart"          "setup.compaction.1D"  "setup.grid.1D"
[13] "setup.grid.2D"       "setup.prop.1D"        "setup.prop.2D"
[16] "tran.1D"             "tran.2D"              "tran.3D"
[19] "tran.cylindrical"    "tran.polar"           "tran.spherical"
[22] "tran.volume.1D"      "tran.volume.2D"       "tran.volume.3D"


> # In the R domain:
>
> install.packages("ReacTran")
Installing package into 'C:/Users/Bert/Documents/R/win-library/3.3'
(as 'lib' is unspecified)
--- Please select a CRAN mirror for use in this session ---

A CRAN mirror is selected.
also installing the dependencies 'rootSolve', 'deSolve'
trying URL 'https://mirrors.tuna.tsinghua.edu.cn/CRAN/bin/windows/contrib/3.3/
rootSolve_1.7.zip'
Content type 'application/zip' length 683839 bytes (667 KB)
downloaded 667 KB
trying URL 'https://mirrors.tuna.tsinghua.edu.cn/CRAN/bin/windows/contrib/3.3/
deSolve_1.20.zip'
Content type 'application/zip' length 2667028 bytes (2.5 MB)
downloaded 2.5 MB
trying URL 'https://mirrors.tuna.tsinghua.edu.cn/CRAN/bin/windows/contrib/3.3/
ReacTran_1.4.3.1.zip'
Content type 'application/zip' length 915996 bytes (894 KB)
downloaded 894 KB
package 'rootSolve' successfully unpacked and MD5 sums checked
package 'deSolve' successfully unpacked and MD5 sums checked
package 'ReacTran' successfully unpacked and MD5 sums checked

The downloaded binary packages are in
C:\Users\Bert\AppData\Local\Temp\RtmpmEkAb2\downloaded_packages
> require(ReacTran)
Loading required package: ReacTran
Loading required package: rootSolve
Loading required package: deSolve
Loading required package: shape
> library(ReacTran)
> ls("package:ReacTran") # Listing all the programs in this package:
```

```
 [1] "advection.1D"       "advection.volume.1D"  "fiadeiro"
 [4] "g.cylinder"         "g.sphere"             "g.spheroid"
 [7] "p.exp"              "p.lin"                "p.sig"
[10] "polar2cart"         "setup.compaction.1D"  "setup.grid.1D"
[13] "setup.grid.2D"      "setup.prop.1D"        "setup.prop.2D"
[16] "tran.1D"            "tran.2D"              "tran.3D"
[19] "tran.cylindrical"   "tran.polar"           "tran.spherical"
[22] "tran.volume.1D"     "tran.volume.2D"       "tran.volume.3D"
```

```
> tran.volume.1D # Outputting the program listing of "tran.volume.1D"

function (C, C.up = C[1], C.down = C[length(C)], C.lat = C, F.up = NULL,
   F.down = NULL, F.lat = NULL, Disp, flow = 0, flow.lat = NULL,
   AFDW = 1, V = NULL, full.check = FALSE, full.output = FALSE)
{
   Rep <- function(x, length.out) if (!is.null(x))
      return(rep(x, length.out = length.out))
   else return(NULL)
   N <- length(C)
   if (N == 0)
      stop("Error: C should be a vector with numeric values")
   if (!is.list(AFDW))
      AFDW <- list(int = AFDW)
   if (!is.list(F.lat))
      F.lat <- list(mid = Rep(F.lat, length.out = N))
   if (!is.list(Disp))
      Disp <- list(int = rep(Disp, length.out = N + 1))
   if (!is.list(flow))
      flow <- list(int = flow)
   if (!is.list(flow.lat))
      flow.lat <- list(mid = flow.lat)
   if (!is.list(V))
      V <- list(mid = rep(V, length.out = N))

   if (!is.list(C.lat))
      C.lat <- list(mid = rep(C.lat, length.out = length(C)))
   if (full.check) {
      gn <- names(AFDW)
      if (!"int" %in% gn)
         stop("error: AFDW should be a list that contains 'int', the AFDW values
at the interface of the grid cells")
      if (is.null(AFDW$int) || length(AFDW$int) == 0)
         stop("error: AFDW should contain (numeric) values")
      if (any(AFDW$int < 0) || any(AFDW$int > 1))
         stop("error: AFDW values should always range between 0 and 1")
      gn <- names(V)
      if (!"mid" %in% gn)
```

```
        stop("error: V should be a list that contains 'mid'")
    if (is.null(V$mid))
        stop("error: the argument V should contain (numeric) values")
    if (any(V$mid < 0))
        stop("error: the volume V should always be larger than zero")
    gn <- names(Disp)
    if (!"int" %in% gn)
      stop("error: Disp should be a list that contains 'int', the dispersion coefficient
at the grid cell interfaces")
    if (is.null(Disp$int))
        stop("error: the argument Disp should be contain (numeric) values")
    if (any(Disp$int < 0))
      stop("error: the bulk dispersion coefficient Disp should always be positive")
    gn <- names(flow)
    if (!"int" %in% gn)
        stop("error: flow should be a list that contains 'int'")
    if (is.null(flow$int))
        stop("error: the argument flow should contain (numeric) values")
    gn <- names(flow.lat)
    if (!"mid" %in% gn)
        stop("error: flow.lat should be a list that contains 'mid'")
    gn <- names(C.lat)
    if (!"mid" %in% gn)
        stop("error: the argument C.lat should be a list that contains 'mid'")
    if (any(C.lat$int < 0))
        stop("error: the concentration C.lat should always be positive")
    gn <- names(F.lat)
   if (!"mid" %in% gn)
        stop("error: the argument F.lat should be a list that contains 'mid'")
  }
  if (is.null(flow.lat$mid)) {
    flow$int <- rep(flow$int, length.out = N + 1)
    flow.lat$mid <- diff(flow$int)
  }
  else {
    if (!is.vector(flow.lat$mid) || length(flow.lat$mid) ==
        1)
        flow.lat$mid <- rep(flow.lat$mid, length.out = N)
    if (length(flow.lat$mid) != N)
        stop("flow.lat should be one number or a vector of length = N")
    if (!(length(flow$int) = 1))
        stop("flow should be of length = 1 if flow.lat is specified")
    f1 <- flow$int[1]
    flow$int <- c(f1, f1 + cumsum(flow.lat$mid))
  }
  if (is.null(F.lat$mid)) {
```

```
    if (is.null(C.lat$mid))
      stop("C.lat and F.lat cannot be both NULL")
    v1 <- flow.lat$mid
    F.lat$mid <- 0
    if (any(flow.lat$mid > 0)) {
      v1[v1 < 0] <- 0
      F.lat$mid <- C.lat$mid * v1
    }
    if (any(flow.lat$mid < 0)) {
      ii <- which(flow.lat$mid < 0)
      F.lat$mid[ii] <- C[ii] * flow.lat$mid[ii]
    }
  }
  else {
    C.lat$mid <- F.lat$mid/flow.lat$mid
  }
  Dif.F <- as.vector(-Disp$int * diff(c(C.up, C, C.down)))
  v1 <- flow$int
  Adv.F <- 0
  if (any(v1 > 0)) {
    v1[v1 < 0] <- 0
    conc <- AFDW$int * c(C.up, C)

if (any(AFDW$int < 1))
      conc <- conc + (1 - AFDW$int) * c(C, C.down)
 Adv.F <- Adv.F + as.vector(v1 * conc)
  }
  else Adv.F <- 0
  if (any(flow$int < 0)) {
    v1 <- flow$int
    v1[v1 > 0] <- 0
    conc <- AFDW$int * c(C, C.down)
    if (any(AFDW$int < 1))
      conc <- conc + (1 - AFDW$int) * c(C.up, C)
    Adv.F <- Adv.F + as.vector(v1 * conc)
  }
  F <- as.vector(Dif.F + Adv.F)
  if (!is.null(F.up))
    F[1] <- F.up
  if (!is.null(F.down))
    F[length(F)] <- F.down
  dC <- -diff(F)/V$mid + F.lat$mid/V$mid
  if (!full.output) {
    return(list(dC = dC, F.up = F[1], F.down = F[length(F)]))
  }
  else {
```

```
    return(list(dC = dC, flow = flow$int, flow.up = flow$int[1],
       flow.down = flow$int[N + 1], flow.lat = flow.lat$mid,
       F = F, F.up = F[1], F.down = F[length(F)], F.lat = F.lat$mid))
  }
}
<environment: namespace:ReacTran>
>
> # Running the program " tran.volume.1D ":
>
>    <environment: namespace:ReacTran>
> q()
>
> ## ========================================
> ## EXAMPLE : organic carbon (OC) decay in a widening estuary
> ## ========================================
> # Two scenarios are simulated: the baseline includes only input
> # of organic matter upstream. The second scenario simulates the
> # input of an important side river half way the estuary.
> #==============#
> # Model formulation      #
> #==============#
> river.model <- function (t = 0, OC, pars = NULL) {
+ tran <- tran.volume.1D(C = OC, F.up = F.OC, F.lat = F.lat,
+ Disp = Disp, flow = flow.up, flow.lat = flow.lat,
+ V = Volume, full.output = TRUE)
+ reac <- - k*OC
+ return(list(dCdt = tran$dC + reac, Flow = tran$flow))
+ }
> #===============#
> # Parameter definition     #
> #===============#
> # Initializing morphology estuary:
> nbox <- 500                         # number of grid cells
> lengthEstuary <- 100000             # length of estuary [m]
> BoxLength <- lengthEstuary/nbox  # [m]
> Distance <- seq(BoxLength/2, by = BoxLength, len =nbox) # [m]
> Int.Distance <- seq(0, by = BoxLength, len = (nbox+1)) # [m]
> # Cross-sectional area: sigmoid function of estuarine distance [m2]
> CrossArea <- 4000 + 72000 * Distance^5 /(Distance^5+50000^5)
> # Volume of boxes (m3)
> Volume <- CrossArea*BoxLength
> # Transport coefficients
> Disp <- 1000                       # m3/s, bulk dispersion coefficient
> flow.up <- 180                     # m3/s, main river upstream inflow
> flow.lat.0 <- 180                  # m3/s, side river inflow
> F.OC <- 180                         # input organic carbon [mol s-1]
```

```
> F.lat.0 <- 180                        # lateral input organic carbon [mol s-1]
> k <- 10/(365*24*3600) # decay constant organic carbon [s-1]

> #===========#
> # Model solution    #
> #===========#
> #scenario 1: without lateral input
> F.lat <- rep(0, length.out = nbox)
> flow.lat <- rep(0, length.out = nbox)
> Conc1 <- steady.1D(runif(nbox), fun = river.model, nspec = 1,
# name = "OC")
> #scenario 2: with lateral input
> F.lat <- F.lat.0 * dnorm(x =Distance/lengthEstuary,
+ mean = Distance[nbox/2]/lengthEstuary,
+ sd = 1/20,
+ log = FALSE)/nbox
> flow.lat <- flow.lat.0 * dnorm(x = Distance/lengthEstuary,
+ mean = Distance[nbox/2]/lengthEstuary,
+ sd = 1/20,
+ log = FALSE)/nbox
> Conc2 <- steady.1D(runif(nbox), fun = river.model, nspec = 1,
+ name = "OC")
>
> #============#
> #  Plotting output    #
> #============#
> # use S3 plot method
> plot(Conc1, Conc2, grid = Distance/1000, which = "OC",
+ mfrow = c(2, 1), lwd = 2, xlab = "distance [km]",
+ main = "Organic carbon decay in the estuary",
+ ylab = "OC Concentration [mM]")
> #
>
> # Outputting: Figure A6-1: tran.volume.1D-1
>
> #
> plot(Conc1, Conc2, grid = Int.Distance/1000, which = "Flow",
+ mfrow = NULL, lwd = 2, xlab = "distance [km]",
+ main = "Longitudinal change in the water flow rate",
+ ylab = "Flow rate [m3 s-1]")
> #
>
> # Outputting: Figure A6-2: tran.volume.1D-2
>
> #
```

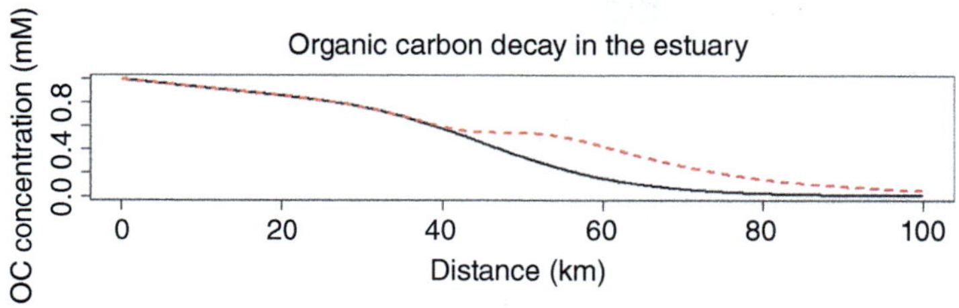

Figure A6-1 tran.volume.1D-1 1D Diffusive transport in an aquatic system rate of organic carbon decay in the estuary away from the source.

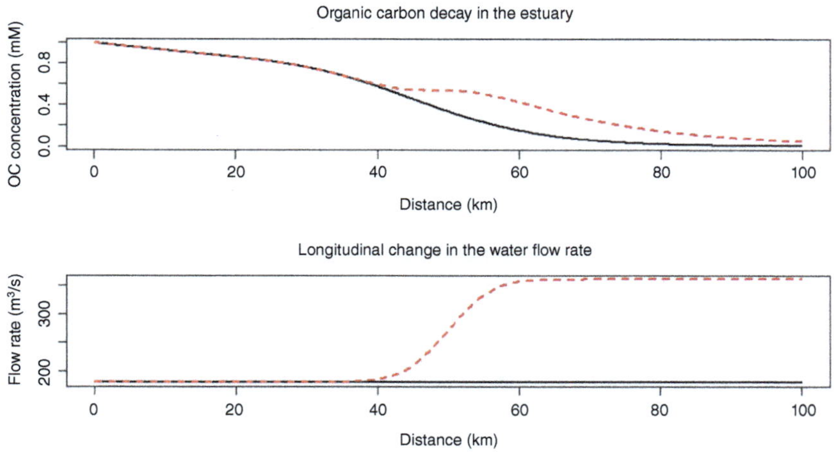

Figure A6-2 tran.volume.1D-2 1D Diffusive transport in an aquatic system rate of organic carbon decay in the estuary compared with longitudinal change in the water flow rate as a function of distance from the source.

```
> legend ("topright", lty = 1:2, col = 1:2, lwd = 2,
+ c("baseline", "+ side river input"))
> #
>
> # Outputting: Figure A6-3 tran.volume.1D-3
>
> #
>
```

Worked Example 4.8 ***Modeling Framework for Cellular Communities in Their Environments***

Version	1.6	
Package	'BacArena'	**May 23, 2017**
Title	Modeling Framework for Cellular Communities in Their Environments	
Version	1.6	
Authors	Eugen Bauer Johannes Zimmermann	

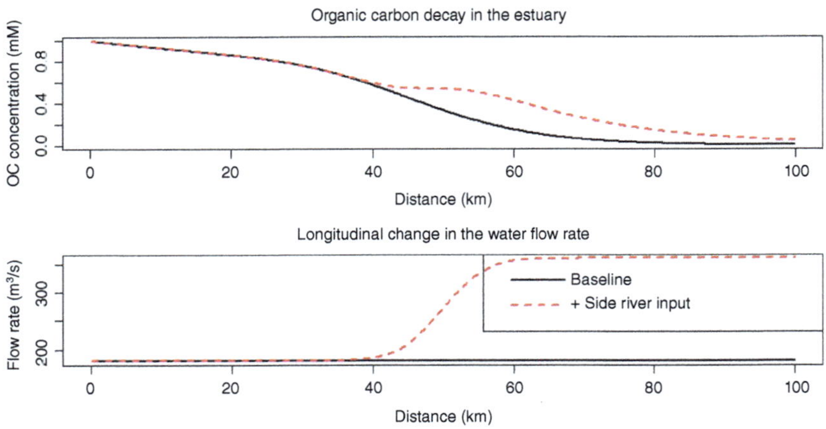

Figure A6-3 tran.volume.1D-3 1D Diffusive transport in an aquatic system rate of organic carbon decay in the estuary compared with longitudinal change in the water flow rate as a function of distance from the source, with added legend.

Maintainer	Johannes Zimmermann <j.zimmermann@iem.uni-kiel.de>
Description	Can be used for simulation of organisms living in communities. Each organism is represented individually and genome-scale metabolic models determine the uptake and release of compounds. Biological processes such as movement, diffusion, chemotaxis, and kinetics are available along with data analysis techniques
URL	https://github.com/euba/BacArena
Bug Reports	https://github.com/euba/BacArena/issues
Depends	R (>= 3.0.0), sybil (>= 1.3.0), ReacTran (>= 1.4.2), deSolve (>= 1.12), Matrix (>= 1.2)
Imports	igraph, methods, utils, stats, graphics, ggplot2, reshape2, glpkAPI, plyr, Rcpp
Suggests	sybilSBML, parallel, knitr, rmarkdown
LinkingTo	Rcpp, RcppArmadillo, RcppEigen
License	GPL-3
VignetteBuilder	knitr
RoxygenNote	6.0.1
LazyData	True
NeedsCompilation	Yes
Repository	CRAN
Date/Publication	2017-05-23 17:28:04 UTC

R **topics demonstrated:**

addDefaultMed 4
addEssentialMed 4
addEval 5

addOrg	6
addSubs	7
Arena-class	8
Arena-constructor	9
Bac-class	10
Bac-Constructor	10
BacArena	11
cellgrowth	11
changeDiff	12
changeFobj	13
changeOrg	14
changeSub	15
checkCorr	16
checkPhen	17
checkPhen_par	18
chemotaxis	18
colpal1	19
colpal2	20
colpal3	20
colpal4	21
colpal5	21
colpal6	22
constrain	22
consume	23
createGradient	24
dat2mat	25
diffuse	26
diffusePDE	26
diffuseR	27
diffuse_par	28
emptyHood	28
Eval-class	29
Eval-constructor	30
evalArena	30
extractMed	31
findFeeding	32
findFeeding2	33
findFeeding3	34
findFeeding3rep	34
findInArena	35
findRxnFlux	35
flushSubs	36
fluxVarSim	37
getArena	38
getCorrM	39

getPhenoMat	40
getPhenotype	41
getSubHist	42
getSublb	42
getVarSubs	43
growExp	44
growLin	44
growth	45
growth_par	46
Human-class	47
Human-constructor	47
lsd	48
lysis	48
minePheno	49
move	50
NemptyHood	51
openArena	52
optimizeLP	52
Organism-class	53
Organism-constructor	54
plotAbundance	55
plotCurves	56
plotCurves2	57
plotFluxVar	58
plotFVA	59
plotGrowthCurve	59
plotInterNum	60
plotPhenCurve	60
plotPhenNum	61
plotReaActivity	61
plotShadowCost	62
plotSpecActivity	62
plotSubCurve	63
plotSubDist	64
plotSubDist2	64
plotSubUsage	65
plotSubVar	65
plotTotFlux	66
redEval	66
reset_screen	67
rmSubs	68
selPheno	68
setKinetics	69
sihumi_test	70
simBac	70

simBac_par	71
simEnv	72
simEnv_par	73
simHum	75
statPheno	76
stirEnv	77
Substance-class	78
Substance-constructor	79
unit_conversion	80
usd	80

Worked Example 4.9 plotCurves ***Function for Plotting the Overall Change as Curves***

Description

The generic function plotCurves() plots the growth curves and concentration changes of substances from simulation steps in an Eval object.

Usage

```
plotCurves(object, medplot = object@mediac, retdata = F, remove = F,
           legend = F, graph = T)
## S4 method for signature 'Eval'

plotCurves(object, medplot = object@mediac, retdata = F, remove = F,
          legend = F, graph = T)
```

Arguments

object	An object of class Eval
medplot	A character vector giving the name of substances which should be plotted
retdata	A boolean variable indicating if the data used to generate the plots should be returned
remove	A boolean variable indicating if substances, which don't change in their concentration should be removed from the plot
legend	Boolean variable indicating if legend should be plotted graph True if graphic should be plotted

Details

The parameter retdata can be used to access the data used for the returned plots to create own custom plots.

Value

Returns two graphs in one plot: the growth curves and the curves of concentration changes. Optional the data to generate the original plots can be returned.

See Also

Eval-class and Arena-class

Examples

```
get Escherichia coli core metabolic model
data(Ec_core, envir = environment())
bac <- Bac(Ec_core,deathrate=0.05,
            minweight=0.05, growtype="exponential") #initialize a bacterium
arena <- Arena(n=20,m=20) #initialize the environment
arena <- addOrg(arena,bac,amount=10) #add 10 organisms
arena <- addSubs(arena,40) #add all possible substances
eval <- simEnv(arena,5)
plotCurves(eval)
```

In the R domain:

R version 3.3.3 (2017-03-06) – "Another Canoe"
Copyright (C) 2017 The R Foundation for Statistical Computing
Platform: x86_64-w64-mingw32/x64 (64-bit)

R is a free software and comes with ABSOLUTELY NO WARRANTY.
You are welcome to redistribute it under certain conditions.
Type 'license()' or 'licence()' for distribution details.
Natural language support but running in an English locale.

R is a collaborative project with many contributors.
Type 'contributors()' for more information and
'citation()' on how to cite R or R packages in publications.

Type 'demo()' for some demos, 'help()' for on-line help, or
'help.start()' for an HTML browser interface to help.
Type 'q()' to quit R.

```
>
> install.packages("BacArena")
Installing package into 'C:/Users/Bert/Documents/R/win-library/3.3'
(as 'lib' is unspecified)
--- Please select a CRAN mirror for use in this session ---
```

A CRAN mirror is selected

```
trying URL 'https://mirrors.tuna.tsinghua.edu.cn/CRAN/bin/windows/contrib/3.3/
BacArena_1.6.zip'
```

```
Content type 'application/zip' length 5091073 bytes (4.9 MB)
downloaded 4.9 MB
```

package 'BacArena' successfully unpacked and MD5 sums checked
The downloaded binary packages are in
C:\Users\Bert\AppData\Local\Temp\Rtmp8sWwS0\downloaded_packages
> library(BacArena)
Loading required package: sybil
Loading required package: Matrix
Loading required package: lattice
Loading required package: ReacTran
Loading required package: rootSolve
Loading required package: deSolve
Loading required package: shape
BacArena paper: https://doi.org/10.1371/journal.pcbi.1005544
Tutorial: https://CRAN.R-project.org/package=BacArena/vignettes/BacArena-Introduction.pdf
Development and help: https://github.com/euba/bacarena

```
> ls("package:BacArena")
 [1] "addDefaultMed"    "addEssentialMed"  "addEval"          "addOrg"
 [5] "addSubs"          "Arena"            "Bac"              "cellgrowth"
 [9] "changeDiff"       "changeFobj"       "changeOrg"        "changeSub"
[13] "checkCorr"        "checkPhen"        "checkPhen_par"    "chemotaxis"
[17] "colpal1"          "colpal2"          "colpal3"          "colpal4"
[21] "colpal5"          "colpal6"          "constrain"        "consume"
[25] "createGradient"   "dat2mat"          "diffuse"          "diffuse_par"
[29] "diffusePDE"       "diffuseR"         "emptyHood"        "Eval"
[33] "evalArena"        "extractMed"       "findFeeding"      "findFeeding2"
[37] "findFeeding3"     "findFeeding3rep"  "findInArena"      "findRxnFlux"
[41] "flushSubs"        "fluxVarSim"       "getArena"         "getCorrM"
[45] "getPhenoMat"      "getPhenotype"     "getSubHist"       "getSublb"
[49] "getVarSubs"       "growExp"          "growLin"          "growth"
[53] "growth_par"       "Human"            "lsd"              "lysis"
[57] "minePheno"        "move"             "NemptyHood"       "openArena"
[61] "optimizeLP"       "Organism"         "plotAbundance"    "plotCurves"
[65] "plotCurves2"      "plotFluxVar"      "plotFVA"          "plotGrowthCurve"
[69] "plotInterNum"     "plotPhenCurve"    "plotPhenNum"      "plotReaActivity"
[73] "plotShadowCost"   "plotSpecActivity" "plotSubCurve"     "plotSubDist"
[77] "plotSubDist2"     "plotSubUsage"     "plotSubVar"       "plotTotFlux"
[81] "redEval"          "reset_screen"     "rmSubs"           "selPheno"
[85] "setKinetics"      "sihumi_test"      "simBac"           "simBac_par"
[89] "simEnv"           "simEnv_par"       "simHum"           "statPheno"
[93] "stirEnv"          "Substance"        "unit_conversion"  "usd"
```

> plotCurves ***# Outputting the listing of the*** R ***program*** plotCurves
nonstandardGenericFunction for "plotCurves" defined from package

```
"BacArena"
function (object, medplot = object@mediac, retdata = F, remove = F,
   legend = F, graph = T)
{
   standardGeneric("plotCurves")
}
<environment: 0x0000000017361628>
Methods may be defined for arguments: object, medplot, retdata, remove, legend,
graph
Use  showMethods("plotCurves")  for currently available ones.
> data(Ec_core, envir = environment()) # get Escherichia coli core
+ # metabolic model
> bac <- Bac(Ec_core,deathrate=0.05,
+ minweight=0.05,growtype="exponential")
# initialize a bacterium
Loading required package: glpkAPI
using GLPK version 4.47
> arena <- Arena(n=20,m=20) #initialize the environment
> arena <- addOrg(arena,bac,amount=10) #add 10 organisms
> arena <- addSubs(arena,40) #add all possible substances
> eval <- simEnv(arena,5)

iteration: 1     organisms: 10   biomass: 5.25312 pg
                 count biomass %
ecoli_core_model    10 5.25312 0
      time total:  0.49       diffusion:  0.22  ( 44.9 %)

iteration: 2     organisms: 15   biomass: 11.59317 pg
                 count  biomass       %
ecoli_core_model    15 11.59317 120.6912
      time total:  0.5       diffusion:  0.21  ( 42 %)

iteration: 3     organisms: 30   biomass: 22.54044 pg
                 count  biomass       %
ecoli_core_model    30 22.54044 94.42863
      time total:  0.63       diffusion:  0.2  ( 31.7 %)

iteration: 4     organisms: 60   biomass: 44.95924 pg
                 count  biomass       %
ecoli_core_model    60 44.95924 99.46032
      time total:  0.93       diffusion:  0.19  ( 20.4 %)

iteration: 5     organisms: 120          biomass: 90.11333 pg
                 count  biomass       %
ecoli_core_model   120 90.11333 100.4334
      time total:  1.46       diffusion:  0.17  ( 11.6 %)
```

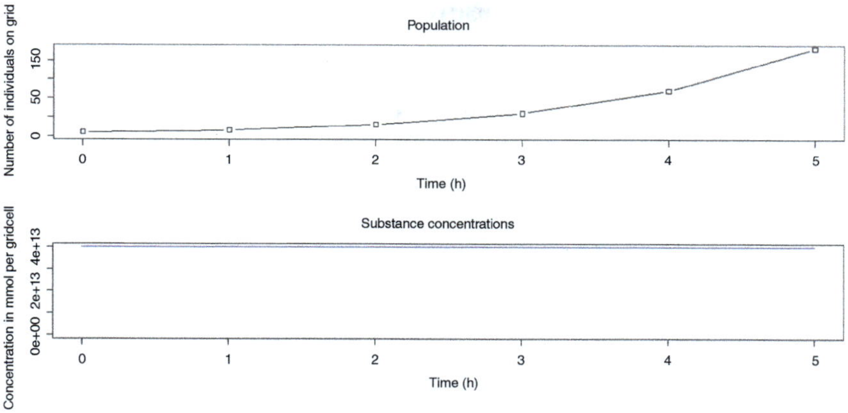

Figure A8-1 plotCurves-1 Diffusion and reaction.

```
> plotCurves(eval)
>
> # Outputting: Figure A8-1 plotCurves-1
>
```

Worked Example 4.10 plotTotFlux ***Function for Plotting the Overall Change in Reaction Activity***

Description

The generic function plotTotFlux() plots the time course of reactions with high variation in activity for an Eval object.

Usage

```
plotTotFlux(object, legendpos = "topright", num = 20)
## S4 method for signature 'Eval'
plotTotFlux(object, legendpos = "topright", num = 20)
```

Arguments

object	An object of class Eval
legendpos	A character variable declaring the position of the legend
num	An integer defining the number of substrates to be plot

Examples

```
data(Ec_core, envir = environment())
#get Escherichia coli core metabolic model
bac <- Bac(Ec_core,deathrate=0.05,
```

```
minweight=0.05,growtype="exponential")
#initialize a bacterium
arena <- Arena(n=20,m=20) #initialize the environment
arena <- addOrg(arena,bac,amount=10) #add 10 organisms
arena <- addSubs(arena,40) #add all possible substances
eval <- simEnv(arena,5)
plotTotFlux(eval)
```

In the R domain:

```
>
> data(Ec_core, envir = environment())
# get Escherichia coli core metabolic model
> bac <- Bac(Ec_core,deathrate=0.05,
+ minweight=0.05,growtype="exponential")
+ #initialize a bacterium
> arena <- Arena(n=20,m=20) #initialize the environment
> arena <- addOrg(arena,bac,amount=10) #add 10 organisms
> arena <- addSubs(arena,40) #add all possible substances
> eval <- simEnv(arena,5)

iteration: 1    organisms: 10   biomass: 4.797873 pg
                count  biomass %
ecoli_core_model    10 4.797873 0
      time total:  0.48      diffusion:  0.22  ( 45.8 %)

iteration: 2    organisms: 15   biomass: 10.76978 pg
                count  biomass        %
ecoli_core_model    15 10.76978 124.4699
      time total:  0.52      diffusion:  0.22  ( 42.3 %)

iteration: 3    organisms: 30   biomass: 21.56735 pg
                count  biomass       %
ecoli_core_model    30 21.56735 100.258
      time total:  2.34      diffusion:  1.91  ( 81.6 %)

iteration: 4    organisms: 60   biomass: 43.73724 pg
                count  biomass        %
ecoli_core_model    60 43.73724 102.7938
time total:  0.96      diffusion:  0.21  ( 21.9 %)
iteration: 5    organisms: 120          biomass: 88.57083 pg
                count  biomass        %
ecoli_core_model   120 88.57083 102.5067
      time total:  1.49      diffusion:  0.21  ( 14.1 %)

> plotTotFlux(eval)
 [1] "ecoli_core_model.EX_glu_L(e)" "ecoli_core_model.GLUt2r"
"ecoli_core_model.EX_fum(e)"
```

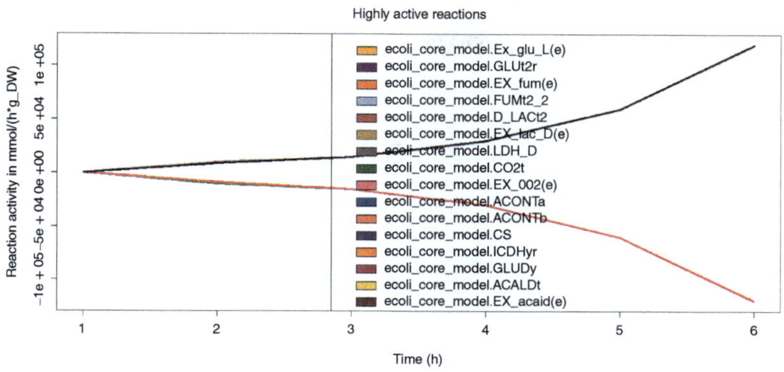

Figure A9-1 plotTotFlux Diffusion and reaction.

```
 [4] "ecoli_core_model.FUMt2_2"    "ecoli_core_model.D_LACt2"
"ecoli_core_model.EX_lac_D(e)"
 [7] "ecoli_core_model.LDH_D"      "ecoli_core_model.CO2t"
"ecoli_core_model.EX_co2(e)"
[10] "ecoli_core_model.ACONTa"    "ecoli_core_model.ACONTb"
"ecoli_core_model.CS"        eaction
[13] "ecoli_core_model.ICDHyr"     "ecoli_core_model.GLUDy"
"ecoli_core_model.ACALDt"
[16] "ecoli_core_model.EX_acald(e)" "ecoli_core_model.EX_pyr(e)"
"ecoli_core_model.PYRt2r"
[19] "ecoli_core_model.EX_nh4(e)"   "ecoli_core_model.NH4t"
>
> # Outputting: Figure A9-1 plotTotFlux Diffusion and Reaction
```

Soetaert, K. and Meysman, F., Centre for Estuarine and Marine Ecology, Netherlands Institute of Ecology, The Netherlands. R-package ReacTran: reactive transport modelling in R – R Project. https://cran.r-project.org/web/packages/ReacTran/vignettes/ReacTran.pdf.

Abstract

The R-package ReacTran (Soetaert and Meysman 2009) contains functions to solve reactive-transport equations, as used, for example, in the environmental sciences.

Essentially, it

(1) Provides functions that subdivide the spatial extent into a number of discrete grid cells.
(2) Approximates the advective–diffusive transport term by finite differences or finite volumes.

The main package vignette (Soetaert and Meysman 2010) explains how ReacTran may be used to model reaction-transport phenomena.

However, the functions from ReacTran may be used to solve more general types of partial differential equations (≤order 2).

This vignette shows how the package may be used to solve partial differential equations of the parabolic, hyperbolic, and elliptic type, providing an example each.

Keywords: Partial Differential Equations, hyperbolic, parabolic, elliptic, R.

4.3 Partial Differential Equations

In partial differential equations (PDE), the function has several independent variables (e.g. time and depth) and may contain their partial derivatives.

A first step to solve PDE is to discretize one or more of the independent variables.

Usually, the independent variable "time" is not discretized, while other variables (e.g. spatial axes) are discretized, so that a set of ODE is obtained, which may be solved with appropriate initial values solvers from the package deSolve (Soetaert, Petzoldt, and Setzer 2010). This technique, the method-of-lines, applies to hyperbolic and parabolic PDEs.

For time-invariant problems, usually all independent variables are discretized, and the resulting algebraic equations solved with root-solving functions from package rootSolve (Soetaert 2009).

Functions tran.1D, tran.2D, and tran.3D from the R package ReacTran implement finite difference approximations of the general diffusive–advective transport equation, which for 1D is:

$$-\left(1/A_x\xi_x\right)\cdot\left[(\partial/\partial x)A_x\cdot\left\{-D\left(\partial\xi_x C/\partial x\right)\right\}-(\partial/\partial x)\left(A_x\cdot v\cdot\xi_x C\right)\right]$$

Here D is the "diffusion coefficient," v is the "advection rate," and A_x and ξ are the surface area and volume fraction, respectively.

Assuming that A, ξ, D, and v are constant along x, one may rewrite this in a more general form: $D\partial^2C/\partial x^2 - u\partial C/\partial x$ in which the first term is a second-order, the second term a first-order derivative.

R-function tran.1D is defined as:

```
tran.1D <- function (C, C.up = C[1], C.down = C[length(C)],
                     flux.up = NULL, flux.down = NULL, a.bl.up = NULL,
                     a.bl.down = NULL, D = 0,
                     v = 0, AFDW = 1, VF = 1, A = 1, dx, ...)
```

where C.up and C.down are the upstream and downstream boundary values, respectively, flux.up and flux.down are the upstream and downstream fluxes, respectively, v and D are the advection and diffusion coefficient, respectively, A is the surface area, x contains the grid, and VF is the volume fraction (ξ).

For the other arguments, see the help file of tran.1D.

A suitable *grid* may be generated with functions setup.grid.1D and setup.grid.2D (there is no corresponding 3D function), while a *property* may be added to this grid using functions setup.prop.1D, and setup.prop.2D. These two latter functions are useful to have the variable surface areas, volume fractions, advection, and diffusion rates being defined at the correct places of the grid.

These functions are defined as:

```
setup.grid.1D <- function(x.up = 0, x.down = NULL, L = NULL, N = NULL,
                          dx.1 = NULL, ...)
setup.grid.2D <- function(x.grid = NULL, y.grid = NULL)
setup.prop.1D <- function (func = NULL, value = NULL, xy = NULL,
                           interpolate = "spline", grid, ...)
setup.prop.2D <- function (func = NULL, value = NULL, grid, y.func = func,
                           y.value = value, ...)
```

4.4 A Parabolic PDE

As an example of the parabolic type, consider the 1D diffusion–reaction model, in spherical, cylindrical, and Cartesian coordinates, defined for r in [0, 10]:

$$\partial C/\partial t = \left(1/r^2\right)(\partial/\partial r)\left\{r^2 \cdot D \cdot (\partial C/\partial r)\right\} - Q$$

$$\partial C/\partial t = (1/r)(\partial/\partial r)\{r \cdot D \cdot (\partial C/\partial r)\} - Q$$

$$\partial C/\partial t = (\partial/\partial r)\{D \cdot (\partial C/\partial r)\} - Q$$

with t the time, r the (radial) distance from the origin, Q, the consumption rate, and with boundary conditions (values at the edges of the model):

$$(\partial C/\partial r)_{r=0} = 0$$

$$C_{r=10} = C_{\text{ext}}$$

To solve this model in R, first the 1D model Grid is defined; it divides 10 cm (L) into 1000 equally sized boxes (N).

```
Grid <- setup.grid.1D(N = 1000, L = 10)
```

Next the properties r and r^2 are defined on this grid:

```
r <- setup.prop.1D(grid = Grid, func = function(r) r)
r2 <- setup.prop.1D(grid = Grid, func = function(r) r^2)
```

The model equation includes a transport term, approximated by ReacTran function tran.1D() and a consumption term (Q); only the downstream boundary condition, prescribed as a concentration (C.down) which has to be specified, as the zero-gradient at the upstream boundary is the default:

```
library(ReacTran)
pde1D <- function(t, C, parms, A = 1) {
   tran <- tran.1D(C = C, A = A, D = D, C.down = Cext,
        dx = Grid)$dC
   list(tran - Q)
}
```

The model parameters are defined:

```
D <- 1
Q <- 1
Cext <- 20
```

4.4.1 Steady-State Solution

In a first application, the model is solved to *steady-state*, which retrieves the condition where the concentrations are invariant, e.g. for the cylindrical coordinate case:

$$0 = \left(1/r^2\right) \cdot (\partial/\partial r) \left\{r^2 \cdot D \cdot (\partial C/\partial r)\right\} - Q$$

The R package ReacTran (**Soetaert and Meysman** 2012) contains functions for ...R-**package** ReacTran: Reactive Transport Modelling in R ... and Herman, **2009**. In R, steady-state conditions may be estimated using functions from the package rootSolve which implement among others a Newton–Raphson algorithm (Press, Teukolsky, Vetterling, and Flannery 1992). For one-dimensional models, steady.1D should be used. The initial estimate of the steady-state solution (y) is unimportant; here one may take simply N random numbers.

Argument nspec = 1 informs the solver that only one component is described.

Although a system of 1000 equations needs to be solved, this takes only a fraction of a second:

```
library(rootSolve)
Cartesian <- steady.1D(y = runif(Grid$N),
   func = pde1D, parms = NULL, nspec = 1, A = 1)
Cylindrical <- steady.1D(y = runif(Grid$N),
   func = pde1D, parms = NULL, nspec = 1, A = r)

print(system.time(
  Spherical <- steady.1D(y = runif(Grid$N),
      func = pde1D, parms = NULL, nspec = 1, A = r2)
))

user   system   elapsed
0.02      0.00        0.00
```

The values of the state variables (y) are plotted against the radial distance, in the middle of the grid cells (Grid$x.mid).

```
plot (Grid$x.mid, Cartesian$y, type = "l", main = "steady-state PDE",
  lwd = 3, xlab = "x", ylab = "C", col = "darkgreen", lty = 1)
lines(Grid$x.mid, Cylindrical$y, lwd = 3, col="blue", lty = 2)
lines(Grid$x.mid, Spherical$y, lwd = 3, col = "red", lty = 3)
legend("bottomright", c("cartesian", "cylindrical", "spherical"),
  col = c("darkgreen", "blue", "red"), lwd = 3, lty = 1:3)
```

The analytical solutions compare well with the numerical approximation for all three cases:

```
max(abs(Q/6/D*(r2$mid - 10^2) + Cext - Spherical$y))
[1] 5.82081e-05
max(abs(Q/4/D*(r2$mid - 10^2) + Cext - Cylindrical$y))
[1] 6.250022e-06
max(abs(Q/2/D*(r2$mid - 10^2) + Cext - Cartesian$y))
[1] 1.250002e-05
```

4.4.2 The Method of Lines

Next, the model (for spherical coordinates) is run dynamically for 100 time units using the deSolve() function ode.1D, and starting with an initially uniform distribution (y = rep(1, Grid$N)):

```
require(deSolve)
times <- seq(0, 100, by = 1)
system.time(
  out <- ode.1D(y = rep(1, Grid$N), times = times, func = pde1D,
    parms = NULL, nspec = 1, A = r2)
)
```

```
user  system  elapsed
0.54    0.04     0.56
```

Here, out is a matrix, whose first column contains the output times, and the next columns the values of the state variables in the different boxes:

```
tail(out[, 1:4], n = 3)
       time        1        2        3
[99,]    98 3.332278 3.332303 3.332366
[100,]   99 3.332383 3.332408 3.332471
[101,]  100 3.332478 3.332503 3.332566
```

Plotting the result using a blue-yellow-red color scheme, and using deSolve's S3 method image:

```
image(out, grid = Grid$x.mid, xlab = "time, days",
      ylab = "Distance, cm", main = "PDE", add.contour = TRUE)
```

The result is shown in Figure A9-2.

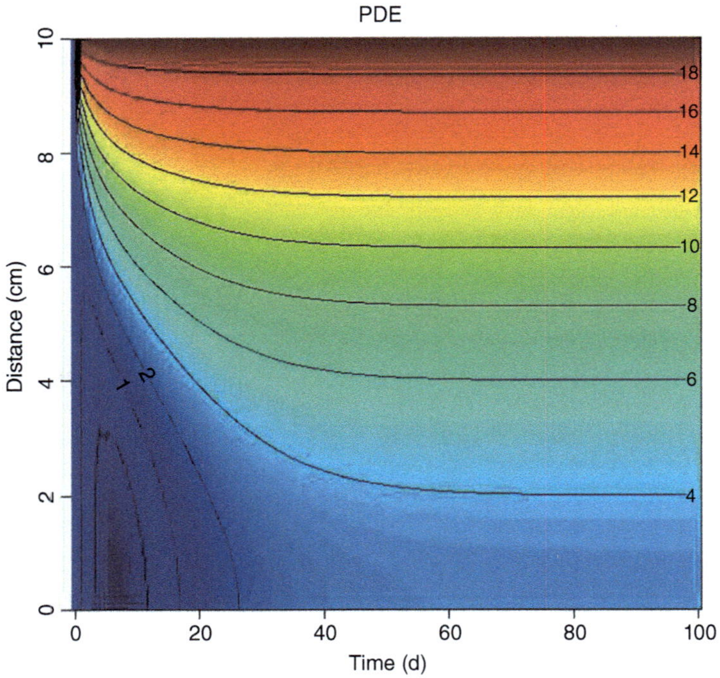

Figure A9-2 Dynamic solution of the 1D diffusion–reaction model.

Further Reading

https://en.wikipedia.org/wiki/Advection

Soetaert, K. and Herman, P.M.J. (2009). *A Practical Guide to Ecological Modelling – Using* R *as a Simulation Platform*. Springer.

5

More Numerical Worked Examples Using R for Simultaneous Mass Transfer and Chemical Reactions

5.1 Introduction

Several numerical examples are chosen to illustrate and highlight the application of the R Program for numerical computations in Simultaneous Mass Transfer and Chemical Reactions in typical transport phenomena in Chemical Engineering Science.

The following three classes of examples are selected to showcase such applications of solving partial differential equations using the R Package: ReacTran

Class A: Simultaneous Mass Transfer and Chemical Reactions Using R
* 1D, 2D, and 3D Rectangular Geometries
* Cylindrical Geometry
* Polar Coordinates
* 1D, 2D, and 3D Volumetric Advective-Diffusive Transport in an Aquatic System

Class B: Diffusive Models of Environmental Transport Using R
* Equations for Diffusion in Environmental Transport

Class C: Numerical Worked Examples of Advection Using R for:
* 1D, 2D, and 3D Rectangular Geometries
* Cylindrical Geometry
* Polar Coordinates

5.2 Advection

In the field of science and engineering, ***advection*** is the transport of a substance by *bulk motion*. The properties of that substance are carried with it. Generally, the majority of the advected substance is a fluid (gas or liquid). The properties that are carried with the advected substance are conserved properties such as energy.

Simultaneous Mass Transfer and Chemical Reactions in Engineering Science, First Edition. Bertram K. C. Chan.

An example of advection is the transport of pollutants or silt in a river by bulk water flowing downstream. Another commonly advected quantity is energy or enthalpy. Here, the fluid may be any material that contains thermal energy, such as air or water. In general, any substance or conserved, extensive quantity may be advected by a fluid that can hold or contain the quantity or substance.

During advection, a fluid transports some conserved quantity or material via bulk motion. The fluid motion may be described mathematically as a vector field, and the transported material is described by a scalar field showing its distribution over space. Advection requires currents in the fluid, and so it cannot happen in rigid solids. It does ***not*** include transport of substances by molecular diffusion. Advection is not to be equated with the process of ***convection*** which is the ***combination*** of advective transport and diffusive transport.

In meteorology and physical oceanography, advection often refers to the transport of some property of the atmosphere or ocean, such as heat, humidity, or salinity. (Advection is important for the formation of orographic clouds and the precipitation of water from clouds, as part of the hydrological cycle.)

5.2.1 Advection vs. Convection

The term *advection* is sometimes considered as a synonym for *convection*, but technically, *convection* covers the sum of transport both by diffusion and by advection. Advective transport describes the movement of some quantity via the bulk flow of a fluid (as in a river or pipeline).

5.2.1.1 Meteorology

In meteorology and physical oceanography, advection often refers to the horizontal transport of some property of the atmosphere or ocean, such as heat, humidity, or salinity, and convection generally refers to vertical transport (vertical advection). Advection is important for the formation of orographic clouds (terrain-forced convection) and the precipitation of water from clouds, as part of the hydrological cycle. Advection also applies if the quantity being advected is represented by a probability density function at each point.

5.2.1.2 The Mathematics of Advection

The **advection equation** is the partial differential equation that governs the motion of a conserved scalar field as it is advected by a known velocity vector field. It is derived from the scalar field's conservation law, together with Gauss's theorem and taking the infinitesimal limit.

A simple visualized example of advection is the transport of ink dumped into a river. As the river flows, ink will move downstream in a "pulse" via advection, as the movement of bulk water itself transports the ink. Note that as it moves downstream, the "pulse" of ink will also spread via diffusion. The *sum* of these processes is called **convection**.

Note: If the ink is added to a lake without significant bulk water flow, the ink would simply disperse outward from its source via diffusion alone, which is ***not*** advection.

5.2.1.3 The Advection Equation

In Cartesian coordinates, the advection *operator* is

$$\mathbf{u} \cdot \mathbf{\nabla} = u_x \frac{\partial}{\partial x} + u_y \frac{\partial}{\partial y} + u_z \frac{\partial}{\partial z} \tag{5.1}$$

where $\mathbf{u} = (u_x, u_y, u_z)$ is the *velocity field*, and $\nabla =$ is the **del** operator.

Note that Cartesian coordinates are used here.

The advection equation for a conserved quantity described by a scalar field ψ is expressed mathematically by a **continuity equation**:

$$\frac{\partial \psi}{\partial t} + \nabla \cdot (\psi \mathbf{u}) = 0 \tag{5.2}$$

where $\mathbf{\nabla}\cdot$ is the divergence operator and **u** is the velocity vector field.

Frequently, it is assumed that the flow is incompressible, that is, the velocity field satisfies the equation

$$\mathbf{\nabla} \cdot \mathbf{u} = 0 \tag{5.3}$$

and **u** is said to be ***solenoidal***. Under such conditions, the above equation may be written as

$$\frac{\partial \psi}{\partial t} + \mathbf{u} \cdot \mathbf{\nabla} \psi = 0 \tag{5.4}$$

In particular, if the flow is steady, then $\partial/\partial t = 0$, and

$$\mathbf{u} \cdot \mathbf{\nabla} \psi = 0 \tag{5.5}$$

which shows that ψ is constant along a *streamline*. Hence,

$$\partial \psi / \partial t = 0 \tag{5.6}$$

so ψ does not vary in time.

If a vector quantity **a** (such as a magnetic field) is being advected by the solenoidal velocity field **u**, the advection equation above becomes:

$$\frac{\partial a}{\partial t} + (\mathbf{u} \cdot \nabla)\mathbf{a} = 0 \tag{5.7}$$

5.2.1.4 Solving the Advection Equation

The advection equation is not simple to solve numerically: the system is a hyperbolic partial differential equation, and interest typically centers on discontinuous "shock" solutions, which are difficult for numerical schemes to solve (Figure 5.1). Even with

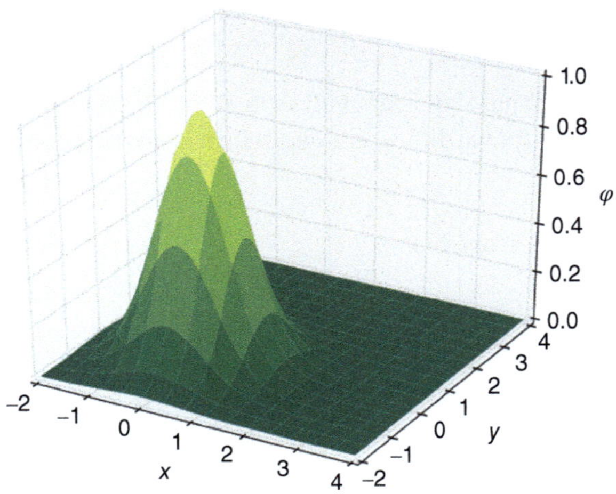

Figure 5.1 A simulation of the advection equation where **u** = (sin *t*, cos *t*) is solenoidal. Source: Shiyu Ji/Wikimedia Commons/CC BY-SA 4.0.

one space dimension and a constant velocity field, the system remains difficult to simulate. The equation becomes

$$\frac{\partial \psi}{\partial t} + u_x \frac{\partial \psi}{\partial x} = 0 \tag{5.8}$$

where $\psi = \psi(x, t)$ is the scalar field being advected and u_x is the x component of the vector $\mathbf{u} = (u_x, 0, 0)$.

5.2.1.5 The Advection Operator in the Incompressible Navier–Stokes Equations

Numerical simulation can be aided by considering the *skew symmetric* form for the advection operator:

$$\frac{1}{2}\mathbf{u} \cdot \nabla \mathbf{u} + \frac{1}{2}\nabla(\mathbf{u}\mathbf{u}) \tag{5.9}$$

where

$$\nabla(\mathbf{u}\mathbf{u}) = [\nabla(\mathbf{u}u_x), \nabla(\mathbf{u}u_y), \nabla(\mathbf{u}u_z)] \tag{5.10}$$

and **u** is the same as above.

Since skew symmetry implies only imaginary eigenvalues, this form reduces the "blow up" and "spectral blocking" often experienced in numerical solutions with sharp discontinuities.

Using vector calculus identities, these operators can also be expressed in other ways, available in more software packages for more coordinate systems.

$$\mathbf{u} \cdot \nabla \mathbf{u} = \nabla \left(\frac{\|\mathbf{u}\|^2}{2} \right) + (\nabla \times \mathbf{u}) \times \mathbf{u} \tag{5.11}$$

$$\frac{1}{2}\mathbf{u} \cdot \nabla \mathbf{u} + \frac{1}{2}\nabla(\mathbf{u}\mathbf{u}) = \nabla \left(\frac{\|\mathbf{u}\|^2}{2} \right) + (\nabla \times \mathbf{u}) \times \mathbf{u} + \frac{1}{2}\mathbf{u}(\nabla \cdot \mathbf{u}) \tag{5.12}$$

This form also makes visible that the ***skew symmetric*** operator introduces error when the velocity field diverges. Solving the advection equation by numerical methods is challenging, and there is a large scientific literature on this subject!

5.3 Solving Partial Differential Equations Using the R Package ReacTran

5.3.1 Worked Examples

Example 5.1 ***Solving Reactive Transport Equations Using*** R

Package	'ReacTran' **August 16, 2017**
Version	1.4.3.1
Title	Reactive Transport Modeling in 1D, 2D, and 3D
Authors	Karline Soetaert <karline.soetaert@nioz.nl>, Filip Meysman <filip.meysman@nioz.nl>
Maintainer	Karline Soetaert <karline.soetaert@nioz.nl>
Depends	R (>= 2.10), rootSolve, deSolve, shape
Imports	stats, graphics
Description	Routines for developing models that describe reaction and advective–diffusive transport in one, two, or three dimensions. Includes transport routines in porous media, in estuaries, and in bodies with variable shape
License	GPL (>= 3)
LazyData	Yes
Repository	CRAN

Repository/R-Forge/Project reactran
Repository/R-Forge/Revision 100
Repository/R-Forge/DateTimeStamp 2017-08-15 06:57:39

Date/Publication	2017-08-15 22:13:03 UTC
Needs Compilation	Yes

R topics documented:

1. ReacTran-package
2. advection.1D
3. fiadeiro
4. g.sphere
5. p.exp
6. setup.compaction.1D
7. setup.grid.1D
8. setup.grid.2D
9. setup.prop.1D
10. setup.prop.2D
11. tran.1D
12. tran.2D
13. tran.3D
14. tran.cylindrical

15. tran.polar
16. tran.volume.1D

ReacTran-Package Reactive Transport Modeling in 1D, 2D, and 3D

Description

R-package ReacTran contains routines that enable the development of reactive transport models in aquatic systems (rivers, lakes, oceans), porous media (floc aggregates, sediments, etc.), and even idealized organisms (spherical cells, cylindrical worms, etc.).

The geometry of the model domain is either one-dimensional, two-dimensional, or three-dimensional.

The package contains:

- Functions to setup a finite-difference grid (1D or 2D)
- Functions to attach parameters and properties to this grid (1D or 2D)
- Functions to calculate the advective–diffusive transport term over the grid (1D, 2D, or 3D)
- Utility functions.

Details

Package	ReacTran
Type	Package
Version	1.4.3
Date	2017-08-14
License	GNU Public License 2 or above

Authors

Karline Soetaert (Maintainer)
Filip Meysman

See Also

Functions ode.1D, ode.2D, ode.3D from package deSolve to integrate the reactive-transport model.

Functions steady.1D, steady.2D, steady.3D from the package rootSolve to find the steady-state solution of the reactive transport model.

tran.1D, tran.2D, tran.3D for a discretization of the general transport equations.

tran.volume.1D for discretization of the 1D transport equations using finite volumes.

tran.cylindrical, tran.spherical for a discretization of 3D transport equations in cylindrical and spherical coordinates.
tran.polar for a discretization of 2D transport equations in polar coordinates.
setup.grid.1D, setup.grid.2D for the creation of grids in 1D and in 2D.
setup.prop.1D, setup.prop.2D for defining properties on these grids.

Examples

```
## show examples (see respective help pages for details)
## 1-dimensional transport
example(tran.1D)
example(tran.volume.1D)

## 2-dimensional transport
example(tran.2D)
example(tran.polar)

## 3-dimensional transport
example(tran.3D)
example(tran.cylindrical)
example(tran.spherical)

## open the directory with documents
browseURL(paste(system.file(package="ReacTran"), "/doc", sep=""))

## open the directory with fortran codes of the transport functions
browseURL(paste(system.file(package="ReacTran"), "/doc/fortran", sep=""))

## show package vignette with how to use ReacTran and how to solve PDEs
## + source code of the vignettes
vignette("ReacTran")
vignette("PDE")
edit(vignette("ReacTran"))

## a directory with fortran implementations of the transport
browseURL(paste(system.file(package="ReacTran"), "/doc/fortran", sep=""))

## End
```

Example 5.2 trans.1D ***General One-Dimensional Advective-Diffusive Transport***

Description

This program estimates the transport term (i.e. the rate of change of a concentration owing to diffusion and advection) in a one-dimensional model of a liquid (volume fraction constant and equal to 1) or in a porous medium (volume fraction variable and lower than 1).

The interfaces between grid cells can have a variable cross-sectional area, e.g. when modelling spherical or cylindrical geometries (see example).

Usage

```
tran.1D(C, C.up = C[1], C.down = C[length(C)],
    flux.up = NULL, flux.down = NULL,
    a.bl.up = NULL, a.bl.down = NULL,
    D = 0, v = 0, AFDW = 1, VF = 1, A = 1, dx,
    full.check = FALSE, full.output = FALSE)
```

Arguments

C	Concentration, expressed per unit of phase volume
C.up	Concentration at upstream boundary. One value $[M/L^3]$
C.down	Concentration at downstream boundary. One value $[M/L^3]$
flux.up	Flux across the upstream boundary, positive = INTO model domain. One value, expressed per unit of total surface $[M/L^2/T]$. If NULL, the boundary is prescribed as a concentration or a convective transfer boundary
flux.down	Flux across the downstream boundary, positive = OUT of model domain. One value, expressed per unit of total surface $[M/L^2/T]$. If NULL, the boundary is prescribed as a concentration or a convective transfer boundary
a.bl.up	Convective transfer coefficient across the upstream boundary layer. Flux = a.bl.up*(C.up-C0). One value [L/T]
a.bl.down	Convective transfer coefficient across the downstream boundary layer (L). Flux = a.bl.down*(CL-C.down). One value [L/T]
D	Diffusion coefficient, defined on grid cell interfaces. One value, a vector of length N+1 $[L^2/T]$, or a 1D property list; the list contains at least the element int (see setup.prop.1D) $[L^2/T]$
v	Advective velocity, defined on the grid cell interfaces. Can be positive (downstream flow) or negative (upstream flow). One value, a vector of length N+1 [L/T], or a 1D property list; the list contains at least the element int (see setup.prop.1D) [L/T]
AFDW	Weight used in the finite difference scheme for advection, defined on grid cell interfaces; backward = 1, centered = 0.5, forward = 0; default is backward. One value, a vector of length N+1, or a 1D property list; the list contains at least the element int (see setup.prop.1D) [–]

VF	Volume fraction defined at the grid cell interfaces. One value, a vector of Length N+1, or a 1D property list; the list contains at least the elements int and mid (see setup.prop.1D) [–]
A	Interface area defined at the grid cell interfaces. One value, a vector of length N+1, or a 1D grid property list; the list contains at least the elements int and mid (see setup.prop.1D) [L^2]
dx	Distance between adjacent cell interfaces (thickness of grid cells). One value, a vector of length N, or a 1D grid list containing at least the elements dx and dx.aux (see setup.grid.1D) [L]
full.check	Logical flag enabling a full check of the consistency of the arguments (default = FALSE; TRUE slows down execution by 50%)
full.output	Logical flag enabling a full return of the output (default = FALSE; TRUE slows down execution by 20%)

Details

The **boundary conditions may take one of the following values:**

(1) zero gradient
(2) fixed concentration
(3) convective boundary layer
(4) fixed flux.

The above order also shows the priority. The default condition is the zero gradient. The fixed concentration condition overrules the zero gradient. The convective boundary layer condition overrules the fixed concentration and zero gradient. The fixed flux overrules all other specifications.

To ensure that the boundary conditions are well defined: for instance, it does not make sense to specify an influx in a boundary cell with the advection velocity pointing outward.

Transport Properties

The *diffusion coefficient* (D), the *advective velocity* (v), the *volume fraction* (VF), the *interface surface* (A), and the *advective finite difference weight* (AFDW) can either be specified as one value, a vector or a 1D property list as generated by `setup.prop.1D`.

When a vector, this vector must be of length $N+1$, defined at all grid cell interfaces, including the upper and lower boundary.

The **finite difference grid** (dx) is specified either as one value, a vector or a 1D grid list, as generated by `setup.grid.1D`.

Value

dC	The rate of change of the concentration C due to transport, defined in the center of each grid cell. The rate of change is expressed per unit of phase volume $[M/L^3/T]$
C.up	Concentration at the upstream interface. One value $[M/L^3]$ only when (full.output = TRUE)
C.down	Concentration at the downstream interface. One value $[M/L^3]$ only when (full.output = TRUE)
dif.flux	Diffusive flux across at the interface of each grid cell. A vector of length N+1 $[M/L^2/T]$ only when (full.output = TRUE)
adv.flux	Advective flux across at the interface of each grid cell. A vector of length N+1 $[M/L^2/T]$ only when (full.output = TRUE)
flux	Total flux across at the interface of each grid cell. A vector of length N+1 $[M/L^2/T]$ only when (full.output = TRUE)
flux.up	Flux across the upstream boundary, positive = INTO model domain. One value $[M/L^2/T]$
flux.down	Flux across the downstream boundary, positive = OUT of model domain. One value $[M/L^2/T]$

Note

The advective equation is not checked for mass conservation. Sometimes, this is not an issue, for instance when `v` represents a sinking velocity of particles or a swimming velocity of organisms. In others cases, however, mass conservation needs to be accounted for. To ensure mass conservation, the advective velocity must obey certain continuity constraints: in essence, the product of the volume fraction (VF), interface surface area (A), and advective velocity (v) should be constant. In sediments, one can use `setup.compaction.1D` to ensure that the advective velocities for the pore water and solid phase meet these constraints.

In terms of the units of concentrations and fluxes, one may follow the convention in the geosciences. The concentration `C`, `C.up`, `C.down` as well at the rate of change of the concentration `dC` are always expressed per unit of phase volume (i.e. per unit volume of solid or liquid).

Total concentrations (e.g. per unit volume of bulk sediment) may be obtained by multiplication with the appropriate volume fraction. In contrast, fluxes are always expressed per unit of total interface area (so here the volume fraction is accounted for).

Authors

Filip Meysman <filip.meysman@nioz.nl>,
Karline Soetaert <karline.soetaert@nioz.nl>.

See Also

tran.volume.1D for a discretization of the transport equation using finite volumes.
tran.2D, tran.3D
advection.1D, for more sophisticated advection schemes.

Examples

```
# ==============================================#
# EXAMPLE 1: O2 and OC consumption in sediments          #
# ==============================================#

# This example uses only the volume fractions
# in the reactive transport term

#==============#
# Model formulation  #
#==============#

# Monod consumption of oxygen (O2)

O2.model <- function (t = 0, O2, pars = NULL) {
 tran <- tran.1D(C = O2, C.up = C.ow.O2, D = D.grid,
          v = v.grid, VF = por.grid, dx = grid)$dC
 reac <- - R.O2*(O2/(Ks+O2))
 return(list(dCdt = tran + reac))
}

# First order consumption of organic carbon (OC)

OC.model <- function (t = 0, OC, pars = NULL) {

  tran <- tran.1D(C = OC, flux.up = F.OC, D = Db.grid,
           v = v.grid, VF = svf.grid, dx = grid)$dC
  reac <- - k*OC
  return(list(dCdt = tran + reac))
}

#============= #
# Parameter definition #
#==============#

# Parameter values
```

```
F.OC   <- 25 # input flux organic carbon [micromol cm-2 yr-1]
C.ow.O2 <- 0.25 # concentration O2 in overlying water [micromol cm-3]
por    <- 0.8 # porosity
D      <- 400 # diffusion coefficient O2 [cm2 yr-1]
Db     <- 10 # mixing coefficient sediment [cm2 yr-1]
v      <- 1 # advective velocity [cm yr-1]
k      <- 1 # decay constant organic carbon [yr-1]
R.O2   <- 10 # O2 consumption rate [micromol cm-3 yr-1]
Ks     <- 0.005 # O2 consumption saturation constant
# Grid definition

L  <- 10 # depth of sediment domain [cm]
N  <- 100 # number of grid layers
grid <- setup.grid.1D(x.up = 0, L = L, N = N)

# Volume fractions

por.grid <- setup.prop.1D(value = por, grid = grid)
svf.grid <- setup.prop.1D(value = (1-por), grid = grid)
D.grid  <- setup.prop.1D(value = D, grid = grid)
Db.grid <- setup.prop.1D(value = Db, grid = grid)
v.grid  <- setup.prop.1D(value = v, grid = grid)

#===========#
# Model solution  #
#===========#
=
# Initial conditions + simulation O2

yini <- rep(0, length.out = N)
O2 <- steady.1D(y = yini, func = O2.model, nspec = 1)

# Initial conditions + simulation OC

yini <- rep(0, length.out = N)
OC <- steady.1D(y = yini, func = OC.model, nspec = 1)

# Plotting output, using S3 plot method of package rootSolve"

plot(O2, grid = grid$x.mid, xyswap = TRUE, main = "O2 concentration",
    ylab = "depth [cm]", xlab = "", mfrow = c(1,2), type = "p", pch = 16)

plot(OC, grid = grid$x.mid, xyswap = TRUE, main = "C concentration",
    ylab = "depth [cm]", xlab = "", mfrow = NULL)
```

```
## ======================================
## EXAMPLE 2: O2 in a cylindrical and spherical organism
## ======================================

# This example uses only the surface areas
# in the reactive transport term

#============#
# Model formulation #
#============#

# The numerical model - rate of change = transport-consumption
Cylinder.Model <- function(time, O2, pars)
return (list(
 tran.1D(C = O2, C.down = BW, D = Da, A = A.cyl, dx = dx)$dC - Q
 ))

Sphere.Model <- function(time, O2, pars)
 return (list(
  tran.1D(C = O2, C.down = BW, D = Da, A = A.sphere, dx = dx)$dC - Q
   ))

#==============#
# Parameter definition #
#==============#

# Parameter values

BW <- 2 # mmol/m3, oxygen conc in surrounding water
Da <- 0.5 # cm2/d effective diffusion coeff in organism
R <- 0.0025 # cm radius of organism
Q <- 250000 # nM/cm3/d oxygen consumption rate/volume/day
L  <- 0.05 # cm length of organism (if a cylinder)

# the numerical model

N <- 40                          # layers in the body
dx <- R/N                         # thickness of each layer
x.mid <- seq(dx/2, by=dx, length.out=N) # distance of center to mid-layer
x.int <- seq(0, by=dx, length.out = N+1) # distance to layer interface

# Cylindrical surfaces
A.cyl <- 2*pi*x.int*L              # surface at mid-layer depth

# Spherical surfaces
A.sphere <- 4*pi*x.int^2           # surface of sphere, at each mid-layer
```

```
#=========== #
# Model solution  #
#===========#

# The analytical solution of cylindrical and spherical model

cylinder <- function(Da, Q, BW, R, r) BW + Q/(4*Da)*(r^2-R^2)

sphere <- function(Da, Q, BW, R, r) BW + Q/(6*Da)*(r^2-R^2)

# solve the model numerically for a cylinder
O2.cyl <- steady.1D (y = runif(N), name = "O2",
      func = Cylinder.Model, nspec = 1, atol = 1e-10)

# solve the model numerically for a sphere
O2.sphere <- steady.1D (y = runif(N), name = "O2",
        func = Sphere.Model, nspec = 1, atol = 1e-10)

#===========#
# Plotting output  #
#===========#

# Analytical solution - "observations"
Ana.cyl <- cbind(x.mid, O2 = cylinder(Da, Q, BW, R, x.mid))
Ana.spher <- cbind(x.mid, O2 = sphere(Da, Q, BW, R, x.mid))
plot(O2.cyl, O2.sphere, grid = x.mid, lwd = 2, lty = 1, col = 1:2,
   xlab = "distance from center, cm",
   ylab = "mmol/m3", main = "tran.1D",
   sub = "diffusion-reaction in a cylinder and sphere",
   obs = list(Ana.cyl, Ana.spher), obspar = list(pch = 16, col =1:2))

legend ("topleft", lty = c(1, NA), pch = c(NA, 18),
     c("numerical approximation", "analytical solution"))

legend ("bottomright", pch = 16, lty = 1, col = 1:2,
     c("cylinder", "sphere"))

#==========================================#
# EXAMPLE 3: O2 consumption in a spherical aggregate              #
#==========================================#

# This example uses both the surface areas and the volume fractions
# in the reactive transport term

#=============#
# Model formulation  #
#=============#
```

```
Aggregate.Model <- function(time, O2, pars) {

 tran <- tran.1D(C = O2, C.down = C.ow.O2,
          D = D.grid, A = A.grid,
          VF = por.grid, dx = grid)$dC

reac <- - R.O2*(O2/(Ks+O2))*(O2>0)
return(list(dCdt = tran + reac, consumption = -reac))
}

#============= #
# Parameter definition #
#============= #

# Parameters

C.ow.O2 <- 0.25    # concentration O2 water [micromol cm-3]
por    <- 0.8    # porosity
D     <- 400    # diffusion coefficient O2 [cm2 yr-1]
v     <- 0    # advective velocity [cm yr-1]
R.O2 <- 1000000   # O2 consumption rate [micromol cm-3 yr-1]
Ks  <- 0.005        # O2 saturation constant [micromol cm-3]

# Grid definition
R <- 0.025        # radius of the aggregate [cm]
N <- 100         # number of grid layers
grid <- setup.grid.1D(x.up = 0, L = R, N = N)

# Volume fractions

por.grid <- setup.prop.1D(value = por, grid = grid)
D.grid  <- setup.prop.1D(value = D, grid = grid)

# Surfaces

A.mid <- 4*pi*grid$x.mid^2 # surface of sphere at middle of grid cells
A.int <- 4*pi*grid$x.int^2 # surface of sphere at interface
A.grid <- list(int = A.int, mid = A.mid)

#===========#
# Model solution  #
#===========#
# Numerical solution: steady state

O2.agg <- steady.1D (runif(N), func = Aggregate.Model, nspec = 1,
            atol = 1e-10, names = "O2")
```

```
#============#
# Plotting output #
#============#
par(mfrow = c(1,1))

plot(grid$x.mid, O2.agg$y, xlab = "distance from center, cm",
   ylab = "mmol/m3",
   main = "Diffusion-reaction of O2 in a spherical aggregate")

legend ("bottomright", pch = c(1, 18), lty = 1, col = "black",
     c("O2 concentration"))

# Similar, using S3 plot method of package rootSolve"
plot(O2.agg, grid = grid$x.mid, which = c("O2", "consumption"),
   xlab = "distance from center, cm", ylab = c("mmol/m3","mmol/m3/d"))
```

In the R domain:

```
>
> install.packages("ReacTran")
Installing package into 'C:/Users/Bert/Documents/R/win-library/3.3'
(as 'lib' is unspecified)
--- Please select a CRAN mirror for use in this session ---
A CRAN mirror is selected.
also installing the dependencies 'rootSolve', 'deSolve'
trying URL 'https://mirrors.tuna.tsinghua.edu.cn/CRAN/bin/windows/contrib/3.3/
rootSolve_1.7.zip'
Content type 'application/zip' length 683839 bytes (667 KB)
downloaded 667 KB
trying URL 'https://mirrors.tuna.tsinghua.edu.cn/CRAN/bin/windows/contrib/3.3/
deSolve_1.20.zip'
Content type 'application/zip' length 2667028 bytes (2.5 MB)
downloaded 2.5 MB
trying URL 'https://mirrors.tuna.tsinghua.edu.cn/CRAN/bin/windows/contrib/3.3/
ReacTran_1.4.3.1.zip'
Content type 'application/zip' length 915996 bytes (894 KB)
downloaded 894 KB

package 'rootSolve' successfully unpacked and MD5 sums checked
package 'deSolve' successfully unpacked and MD5 sums checked
package 'ReacTran' successfully unpacked and MD5 sums checked

The downloaded binary packages are in
C:\Users\Bert\AppData\Local\Temp\RtmpmEkAb2\downloaded_packages
> require(ReacTran)
```

```
Loading required package: ReacTran
Loading required package: rootSolve
Loading required package: deSolve
Loading required package: shape
> library(ReacTran)
> ls("package:ReacTran") # Listing all the programs in this package:
 [1] "advection.1D"      "advection.volume.1D"  "fiadeiro"
 [4] "g.cylinder"        "g.sphere"             "g.spheroid"
 [7] "p.exp"             "p.lin"                "p.sig"
[10] "polar2cart"        "setup.compaction.1D"  "setup.grid.1D"
[13] "setup.grid.2D"     "setup.prop.1D"        "setup.prop.2D"
[16] "tran.1D"           "tran.2D"              "tran.3D"
[19] "tran.cylindrical"  "tran.polar"           "tran.spherical"
[22] "tran.volume.1D"    "tran.volume.2D"       "tran.volume.3D"
> tran.1D # Outputting the program listing of "tran.1D"
function (C, C.up = C[1], C.down = C[length(C)], flux.up = NULL,
  flux.down = NULL, a.bl.up = NULL, a.bl.down = NULL, D = 0,
  v = 0, AFDW = 1, VF = 1, A = 1, dx, full.check = FALSE, full.output = FALSE)
{
  if (is.null(dx))
    stop("error: dx should be inputted ")
  N <- length(C)
  if (N == 0)
    stop("C should be a vector with numeric values")
  if (is.null(C.up))
    C.up <- C[1]
  if (is.null(C.down))
    C.down <- C[N]
  if (!is.list(AFDW))
    AFDW <- list(int = rep_len(AFDW, N + 1))
  if (!is.list(D))
    D <- list(int = rep_len(D, N + 1))
  if (!is.list(v))
    v <- list(int = rep_len(v, N + 1))
  if (!is.list(VF))
    VF <- list(int = rep_len(VF, N + 1), mid = 0.5 * (rep_len(VF,
      N + 1)[1:N] + rep_len(VF, N + 1)[2:(N + 1)]))
  if (!is.list(A))
    A <- list(int = rep_len(A, N + 1), mid = 0.5 * (rep_len(A,
      N + 1)[1:N] + rep_len(A, N + 1)[2:(N + 1)]))
  if (is.list(dx))
    grid <- dx
  if (!is.list(dx))
    grid <- list(dx = rep_len(dx, N), dx.aux = 0.5 * (c(0,
      rep_len(dx, N)) + c(rep_len(dx, N), 0)))
  if (full.check) {
```

```
  gn <- names(AFDW)
  if (!"int" %in% gn)
    stop("error: AFDW should be a list that contains 'int', the AFDW values
at the interface of the grid cell ")
  if (is.null(AFDW$int))
    stop("error: AFDW is NULL, should contain (numeric) values")
  if (!is.null(AFDW$int)) {
    if (!((length(AFDW$int) == 1) || (length(AFDW$int) ==
      (N + 1))))
      stop("error: AFDW should be a vector of length 1 or N+1")
  }
  if (any(AFDW$int < 0) || any(AFDW$int > 1))
    stop("error: the AFDW should always range between 0 and 1")
  gn <- names(D)
  if (!"int" %in% gn)
    stop("error: D should be a list that contains 'int', the D values at the interface
of the grid cell ")
  if (is.null(D$int))
    stop("error: D is NULL, should contain (numeric) values")
   if (!is.null(D$int)) {
    if (!((length(D$int) == 1) || (length(D$int) == (N +
      1))))
      stop("error: D should be a vector of length 1 or N+1")
  }
  if (any(D$int < 0))
    stop("error: the diffusion coefficient should always be positive")
  gn <- names(v)
  if (!"int" %in% gn)
    stop("error: v should be a list that contains 'int', the v values at the interface
of the grid cell ")
  if (is.null(v$int))
   stop("error: the advective velocity v is NULL, should contain (numeric) values")
  if (!is.null(v$int)) {
    if (!((length(v$int) == 1) || (length(v$int) == (N +
      1))))
      stop("error: v should be a vector of length 1 or N+1")
  }
  gn <- names(VF)
  if (!"int" %in% gn)
   stop("error: VF should be a list that contains 'int', the area values at the interface
of the grid cell ")
  if (!"mid" %in% gn)
    stop("error: VF should be a list that contains 'mid', the area at the middle
of the grid cells")
  if (is.null(VF$int) || is.null(VF$mid))
    stop("error: the volume fraction VF should contain (numeric) values")
```

```
  if (!is.null(VF$int)) {
    if (!((length(VF$int) == 1) || (length(VF$int) ==
      (N + 1))))
      stop("error: VF$int should be a vector of length 1 or N+1")
  }
  if (!is.null(VF$mid)) {
    if (!((length(VF$mid) == 1) || (length(VF$mid) ==
      (N))))
      stop("error: VF$mid should be a vector of length 1 or N")
  }
  if (any(VF$int < 0) || any(VF$mid < 0) || any(VF$int >
    1) || any(VF$mid > 1))
    stop("error: the volume fraction should range between 0 and 1")
  gn <- names(A)
  if (!"int" %in% gn)
   stop("error: A should be a list that contains 'int', the area values at the interface
of the grid cell ")
  if (!"mid" %in% gn)
    stop("error: A should be a list that contains 'mid', the area at the middle
of the grid cells")
  if (is.null(A$int) || is.null(A$mid))
    stop("error: the surface area A is NULL, should contain (numeric) values")
  if (!is.null(A$int)) {
    if (!((length(A$int) == 1) || (length(A$int) == (N +
      1))))
      stop("error: A$int should be a vector of length 1 or N+1")
  }
  if (!is.null(A$mid)) {
    if (!((length(A$mid) == 1) || (length(A$mid) == (N))))
      stop("error: A$mid should be a vector of length 1 or N")
  }
  if (any(A$int < 0) || any(A$mid < 0))
    stop("error: the area A should always be positive")
  gn <- names(grid)
  if (!"dx" %in% gn)
    stop("error: grid should be a list that contains 'dx' ")
  if (!"dx.aux" %in% gn)
    stop("error: grid should be a list that contains 'dx.aux' ")
  if (is.null(grid$dx) || is.null(grid$dx.aux))
   stop("error: the grid should be a list with (numeric) values for 'dx' and 'dx.aux' ")
  if (!is.null(grid$dx)) {
    if (!((length(grid$dx) == 1) || (length(grid$dx) ==
      N)))
      stop("error: dx should be a vector of length 1 or N")
  }
  if (!is.null(grid$dx.aux)) {
```

```
    if (!((length(grid$dx.aux) == 1) || (length(grid$dx.aux) ==
      (N + 1))))
      stop("error: dx.aux should be a vector of length 1 or N+1")
  }
  if (any(grid$dx <= 0) || any(grid$dx.aux <= 0))
    stop("error: the grid distances dx and dx.aux should always be positive")
  if (!is.null(a.bl.up) & !is.null(C.up)) {
    if (a.bl.up < 0)
      stop("error: the boundary layer transfer coefficient should be positive")
  }
  if (!is.null(a.bl.down) & !is.null(C.down)) {
    if (a.bl.down < 0)
      stop("error: the boundary layer transfer coefficient should be positive")
  }
}
if (full.output) {
  if (!is.null(flux.up)) {
    if (v$int[1] >= 0) {
      nom <- flux.up + VF$int[1] * (D$int[1]/grid$dx.aux[1] +
       (1 - AFDW$int[1]) * v$int[1]) * C[1]
      denom <- VF$int[1] * (D$int[1]/grid$dx.aux[1] +
       AFDW$int[1] * v$int[1])
    }
    else {
      nom <- flux.up + VF$int[1] * (D$int[1]/grid$dx.aux[1] +
       AFDW$int[1] * v$int[1]) * C[1]
      denom <- VF$int[1] * (D$int[1]/grid$dx.aux[1] +
       (1 - AFDW$int[1]) * v$int[1])
    }
    C.up <- nom/denom
  }
  if (!is.null(flux.down)) {
    if (v$int[N + 1] >= 0) {
      nom <- flux.down - VF$int[N + 1] * (D$int[N +
       1]/grid$dx.aux[N + 1] + AFDW$int[N + 1] * v$int[N +
       1]) * C[N]
      denom <- -VF$int[N + 1] * (D$int[N + 1]/grid$dx.aux[N +
       1] + (1 - AFDW$int[N + 1]) * v$int[N + 1])
    }
    else {
      nom <- flux.down - VF$int[N + 1] * (D$int[N +
       1]/grid$dx.aux[N + 1] + (1 - AFDW$int[N + 1]) *
       v$int[N + 1]) * C[N]
      denom <- -VF$int[N + 1] * (D$int[N + 1]/grid$dx.aux[N +
       1] + AFDW$int[N + 1] * v$int[N + 1])
    }
```

```
  C.down <- nom/denom
}
if (!is.null(a.bl.up) & !is.null(C.up)) {
  if (v$int[1] >= 0) {
    nom <- a.bl.up * C.up + VF$int[1] * (D$int[1]/grid$dx.aux[1] +
     (1 - AFDW$int[1]) * v$int[1]) * C[1]
    denom <- a.bl.up + VF$int[1] * (D$int[1]/grid$dx.aux[1] +
     AFDW$int[1] * v$int[1])
  }
  else {
    nom <- a.bl.up * C.up + VF$int[1] * (D$int[1]/grid$dx.aux[1] +
     AFDW$int[1] * v$int[1]) * C[1]
    denom <- a.bl.up + VF$int[1] * (D$int[1]/grid$dx.aux[1] +
     (1 - AFDW$int[1]) * v$int[1])
  }
  C.up <- nom/denom
}
if (!is.null(a.bl.down) & !is.null(C.down)) {
  if (v$int[N + 1] >= 0) {
    nom <- a.bl.down * C.down + VF$int[N + 1] * (D$int[N +
     1]/grid$dx.aux[N + 1] + (1 - AFDW$int[N + 1]) *
     v$int[N + 1]) * C[N]
    denom <- a.bl.down + VF$int[N + 1] * (D$int[N +
     1]/grid$dx.aux[N + 1] + AFDW$int[N + 1] * v$int[N +
     1])
  }
  else {
    nom <- a.bl.down * C.down + VF$int[N + 1] * (D$int[N +
     1]/grid$dx.aux[N + 1] + AFDW$int[N + 1] * v$int[N +
     1]) * C[N]
    denom <- a.bl.down + VF$int[N + 1] * (D$int[N +
     1]/grid$dx.aux[N + 1] + (1 - AFDW$int[N + 1]) *
     v$int[N + 1])
 }
  C.down <- nom/denom
}
dif.flux <- as.vector(-VF$int * D$int * diff(c(C.up,
  C, C.down))/grid$dx.aux)
adv.flux <- rep(0, length.out = length(dif.flux))
if (any(v$int > 0)) {
  vv <- v$int
  vv[v$int < 0] <- 0
  conc <- AFDW$int * c(C.up, C)
  if (any(AFDW$int < 1))
    conc <- conc + (1 - AFDW$int) * c(C, C.down)
  adv.flux <- adv.flux + as.vector(VF$int * vv * conc)
```

```
    }
    if (any(v$int < 0)) {
      vv <- v$int
      vv[v$int > 0] <- 0
      conc <- AFDW$int * c(C, C.down)
      if (any(AFDW$int < 1))
        conc <- conc + (1 - AFDW$int) * c(C.up, C)
      adv.flux <- adv.flux + as.vector(VF$int * vv * conc)
    }
    flux <- dif.flux + adv.flux
  }
  else {
    if (!is.null(a.bl.up) & !is.null(C.up)) {
      if (v$int[1] >= 0) {
        nom <- a.bl.up * C.up + VF$int[1] * (D$int[1]/grid$dx.aux[1] +
         (1 - AFDW$int[1]) * v$int[1]) * C[1]
        denom <- a.bl.up + VF$int[1] * (D$int[1]/grid$dx.aux[1] +
         AFDW$int[1] * v$int[1])
      }
      else {
        nom <- a.bl.up * C.up + VF$int[1] * (D$int[1]/grid$dx.aux[1] +
         AFDW$int[1] * v$int[1]) * C[1]
        denom <- a.bl.up + VF$int[1] * (D$int[1]/grid$dx.aux[1] +
         (1 - AFDW$int[1]) * v$int[1])
      }
      C.up <- nom/denom
    }
    if (!is.null(a.bl.down) & !is.null(C.down)) {
      if (v$int[N + 1] >= 0) {
       nom <- a.bl.down * C.down + VF$int[N + 1] * (D$int[N +
         1]/grid$dx.aux[N + 1] + (1 - AFDW$int[N + 1]) *
         v$int[N + 1]) * C[N]
        denom <- a.bl.down + VF$int[N + 1] * (D$int[N +
         1]/grid$dx.aux[N + 1] + AFDW$int[N + 1] * v$int[N +
         1])
      }
      else {
        nom <- a.bl.down * C.down + VF$int[N + 1] * (D$int[N +
         1]/grid$dx.aux[N + 1] + AFDW$int[N + 1] * v$int[N +
         1]) * C[N]
        denom <- a.bl.down + VF$int[N + 1] * (D$int[N +
         1]/grid$dx.aux[N + 1] + (1 - AFDW$int[N + 1]) *
         v$int[N + 1])
      }
      C.down <- nom/denom
    }
```

```
    flux <- as.vector(-(VF$int) * D$int * diff(c(C.up, C,
      C.down))/grid$dx.aux)
    if (any(v$int > 0)) {
      vv <- v$int
      vv[v$int < 0] <- 0
      conc <- AFDW$int * c(C.up, C)
      if (any(AFDW$int < 1))
        conc <- conc + (1 - AFDW$int) * c(C, C.down)
      flux <- flux + as.vector(VF$int * vv * conc)
    }

    if (any(v$int < 0)) {
      vv <- v$int
      vv[v$int > 0] <- 0
      conc <- AFDW$int * c(C, C.down)
      if (any(AFDW$int < 1))
        conc <- conc + (1 - AFDW$int) * c(C.up, C)
      flux <- flux + as.vector(VF$int * vv * conc)
    }
  }
  if (!is.null(flux.up))
    flux[1] <- flux.up
  if (!is.null(flux.down))
    flux[N + 1] <- flux.down
  dC <- -diff(A$int * flux)/A$mid/VF$mid/grid$dx
  if (!full.output) {
    return(list(dC = dC, flux.up = flux[1], flux.down = flux[length(flux)]))
  }
  else {
    return(list(dC = dC, C.up = C.up, C.down = C.down, dif.flux = dif.flux,
      adv.flux = adv.flux, flux = flux, flux.up = flux[1],
      flux.down = flux[length(flux)]))
  }
}
<environment: namespace:ReacTran>
>

> install.packages("ReacTran")
Installing package into 'C:/Users/Bert/Documents/R/win-library/3.3'
(as 'lib' is unspecified)
--- Please select a CRAN mirror for use in this session ---
A CRAN mirror is selected.
trying URL 'https://mirrors.tuna.tsinghua.edu.cn/CRAN/bin/windows/contrib/3.3/
ReacTran_1.4.3.1.zip'

Content type 'application/zip' length 915996 bytes (894 KB)
```

```
downloaded 894 KB
package 'ReacTran' successfully unpacked and MD5 sums checked
The downloaded binary packages are in
C:\Users\Bert\AppData\Local\Temp\RtmpCiBQxG\downloaded_packages
> require(ReacTran)
Loading required package: ReacTran
Loading required package: rootSolve
Loading required package: deSolve
Loading required package: shape
> library(ReacTran)
> ls("package:ReacTran")
 [1] "advection.1D"      "advection.volume.1D"  "fiadeiro"
 [4] "g.cylinder"        "g.sphere"             "g.spheroid"
 [7] "p.exp"             "p.lin"                "p.sig"
[10] "polar2cart"        "setup.compaction.1D"  "setup.grid.1D"
[13] "setup.grid.2D"     "setup.prop.1D"        "setup.prop.2D"
[16] "tran.1D"           "tran.2D"              "tran.3D"
[19] "tran.cylindrical"  "tran.polar"           "tran.spherical"
[22] "tran.volume.1D"    "tran.volume.2D"       "tran.volume.3D"
>
> tran.1D # Listing the program "tran.1D"
function (C, C.up = C[1], C.down = C[length(C)], flux.up = NULL,
  flux.down = NULL, a.bl.up = NULL, a.bl.down = NULL, D = 0,
  v = 0, AFDW = 1, VF = 1, A = 1, dx, full.check = FALSE, full.output = FALSE)
{
  if (is.null(dx))
    stop("error: dx should be inputted ")
  N <- length(C)
  if (N == 0)
    stop("C should be a vector with numeric values")
  if (is.null(C.up))
    C.up <- C[1]
  if (is.null(C.down))
    C.down <- C[N]
  if (!is.list(AFDW))
    AFDW <- list(int = rep_len(AFDW, N + 1))
  if (!is.list(D))
    D <- list(int = rep_len(D, N + 1))
  if (!is.list(v))
    v <- list(int = rep_len(v, N + 1))
  if (!is.list(VF))
    VF <- list(int = rep_len(VF, N + 1), mid = 0.5 * (rep_len(VF,
      N + 1)[1:N] + rep_len(VF, N + 1)[2:(N + 1)]))
  if (!is.list(A))
    A <- list(int = rep_len(A, N + 1), mid = 0.5 * (rep_len(A,
      N + 1)[1:N] + rep_len(A, N + 1)[2:(N + 1)]))
```

```
  if (is.list(dx))
    grid <- dx
if (!is.list(dx))
    grid <- list(dx = rep_len(dx, N), dx.aux = 0.5 * (c(0,
      rep_len(dx, N)) + c(rep_len(dx, N), 0)))
  if (full.check) {
    gn <- names(AFDW)
    if (!"int" %in% gn)
      stop("error: AFDW should be a list that contains 'int', the AFDW values at the
interface of the grid cell ")
    if (is.null(AFDW$int))
      stop("error: AFDW is NULL, should contain (numeric) values")
    if (!is.null(AFDW$int)) {
      if (!((length(AFDW$int) == 1) || (length(AFDW$int) ==
        (N + 1))))
        stop("error: AFDW should be a vector of length 1 or N+1")
    }
    if (any(AFDW$int < 0) || any(AFDW$int > 1))
      stop("error: the AFDW should always range between 0 and 1")
    gn <- names(D)
    if (!"int" %in% gn)
      stop("error: D should be a list that contains 'int', the D values at the interface
of the grid cell ")
    if (is.null(D$int))
      stop("error: D is NULL, should contain (numeric) values")
    if (!is.null(D$int)) {
      if (!((length(D$int) == 1) || (length(D$int) == (N +
        1))))
       stop("error: D should be a vector of length 1 or N+1")
    }
    if (any(D$int < 0))
      stop("error: the diffusion coefficient should always be positive")
    gn <- names(v)
    if (!"int" %in% gn)
      stop("error: v should be a list that contains 'int', the v values at the interface
of the grid cell ")
    if (is.null(v$int))
     stop("error: the advective velocity v is NULL, should contain (numeric) values")
    if (!is.null(v$int)) {
      if (!((length(v$int) == 1) || (length(v$int) == (N +
        1))))
        stop("error: v should be a vector of length 1 or N+1")
    }
    gn <- names(VF)
    if (!"int" %in% gn)
```

```
    stop("error: VF should be a list that contains 'int', the area values at the interface
of the grid cell ")
    if (!"mid" %in% gn)
      stop("error: VF should be a list that contains 'mid', the area at the middle
of the grid cells")
    if (is.null(VF$int) || is.null(VF$mid))
      stop("error: the volume fraction VF should contain (numeric) values")
    if (!is.null(VF$int)) {
      if (!((length(VF$int) == 1) || (length(VF$int) ==
        (N + 1))))
        stop("error: VF$int should be a vector of length 1 or N+1")
    }
    if (!is.null(VF$mid)) {
      if (!((length(VF$mid) == 1) || (length(VF$mid) ==
        (N))))
        stop("error: VF$mid should be a vector of length 1 or N")
    }
    if (any(VF$int < 0) || any(VF$mid < 0) || any(VF$int >
      1) || any(VF$mid > 1))
      stop("error: the volume fraction should range between 0 and 1")
    gn <- names(A)
    if (!"int" %in% gn)
     stop("error: A should be a list that contains 'int', the area values at the interface
of the grid cell ")
    if (!"mid" %in% gn)
      stop("error: A should be a list that contains 'mid', the area at the middle of
the grid cells")
    if (is.null(A$int) || is.null(A$mid))
      stop("error: the surface area A is NULL, should contain (numeric) values")
    if (!is.null(A$int)) {
      if (!((length(A$int) == 1) || (length(A$int) == (N +
        1))))
        stop("error: A$int should be a vector of length 1 or N+1")
    }
   if (!is.null(A$mid)) {
      if (!((length(A$mid) == 1) || (length(A$mid) == (N))))
        stop("error: A$mid should be a vector of length 1 or N")
    }
    if (any(A$int < 0) || any(A$mid < 0))
      stop("error: the area A should always be positive")
    gn <- names(grid)
    if (!"dx" %in% gn)
      stop("error: grid should be a list that contains 'dx' ")
    if (!"dx.aux" %in% gn)
      stop("error: grid should be a list that contains 'dx.aux' ")
    if (is.null(grid$dx) || is.null(grid$dx.aux))
```

```
    stop("error: the grid should be a list with (numeric) values for 'dx' and 'dx.aux' ")
    if (!is.null(grid$dx)) {
      if (!((length(grid$dx) == 1) || (length(grid$dx) ==
        N)))
        stop("error: dx should be a vector of length 1 or N")
    }
    if (!is.null(grid$dx.aux)) {
      if (!((length(grid$dx.aux) == 1) || (length(grid$dx.aux) ==
        (N + 1))))
        stop("error: dx.aux should be a vector of length 1 or N+1")
    }
    if (any(grid$dx <= 0) || any(grid$dx.aux <= 0))
      stop("error: the grid distances dx and dx.aux should always be positive")
if (!is.null(a.bl.up) & !is.null(C.up)) {
      if (a.bl.up < 0)
        stop("error: the boundary layer transfer coefficient should be positive")
    }
    if (!is.null(a.bl.down) & !is.null(C.down)) {
      if (a.bl.down < 0)
        stop("error: the boundary layer transfer coefficient should be positive")
    }
  }
  if (full.output) {
    if (!is.null(flux.up)) {
      if (v$int[1] >= 0) {
        nom <- flux.up + VF$int[1] * (D$int[1]/grid$dx.aux[1] +
         (1 - AFDW$int[1]) * v$int[1]) * C[1]
        denom <- VF$int[1] * (D$int[1]/grid$dx.aux[1] +
         AFDW$int[1] * v$int[1])
      }
      else {
        nom <- flux.up + VF$int[1] * (D$int[1]/grid$dx.aux[1] +
         AFDW$int[1] * v$int[1]) * C[1]
        denom <- VF$int[1] * (D$int[1]/grid$dx.aux[1] +
         (1 - AFDW$int[1]) * v$int[1])
      }
      C.up <- nom/denom
    }
 if (!is.null(flux.down)) {
      if (v$int[N + 1] >= 0) {
        nom <- flux.down - VF$int[N + 1] * (D$int[N +
         1]/grid$dx.aux[N + 1] + AFDW$int[N + 1] * v$int[N +
         1]) * C[N]
        denom <- -VF$int[N + 1] * (D$int[N + 1]/grid$dx.aux[N +
         1] + (1 - AFDW$int[N + 1]) * v$int[N + 1])
      }
```

```
    else {
      nom <- flux.down - VF$int[N + 1] * (D$int[N +
       1]/grid$dx.aux[N + 1] + (1 - AFDW$int[N + 1]) *
       v$int[N + 1]) * C[N]
      denom <- -VF$int[N + 1] * (D$int[N + 1]/grid$dx.aux[N +
       1] + AFDW$int[N + 1] * v$int[N + 1])
    }
    C.down <- nom/denom
  }
  if (!is.null(a.bl.up) & !is.null(C.up)) {
    if (v$int[1] >= 0) {
      nom <- a.bl.up * C.up + VF$int[1] * (D$int[1]/grid$dx.aux[1] +
       (1 - AFDW$int[1]) * v$int[1]) * C[1]
      denom <- a.bl.up + VF$int[1] * (D$int[1]/grid$dx.aux[1] +
       AFDW$int[1] * v$int[1])
    }
    else {
      nom <- a.bl.up * C.up + VF$int[1] * (D$int[1]/grid$dx.aux[1] +
       AFDW$int[1] * v$int[1]) * C[1]
      denom <- a.bl.up + VF$int[1] * (D$int[1]/grid$dx.aux[1] +
       (1 - AFDW$int[1]) * v$int[1])
    }
    C.up <- nom/denom
  }
  if (!is.null(a.bl.down) & !is.null(C.down)) {
    if (v$int[N + 1] >= 0) {
      nom <- a.bl.down * C.down + VF$int[N + 1] * (D$int[N +
       1]/grid$dx.aux[N + 1] + (1 - AFDW$int[N + 1]) *
       v$int[N + 1]) * C[N]
      denom <- a.bl.down + VF$int[N + 1] * (D$int[N +
       1]/grid$dx.aux[N + 1] + AFDW$int[N + 1] * v$int[N +
       1])
    }
    else {
      nom <- a.bl.down * C.down + VF$int[N + 1] * (D$int[N +
       1]/grid$dx.aux[N + 1] + AFDW$int[N + 1] * v$int[N +
       1]) * C[N]
      denom <- a.bl.down + VF$int[N + 1] * (D$int[N +
       1]/grid$dx.aux[N + 1] + (1 - AFDW$int[N + 1]) *
       v$int[N + 1])
    }
    C.down <- nom/denom
  }
  dif.flux <- as.vector(-VF$int * D$int * diff(c(C.up,
   C, C.down))/grid$dx.aux)
  adv.flux <- rep(0, length.out = length(dif.flux))
```

```
  if (any(v$int > 0)) {
   vv <- v$int
   vv[v$int < 0] <- 0
   conc <- AFDW$int * c(C.up, C)
   if (any(AFDW$int < 1))
    conc <- conc + (1 - AFDW$int) * c(C, C.down)
   adv.flux <- adv.flux + as.vector(VF$int * vv * conc)
  }
  if (any(v$int < 0)) {
   vv <- v$int
   vv[v$int > 0] <- 0
   conc <- AFDW$int * c(C, C.down)
   if (any(AFDW$int < 1))
    conc <- conc + (1 - AFDW$int) * c(C.up, C)
   adv.flux <- adv.flux + as.vector(VF$int * vv * conc)
  }
  flux <- dif.flux + adv.flux
 }
 else {
  if (!is.null(a.bl.up) & !is.null(C.up)) {
   if (v$int[1] >= 0) {
    nom <- a.bl.up * C.up + VF$int[1] * (D$int[1]/grid$dx.aux[1] +
     (1 - AFDW$int[1]) * v$int[1]) * C[1]
    denom <- a.bl.up + VF$int[1] * (D$int[1]/grid$dx.aux[1] +
     AFDW$int[1] * v$int[1])
   }
   else {
    nom <- a.bl.up * C.up + VF$int[1] * (D$int[1]/grid$dx.aux[1] +
     AFDW$int[1] * v$int[1]) * C[1]
    denom <- a.bl.up + VF$int[1] * (D$int[1]/grid$dx.aux[1] +
     (1 - AFDW$int[1]) * v$int[1])
   }
   C.up <- nom/denom
  }
  if (!is.null(a.bl.down) & !is.null(C.down)) {
   if (v$int[N + 1] >= 0) {
    nom <- a.bl.down * C.down + VF$int[N + 1] * (D$int[N +
     1]/grid$dx.aux[N + 1] + (1 - AFDW$int[N + 1]) *
     v$int[N + 1]) * C[N]
    denom <- a.bl.down + VF$int[N + 1] * (D$int[N +
     1]/grid$dx.aux[N + 1] + AFDW$int[N + 1] * v$int[N +
     1])
   }
   else {
    nom <- a.bl.down * C.down + VF$int[N + 1] * (D$int[N +
     1]/grid$dx.aux[N + 1] + AFDW$int[N + 1] * v$int[N +
```

```
        1]) * C[N]
      denom <- a.bl.down + VF$int[N + 1] * (D$int[N +
        1]/grid$dx.aux[N + 1] + (1 - AFDW$int[N + 1]) *
        v$int[N + 1])
    }
    C.down <- nom/denom
  }
  flux <- as.vector(-(VF$int) * D$int * diff(c(C.up, C,
    C.down))/grid$dx.aux)
  if (any(v$int > 0)) {
    vv <- v$int
    vv[v$int < 0] <- 0
    conc <- AFDW$int * c(C.up, C)
    if (any(AFDW$int < 1))
      conc <- conc + (1 - AFDW$int) * c(C, C.down)
    flux <- flux + as.vector(VF$int * vv * conc)
  }
  if (any(v$int < 0)) {
    vv <- v$int
    vv[v$int > 0] <- 0
    conc <- AFDW$int * c(C, C.down)
    if (any(AFDW$int < 1))
      conc <- conc + (1 - AFDW$int) * c(C.up, C)
    flux <- flux + as.vector(VF$int * vv * conc)
  }
 }
 if (!is.null(flux.up))
   flux[1] <- flux.up
 if (!is.null(flux.down))
   flux[N + 1] <- flux.down
dC <- -diff(A$int * flux)/A$mid/VF$mid/grid$dx
 if (!full.output) {
   return(list(dC = dC, flux.up = flux[1], flux.down = flux[length(flux)]))
 }
 else {
   return(list(dC = dC, C.up = C.up, C.down = C.down, dif.flux = dif.flux,
     adv.flux = adv.flux, flux = flux, flux.up = flux[1],
     flux.down = flux[length(flux)]))
 }
}
<environment: namespace:ReacTran
```

 # Computing EXAMPLE 1 in the R ***environment:***

```
> ## ====================================================
> ## EXAMPLE 1: O2 and OC (Organic Carbon) consumption in sediments =
> ## ====================================================
```

```
> # This example uses only the volume fractions
> # in the reactive transport term
> #============#
> # Model formulation #
> #============#
> # Monod consumption of oxygen (O2)
> O2.model <- function (t = 0, O2, pars = NULL) {
+ tran <- tran.1D(C = O2, C.up = C.ow.O2, D = D.grid,
+ v = v.grid, VF = por.grid, dx = grid)$dC
+ reac <- - R.O2*(O2/(Ks+O2))
+ return(list(dCdt = tran + reac))
+ }
> # First order consumption of organic carbon (OC)
> OC.model <- function (t = 0, OC, pars = NULL) {
+ tran <- tran.1D(C = OC, flux.up = F.OC, D = Db.grid,
+ v = v.grid, VF = svf.grid, dx = grid)$dC
+ reac <- - k*OC
+ return(list(dCdt = tran + reac))
+ }
> #==============#
> # Parameter definition #
> #==============#
> # Parameter values
> F.OC <- 25 # input flux organic carbon [micromol cm-2 yr-1]
> C.ow.O2 <- 0.25 # concentration O2 in overlying water [micromol cm-3]
> por <- 0.8 # porosity
> D <- 400 # diffusion coefficient O2 [cm2 yr-1]
> Db <- 10 # mixing coefficient sediment [cm2 yr-1]
> v <- 1 # advective velocity [cm yr-1]
> k <- 1 # decay constant organic carbon [yr-1]
> R.O2 <- 10 # O2 consumption rate [micromol cm-3 yr-1]
> Ks <- 0.005 # O2 consumption saturation constant
> # Grid definition
> L <- 10 # depth of sediment domain [cm]
> N <- 100 # number of grid layers
> grid <- setup.grid.1D(x.up = 0, L = L, N = N)
> # Volume fractions
> por.grid <- setup.prop.1D(value = por, grid = grid)
> svf.grid <- setup.prop.1D(value = (1-por), grid = grid)
> D.grid <- setup.prop.1D(value = D, grid = grid)
> Db.grid <- setup.prop.1D(value = Db, grid = grid)
> v.grid <- setup.prop.1D(value = v, grid = grid)
> #==========#
> # Model solution #
> #==========#
> # Initial conditions + simulation O2
```

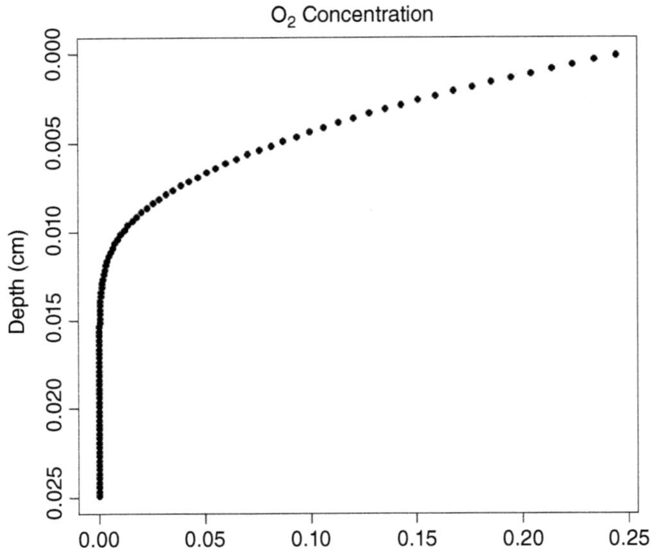

Figure A1-1 trans.1D 1 Reactive transport modeling in 1D.

```
> yini <- rep(0, length.out = N)
> O2 <- steady.1D(y = yini, func = O2.model, nspec = 1)
> # Initial conditions + simulation OC
> yini <- rep(0, length.out = N)
> OC <- steady.1D(y = yini, func = OC.model, nspec = 1)
> # Plotting output, using S3 plot method of package rootSolve"
> plot(O2, grid = grid$x.mid, xyswap = TRUE, main = "O2 concentration",
+ ylab = "depth [cm]", xlab = "", mfrow = c(1,2), type = "p", pch = 16)
> #
> # Outputting: Figure A1-1: tran.1D-1
> #
> plot(OC, grid = grid$x.mid, xyswap = TRUE, main = "C concentration",
+ ylab = "depth [cm]", xlab = "", mfrow = NULL)
> #
> # Outputting: Figure A1-2: tran.1D-2
> #
> ## =====================================
> ## EXAMPLE 2: O2 in a cylindrical and spherical organism =
> ## =====================================
> # This example uses only the surface areas
> # in the reactive transport term
> #=============#
> # Model formulation #
> #=============#
> # The numerical model - rate of change = transport-consumption
> Cylinder.Model <- function(time, O2, pars)
+ return (list(
```

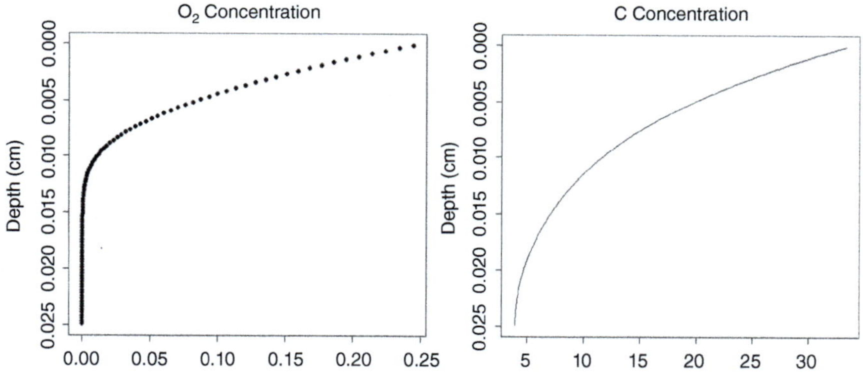

Figure A1-2 trans.1D 2 Reactive transport modeling in 1D.

```
+ tran.1D(C = O2, C.down = BW, D = Da, A = A.cyl, dx = dx)$dC - Q
+ ))
> Sphere.Model <- function(time, O2, pars)
+ return (list(
+ tran.1D(C = O2, C.down = BW, D = Da, A = A.sphere, dx = dx)$dC - Q
+ ))
> #==============#
> # Parameter definition #
> #==============#
> # Parameter values
> BW <- 2 # mmol/m3, oxygen conc in surrounding water
> Da <- 0.5 # cm2/d effective diffusion coeff in organism
> R <- 0.0025 # cm radius of organism
> Q <- 250000 # nM/cm3/d oxygen consumption rate/volume/day
> L <- 0.05 # cm length of organism (if a cylinder)
> # the numerical model
> N <- 40 # layers in the body
> dx <- R/N # thickness of each layer
> x.mid <- seq(dx/2, by = dx, length.out = N) # distance of center to mid-layer
> x.int <- seq(0, by = dx, length.out = N+1) # distance to layer interface
> # Cylindrical surfaces
> A.cyl <- 2*pi*x.int*L # surface at mid-layer depth
> # Spherical surfaces
> A.sphere <- 4*pi*x.int^2 # surface of sphere, at each mid-layer
> #==========#
> # Model solution #
> #==========#
> # The analytical solution of cylindrical and spherical model
> cylinder <- function(Da, Q, BW, R, r) BW + Q/(4*Da)*(r^2-R^2)
> sphere <- function(Da, Q, BW, R, r) BW + Q/(6*Da)*(r^2-R^2)
> # solve the model numerically for a cylinder
> O2.cyl <- steady.1D (y = runif(N), name = "O2",
```

```
+ func = Cylinder.Model, nspec = 1, atol = 1e-10)
> # solve the model numerically for a sphere
> O2.sphere <- steady.1D (y = runif(N), name = "O2",
+ func = Sphere.Model, nspec = 1, atol = 1e-10)
> #==========#
> # Plotting output #
> #==========#
> # Analytical solution - "observations"
> Ana.cyl <- cbind(x.mid, O2 = cylinder(Da, Q, BW, R, x.mid))
> Ana.spher <- cbind(x.mid, O2 = sphere(Da, Q, BW, R, x.mid))
> plot(O2.cyl, O2.sphere, grid = x.mid, lwd = 2, lty = 1, col = 1:2,
+ xlab = "distance from center, cm",
+ ylab = "mmol/m3", main = "tran.1D",
+ sub = "diffusion-reaction in a cylinder and sphere",
+ obs = list(Ana.cyl, Ana.spher), obspar = list(pch = 16, col =1:2))
> #
> # Outputting: Figure A1-3: tran.1D-3
> #
> legend ("topleft", lty = c(1, NA), pch = c(NA, 18),
+ c("numerical approximation", "analytical solution"))
> #
> # Outputting: Figure A1-4: tran.1D-4
> #
> legend ("bottomright", pch = 16, lty = 1, col = 1:2,
+ c("cylinder", "sphere"))
> #
> # Outputting: Figure A1-5: tran.1D-5
> #
>===================================
> ## EXAMPLE 3: O2 consumption in a spherical aggregate =
> ## ================================ =
> # This example uses both the surface areas and the volume fractions
> # in the reactive transport term
> #==============#
> # Model formulation  #
> #==============#
> Aggregate.Model <- function(time, O2, pars) {
+ tran <- tran.1D(C = O2, C.down = C.ow.O2,
+ D = D.grid, A = A.grid,
+ VF = por.grid, dx = grid )$dC
+ reac <- - R.O2*(O2/(Ks+O2))*(O2>0)
+ return(list(dCdt = tran + reac, consumption = -reac))
+ }
> #==============#
> # Parameter definition #
> #==============#
```

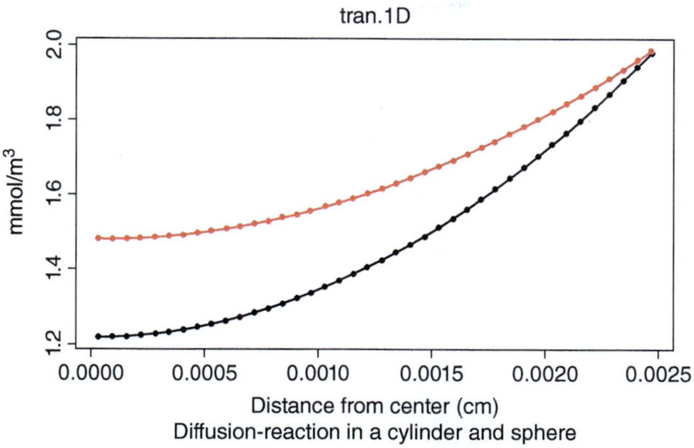

Figure A1-3 trans.1D 3 Reactive transport modeling in 1D.

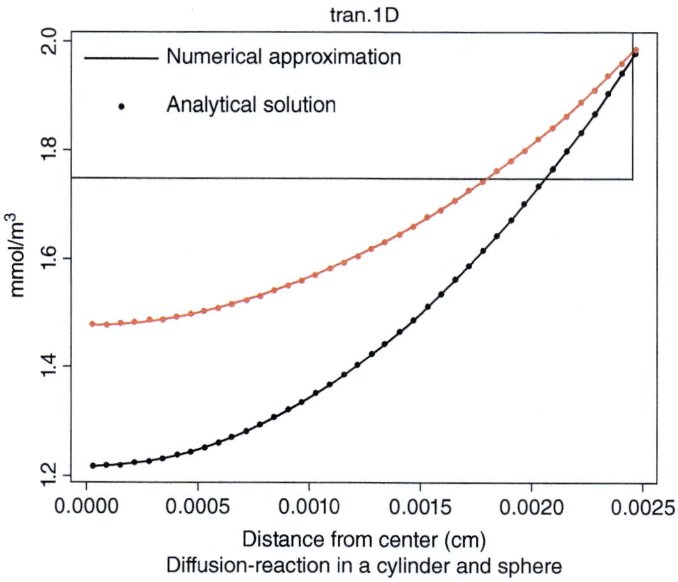

Figure A1-4 trans.1D 4 Reactive transport modeling in 1D.

```
>
> # Parameters
> C.ow.O2 <- 0.25 # concentration O2 water [micromol cm-3]
> por <- 0.8 # porosity
> D <- 400 # diffusion coefficient O2 [cm2 yr-1]
> v <- 0 # advective velocity [cm yr-1]
> R.O2 <- 1000000 # O2 consumption rate [micromol cm-3 yr-1]
> Ks <- 0.005 # O2 saturation constant [micromol cm-3]
> # Grid definition
```

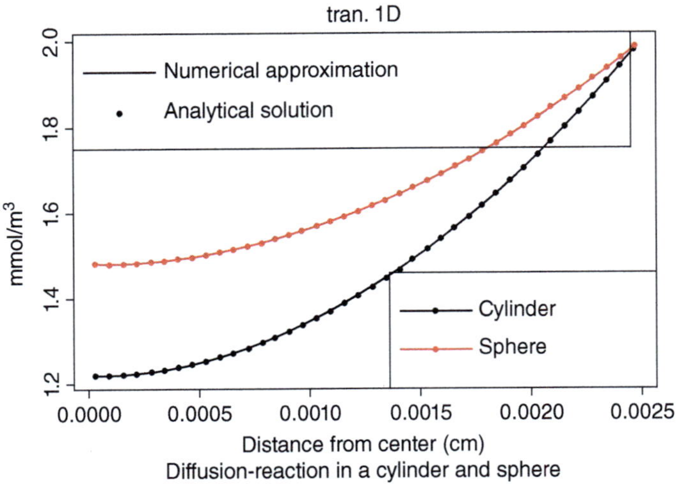

Figure A1-5 trans.1D 5 Reactive transport modeling in 1D.

```
> R <- 0.025 # radius of the aggregate [cm]
> N <- 100 # number of grid layers
> grid <- setup.grid.1D(x.up = 0, L = R, N = N)
> # Volume fractions
> por.grid <- setup.prop.1D(value = por, grid = grid)
> D.grid <- setup.prop.1D(value = D, grid = grid)
> # Surfaces
> A.mid <- 4*pi*grid$x.mid^2 # surface of sphere at middle of grid cells
> A.int <- 4*pi*grid$x.int^2 # surface of sphere at interface
> A.grid <- list(int = A.int, mid = A.mid)
> #==========#
> # Model solution #
> #==========#
> # Numerical solution: steady state
> O2.agg <- steady.1D (runif(N), func = Aggregate.Model, nspec = 1,
+ atol = 1e-10, names = "O2")
> #===========#
> # Plotting output  #
> #===========#
> par(mfrow = c(1,1))
> plot(grid$x.mid, O2.agg$y, xlab = "distance from center, cm",
+ ylab = "mmol/m3",
+ main = "Diffusion-reaction of O2 in a spherical aggregate")
> #
> # Outputting: Figure A1-6: tran.1D-6
> #
```

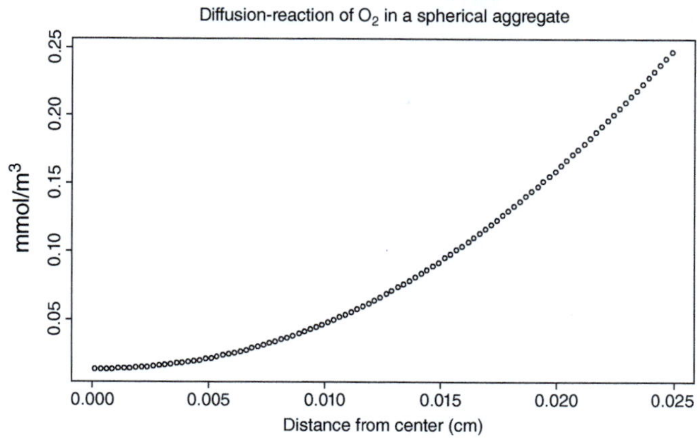

Figure A1-6 trans.1D 6 Reactive transport modeling in 1D.

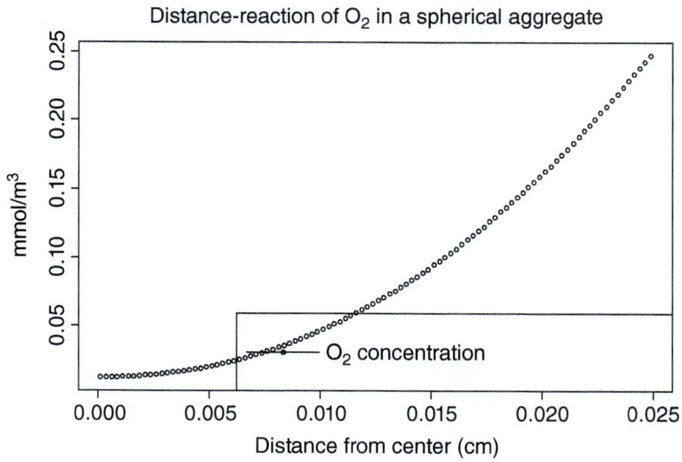

Figure A1-7 trans.1D 7 Reactive transport modeling in 1D.

```
> legend ("bottomright", pch = c(1, 18), lty = 1, col = "black",
+ c("O2 concentration"))
> #
> # Outputting: Figure A1-7: tran.1D-7
> #
> # Similar, using S3 plot method of package rootSolve"
> plot(O2.agg, grid = grid$x.mid, which = c("O2", "consumption"),
+ xlab = "distance from center, cm", ylab = c("mmol/m3","mmol/m3/d"))
> #
> # Outputting: Figure A1-8: tran.1D-8
> #
```

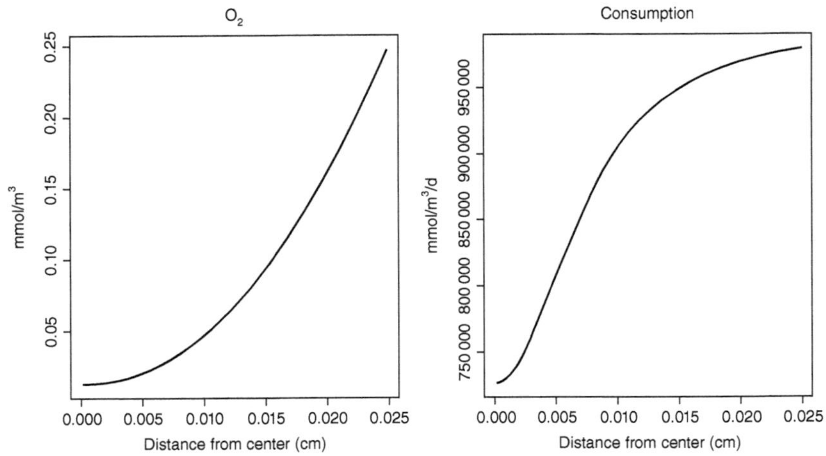

Figure A1-8 trans.1D 8 Reactive transport modeling in 1D.

Example 5.3 tran.2D *General Two-Dimensional Advective–Diffusive Transport*

Description

This program estimates the transport term (i.e. the rate of change of a concentration owing to diffusion and advection) in a two-dimensional model domain.

Usage

```
tran.2D (C, C.x.up = C[1,], C.x.down = C[nrow(C),],
    C.y.up = C[,1], C.y.down = C[ ,ncol(C)],
    flux.x.up = NULL, flux.x.down = NULL,
    flux.y.up = NULL, flux.y.down = NULL,
    a.bl.x.up = NULL, a.bl.x.down = NULL,
    a.bl.y.up = NULL, a.bl.y.down = NULL,
    D.grid = NULL, D.x = NULL, D.y = D.x,
    v.grid = NULL, v.x = 0, v.y = 0,
    AFDW.grid = NULL, AFDW.x = 1, AFDW.y = AFDW.x,
    VF.grid = NULL, VF.x = 1, VF.y = VF.x,
    A.grid = NULL, A.x = 1, A.y = 1,
    grid = NULL, dx = NULL, dy = NULL,
    full.check = FALSE, full.output = FALSE)
```

Arguments

C	Concentration, expressed per unit volume, defined at the center of each grid cell; Nx*Ny matrix $[M/L^3]$
C.x.up	Concentration at upstream boundary in x-direction; vector of length Ny $[M/L^3]$

C.x.down	Concentration at downstream boundary in x-direction; vector of length Ny [M/L^3]
C.y.up	Concentration at upstream boundary in y-direction; vector of length Nx [M/L^3]
C.y.down	Concentration at downstream boundary in y-direction; vector of length Nx [M/L^3]
flux.x.up	Flux across the upstream boundary in x-direction, positive = INTO model domain; vector of length Ny [M/L^2/T]
flux.x.down	Flux across the downstream boundary in x-direction, positive = OUT of model domain; vector of length Ny [M/L^2/T]
flux.y.up	Flux across the upstream boundary in y-direction, positive = INTO model domain; vector of length Nx [M/L^2/T]
flux.y.down	Flux across the downstream boundary in y-direction, positive = OUT of model domain; vector of length Nx [M/L^2/T]
a.bl.x.up	Transfer coefficient across the upstream boundary layer in x-direction; Flux=a.bl.x.up*(C.x.up-C[1,]). One value [L/T]
a.bl.x.down	Transfer coefficient across the downstream boundary layer in x-direction; Flux=a.bl.x.down*(C[Nx,]-C.x.down). One value [L/T]
a.bl.y.up	Transfer coefficient across the upstream boundary layer in y-direction; Flux=a.bl.y.up*(C.y.up-C[,1]). One value [L/T]
a.bl.y.down	Transfer coefficient across the downstream boundary layer in y-direction; Flux=a.bl.y.down*(C[,Ny]-C.y.down). One value [L/T]
D.grid	Diffusion coefficient defined on all grid cell interfaces. A prop.2D list created by setup.prop.2D [L^2/T]. See last example for creating spatially-varying diffusion coefficients
D.x	Diffusion coefficient in x-direction, defined on grid cell interfaces. One value, a vector of length (Nx+1), a prop.1D list created by setup.prop.1D, or a (Nx+1)* Ny matrix [L^2/T]
D.y	Diffusion coefficient in y-direction, defined on grid cell interfaces. One value, a vector of length (Ny+1), a prop.1D list created by setup.prop.1D, or a Nx*(Ny+1) matrix [L^2/T]
v.grid	Advective velocity defined on all grid cell interfaces. Can be positive (downstream flow) or negative (upstream flow). A prop.2D list created by setup.prop.2D [L/T]
v.x	Advective velocity in the x-direction, defined on grid cell interfaces. Can be positive (downstream flow) or negative (upstream flow). One value, a vector of length (Nx+1), a prop.1D list created by `setup.prop.1D`, or a (Nx+1)*Ny matrix [L/T]

v.y	Advective velocity in the y-direction, defined on grid cell interfaces. Can be positive (downstream flow) or negative (upstream flow). One value, a vector of length (Ny+1), a prop.1D list created by `setup.prop.1D`, or a Nx*(Ny+1) matrix [L/T]
AFDW.grid	Weight used in the finite difference scheme for advection in the x- and y-direction, defined on grid cell interfaces; backward = 1, centered = 0.5, forward = 0; default is backward. A `prop.2D` list created by `setup.prop.2D` [–]
AFDW.x	Weight used in the finite difference scheme for advection in the x-direction, defined on grid cell interfaces; backward = 1, centered = 0.5, forward = 0; default is backward. One value, a vector of length (Nx+1), a prop.1D list created by `setup.prop.1D`, or a (Nx+1)*Ny matrix [–]
AFDW.y	Weight used in the finite difference scheme for advection in the y-direction, defined on grid cell interfaces; backward = 1, centered = 0.5, forward = 0; default is backward. One value, a vector of length (Ny+1), a prop.1D list created by `setup.prop.1D`, or a Nx*(Ny+1) matrix [–]
VF.grid	Volume fraction. A prop.2D list created by setup.prop.2D [–]
VF.x	Volume fraction at the grid cell interfaces in the x-direction. One value, a vector of length (Nx+1), a prop.1D list created by `setup.prop.1D`, or a (Nx+1)*Ny matrix [–]
VF.y	Volume fraction at the grid cell interfaces in the y-direction. One value, a vector of length (Ny+1), a prop.1D list created by `setup.prop.1D`, or a Nx*(Ny+1) matrix [–]
A.grid	Interface area. A prop.2D list created by setup.prop.2D [L^2]
A.x	Interface area defined at the grid cell interfaces in the x-direction. One value, a vector of length (Nx+1), a prop.1D list created by `setup.prop.1D`, or a (Nx+1)*Ny matrix [L^2]
A.y	Interface area defined at the grid cell interfaces in the y-direction. One value, a vector of length (Ny+1), a prop.1D list created by `setup.prop.1D`, or a Nx*(Ny+1) matrix [L^2]
dx	Distance between adjacent cell interfaces in the x-direction (thickness of grid cells). One value or vector of length Nx [L]
dy	Distance between adjacent cell interfaces in the y-direction (thickness of grid cells). One value or vector of length Ny [L]
grid	Discretization grid, a list containing at least elements dx, dx.aux, dy, dy.aux (see `setup.grid.2D`) [L]
full.check	Logical flag enabling a full check of the consistency of the arguments (`default = FALSE; TRUE` slows down execution by 50%)
full.output	Logical flag enabling a full return of the output (default = `FALSE; TRUE` slows down execution by 20%)

Details

The boundary conditions are either

(1) zero-gradient
(2) fixed concentration
(3) convective boundary layer
(4) fixed flux.

This is also the order of priority. The zero gradient is the default, the fixed flux overrules all other.

Value

A list containing:

dC	The rate of change of the concentration *C* due to transport, defined in the center of each grid cell, a Nx*Ny matrix [M/L^3/T]
C.x.up	Concentration at the upstream interface in *x*-direction. A vector of length Ny [M/L^3]. Only when full.output = TRUE
C.x.down	Concentration at the downstream interface in *x*-direction. A vector of length Ny [M/L^3]. Only when full.output = TRUE
C.y.up	Concentration at the upstream interface in *y*-direction. A vector of length Nx [M/L^3]. Only when full.output = TRUE
C.y.down	Concentration at the downstream interface in *y*-direction. A vector of length Nx [M/L^3]. Only when full.output = TRUE
x.flux	Flux across the interfaces in *x*-direction of the grid cells. A (Nx+1)*Ny matrix [M/L^2/T]. Only when full.output = TRUE
y.flux	Flux across the interfaces in *y*-direction of the grid cells. A Nx*(Ny+1) matrix [M/L^2/T]. Only when full.output = TRUE
flux.x.up	Flux across the upstream boundary in *x*-direction, positive = INTO model domain. A vector of length Ny [M/L^2/T]
flux.x.down	Flux across the downstream boundary in *x*-direction, positive = OUT of model domain. A vector of length Ny [M/L^2/T]
flux.y.up	Flux across the upstream boundary in *y*-direction, positive = INTO model domain. A vector of length Nx [M/L^2/T]
flux.y.down	Flux across the downstream boundary in *y*-direction, positive = OUT of model domain. A vector of length Nx [M/L^2/T]

Note

It is much more efficient to use the *grid* input rather than vectors or single numbers.

Thus: to optimize the code, use setup.grid.2D to create the grid, and use setup.prop.2D to create D.grid, v.grid, AFDW.grid, VF.grid, and A.grid, even if the values are 1 or remain constant.

There is no provision (yet) to deal with *cross-diffusion*. Set `D.x` and `D.y` different only if cross-diffusion effects are unimportant.

Authors

Filip Meysman <filip.meysman@nioz.nl>,
Karline Soetaert <karline.soetaert@nioz.nl>.

See Also

`tran.polar` for a discretization of 2D transport equations in polar coordinates.
`tran.1D`, `tran.3D`

Examples

```
## ============================================
## Testing the functions
## ============================================
# Parameters
F     <- 100      # input flux [micromol cm-2 yr-1]
por   <- 0.8      # constant porosity
D     <- 400      # mixing coefficient [cm2 yr-1]
v     <- 1        # advective velocity [cm yr-1]

# Grid definition
x.N <- 4   # number of cells in x-direction
y.N <- 6   # number of cells in y-direction
x.L <- 8   # domain size x-direction [cm]
y.L <- 24  # domain size y-direction [cm]
dx <- x.L/x.N        # cell size x-direction [cm]
dy <- y.L/y.N        # cell size y-direction [cm]

# Initial conditions
C <- matrix(nrow = x.N, ncol = y.N, data = 0, byrow = FALSE)

# Boundary conditions: fixed concentration
C.x.up     <- rep(1, times = y.N)
C.x.down   <- rep(0, times = y.N)
C.y.up     <- rep(1, times = x.N)
C.y.down   <- rep(0, times = x.N)

# Only diffusion
tran.2D(C = C, D.x = D, D.y = D, v.x = 0, v.y = 0,
```

```
  VF.x = por, VF.y = por, dx = dx, dy = dy,
  C.x.up = C.x.up, C.x.down = C.x.down,
  C.y.up = C.y.up, C.y.down = C.y.down, full.output = TRUE)

# Strong advection, backward (default), central and forward
#finite difference schemes
tran.2D(C = C, D.x = D, v.x = 100*v,
  VF.x = por, dx = dx, dy = dy,
  C.x.up = C.x.up, C.x.down = C.x.down,
  C.y.up = C.y.up, C.y.down = C.y.down)

tran.2D(AFDW.x = 0.5, C = C, D.x = D, v.x = 100*v,
  VF.x = por, dx = dx, dy = dy,
  C.x.up = C.x.up, C.x.down = C.x.down,
  C.y.up = C.y.up, C.y.down = C.y.down)

tran.2D(AFDW.x = 0, C = C, D.x = D, v.x = 100*v,
  VF.x = por, dx = dx, dy = dy,
  C.x.up = C.x.up, C.x.down = C.x.down,
  C.y.up = C.y.up, C.y.down = C.y.down)

# Boundary conditions: fixed fluxes

flux.x.up <- rep(200, times = y.N)
flux.x.down <- rep(-200, times = y.N)

flux.y.up <- rep(200, times = x.N)
flux.y.down <- rep(-200, times = x.N)
tran.2D(C = C, D.x = D, v.x = 0,
  VF.x = por, dx = dx, dy = dy,
  flux.x.up = flux.x.up, flux.x.down = flux.x.down,
  flux.y.up = flux.y.up, flux.y.down = flux.y.down)

# Boundary conditions: convective boundary layer on all sides

a.bl <- 800 # transfer coefficient
C.x.up <- rep(1, times = (y.N)) # fixed conc at boundary layer
C.y.up <- rep(1, times = (x.N)) # fixed conc at boundary layer
tran.2D(full.output = TRUE, C = C, D.x = D, v.x = 0,
  VF.x = por, dx = dx, dy = dy,
  C.x.up = C.x.up, a.bl.x.up = a.bl,
  C.x.down = C.x.up, a.bl.x.down = a.bl,
  C.y.up = C.y.up, a.bl.y.up = a.bl,
  C.y.down = C.y.up, a.bl.y.down = a.bl)
```

```
# Runtime test with and without argument checking

n.iterate <-500

test1 <- function() {
 for (i in 1:n.iterate)
  ST <- tran.2D(full.check = TRUE, C = C, D.x = D,
    v.x = 0, VF.x = por, dx = dx, dy = dy,
    C.x.up = C.x.up, a.bl.x.up = a.bl, C.x.down = C.x.down)
}
system.time(test1())
test2 <- function() {
 for (i in 1:n.iterate )
  ST <- tran.2D(full.output = TRUE, C = C, D.x = D,
    v.x = 0, VF.x = por, dx = dx, dy = dy,
    C.x.up = C.x.up, a.bl.x.up = a.bl, C.x.down = C.x.down)
}
system.time(test2())

test3 <- function() {
 for (i in 1:n.iterate )
  ST <- tran.2D(full.output = TRUE, full.check = TRUE, C = C,
    D.x = D, v.x = 0, VF.x = por, dx = dx, dy = dy,
    C.x.up = C.x.up, a.bl.x.up = a.bl, C.x.down = C.x.down)
}
system.time(test3())

## =============================================
## A 2D model with diffusion in x- and y-direction and first-order
## consumption - inefficient implementation
## =============================================

N  <- 51              # number of grid cells
XX <- 10              # total size
dy <- dx <- XX/N      # grid size
Dy <- Dx <- 0.1       # diffusion coeff, X- and Y-direction
r  <- 0.005           # consumption rate
ini <- 1              # initial value at x=0

N2 <- ceiling(N/2)
X  <- seq (dx, by = dx, len = (N2-1))
X  <- c(-rev(X), 0, X)

# The model equations

Diff2D <- function (t, y, parms) {
```

```
CONC <- matrix(nrow = N, ncol = N, y)
dCONC <- tran.2D(CONC, D.x = Dx, D.y=Dy, dx=dx, dy=dy)$dC + r*CONC
return (list(dCONC))
}

# initial condition: 0 everywhere, except in central point

y <- matrix(nrow = N, ncol = N, data = 0)
y[N2, N2] <- ini # initial concentration in the central point...

# solve for 10 time units
times <- 0:10
out <- ode.2D (y = y, func = Diff2D, t = times, parms = NULL,
        dim = c(N,N), lrw = 160000)

pm <- par (mfrow = c(2, 2))

# Compare solution with analytical solution...
for (i in seq(2, 11, by = 3)) {
 tt <- times[i]
  mat <- matrix(nrow = N, ncol = N,
          data = subset(out, time == tt))
  plot(X, mat[N2,], type = "l", main = paste("time=", times[i]),
    ylab = "Conc", col = "red")
  ana <- ini*dx^2/(4*pi*Dx*tt)*exp(r*tt-X^2/(4*Dx*tt))
  points(X, ana, pch = "+")
}

legend ("bottom", col = c("red","black"), lty = c(1, NA),
  pch = c(NA, "+"), c("tran.2D", "exact"))
par("mfrow" = pm )

## ==========================================
## A 2D model with diffusion in x- and y-direction and first-order
## consumption - more efficient implementation, specifying ALL 2D grids
## ==========================================

N   <- 51             # number of grid cells
Dy  <- Dx <- 0.1      # diffusion coeff, X- and Y-direction
r   <- 0.005          # consumption rate
ini  <- 1             # initial value at x=0

x.grid   <- setup.grid.1D(x.up = -5, x.down = 5, N = N)
y.grid   <- setup.grid.1D(x.up = -5, x.down = 5, N = N)
grid2D   <- setup.grid.2D(x.grid, y.grid)
```

```
D.grid    <- setup.prop.2D(value = Dx, y.value = Dy, grid = grid2D)
v.grid    <- setup.prop.2D(value = 0, grid = grid2D)
A.grid    <- setup.prop.2D(value = 1, grid = grid2D)
AFDW.grid <- setup.prop.2D(value = 1, grid = grid2D)
VF.grid   <- setup.prop.2D(value = 1, grid = grid2D)
# The model equations - using the grids

Diff2Db <- function (t, y, parms) {
  CONC <- matrix(nrow = N, ncol = N, data = y)
  dCONC <- tran.2D(CONC, grid = grid2D, D.grid = D.grid,
     A.grid = A.grid, VF.grid = VF.grid, AFDW.grid = AFDW.grid,
     v.grid = v.grid)$dC + r * CONC
 return (list(dCONC))
}

# initial condition: 0 everywhere, except in central point
y <- matrix(nrow = N, ncol = N, data = 0)
y[N2,N2] <- ini # initial concentration in the central point...

# solve for 8 time units
Times <- 0:8
outb <- ode.2D (y = y, func = Diff2Db, t = times, parms = NULL,
                dim = c(N, N), lrw = 160000)

outtimes <- c(1, 3, 5, 7)
image(outc, ask = FALSE, mfrow = c(2, 2), main = paste("time", outtimes),
   legend = TRUE, add.contour = TRUE, subset = time %in% outtimes)
```

```
# In the R domain:
> install.packages("ReacTran")
Installing package into 'C:/Users/Bert/Documents/R/win-library/3.3'
(as 'lib' is unspecified)
--- Please select a CRAN mirror for use in this session ---
A CRAN mirror is selected.
trying URL 'https://mirrors.tuna.tsinghua.edu.cn/CRAN/bin/windows/contrib/3.3/
ReacTran_1.4.3.1.zip'
Content type 'application/zip' length 915996 bytes (894 KB)
downloaded 894 KB

package 'ReacTran' successfully unpacked and MD5 sums checked

The downloaded binary packages are in
C:\Users\Bert\AppData\Local\Temp\RtmpEvJcoa\downloaded_packages
> require(ReacTran)
Loading required package: ReacTran
```

```
Loading required package: rootSolve
Loading required package: deSolve
Loading required package: shape
> library(ReacTran)
> ls("package:ReacTran")
 [1] "advection.1D"   "advection.volume.1D"  "fiadeiro"
 [4] "g.cylinder"     "g.sphere"             "g.spheroid"
 [7] "p.exp"          "p.lin"                "p.sig"
[10] "polar2cart"     "setup.compaction.1D"  "setup.grid.1D"
[13] "setup.grid.2D"  "setup.prop.1D"        "setup.prop.2D"
[16] "tran.1D"        "tran.2D"              "tran.3D"
[19] "tran.cylindrical"  "tran.polar"        "tran.spherical"
[22] "tran.volume.1D" "tran.volume.2D"       "tran.volume.3D"

> tran.2D # Outputting the program listing of tran.2D
function (C, C.x.up = C[1, ], C.x.down = C[nrow(C), ], C.y.up = C[,
  1], C.y.down = C[, ncol(C)], flux.x.up = NULL, flux.x.down = NULL,
  flux.y.up = NULL, flux.y.down = NULL, a.bl.x.up = NULL, a.bl.x.down =
  NULL,
  a.bl.y.up = NULL, a.bl.y.down = NULL, D.grid = NULL, D.x = NULL,
  D.y = D.x, v.grid = NULL, v.x = 0, v.y = 0, AFDW.grid = NULL,
  AFDW.x = 1, AFDW.y = AFDW.x, VF.grid = NULL, VF.x = 1, VF.y = VF.x,
  A.grid = NULL, A.x = 1, A.y = 1, grid = NULL, dx = NULL,
dy = NULL, full.check = FALSE, full.output = FALSE)
{
  if (is.null(grid))
    if (is.null(dx) | is.null(dy))
      stop("error: either grid or dx and dy should be specified ")
  Nx <- nrow(C)
  Ny <- ncol(C)
  if (is.null(grid)) {
    DX <- if (is.list(dx))
      dx$dx
    else rep(dx, length.out = Nx)
    DXaux <- if (is.list(dx))
      dx$dx.aux
    else 0.5 * (c(0, rep(dx, length.out = Nx)) + c(rep(dx,
      length.out = Nx), 0))
    DY <- if (is.list(dy))
      dy$dx
    else rep(dy, length.out = Ny)
    DYaux <- if (is.list(dy))
      dy$dx.aux
    else 0.5 * (c(0, rep(dy, length.out = Ny)) + c(rep(dy,
      length.out = Ny), 0))
    grid <- list(dx = matrix(nrow = Nx, ncol = Ny, data = DX),
```

```
      dx.aux = matrix(nrow = Nx + 1, ncol = Ny, data = DXaux),
      dy = matrix(nrow = Nx, ncol = Ny, data = DY, byrow = TRUE),
      dy.aux = matrix(nrow = Nx, ncol = Ny + 1, data = DYaux,
               byrow = TRUE))
  }
  else if (class(grid) != "grid.2D") {
    if (length(grid$dx) != Nx * Ny)
      grid$dx <- matrix(nrow = Nx, ncol = Ny, grid$dx)
    if (length(grid$dx.aux) != (Nx + 1) * Ny)
      grid$dx.aux <- matrix(nrow = Nx + 1, ncol = Ny, grid$dx.aux)
    if (length(grid$dy) != Nx * Ny)
      grid$dy <- matrix(nrow = Nx, ncol = Ny, data = grid$dy,
        byrow = TRUE)
    if (length(grid$dy.aux) != Nx * (Ny + 1))
      grid$dy.aux <- matrix(nrow = Nx, ncol = Ny + 1, grid$dy.aux,
        byrow = TRUE)
  }
  gridFill <- function(G.x, G.y, Name) {
    if (is.null(G.x) | is.null(G.y))
      stop((paste("error: ", Name, "and (", Name, ".x and",
        Name, ".y) cannot be NULL at the same time",
        del = "")))
    G.grid <- list()
    if (is.matrix(G.x)) {
      if (sum(abs(dim(G.x) - c(Nx + 1, Ny))) != 0)
        stop(paste("error: ", Name, ".x matrix not of correct (Nx+1) Ny dimensions",
         del = ""))
      G.grid$x.int <- G.x
G.grid$x.mid <- 0.5 * (G.x[1:Nx, ] + G.x[2:(Nx +
              1), ])
    }
    else if (class(G.x) == "prop.1D") {
      G.grid$x.int <- matrix(data = G.x$int, nrow = (Nx +
        1), ncol = Ny)
      G.grid$x.mid <- matrix(data = G.x$mid, nrow = Nx,
        ncol = Ny)
    }
    else if (length(G.x) == 1) {
      G.grid$x.int <- matrix(data = G.x, nrow = (Nx + 1),
        ncol = Ny)
      G.grid$x.mid <- matrix(data = G.x, nrow = Nx, ncol = Ny)
    }
    else if (length(G.x) != Nx + 1) {
      stop(paste("error: ", Name, ".x should be a vector of length 1 or Nx+1",
        del = ""))
    }
```

```
  else {
    G.grid$x.int <- matrix(data = G.x, nrow = (Nx + 1),
      ncol = Ny)
    G.grid$x.mid <- matrix(data = 0.5 * (G.x[1:Nx] +
      G.x[2:(Nx + 1)]), nrow = Nx, ncol = Ny)
  }
    if (is.matrix(G.y)) {
    if (sum(abs(dim(G.y) - c(Nx, Ny + 1))) != 0)
      stop(paste("error: ", Name, ".y matrix not of correct Nx(Ny+1)dimensions",
       del = ""))
    G.grid$y.int <- G.y
    G.grid$y.mid <- 0.5 * (G.y[, 1:Ny] + G.y[, 2:(Ny +
      1)])
  }
  else if (class(G.y) == "prop.1D") {
    G.grid$y.int <- matrix(data = G.y$int, nrow = Nx,
      ncol = (Ny + 1))
    G.grid$y.mid <- matrix(data = G.y$mid, nrow = Nx,
      ncol = Ny)
  }
  else if (length(G.y) == 1) {
    G.grid$y.int <- matrix(data = G.y, nrow = Nx, ncol = (Ny +
      1))
    G.grid$y.mid <- matrix(data = G.y, nrow = Nx, ncol = Ny)
  }
  else if (length(G.y) != Ny + 1) {
    stop(paste("error: ", Name, ".y should be a vector of length 1 or Ny+1",
      del = ""))
  }
  else {
    G.grid$y.int <- matrix(data = G.y, nrow = Nx, ncol = (Ny +
     1))
    G.grid$y.mid <- matrix(data = 0.5 * (G.y[1:Nx] +
      G.y[2:(Nx + 1)]), nrow = Nx, ncol = Ny)
  }
  G.grid
}
if (is.null(VF.grid))
  VF.grid <- gridFill(VF.x, VF.y, "VF")
if (is.null(A.grid))
  A.grid <- gridFill(A.x, A.y, "A")
gridInt <- function(G.x, G.y, Name) {
  if (is.null(G.x) | is.null(G.y))
    stop((paste("error: ", Name, "and (", Name, ".x and",
      Name, ".y) cannot be NULL at the same time",
      del = "")))
```

```
G.grid <- list()
if (is.matrix(G.x)) {
  if (sum(abs(dim(G.x) - c(Nx + 1, Ny))) != 0)
    stop(paste("error: ", Name, ".x matrix not of correct (Nx+1) Ny dimensions",
     del = ""))
  G.grid$x.int <- G.x
}
else if (class(G.x) == "prop.1D") {
  G.grid$x.int <- matrix(data = G.x$int, nrow = (Nx +
    1), ncol = Ny)
}
else if (length(G.x) == 1) {
  G.grid$x.int <- matrix(data = G.x, nrow = (Nx + 1),
    ncol = Ny)
}
else if (length(G.x) != Nx + 1) {
  stop(paste("error: ", Name, ".x should be a vector of length 1 or Nx+1",
    del = ""))
}
else {
  G.grid$x.int <- matrix(data = G.x, nrow = (Nx + 1),
    ncol = Ny)
}
if (is.matrix(G.y)) {
  if (sum(abs(dim(G.y) - c(Nx, Ny + 1))) != 0)
    stop(paste("error: ", Name, ".y matrix not of correct Nx(Ny+1)dimensions",
     del = ""))
  G.grid$y.int <- G.y
}
else if (class(G.y) == "prop.1D") {
  G.grid$y.int <- t(matrix(data = G.y$int, ncol = Nx,
    nrow = (Ny + 1)))
}
else if (length(G.y) == 1) {
  G.grid$y.int <- matrix(data = G.y, nrow = Nx, ncol = (Ny +
   1))
}
else if (length(G.y) != Ny + 1) {
  stop(paste("error: ", Name, ".y should be a vector of length 1 or
Ny+1",
    del = ""))
}
else {
  G.grid$y.int <- matrix(data = G.y, nrow = Nx, ncol = (Ny +
    1))
}
```

```
  G.grid
}
if (is.null(AFDW.grid))
  AFDW.grid <- gridInt(AFDW.x, AFDW.y, "AFDW")
if (is.null(D.grid))
  D.grid <- gridInt(D.x, D.y, "D")
if (is.null(v.grid))
  v.grid <- gridInt(v.x, v.y, "v")
if (full.check) {
  if (!is.null(C.x.up)) {
    if (!((length(C.x.up) == 1) || (length(C.x.up) ==
      (Ny))))
      stop("error: C.x.up should be a vector of length 1 or ncol(C)")
  }
  if (!is.null(C.x.down)) {
   if (!((length(C.x.down) == 1) || (length(C.x.down) ==
      (Ny))))
      stop("error: C.x.down should be a vector of length 1 or ncol(C)")
  }
  if (!is.null(C.y.up)) {
    if (!((length(C.y.up) == 1) || (length(C.y.up) ==
      (Nx))))
      stop("error: C.y.up should be a vector of length 1 or nrow(C)")
  }
  if (!is.null(C.y.down)) {
    if (!((length(C.y.down) == 1) || (length(C.y.down) ==
      (Nx))))
      stop("error: C.y.down should be a vector of length 1 or nrow(C)")
  }
  if (!is.null(flux.x.up)) {
    if (!((length(flux.x.up) == 1) || (length(flux.x.up) ==
      (Ny))))
      stop("error: flux.x.up should be a vector of length 1 or ncol(C)")
  }
  if (!is.null(flux.x.down)) {
    if (!((length(flux.x.down) == 1) || (length(flux.x.down) ==
      (Ny))))
      stop("error: flux.x.down should be a vector of length 1 or
          ncol(C)")
  }
  if (!is.null(flux.y.up)) {
    if (!((length(flux.y.up) == 1) || (length(flux.y.up) ==
      (Nx))))
      stop("error: flux.y.up should be a vector of length 1 or nrow(C)")
  }
  if (!is.null(flux.y.down)) {
```

```
  if (!((length(flux.y.down) == 1) || (length(flux.y.down) ==
    (Nx))))
    stop("error: flux.y.down should be a vector of length 1 or
       nrow(C)")
}
if (is.null(dx) && is.null(dy) && is.null(grid))
  stop("error: dx, dy, and grid cannot be NULL at the same time")
gn <- names(grid)
if (!"dx" %in% gn)
  stop("error: grid should be a list that contains 'dx' ")
if (!"dx.aux" %in% gn)
  stop("error: grid should be a list that contains 'dx.aux' ")
if (!"dy" %in% gn)
  stop("error: grid should be a list that contains 'dy' ")
if (!"dy.aux" %in% gn)
  stop("error: grid should be a list that contains 'dy.aux' ")
if (is.null(grid$dx) || is.null(grid$dx.aux))
  stop("error: the grid should be a list with (numeric) values for 'dx'
     and 'dx.aux' ")
if (is.null(grid$dy) || is.null(grid$dy.aux))
  stop("error: the grid should be a list with (numeric) values for 'dy'
    and 'dy.aux' ")
if (any(grid$dx <= 0) || any(grid$dx.aux <= 0))
  stop("error: the grid distances dx and dx.aux should always be
     positive")
if (any(grid$dy <= 0) || any(grid$dy.aux <= 0))
  stop("error: the grid distances dy and dy.aux should always be
     positive")
gn <- names(AFDW.grid)
if (!"x.int" %in% gn)
  stop("error: AFDW.grid should be a list that contains 'x.int', the AFDW values
     at the interfaces of the grid cells in x-direction")
if (!"y.int" %in% gn)
  stop("error: AFDW.grid should be a list that contains 'y.int', the
     AFDW values at the interfaces of the grid cells in y-direction")
if (is.null(AFDW.grid$x.int))
  stop("error: AFDW.grid$x.int should be a list with (numeric) values")
if (is.null(AFDW.grid$y.int))
  stop("error: AFDW.grid$y.int should be a list with (numeric)
     values")
if (any(AFDW.grid$x.int < 0) || any(AFDW.grid$x.int >
  1))
  stop("error: the AFDW should range between 0 and 1")
if (any(AFDW.grid$y.int < 0) || any(AFDW.grid$y.int >
  1))
  stop("error: the AFDW should range between 0 and 1")
```

```
  gn <- names(D.grid)
if (!"x.int" %in% gn)
    stop("error: D.grid should be a list that contains 'x.int', the D values
       at the interfaces of the grid cells in x-direction")
  if (!"y.int" %in% gn)
    stop("error: D.grid should be a list that contains 'y.int', the D values
       at the interfaces of the grid cells in y-direction")
  if (is.null(D.grid$x.int))
    stop("error: D.grid$x.int should be a list with (numeric) values")
  if (is.null(D.grid$y.int))
    stop("error: D.grid$y.int should be a list with (numeric) values")
  if (any(D.grid$x.int < 0) || any(D.grid$y.int < 0))
    stop("error: the diffusion coefficient should always be positive")
  gn <- names(v.grid)
  if (!"x.int" %in% gn)
    stop("error: v.grid should be a list that contains 'x.int', the velocity
       values at the interfaces of the grid cells in x-direction")
  if (!"y.int" %in% gn)
    stop("error: v.grid should be a list that contains 'y.int', the velocity
       values at the interfaces of the grid cells in y-direction")
  if (is.null(v.grid$x.int))
    stop("error: the advective velocity v.grid$x.int should be a list with
       (numeric) values")
  if (is.null(v.grid$y.int))
    stop("error: the advective velocity v.grid$y.int should be a list with
       (numeric) values")
 gn <- names(VF.grid)
  if (!"x.int" %in% gn)
    stop("error: VF.grid should be a list that contains 'x.int'")
  if (!"y.int" %in% gn)
    stop("error: VF.grid should be a list that contains 'y.int'")
  if (!"x.mid" %in% gn)
    stop("error: VF.grid should be a list that contains 'x.mid'")
  if (!"y.mid" %in% gn)
    stop("error: VF.grid should be a list that contains 'y.mid'")
  if (is.null(VF.grid$x.int) || is.null(VF.grid$y.int) ||
    is.null(VF.grid$x.mid) || is.null(VF.grid$y.mid))
    stop("error: VF should contain (numeric) values")
  if (any(VF.grid$x.int < 0) || any(VF.grid$y.int < 0) ||
    any(VF.grid$x.mid < 0) || any(VF.grid$y.mid < 0))
    stop("error: the VF values should always be positive")
  gn <- names(A.grid)
  if (!"x.int" %in% gn)
    stop("error: A.grid should be a list that contains 'x.int'")
  if (!"y.int" %in% gn)
    stop("error: A.grid should be a list that contains 'y.int'")
```

```
    if (!"x.mid" %in% gn)
      stop("error: A.grid should be a list that contains 'x.mid'")
    if (!"y.mid" %in% gn)
      stop("error: A.grid should be a list that contains 'y.mid'")
    if (is.null(A.grid$x.int) || is.null(A.grid$y.int) ||
      is.null(A.grid$x.mid) || is.null(A.grid$y.mid))
stop("error: the VF should contain (numeric) values")
    if (any(A.grid$x.int < 0) || any(A.grid$y.int < 0) ||
      any(A.grid$x.mid < 0) || any(A.grid$y.mid < 0))
      stop("error: the A values should always be positive")
  }
  if (!is.null(flux.x.up[1])) {
    nom <- flux.x.up + VF.grid$x.int[1, ] * (D.grid$x.int[1,
      ]/grid$dx.aux[1, ] + (1 - AFDW.grid$x.int[1, ]) *
      v.grid$x.int[1, ]) * C[1, ]
    denom <- VF.grid$x.int[1, ] * (D.grid$x.int[1, ]/grid$dx.aux[1,
      ] + AFDW.grid$x.int[1, ] * v.grid$x.int[1, ])
    C.x.up <- nom/denom
  }
  if (!is.null(flux.x.down[1])) {
    nom <- flux.x.down - VF.grid$x.int[(Nx + 1), ] * (D.grid$x.int[(Nx +
      1), ]/grid$dx.aux[Nx + 1, ] + AFDW.grid$x.int[(Nx +
      1), ] * v.grid$x.int[(Nx + 1), ]) * C[Nx, ]
    denom <- -VF.grid$x.int[(Nx + 1), ] * (D.grid$x.int[(Nx +
      1), ]/grid$dx.aux[Nx + 1, ] + (1 - AFDW.grid$x.int[(Nx +
      1), ]) * v.grid$x.int[(Nx + 1), ])
    C.x.down <- nom/denom
  }
  if (!is.null(flux.y.up[1])) {
    nom <- flux.y.up + VF.grid$y.int[, 1] * (D.grid$y.int[,
      1]/grid$dy.aux[, 1] + (1 - AFDW.grid$y.int[, 1]) *
      v.grid$y.int[, 1]) * C[, 1]
denom <- VF.grid$y.int[, 1] * (D.grid$y.int[, 1]/grid$dy.aux[,
      1] + AFDW.grid$y.int[, 1] * v.grid$y.int[, 1])
    C.y.up <- nom/denom
  }
  if (!is.null(flux.y.down[1])) {
    nom <- flux.y.down - VF.grid$y.int[, (Ny + 1)] * (D.grid$y.int[,
      (Ny + 1)]/grid$dy.aux[, (Ny + 1)] + AFDW.grid$y.int[,
      (Ny + 1)] * v.grid$y.int[, (Ny + 1)]) * C[, Ny]
    denom <- -VF.grid$y.int[, (Ny + 1)] * (D.grid$y.int[,
      (Ny + 1)]/grid$dy.aux[, Ny + 1] + (1 - AFDW.grid$y.int[,
      (Ny + 1)]) * v.grid$y.int[, (Ny + 1)])
    C.y.down <- nom/denom
  }
```

```
  if (!is.null(a.bl.x.up) & !is.null(C.x.up[1])) {
    nom <- a.bl.x.up * C.x.up + VF.grid$x.int[1, ] * (D.grid$x.int[1,
      ]/grid$dx.aux[1, ] + (1 - AFDW.grid$x.int[1, ]) *
      v.grid$x.int[1, ]) * C[1, ]
    denom <- a.bl.x.up + VF.grid$x.int[1, ] * (D.grid$x.int[1,
      ]/grid$dx.aux[1, ] + AFDW.grid$x.int[1, ] * v.grid$x.int[1,
      ])
    C.x.up <- nom/denom
  }
  if (!is.null(a.bl.x.down) & !is.null(C.x.down[1])) {
    nom <- a.bl.x.down * C.x.down + VF.grid$x.int[(Nx + 1),
      ] * (D.grid$x.int[(Nx + 1), ]/grid$dx.aux[(Nx + 1),
      ] + (1 - AFDW.grid$x.int[(Nx + 1), ]) * v.grid$x.int[(Nx +
     1), ]) * C[Nx, ]
    denom <- a.bl.x.down + VF.grid$x.int[(Nx + 1), ] * (D.grid$x.int[(Nx +
      1), ]/grid$dx.aux[(Nx + 1), ] + AFDW.grid$x.int[(Nx +
      1), ] * v.grid$x.int[(Nx + 1), ])
    C.x.down <- nom/denom
  }
  if (!is.null(a.bl.y.up) & !is.null(C.y.up[1])) {
    nom <- a.bl.y.up * C.y.up + VF.grid$y.int[, 1] * (D.grid$y.int[,
      1]/grid$dy.aux[, 1] + (1 - AFDW.grid$y.int[, 1]) *
      v.grid$y.int[, 1]) * C[, 1]
    denom <- a.bl.y.up + VF.grid$y.int[, 1] * (D.grid$y.int[,
      1]/grid$dy.aux[, 1] + AFDW.grid$y.int[, 1] * v.grid$y.int[,
      1])
    C.y.up <- nom/denom
  }
  if (!is.null(a.bl.y.down) & !is.null(C.y.down[1])) {
    nom <- a.bl.y.down * C.y.down + VF.grid$y.int[, (Ny +
      1)] * (D.grid$y.int[, (Ny + 1)]/grid$dy.aux[, (Ny +
      1)] + (1 - AFDW.grid$y.int[, (Ny + 1)]) * v.grid$y.int[,
      (Ny + 1)]) * C[, Ny]
    denom <- a.bl.y.down + VF.grid$y.int[, (Ny + 1)] * (D.grid$y.int[,
      (Ny + 1)]/grid$dy.aux[, (Ny + 1)] + AFDW.grid$y.int[,
      (Ny + 1)] * v.grid$y.int[, (Ny + 1)])
    C.y.down <- nom/denom
  }
x.Dif.flux <- as.matrix(-VF.grid$x.int * D.grid$x.int * diff(rbind(C.x.up,
C, C.x.down, deparse.level = 0))/grid$dx.aux)
  y.Dif.flux <- as.matrix(-VF.grid$y.int * D.grid$y.int *
              t(diff(t(cbind(C.y.up,
    C, C.y.down, deparse.level = 0))))/grid$dy.aux)
  x.Adv.flux <- 0
  if (any(v.grid$x.int > 0)) {
```

```
    vv <- v.grid$x.int
    vv[vv < 0] <- 0
    x.Adv.flux <- x.Adv.flux + as.matrix(VF.grid$x.int *
      vv * (AFDW.grid$x.int * rbind(C.x.up, C, deparse.level = 0) +
      (1 - AFDW.grid$x.int) * rbind(C, C.x.down, deparse.level = 0)))
  }
  if (any(v.grid$x.int < 0)) {
    vv <- v.grid$x.int
    vv[vv > 0] <- 0
    x.Adv.flux <- x.Adv.flux + as.matrix(VF.grid$x.int *
      vv * ((1 - AFDW.grid$x.int) * rbind(C.x.up, C, deparse.level = 0) +
      AFDW.grid$x.int * rbind(C, C.x.down, deparse.level = 0)))
  }
  y.Adv.flux <- 0
  if (any(v.grid$y.int > 0)) {
    vv <- v.grid$y.int
    vv[vv < 0] <- 0
    y.Adv.flux <- y.Adv.flux + as.matrix(VF.grid$y.int *
      vv * (AFDW.grid$y.int * cbind(C.y.up, C, deparse.level = 0) +
    (1 - AFDW.grid$y.int) * cbind(C, C.y.down, deparse.level = 0)))
  }
  if (any(v.grid$y.int < 0)) {
    vv <- v.grid$y.int
    vv[vv > 0] <- 0
    y.Adv.flux <- y.Adv.flux + as.matrix(VF.grid$y.int *
      vv * ((1 - AFDW.grid$y.int) * cbind(C.y.up, C, deparse.level = 0) +
      AFDW.grid$y.int * cbind(C, C.y.down, deparse.level = 0)))
  }
  x.flux <- x.Dif.flux + x.Adv.flux
  y.flux <- y.Dif.flux + y.Adv.flux
  if (!is.null(flux.x.up[1]))
    x.flux[1, ] <- flux.x.up
  if (!is.null(flux.x.down[1]))
    x.flux[nrow(x.flux), ] <- flux.x.down
  if (!is.null(flux.y.up[1]))
    y.flux[, 1] <- flux.y.up
  if (!is.null(flux.y.down[1]))
    y.flux[, ncol(y.flux)] <- flux.y.down
  dFdx <- -(diff(A.grid$x.int * x.flux)/A.grid$x.mid)/grid$dx/VF.grid$x.mid
  dFdy <- -t(diff(t(A.grid$y.int * y.flux))/t(A.grid$y.mid))/grid$dy/VF.grid$y.mid
  if (!full.output) {
    return(list(dC = dFdx + dFdy, flux.x.up = x.flux[1, ],
      flux.x.down = x.flux[nrow(x.flux), ], flux.y.up = y.flux[,
        1], flux.y.down = y.flux[, ncol(y.flux)]))
  }
```

```
  else {
   return(list(dC = dFdx + dFdy, C.x.up = C.x.up, C.x.down = C.x.down,
     C.y.up = C.y.up, C.y.down = C.y.down, x.flux = x.flux,
     y.flux = y.flux, flux.x.up = x.flux[1, ], flux.x.down = x.flux[nrow(x.flux),
       ], flux.y.up = y.flux[, 1], flux.y.down = y.flux[,
       ncol(y.flux)]))
  }
}
<environment: namespace:ReacTran>

> ## ==============================================
> ## Testing the functions
> ## ==============================================
> # Parameters
> F <- 100 # input flux [micromol cm-2 yr-1]
> por <- 0.8 # constant porosity
> D <- 400 # mixing coefficient [cm2 yr-1]
> v <- 1 # advective velocity [cm yr-1]
> # Grid definition
> x.N <- 4 # number of cells in x-direction
> y.N <- 6 # number of cells in y-direction
> x.L <- 8 # domain size x-direction [cm]
> y.L <- 24 # domain size y-direction [cm]
> dx <- x.L/x.N # cell size x-direction [cm]
> dy <- y.L/y.N # cell size y-direction [cm]
> # Initial conditions
> C <- matrix(nrow = x.N, ncol = y.N, data = 0, byrow = FALSE)
> # Boundary conditions: fixed concentration
> C.x.up <- rep(1, times = y.N)
> C.x.down <- rep(0, times = y.N)
> C.y.up <- rep(1, times = x.N)
> C.y.down <- rep(0, times = x.N)
> # Only diffusion
> tran.2D(C = C, D.x = D, D.y = D, v.x = 0, v.y = 0,
+ VF.x = por, VF.y = por, dx = dx, dy = dy,
+ C.x.up = C.x.up, C.x.down = C.x.down,
+ C.y.up = C.y.up, C.y.down = C.y.down, full.output = TRUE)
```

 # Outputting:

```
 $dC
      [,1] [,2] [,3] [,4] [,5] [,6]
[1,] 250  200  200  200  200  200
[2,]  50    0    0    0    0    0
[3,]  50    0    0    0    0    0
[4,]  50    0    0    0    0    0
```

```
$C.x.up

[1] 1 1 1 1 1 1

$C.x.down
[1] 0 0 0 0 0 0

$C.y.up
[1] 1 1 1 1
$C.y.down
[1] 0 0 0 0
$x.flux
     [,1] [,2] [,3] [,4] [,5] [,6]
[1,]  320  320  320  320  320  320
[2,]    0    0    0    0    0    0
[3,]    0    0    0    0    0    0
[4,]    0    0    0    0    0    0
[5,]    0    0    0    0    0    0
$y.flux
     [,1] [,2] [,3] [,4] [,5] [,6] [,7]
[1,]  160    0    0    0    0    0    0
[2,]  160    0    0    0    0    0    0
[3,]  160    0    0    0    0    0    0
[4,]  160    0    0    0    0    0    0

$flux.x.up
[1] 320 320 320 320 320 320

$flux.x.down
[1] 0 0 0 0 0 0

$flux.y.up
[1] 160 160 160 160

$flux.y.down
[1] 0 0 0 0

> # Strong advection, backward (default), central and forward
> #finite difference schemes
> tran.2D(C = C, D.x = D, v.x = 100*v,
+ VF.x = por, dx = dx, dy = dy,
+ C.x.up = C.x.up, C.x.down = C.x.down,
+ C.y.up = C.y.up, C.y.down = C.y.down)
```

 # Outputting:

```
$dC
```

```
     [,1] [,2] [,3] [,4] [,5] [,6]
[1,] 300  250  250  250  250  250
[2,]  50    0    0    0    0    0
[3,]  50    0    0    0    0    0
[4,]  50    0    0    0    0    0

$flux.x.up
[1] 400 400 400 400 400 400

$flux.x.down
[1] 0 0 0 0 0 0

$flux.y.up
[1] 160 160 160 160

$flux.y.down
[1] 0 0 0 0

> tran.2D(AFDW.x = 0.5, C = C, D.x = D, v.x = 100*v,
+ VF.x = por, dx = dx, dy = dy,
+ C.x.up = C.x.up, C.x.down = C.x.down,
+ C.y.up = C.y.up, C.y.down = C.y.down)

# Outputting:
$dC

[1,] 275 225 225 225 225 225
[2,]  50   0   0   0   0   0
[3,]  50   0   0   0   0   0
[4,]  50   0   0   0   0   0

$flux.x.up
[1] 360 360 360 360 360 360

$flux.x.down
[1] 0 0 0 0 0 0

$flux.y.up
[1] 160 160 160 160

$flux.y.down
[1] 0 0 0 0

> tran.2D(AFDW.x = 0, C = C, D.x = D, v.x = 100*v,
+ VF.x = por, dx = dx, dy = dy,
+ C.x.up = C.x.up, C.x.down = C.x.down,
+ C.y.up = C.y.up, C.y.down = C.y.down)
```

```
# Outputting:
$dC
     [,1] [,2] [,3] [,4] [,5] [,6]
[1,]  250  200  200  200  200  200
[2,]   50    0    0    0    0    0
[3,]   50    0    0    0    0    0
[4,]   50    0    0    0    0    0

$flux.x.up
[1] 320 320 320 320 320 320

$flux.x.down
[1] 0 0 0 0 0 0

$flux.y.up
[1] 160 160 160 160

$flux.y.down
[1] 0 0 0 0

> # Boundary conditions: fixed fluxes
> flux.x.up <- rep(200, times = y.N)
> flux.x.down <- rep(-200, times = y.N)
> flux.y.up <- rep(200, times = x.N)
> flux.y.down <- rep(-200, times = x.N)
> tran.2D(C = C, D.x = D, v.x = 0,
+ VF.x = por, dx = dx, dy = dy,
+ flux.x.up = flux.x.up, flux.x.down = flux.x.down,
+ flux.y.up = flux.y.up, flux.y.down = flux.y.down)

# Outputting:
$dC
      [,1] [,2] [,3] [,4] [,5]  [,6]
[1,] 187.5  125  125  125  125 187.5
[2,]  62.5    0    0    0    0  62.5
[3,]  62.5    0    0    0    0  62.5
[4,] 187.5  125  125  125  125 187.5

$flux.x.up
[1] 200 200 200 200 200 200

$flux.x.down
[1] -200 -200 -200 -200 -200 -200

$flux.y.up
[1] 200 200 200 200
```

```
$flux.y.down
[1] -200 -200 -200 -200

> # Boundary conditions: convective boundary layer on all sides
> a.bl <- 800 # transfer coefficient
> C.x.up <- rep(1, times = (y.N)) # fixed conc at boundary layer
> C.y.up <- rep(1, times = (x.N)) # fixed conc at boundary layer
> tran.2D(full.output = TRUE, C = C, D.x = D, v.x = 0,
+ VF.x = por, dx = dx, dy = dy,
+ C.x.up = C.x.up, a.bl.x.up = a.bl,
+ C.x.down = C.x.up, a.bl.x.down = a.bl,
+ C.y.up = C.y.up, a.bl.y.up = a.bl,
+ C.y.down = C.y.up, a.bl.y.down = a.bl)
```

 # Outputting:

```
$dC
          [,1]     [,2]     [,3]     [,4]     [,5]      [,6]
[1,] 184.52381 142.8571 142.8571 142.8571 142.8571 184.52381
[2,]  41.66667   0.0000   0.0000   0.0000   0.0000  41.66667
[3,]  41.66667   0.0000   0.0000   0.0000   0.0000  41.66667
[4,] 184.52381 142.8571 142.8571 142.8571 142.8571 184.52381

$C.x.up
[1] 0.7142857 0.7142857 0.7142857 0.7142857 0.7142857 0.7142857

$C.x.down
[1] 0.7142857 0.7142857 0.7142857 0.7142857 0.7142857 0.7142857

$C.y.up
[1] 0.8333333 0.8333333 0.8333333 0.8333333
$C.y.down
[1] 0.8333333 0.8333333 0.8333333 0.8333333

$x.flux
          [,1]      [,2]      [,3]      [,4]      [,5]      [,6]
[1,]  228.5714  228.5714  228.5714  228.5714  228.5714  228.5714
[2,]    0.0000    0.0000    0.0000    0.0000    0.0000    0.0000
[3,]    0.0000    0.0000    0.0000    0.0000    0.0000    0.0000
[4,]    0.0000    0.0000    0.0000    0.0000    0.0000    0.0000
[5,] -228.5714 -228.5714 -228.5714 -228.5714 -228.5714 -228.5714

         [,1] [,2] [,3] [,4] [,5] [,6]      [,7]
[1,] 133.3333    0    0    0    0    0 -133.3333
[2,] 133.3333    0    0    0    0    0 -133.3333
[3,] 133.3333    0    0    0    0    0 -133.3333
[4,] 133.3333    0    0    0    0    0 -133.3333
```

```
$flux.x.up
[1] 228.5714 228.5714 228.5714 228.5714 228.5714 228.5714

$flux.x.down
[1] -228.5714 -228.5714 -228.5714 -228.5714 -228.5714 -228.5714

$flux.y.up
[1] 133.3333 133.3333 133.3333 133.3333
$flux.y.down
[1] -133.3333 -133.3333 -133.3333 -133.3333

> # Runtime test with and without argument checking
> n.iterate <-500
> test1 <- function() {
+ for (i in 1:n.iterate )
+ ST <- tran.2D(full.check = TRUE, C = C, D.x = D,
+ v.x = 0, VF.x = por, dx = dx, dy = dy,
+ C.x.up = C.x.up, a.bl.x.up = a.bl, C.x.down = C.x.down)
+ }
> system.time(test1())

# Outputting:
 user  system  elapsed
 0.35   0.00     0.35
> test2 <- function() {
+ for (i in 1:n.iterate )
+ ST <- tran.2D(full.output = TRUE, C = C, D.x = D,
+ v.x = 0, VF.x = por, dx = dx, dy = dy,
+ C.x.up = C.x.up, a.bl.x.up = a.bl, C.x.down = C.x.down)
+ }
> system.time(test2())

# Outputting:
 user system elapsed
 0.25  0.00  0.24
> test3 <- function() {
+ for (i in 1:n.iterate )
+ ST <- tran.2D(full.output = TRUE, full.check = TRUE, C = C,
+ D.x = D, v.x = 0, VF.x = por, dx = dx, dy = dy,
+ C.x.up = C.x.up, a.bl.x.up = a.bl, C.x.down = C.x.down)
+ }
> system.time(test3())

# Outputting:
 user  system  elapsed
 0.36   0.00     0.36
```

```
> ## ==========================================
> ## A 2D model with diffusion in x- and y-direction and first-order    =
> ## consumption - inefficient implementation                           =
> ## ==========================================
> N <- 51 # number of grid cells
> XX <- 10 # total size
> dy <- dx <- XX/N # grid size
> Dy <- Dx <- 0.1 # diffusion coeff, X- and Y-direction
> r <- 0.005 # consumption rate
> ini <- 1 # initial value at x=0
> N2 <- ceiling(N/2)
> X <- seq (dx, by = dx, len = (N2-1))
> X <- c(-rev(X), 0, X)
> # The model equations
> Diff2D <- function (t, y, parms) {
+ CONC <- matrix(nrow = N, ncol = N, y)
+ dCONC <- tran.2D(CONC, D.x = Dx, D.y = Dy, dx = dx, dy = dy)$dC + r
+ * CONC
+ return (list(dCONC))
+ }
> # initial condition: 0 everywhere, except in central point
> y <- matrix(nrow = N, ncol = N, data = 0)
> y[N2, N2] <- ini # initial concentration in the central point...
> # solve for 10 time units
> times <- 0:10
> out <- ode.2D (y = y, func = Diff2D, t = times, parms = NULL,
+ dim = c(N,N), lrw = 160000)
> pm <- par (mfrow = c(2, 2))
> # Compare solution with analytical solution...
> for (i in seq(2, 11, by = 3)) {
+ tt <- times[i]
+ mat <- matrix(nrow = N, ncol = N,
+ data = subset(out, time == tt))
+ plot(X, mat[N2,], type = "l", main = paste("time=", times[i]),
+ ylab = "Conc", col = "red")
+ ana <- ini*dx^2/(4*pi*Dx*tt)*exp(r*tt-X^2/(4*Dx*tt))
+ points(X, ana, pch = "+")
+ }
> legend ("bottom", col = c("red","black"), lty = c(1, NA),
+ pch = c(NA, "+"), c("tran.2D", "exact"))
> par("mfrow" = pm )
> #=============================================
> # A 2D model with diffusion in x- and y-direction and first-order          =
> # consumption - more efficient implementation, specifying ALL 2D grids     =
> # =============================================
> N <- 51 # number of grid cells
```

```
> Dy <- Dx <- 0.1 # diffusion coeff, X- and Y-direction
> r <- 0.005 # consumption rate
> ini <- 1 # initial value at x=0
> x.grid <- setup.grid.1D(x.up = -5, x.down = 5, N = N)
> y.grid <- setup.grid.1D(x.up = -5, x.down = 5, N = N)
> grid2D <- setup.grid.2D(x.grid, y.grid)
> D.grid <- setup.prop.2D(value = Dx, y.value = Dy, grid = grid2D)
> v.grid <- setup.prop.2D(value = 0, grid = grid2D)
> A.grid <- setup.prop.2D(value = 1, grid = grid2D)
> AFDW.grid <- setup.prop.2D(value = 1, grid = grid2D)
> VF.grid <- setup.prop.2D(value = 1, grid = grid2D)
>
> # The model equations - using the grids
> Diff2Db <- function (t, y, parms) {
+ CONC <- matrix(nrow = N, ncol = N, data = y)
+ dCONC <- tran.2D(CONC, grid = grid2D, D.grid = D.grid,
+ A.grid = A.grid, VF.grid = VF.grid, AFDW.grid = AFDW.grid,
+ v.grid = v.grid)$dC + r * CONC
+ return (list(dCONC))
+ }
> # initial condition: 0 everywhere, except in central point
> y <- matrix(nrow = N, ncol = N, data = 0)
> y[N2,N2] <- ini # initial concentration in the central point...
> # solve for 8 time units
> times <- 0:8
> outb <- ode.2D (y = y, func = Diff2Db, t = times, parms = NULL,
+ dim = c(N, N), lrw = 160000)
> image(outb, ask = FALSE, mfrow = c(3, 3), main = paste("time", times))
>
> #==============================================
> # Same 2D model, but now with spatially variable diffusion coefficients  =
> #==============================================
> N <- 51 # number of grid cells
> r <- 0.005 # consumption rate
> ini <- 1 # initial value at x=0
> N2 <- ceiling(N/2)
> D.grid <- list()
> # Diffusion on x-interfaces
> D.grid$x.int <- matrix(nrow = N+1, ncol = N, data = runif(N*(N+1)))
> # Diffusion on y-interfaces
> D.grid$y.int <- matrix(nrow = N, ncol = N+1, data = runif(N*(N+1)))
> dx <- 10/N
> dy <- 10/N
> # The model equations
> Diff2Dc <- function (t, y, parms) {
+ CONC <- matrix(nrow = N, ncol = N, data = y)
```

```
+ dCONC <- tran.2D(CONC, dx = dx, dy = dy, D.grid = D.grid)$dC + r *
+ CONC
+ return (list(dCONC))
+ }
> # initial condition: 0 everywhere, except in central point
> y <- matrix(nrow = N, ncol = N, data = 0)
> y[N2, N2] <- ini # initial concentration in the central point...
> # solve for 8 time units
> times <- 0:8
> outc <- ode.2D (y = y, func = Diff2Dc, t = times, parms = NULL,
+ dim = c(N, N), lrw = 160000)
> outtimes <- c(1, 3, 5, 7)
> image(outc, ask = FALSE, mfrow = c(2, 2), main = paste("time", outtimes),
+ legend = TRUE, add.contour = TRUE, subset = time %in% outtimes)
>
> # In the R domain:
>
> install.packages("ReacTran")
Installing package into 'C:/Users/Bert/Documents/R/win-library/3.3'
(as 'lib' is unspecified)
--- Please select a CRAN mirror for use in this session ---
A CRAN mirror is selected.
also installing the dependencies 'rootSolve', 'deSolve'

trying URL 'https://mirrors.tuna.tsinghua.edu.cn/CRAN/bin/windows/contrib/3.3/
rootSolve_1.7.zip'
Content type 'application/zip' length 683839 bytes (667 KB)
downloaded 667 KB

trying URL 'https://mirrors.tuna.tsinghua.edu.cn/CRAN/bin/windows/contrib/3.3/
deSolve_1.20.zip'
Content type 'application/zip' length 2667028 bytes (2.5 MB)
downloaded 2.5 MB

trying URL 'https://mirrors.tuna.tsinghua.edu.cn/CRAN/bin/windows/contrib/3.3/
ReacTran_1.4.3.1.zip'
Content type 'application/zip' length 915996 bytes (894 KB)
downloaded 894 KB

package 'rootSolve' successfully unpacked and MD5 sums checked
package 'deSolve' successfully unpacked and MD5 sums checked
package 'ReacTran' successfully unpacked and MD5 sums checked

The downloaded binary packages are in
C:\Users\Bert\AppData\Local\Temp\RtmpmEkAb2\downloaded_packages
> require(ReacTran)
```

```
Loading required package: ReacTran
Loading required package: rootSolve
Loading required package: deSolve
Loading required package: shape
> library(ReacTran)
> ls("package:ReacTran") # Listing all the programs in this package:
 [1] "advection.1D"     "advection.volume.1D" "fiadeiro"
 [4] "g.cylinder"       "g.sphere"             "g.spheroid"
 [7] "p.exp"            "p.lin"                "p.sig"
[10] "polar2cart"       "setup.compaction.1D" "setup.grid.1D"
[13] "setup.grid.2D"    "setup.prop.1D"        "setup.prop.2D"
[16] "tran.1D"          "tran.2D"              "tran.3D"
[19] "tran.cylindrical" "tran.polar"           "tran.spherical"
[22] "tran.volume.1D"  "tran.volume.2D"       "tran.volume.3D"
> tran.2D # Outputting the program listing of "tran.2D"
> tran.2D
function (C, C.x.up = C[1, ], C.x.down = C[nrow(C), ], C.y.up = C[,
  1], C.y.down = C[, ncol(C)], flux.x.up = NULL, flux.x.down = NULL,
  flux.y.up = NULL, flux.y.down = NULL, a.bl.x.up = NULL, a.bl.x.down = NULL,
  a.bl.y.up = NULL, a.bl.y.down = NULL, D.grid = NULL, D.x = NULL,
  D.y = D.x, v.grid = NULL, v.x = 0, v.y = 0, AFDW.grid = NULL,
  AFDW.x = 1, AFDW.y = AFDW.x, VF.grid = NULL, VF.x = 1, VF.y = VF.x,
  A.grid = NULL, A.x = 1, A.y = 1, grid = NULL, dx = NULL,
  dy = NULL, full.check = FALSE, full.output = FALSE)
{
  if (is.null(grid))
    if (is.null(dx) | is.null(dy))
      stop("error: either grid or dx and dy should be specified ")
  Nx <- nrow(C)
  Ny <- ncol(C)
  if (is.null(grid)) {
    DX <- if (is.list(dx))
      dx$dx
    else rep(dx, length.out = Nx)
    DXaux <- if (is.list(dx))
      dx$dx.aux
    else 0.5 * (c(0, rep(dx, length.out = Nx)) + c(rep(dx,
      length.out = Nx), 0))
    DY <- if (is.list(dy))
      dy$dx
    else rep(dy, length.out = Ny)
    DYaux <- if (is.list(dy))
      dy$dx.aux
    else 0.5 * (c(0, rep(dy, length.out = Ny)) + c(rep(dy,
      length.out = Ny), 0))
    grid <- list(dx = matrix(nrow = Nx, ncol = Ny, data = DX),
```

```
      dx.aux = matrix(nrow = Nx + 1, ncol = Ny, data = DXaux),
      dy = matrix(nrow = Nx, ncol = Ny, data = DY, byrow = TRUE),
      dy.aux = matrix(nrow = Nx, ncol = Ny + 1, data = DYaux,
       byrow = TRUE))
}
  else if (class(grid) != "grid.2D") {
    if (length(grid$dx) != Nx * Ny)
      grid$dx <- matrix(nrow = Nx, ncol = Ny, grid$dx)
    if (length(grid$dx.aux) != (Nx + 1) * Ny)
      grid$dx.aux <- matrix(nrow = Nx + 1, ncol = Ny, grid$dx.aux)
    if (length(grid$dy) != Nx * Ny)
      grid$dy <- matrix(nrow = Nx, ncol = Ny, data = grid$dy,
       byrow = TRUE)
    if (length(grid$dy.aux) != Nx * (Ny + 1))
      grid$dy.aux <- matrix(nrow = Nx, ncol = Ny + 1, grid$dy.aux,
       byrow = TRUE)
  }
  gridFill <- function(G.x, G.y, Name) {
    if (is.null(G.x) | is.null(G.y))
      stop((paste("error: ", Name, "and (", Name, ".x and",
        Name, ".y) cannot be NULL at the same time",
        del = "")))
    G.grid <- list()
    if (is.matrix(G.x)) {
      if (sum(abs(dim(G.x) - c(Nx + 1, Ny))) != 0)
       stop(paste("error: ", Name, ".x matrix not of correct (Nx+1) Ny dimensions",
        del = ""))
      G.grid$x.int <- G.x
      G.grid$x.mid <- 0.5 * (G.x[1:Nx, ] + G.x[2:(Nx +

    1), ])
    }
    else if (class(G.x) == "prop.1D") {
      G.grid$x.int <- matrix(data = G.x$int, nrow = (Nx +
        1), ncol = Ny)
      G.grid$x.mid <- matrix(data = G.x$mid, nrow = Nx,
        ncol = Ny)
    }
    else if (length(G.x) == 1) {
      G.grid$x.int <- matrix(data = G.x, nrow = (Nx + 1),
        ncol = Ny)
      G.grid$x.mid <- matrix(data = G.x, nrow = Nx, ncol = Ny)
    }
    else if (length(G.x) != Nx + 1) {
      stop(paste("error: ", Name, ".x should be a vector of length 1 or Nx+1",
        del = ""))
```

```
}
else {
  G.grid$x.int <- matrix(data = G.x, nrow = (Nx + 1),
    ncol = Ny)
  G.grid$x.mid <- matrix(data = 0.5 * (G.x[1:Nx] +
    G.x[2:(Nx + 1)]), nrow = Nx, ncol = Ny)
}
if (is.matrix(G.y)) {
   if (sum(abs(dim(G.y) - c(Nx, Ny + 1))) != 0)
    stop(paste("error: ", Name, ".y matrix not of correct Nx(Ny+1)dimensions",
     del = ""))
  G.grid$y.int <- G.y
  G.grid$y.mid <- 0.5 * (G.y[, 1:Ny] + G.y[, 2:(Ny +
    1)])
}
else if (class(G.y) == "prop.1D") {
  G.grid$y.int <- matrix(data = G.y$int, nrow = Nx,
    ncol = (Ny + 1))
  G.grid$y.mid <- matrix(data = G.y$mid, nrow = Nx,
    ncol = Ny)
}
else if (length(G.y) == 1) {
  G.grid$y.int <- matrix(data = G.y, nrow = Nx, ncol = (Ny +
    1))
  G.grid$y.mid <- matrix(data = G.y, nrow = Nx, ncol = Ny)
}
else if (length(G.y) != Ny + 1) {
  stop(paste("error: ", Name, ".y should be a vector of length 1 or Ny+1",
    del = ""))
}
else {
  G.grid$y.int <- matrix(data = G.y, nrow = Nx, ncol = (Ny +
   1))
  G.grid$y.mid <- matrix(data = 0.5 * (G.y[1:Nx] +
    G.y[2:(Nx + 1)]), nrow = Nx, ncol = Ny)
}
G.grid
}
if (is.null(VF.grid))
  VF.grid <- gridFill(VF.x, VF.y, "VF")
if (is.null(A.grid))
  A.grid <- gridFill(A.x, A.y, "A")
gridInt <- function(G.x, G.y, Name) {
  if (is.null(G.x) | is.null(G.y))
    stop((paste("error: ", Name, "and (", Name, ".x and",
      Name, ".y) cannot be NULL at the same time",
```

```
    del = "")))
  G.grid <- list()
  if (is.matrix(G.x)) {
    if (sum(abs(dim(G.x) - c(Nx + 1, Ny))) != 0)
      stop(paste("error: ", Name, ".x matrix not of correct (Nx+1) Ny dimensions",
       del = ""))
    G.grid$x.int <- G.x
  }
  else if (class(G.x) == "prop.1D") {
    G.grid$x.int <- matrix(data = G.x$int, nrow = (Nx +
      1), ncol = Ny)
   }
  else if (length(G.x) == 1) {
    G.grid$x.int <- matrix(data = G.x, nrow = (Nx + 1),
      ncol = Ny)
  }
  else if (length(G.x) != Nx + 1) {
    stop(paste("error: ", Name, ".x should be a vector of length 1 or Nx+1",
      del = ""))
  }
  else {
    G.grid$x.int <- matrix(data = G.x, nrow = (Nx + 1),
      ncol = Ny)
  }
  if (is.matrix(G.y)) {
    if (sum(abs(dim(G.y) - c(Nx, Ny + 1))) != 0)
      stop(paste("error: ", Name, ".y matrix not of correct Nx(Ny+1)dimensions",
       del = ""))
    G.grid$y.int <- G.y
  }
  else if (class(G.y) == "prop.1D") {
    G.grid$y.int <- t(matrix(data = G.y$int, ncol = Nx,
      nrow = (Ny + 1)))
  }
  else if (length(G.y) == 1) {
    G.grid$y.int <- matrix(data = G.y, nrow = Nx, ncol = (Ny +
  1))
  }
  else if (length(G.y) != Ny + 1) {
    stop(paste("error: ", Name, ".y should be a vector of length 1 or Ny+1",
      del = ""))
  }
  else {
    G.grid$y.int <- matrix(data = G.y, nrow = Nx, ncol = (Ny +
      1))
  }
```

```
    G.grid
  }
  if (is.null(AFDW.grid))
    AFDW.grid <- gridInt(AFDW.x, AFDW.y, "AFDW")
  if (is.null(D.grid))
    D.grid <- gridInt(D.x, D.y, "D")
  if (is.null(v.grid))
    v.grid <- gridInt(v.x, v.y, "v")
  if (full.check) {
    if (!is.null(C.x.up)) {
      if (!((length(C.x.up) == 1) || (length(C.x.up) ==
        (Ny))))
        stop("error: C.x.up should be a vector of length 1 or ncol(C)")
    }
    if (!is.null(C.x.down)) {
      if (!((length(C.x.down) == 1) || (length(C.x.down) ==
        (Ny))))
        stop("error: C.x.down should be a vector of length 1 or ncol(C)")
    }
    if (!is.null(C.y.up)) {
      if (!((length(C.y.up) == 1) || (length(C.y.up) ==
        (Nx))))
        stop("error: C.y.up should be a vector of length 1 or nrow(C)")
    }
    if (!is.null(C.y.down)) {
      if (!((length(C.y.down) == 1) || (length(C.y.down) ==
        (Nx))))
        stop("error: C.y.down should be a vector of length 1 or nrow(C)")
    }
    if (!is.null(flux.x.up)) {
      if (!((length(flux.x.up) == 1) || (length(flux.x.up) ==
        (Ny))))
        stop("error: flux.x.up should be a vector of length 1 or ncol(C)")
    }
    if (!is.null(flux.x.down)) {
      if (!((length(flux.x.down) == 1) || (length(flux.x.down) ==
        (Ny))))
        stop("error: flux.x.down should be a vector of length 1 or ncol(C)")
    }
    if (!is.null(flux.y.up)) {
      if (!((length(flux.y.up) == 1) || (length(flux.y.up) ==
        (Nx))))
      stop("error: flux.y.up should be a vector of length 1 or nrow(C)")
    }
    if (!is.null(flux.y.down)) {
      if (!((length(flux.y.down) == 1) || (length(flux.y.down) ==
```

```
      (Nx))))
      stop("error: flux.y.down should be a vector of length 1 or nrow(C)")
   }
   if (is.null(dx) && is.null(dy) && is.null(grid))
     stop("error: dx, dy, and grid cannot be NULL at the same time")
   gn <- names(grid)
   if (!"dx" %in% gn)
     stop("error: grid should be a list that contains 'dx' ")
   if (!"dx.aux" %in% gn)
     stop("error: grid should be a list that contains 'dx.aux' ")
   if (!"dy" %in% gn)
     stop("error: grid should be a list that contains 'dy' ")
   if (!"dy.aux" %in% gn)
     stop("error: grid should be a list that contains 'dy.aux' ")
   if (is.null(grid$dx) || is.null(grid$dx.aux))
    stop("error: the grid should be a list with (numeric) values for 'dx' and 'dx.aux' ")
   if (is.null(grid$dy) || is.null(grid$dy.aux))
    stop("error: the grid should be a list with (numeric) values for 'dy' and 'dy.aux' ")
   if (any(grid$dx <= 0) || any(grid$dx.aux <= 0))
     stop("error: the grid distances dx and dx.aux should always be positive")
   if (any(grid$dy <= 0) || any(grid$dy.aux <= 0))
     stop("error: the grid distances dy and dy.aux should always be positive")
   gn <- names(AFDW.grid)
   if (!"x.int" %in% gn)
     stop("error: AFDW.grid should be a list that contains 'x.int', the AFDW values
at the interfaces of the grid cells in x-direction")
   if (!"y.int" %in% gn)
     stop("error: AFDW.grid should be a list that contains 'y.int', the AFDW values
at the interfaces of the grid cells in y-direction")
   if (is.null(AFDW.grid$x.int))
     stop("error: AFDW.grid$x.int should be a list with (numeric) values")
   if (is.null(AFDW.grid$y.int))
     stop("error: AFDW.grid$y.int should be a list with (numeric) values")
   if (any(AFDW.grid$x.int < 0) || any(AFDW.grid$x.int >
     1))
     stop("error: the AFDW should range between 0 and 1")
   if (any(AFDW.grid$y.int < 0) || any(AFDW.grid$y.int >
     1))
     stop("error: the AFDW should range between 0 and 1")
   gn <- names(D.grid)
   if (!"x.int" %in% gn)
     stop("error: D.grid should be a list that contains 'x.int', the D values at the
interfaces of the grid cells in x-direction")
   if (!"y.int" %in% gn)
     stop("error: D.grid should be a list that contains 'y.int', the D values at the
interfaces of the grid cells in y-direction")
```

```
  if (is.null(D.grid$x.int))
    stop("error: D.grid$x.int should be a list with (numeric) values")
  if (is.null(D.grid$y.int))
    stop("error: D.grid$y.int should be a list with (numeric) values")
  if (any(D.grid$x.int < 0) || any(D.grid$y.int < 0))
    stop("error: the diffusion coefficient should always be positive")
  gn <- names(v.grid)
  if (!"x.int" %in% gn)
    stop("error: v.grid should be a list that contains 'x.int', the velocity values
at the interfaces of the grid cells in x-direction")
  if (!"y.int" %in% gn)
    stop("error: v.grid should be a list that contains 'y.int', the velocity values
at the interfaces of the grid cells in y-direction")
  if (is.null(v.grid$x.int))
    stop("error: the advective velocity v.grid$x.int should be a list with (numeric)
values")
  if (is.null(v.grid$y.int))
    stop("error: the advective velocity v.grid$y.int should be a list with (numeric)
values")
  gn <- names(VF.grid)
  if (!"x.int" %in% gn)
    stop("error: VF.grid should be a list that contains 'x.int'")
  if (!"y.int" %in% gn)
    stop("error: VF.grid should be a list that contains 'y.int'")
  if (!"x.mid" %in% gn)
    stop("error: VF.grid should be a list that contains 'x.mid'")
  if (!"y.mid" %in% gn)
    stop("error: VF.grid should be a list that contains 'y.mid'")
  if (is.null(VF.grid$x.int) || is.null(VF.grid$y.int) ||
    is.null(VF.grid$x.mid) || is.null(VF.grid$y.mid))
    stop("error: VF should contain (numeric) values")
  if (any(VF.grid$x.int < 0) || any(VF.grid$y.int < 0) ||
    any(VF.grid$x.mid < 0) || any(VF.grid$y.mid < 0))
    stop("error: the VF values should always be positive")
  gn <- names(A.grid)
  if (!"x.int" %in% gn)
    stop("error: A.grid should be a list that contains 'x.int'")
  if (!"y.int" %in% gn)
    stop("error: A.grid should be a list that contains 'y.int'")
  if (!"x.mid" %in% gn)
    stop("error: A.grid should be a list that contains 'x.mid'")
  if (!"y.mid" %in% gn)
    stop("error: A.grid should be a list that contains 'y.mid'")
  if (is.null(A.grid$x.int) || is.null(A.grid$y.int) ||
    is.null(A.grid$x.mid) || is.null(A.grid$y.mid))
    stop("error: the VF should contain (numeric) values")
```

```
  if (any(A.grid$x.int < 0) || any(A.grid$y.int < 0) ||
    any(A.grid$x.mid < 0) || any(A.grid$y.mid < 0))
    stop("error: the A values should always be positive")
}
if (!is.null(flux.x.up[1])) {
    nom <- flux.x.up + VF.grid$x.int[1, ] * (D.grid$x.int[1,
    ]/grid$dx.aux[1, ] + (1 - AFDW.grid$x.int[1, ]) *
    v.grid$x.int[1, ]) * C[1, ]
  denom <- VF.grid$x.int[1, ] * (D.grid$x.int[1, ]/grid$dx.aux[1,
    ] + AFDW.grid$x.int[1, ] * v.grid$x.int[1, ])
  C.x.up <- nom/denom
}
if (!is.null(flux.x.down[1])) {
  nom <- flux.x.down - VF.grid$x.int[(Nx + 1), ] * (D.grid$x.int[(Nx +
    1), ]/grid$dx.aux[Nx + 1, ] + AFDW.grid$x.int[(Nx +
    1), ] * v.grid$x.int[(Nx + 1), ]) * C[Nx, ]
  denom <- -VF.grid$x.int[(Nx + 1), ] * (D.grid$x.int[(Nx +
    1), ]/grid$dx.aux[Nx + 1, ] + (1 - AFDW.grid$x.int[(Nx +
    1), ]) * v.grid$x.int[(Nx + 1), ])
  C.x.down <- nom/denom
}
if (!is.null(flux.y.up[1])) {
  nom <- flux.y.up + VF.grid$y.int[, 1] * (D.grid$y.int[,
    1]/grid$dy.aux[, 1] + (1 - AFDW.grid$y.int[, 1]) *
    v.grid$y.int[, 1]) * C[, 1]
  denom <- VF.grid$y.int[, 1] * (D.grid$y.int[, 1]/grid$dy.aux[,
    1] + AFDW.grid$y.int[, 1] * v.grid$y.int[, 1])
  C.y.up <- nom/denom
}
if (!is.null(flux.y.down[1])) {
  nom <- flux.y.down - VF.grid$y.int[, (Ny + 1)] * (D.grid$y.int[,
    (Ny + 1)]/grid$dy.aux[, (Ny + 1)] + AFDW.grid$y.int[,
    (Ny + 1)] * v.grid$y.int[, (Ny + 1)]) * C[, Ny]
  denom <- -VF.grid$y.int[, (Ny + 1)] * (D.grid$y.int[,
    (Ny + 1)]/grid$dy.aux[, Ny + 1] + (1 - AFDW.grid$y.int[,
    (Ny + 1)]) * v.grid$y.int[, (Ny + 1)])
  C.y.down <- nom/denom
}
if (!is.null(a.bl.x.up) & !is.null(C.x.up[1])) {
  nom <- a.bl.x.up * C.x.up + VF.grid$x.int[1, ] * (D.grid$x.int[1,
    ]/grid$dx.aux[1, ] + (1 - AFDW.grid$x.int[1, ]) *
    v.grid$x.int[1, ]) * C[1, ]
  denom <- a.bl.x.up + VF.grid$x.int[1, ] * (D.grid$x.int[1,
    ]/grid$dx.aux[1, ] + AFDW.grid$x.int[1, ] * v.grid$x.int[1,
    ])
  C.x.up <- nom/denom
```

```
}
if (!is.null(a.bl.x.down) & !is.null(C.x.down[1])) {
  nom <- a.bl.x.down * C.x.down + VF.grid$x.int[(Nx + 1),
    ] * (D.grid$x.int[(Nx + 1), ]/grid$dx.aux[(Nx + 1),
    ] + (1 - AFDW.grid$x.int[(Nx + 1), ]) * v.grid$x.int[(Nx +
    1), ]) * C[Nx, ]
  denom <- a.bl.x.down + VF.grid$x.int[(Nx + 1), ] * (D.grid$x.int[(Nx +
    1), ]/grid$dx.aux[(Nx + 1), ] + AFDW.grid$x.int[(Nx +
    1), ] * v.grid$x.int[(Nx + 1), ])
  C.x.down <- nom/denom
}
if (!is.null(a.bl.y.up) & !is.null(C.y.up[1])) {
  nom <- a.bl.y.up * C.y.up + VF.grid$y.int[, 1] * (D.grid$y.int[,
    1]/grid$dy.aux[, 1] + (1 - AFDW.grid$y.int[, 1]) *
    v.grid$y.int[, 1]) * C[, 1]
  denom <- a.bl.y.up + VF.grid$y.int[, 1] * (D.grid$y.int[,
    1]/grid$dy.aux[, 1] + AFDW.grid$y.int[, 1] * v.grid$y.int[,
    1])
  C.y.up <- nom/denom
}
if (!is.null(a.bl.y.down) & !is.null(C.y.down[1])) {
  nom <- a.bl.y.down * C.y.down + VF.grid$y.int[, (Ny +
    1)] * (D.grid$y.int[, (Ny + 1)]/grid$dy.aux[, (Ny +
    1)] + (1 - AFDW.grid$y.int[, (Ny + 1)]) * v.grid$y.int[,
    (Ny + 1)]) * C[, Ny]
  denom <- a.bl.y.down + VF.grid$y.int[, (Ny + 1)] * (D.grid$y.int[,
    (Ny + 1)]/grid$dy.aux[, (Ny + 1)] + AFDW.grid$y.int[,
    (Ny + 1)] * v.grid$y.int[, (Ny + 1)])
  C.y.down <- nom/denom
}
x.Dif.flux <- as.matrix(-VF.grid$x.int * D.grid$x.int * diff(rbind(C.x.up,
  C, C.x.down, deparse.level = 0))/grid$dx.aux)
y.Dif.flux <- as.matrix(-VF.grid$y.int * D.grid$y.int * t(diff(t(cbind(C.y.up,
  C, C.y.down, deparse.level = 0))))/grid$dy.aux)
  x.Adv.flux <- 0
 if (any(v.grid$x.int > 0)) {
  vv <- v.grid$x.int
  vv[vv < 0] <- 0
  x.Adv.flux <- x.Adv.flux + as.matrix(VF.grid$x.int *
    vv * (AFDW.grid$x.int * rbind(C.x.up, C, deparse.level = 0) +
    (1 - AFDW.grid$x.int) * rbind(C, C.x.down, deparse.level = 0)))
}
if (any(v.grid$x.int < 0)) {
  vv <- v.grid$x.int
  vv[vv > 0] <- 0
  x.Adv.flux <- x.Adv.flux + as.matrix(VF.grid$x.int *
```

```
      vv * ((1 - AFDW.grid$x.int) * rbind(C.x.up, C, deparse.level = 0) +
      AFDW.grid$x.int * rbind(C, C.x.down, deparse.level = 0)))
  }
  y.Adv.flux <- 0
  if (any(v.grid$y.int > 0)) {
    vv <- v.grid$y.int
    vv[vv < 0] <- 0
    y.Adv.flux <- y.Adv.flux + as.matrix(VF.grid$y.int *
      vv * (AFDW.grid$y.int * cbind(C.y.up, C, deparse.level = 0) +
      (1 - AFDW.grid$y.int) * cbind(C, C.y.down, deparse.level = 0)))
  }
  if (any(v.grid$y.int < 0)) {
    vv <- v.grid$y.int
    vv[vv > 0] <- 0
      y.Adv.flux <- y.Adv.flux + as.matrix(VF.grid$y.int *
      vv * ((1 - AFDW.grid$y.int) * cbind(C.y.up, C, deparse.level = 0) +
      AFDW.grid$y.int * cbind(C, C.y.down, deparse.level = 0)))
  }
  x.flux <- x.Dif.flux + x.Adv.flux
  y.flux <- y.Dif.flux + y.Adv.flux
  if (!is.null(flux.x.up[1]))
    x.flux[1, ] <- flux.x.up
  if (!is.null(flux.x.down[1]))
    x.flux[nrow(x.flux), ] <- flux.x.down
  if (!is.null(flux.y.up[1]))
    y.flux[, 1] <- flux.y.up
  if (!is.null(flux.y.down[1]))
    y.flux[, ncol(y.flux)] <- flux.y.down
  dFdx <- -(diff(A.grid$x.int * x.flux)/A.grid$x.mid)/grid$dx/VF.grid$x.mid
  dFdy <- -t(diff(t(A.grid$y.int * y.flux))/t(A.grid$y.mid))/grid$dy/VF.grid$y.mid
  if (!full.output) {
    return(list(dC = dFdx + dFdy, flux.x.up = x.flux[1, ],
      flux.x.down = x.flux[nrow(x.flux), ], flux.y.up = y.flux[,
        1], flux.y.down = y.flux[, ncol(y.flux)]))
  }
  else {
    return(list(dC = dFdx + dFdy, C.x.up = C.x.up, C.x.down = C.x.down,
      C.y.up = C.y.up, C.y.down = C.y.down, x.flux = x.flux,
      y.flux = y.flux, flux.x.up = x.flux[1, ], flux.x.down = x.flux[nrow(x.flux),
        ], flux.y.up = y.flux[, 1], flux.y.down = y.flux[,
        ncol(y.flux)]))
  }
}
<environment: namespace:ReacTran>
((((
<environment: namespace:ReacTran>
```

```
>##
> # Running the program tran.2D

> ##================
> ## Testing the functions =
> ## ===============
> # Parameters
> F <- 100 # input flux [micromol cm-2 yr-1]
> por <- 0.8 # constant porosity
> D <- 400 # mixing coefficient [cm2 yr-1]
> v <- 1 # advective velocity [cm yr-1]
> # Grid definition
> x.N <- 4 # number of cells in x-direction
> y.N <- 6 # number of cells in y-direction
> x.L <- 8 # domain size x-direction [cm]
> y.L <- 24 # domain size y-direction [cm]
> dx <- x.L/x.N # cell size x-direction [cm]
> dy <- y.L/y.N # cell size y-direction [cm]
> # Initial conditions
> C <- matrix(nrow = x.N, ncol = y.N, data = 0, byrow = FALSE)
> # Boundary conditions: fixed concentration
> C.x.up <- rep(1, times = y.N)
> C.x.down <- rep(0, times = y.N)
> C.y.up <- rep(1, times = x.N)
> C.y.down <- rep(0, times = x.N)
> # Only diffusion
> tran.2D(C = C, D.x = D, D.y = D, v.x = 0, v.y = 0,
+ VF.x = por, VF.y = por, dx = dx, dy = dy,
+ C.x.up = C.x.up, C.x.down = C.x.down,
+ C.y.up = C.y.up, C.y.down = C.y.down, full.output = TRUE)
$dC
     [,1] [,2] [,3] [,4] [,5] [,6]
[1,]  250  200  200  200  200  200
[2,]   50    0    0    0    0    0
[3,]   50    0    0    0    0    0
[4,]   50    0    0    0    0    0

$C.x.up
[1] 1 1 1 1 1 1

$C.x.down
[1] 0 0 0 0 0 0

$C.y.up
[1] 1 1 1 1

$C.y.down
[1] 0 0 0 0
```

```
$x.flux
     [,1] [,2] [,3] [,4] [,5] [,6]
[1,]  320  320  320  320  320  320
[2,]    0    0    0    0    0    0
[3,]    0    0    0    0    0    0
[4,]    0    0    0    0    0    0
[5,]    0    0    0    0    0    0

$y.flux
     [,1] [,2] [,3] [,4] [,5] [,6] [,7]
[1,]  160    0    0    0    0    0    0
[2,]  160    0    0    0    0    0    0
[3,]  160    0    0    0    0    0    0
[4,]  160    0    0    0    0    0    0

$flux.x.up
[1] 320 320 320 320 320 320

$flux.x.down
[1] 0 0 0 0 0 0
$flux.y.up
[1] 160 160 160 160

$flux.y.down
[1] 0 0 0 0

> # Strong advection, backward (default), central and forward
> #finite difference schemes
> tran.2D(C = C, D.x = D, v.x = 100*v,
+ VF.x = por, dx = dx, dy = dy,
+ C.x.up = C.x.up, C.x.down = C.x.down,
+ C.y.up = C.y.up, C.y.down = C.y.down)
$dC
     [,1] [,2] [,3] [,4] [,5] [,6]
[1,]  300  250  250  250  250  250
[2,]   50    0    0    0    0    0
[3,]   50    0    0    0    0    0
[4,]   50    0    0    0    0    0

$flux.x.up
[1] 400 400 400 400 400 400

$flux.x.down
[1] 0 0 0 0 0 0

$flux.y.up
[1] 160 160 160 160
```

```
$flux.y.down
[1] 0 0 0 0

> tran.2D(AFDW.x = 0.5, C = C, D.x = D, v.x = 100*v,
+ VF.x = por, dx = dx, dy = dy,
+ C.x.up = C.x.up, C.x.down = C.x.down,
+ C.y.up = C.y.up, C.y.down = C.y.down)
$dC
     [,1] [,2] [,3] [,4] [,5] [,6]
[1,] 275  225  225  225  225  225
[2,]  50    0    0    0    0    0
[3,]  50    0    0    0    0    0
[4,]  50    0    0    0    0    0

$flux.x.up
[1] 360 360 360 360 360 360

$flux.x.down
[1] 0 0 0 0 0 0

$flux.y.up
[1] 160 160 160 160

$flux.y.down
[1] 0 0 0 0

> tran.2D(AFDW.x = 0, C = C, D.x = D, v.x = 100*v,
+ VF.x = por, dx = dx, dy = dy,
+ C.x.up = C.x.up, C.x.down = C.x.down,
+ C.y.up = C.y.up, C.y.down = C.y.down)
$dC
     [,1] [,2] [,3] [,4] [,5] [,6]
[1,]  250  200  200  200  200  200
[2,]   50    0    0    0    0    0
[3,]   50    0    0    0    0    0
[4,]   50    0    0    0    0    0

$flux.x.up
[1] 320 320 320 320 320 320

$flux.x.down
[1] 0 0 0 0 0 0

$flux.y.up
[1] 160 160 160 160
```

```
$flux.y.down
[1] 0 0 0 0

> # Boundary conditions: fixed fluxes
> flux.x.up <- rep(200, times = y.N)
> flux.x.down <- rep(-200, times = y.N)
> flux.y.up <- rep(200, times = x.N)
> flux.y.down <- rep(-200, times = x.N)
> tran.2D(C = C, D.x = D, v.x = 0,
+ VF.x = por, dx = dx, dy = dy,
+ flux.x.up = flux.x.up, flux.x.down = flux.x.down,
+ flux.y.up = flux.y.up, flux.y.down = flux.y.down)
$dC
      [,1] [,2] [,3] [,4] [,5]  [,6]
[1,] 187.5  125  125  125  125 187.5
[2,]  62.5    0    0    0    0  62.5
[3,]  62.5    0    0    0    0  62.5
[4,] 187.5  125  125  125  125 187.5

$flux.x.up
[1] 200 200 200 200 200 200

$flux.x.down
[1] -200 -200 -200 -200 -200 -200

$flux.y.up
[1] 200 200 200 200
```

Similarly

```
$flux.y.down
[1] -200 -200 -200 -200

> # Boundary conditions: convective boundary layer on all sides
> a.bl <- 800 # transfer coefficient
> C.x.up <- rep(1, times = (y.N)) # fixed conc at boundary layer
> C.y.up <- rep(1, times = (x.N)) # fixed conc at boundary layer
> tran.2D(full.output = TRUE, C = C, D.x = D, v.x = 0,
+ VF.x = por, dx = dx, dy = dy,
+ C.x.up = C.x.up, a.bl.x.up = a.bl,
+ C.x.down = C.x.up, a.bl.x.down = a.bl,
+ C.y.up = C.y.up, a.bl.y.up = a.bl,
+ C.y.down = C.y.up, a.bl.y.down = a.bl)
$dC
          [,1]     [,2]     [,3]     [,4]     [,5]      [,6]
[1,] 184.52381 142.8571 142.8571 142.8571 142.8571 184.52381
[2,]  41.66667   0.0000   0.0000   0.0000   0.0000  41.66667
```

```
[3,]  41.66667   0.0000   0.0000   0.0000   0.0000  41.66667
[4,] 184.52381 142.8571 142.8571 142.8571 142.8571 184.52381

$C.x.up
[1] 0.7142857 0.7142857 0.7142857 0.7142857 0.7142857 0.7142857

$C.x.down
[1] 0.7142857 0.7142857 0.7142857 0.7142857 0.7142857 0.7142857

$C.y.up
[1] 0.8333333 0.8333333 0.8333333 0.8333333
$C.y.down
[1] 0.8333333 0.8333333 0.8333333 0.8333333

$x.flux
          [,1]      [,2]      [,3]      [,4]      [,5]      [,6]
[1,]  228.5714  228.5714  228.5714  228.5714  228.5714  228.5714
[2,]    0.0000    0.0000    0.0000    0.0000    0.0000    0.0000
[3,]    0.0000    0.0000    0.0000    0.0000    0.0000    0.0000
[4,]    0.0000    0.0000    0.0000    0.0000    0.0000    0.0000
[5,] -228.5714 -228.5714 -228.5714 -228.5714 -228.5714 -228.5714

$y.flux
         [,1] [,2] [,3] [,4] [,5] [,6]      [,7]
[1,] 133.3333    0    0    0    0    0 -133.3333
[2,] 133.3333    0    0    0    0    0 -133.3333
[3,] 133.3333    0    0    0    0    0 -133.3333
[4,] 133.3333    0    0    0    0    0 -133.3333

$flux.x.up
[1] 228.5714 228.5714 228.5714 228.5714 228.5714 228.5714

$flux.x.down
[1] -228.5714 -228.5714 -228.5714 -228.5714 -228.5714 -228.5714

$flux.y.up
[1] 133.3333 133.3333 133.3333 133.3333

$flux.y.down
[1] -133.3333 -133.3333 -133.3333 -133.3333

> # Runtime test with and without argument checking
> n.iterate <-500
> test1 <- function() {
+ for (i in 1:n.iterate )
+ ST <- tran.2D(full.check = TRUE, C = C, D.x = D,
```

```
+ v.x = 0, VF.x = por, dx = dx, dy = dy,
+ C.x.up = C.x.up, a.bl.x.up = a.bl, C.x.down = C.x.down)
+ }
> system.time(test1())
 user system elapsed
 0.35 0.00 0.35
> test2 <- function() {
+ for (i in 1:n.iterate )
+ ST <- tran.2D(full.output = TRUE, C = C, D.x = D,
+ v.x = 0, VF.x = por, dx = dx, dy = dy,
+ C.x.up = C.x.up, a.bl.x.up = a.bl, C.x.down = C.x.down)
+ }

> system.time(test2())
 user system elapsed
 0.25 0.00 0.24

> test3 <- function() {
+ for (i in 1:n.iterate )
+ ST <- tran.2D(full.output = TRUE, full.check = TRUE, C = C,
+ D.x = D, v.x = 0, VF.x = por, dx = dx, dy = dy,
+ C.x.up = C.x.up, a.bl.x.up = a.bl, C.x.down = C.x.down)
+ }
> system.time(test3())
 user system elapsed
 0.36 0.00 0.36
> ## ========================================
> ## A 2D model with diffusion in x- and y-direction and first-order    =
> ## consumption - inefficient implementation                          =
> ## ========================================
> N <- 51 # number of grid cells
> XX <- 10 # total size
> dy <- dx <- XX/N # grid size
> Dy <- Dx <- 0.1 # diffusion coeff, X- and Y-direction
> r <- 0.005 # consumption rate
> ini <- 1 # initial value at x=0
> N2 <- ceiling(N/2)
> X <- seq (dx, by = dx, len = (N2-1))
> X <- c(-rev(X), 0, X)
> # The model equations
> Diff2D <- function (t, y, parms) {
+ CONC <- matrix(nrow = N, ncol = N, y)
+ dCONC <- tran.2D(CONC, D.x = Dx, D.y = Dy, dx = dx, dy = dy)$dC + r
* CONC

141
+ return (list(dCONC))
```

```
+ }
> # initial condition: 0 everywhere, except in central point
> y <- matrix(nrow = N, ncol = N, data = 0)
> y[N2, N2] <- ini # initial concentration in the central point...
> # solve for 10 time units
> times <- 0:10
> out <- ode.2D (y = y, func = Diff2D, t = times, parms = NULL,
+ dim = c(N,N), lrw = 160000)
> pm <- par (mfrow = c(2, 2))
> # Compare solution with analytical solution...
> for (i in seq(2, 11, by = 3)) {
+ tt <- times[i]
+ mat <- matrix(nrow = N, ncol = N,
+ data = subset(out, time == tt))
+ #
+ # Fig 1
+ #
+ plot(X, mat[N2,], type = "l", main = paste("time=", times[i]),
+ ylab = "Conc", col = "red")
+ ana <- ini*dx^2/(4*pi*Dx*tt)*exp(r*tt-X^2/(4*Dx*tt))
+ points(X, ana, pch = "+")
+ }
> #
> # Fig 2
> #
> legend ("bottom", col = c("red","black"), lty = c(1, NA),
+ pch = c(NA, "+"), c("tran.2D", "exact"))
> #
> # Fig 3
> #
> par("mfrow" = pm )
> #
> # Fig 4
> #
> ## =============================================================
> ## A 2D model with diffusion in x- and y-direction and first-order          =
> ## consumption - more efficient implementation, specifying ALL 2D grids     =
> ## =============================================================
> N <- 51 # number of grid cells
> Dy <- Dx <- 0.1 # diffusion coeff, X- and Y-direction
> r <- 0.005 # consumption rate
> ini <- 1 # initial value at x=0
> x.grid <- setup.grid.1D(x.up = -5, x.down = 5, N = N)
> y.grid <- setup.grid.1D(x.up = -5, x.down = 5, N = N)
> grid2D <- setup.grid.2D(x.grid, y.grid)
> D.grid <- setup.prop.2D(value = Dx, y.value = Dy, grid = grid2D)
```

```
> v.grid <- setup.prop.2D(value = 0, grid = grid2D)
> A.grid <- setup.prop.2D(value = 1, grid = grid2D)
> AFDW.grid <- setup.prop.2D(value = 1, grid = grid2D)
> VF.grid <- setup.prop.2D(value = 1, grid = grid2D)
>
> # The model equations - using the grids
> Diff2Db <- function (t, y, parms) {
+ CONC <- matrix(nrow = N, ncol = N, data = y)
+ dCONC <- tran.2D(CONC, grid = grid2D, D.grid = D.grid,
+ A.grid = A.grid, VF.grid = VF.grid, AFDW.grid = AFDW.grid,
+ v.grid = v.grid)$dC + r * CONC
+ return (list(dCONC))
+ }
> # initial condition: 0 everywhere, except in central point
> y <- matrix(nrow = N, ncol = N, data = 0)
> y[N2,N2] <- ini # initial concentration in the central point...
> # solve for 8 time units
> times <- 0:8
> outb <- ode.2D (y = y, func = Diff2Db, t = times, parms = NULL,
+ dim = c(N, N), lrw = 160000)
> image(outb, ask = FALSE, mfrow = c(3, 3), main = paste("time", times))
> #
> # Fig 5
> #
>
> ## ===========================================================
> ## Same 2D model, but now with spatially variable diffusion coefficients     =
> ## ===========================================================
> N <- 51 # number of grid cells
> r <- 0.005 # consumption rate
> ini <- 1 # initial value at x=0
> N2 <- ceiling(N/2)
> D.grid <- list()
> # Diffusion on x-interfaces
> D.grid$x.int <- matrix(nrow = N+1, ncol = N, data = runif(N*(N+1)))
> # Diffusion on y-interfaces
> D.grid$y.int <- matrix(nrow = N, ncol = N+1, data = runif(N*(N+1)))
> dx <- 10/N
> dy <- 10/N
> # The model equations
> Diff2Dc <- function (t, y, parms) {
+ CONC <- matrix(nrow = N, ncol = N, data = y)
+ dCONC <- tran.2D(CONC, dx = dx, dy = dy, D.grid = D.grid)$dC + r * CONC
+ return (list(dCONC))
+ }
> # initial condition: 0 everywhere, except in central point
```

```
> y <- matrix(nrow = N, ncol = N, data = 0)
> y[N2, N2] <- ini # initial concentration in the central point...
> # solve for 8 time units
> times <- 0:8
> outc <- ode.2D (y = y, func = Diff2Dc, t = times, parms = NULL,
+ dim = c(N, N), lrw = 160000)
> outtimes <- c(1, 3, 5, 7)
> image(outc, ask = FALSE, mfrow = c(2, 2), main = paste("time", outtimes),
+ legend = TRUE, add.contour = TRUE, subset = time %in% outtimes)
> #
> # Fig 6
> #
>
> # Outputting Figure A2-1: tran.2D 1
> #
> #
> legend ("bottom", col = c("red","black"), lty = c(1, NA),
+ pch = c(NA, "+"), c("tran.2D", "exact"))
>
> # Outputting Figure A2-2: tran.2D 2
> #


> ##=====================================================
> ## A 2D model with diffusion in x- and y-direction and first-order          =
> ## consumption - more efficient implementation, specifying ALL 2D grids     =
> ## =====================================================
> N <- 51 # number of grid cells
> Dy <- Dx <- 0.1 # diffusion coeff, X- and Y-direction
> r <- 0.005 # consumption rate
```

Figure A2-1 tran.2D 1 Preparation of space for subsequent graphical outputs.

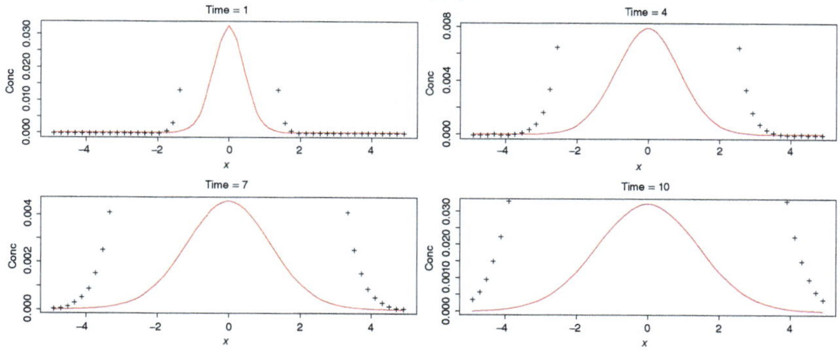

Figure A2-2 tran.2D 2 ini # initial concentration in the central point.

```
> ini <- 1 # initial value at x=0
> x.grid <- setup.grid.1D(x.up = -5, x.down = 5, N = N)
> y.grid <- setup.grid.1D(x.up = -5, x.down = 5, N = N)
> grid2D <- setup.grid.2D(x.grid, y.grid)
> D.grid <- setup.prop.2D(value = Dx, y.value = Dy, grid = grid2D)
> v.grid <- setup.prop.2D(value = 0, grid = grid2D)
> A.grid <- setup.prop.2D(value = 1, grid = grid2D)
> AFDW.grid <- setup.prop.2D(value = 1, grid = grid2D)
> VF.grid <- setup.prop.2D(value = 1, grid = grid2D)
> # The model equations - using the grids
> Diff2Db <- function (t, y, parms) {
+ CONC <- matrix(nrow = N, ncol = N, data = y)
+ dCONC <- tran.2D(CONC, grid = grid2D, D.grid = D.grid,
+ A.grid = A.grid, VF.grid = VF.grid, AFDW.grid = AFDW.grid,
+ v.grid = v.grid)$dC + r * CONC
+ return (list(dCONC))
+ }
> # initial condition: 0 everywhere, except in central point
> y <- matrix(nrow = N, ncol = N, data = 0)
> y[N2,N2] <- ini # initial concentration in the central point...
> # solve for 8 time units
> times <- 0:8
> outb <- ode.2D (y = y, func = Diff2Db, t = times, parms = NULL,
+ dim = c(N, N), lrw = 160000)
> image(outb, ask = FALSE, mfrow = c(3, 3), main = paste("time", times))
> #
> # Outputting: Figure A2-3: tran.2D 3
> #
> #
> ## ===========================================
> ## Same 2D model, but now with spatially variable diffusion coefficients    =
> ## ===========================================
> N <- 51 # number of grid cells
```

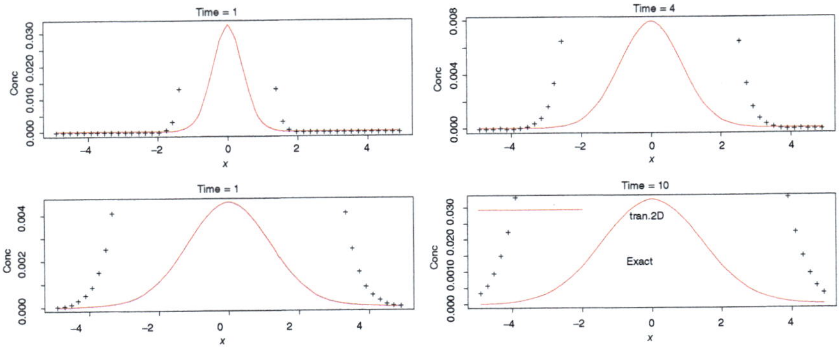

Figure A2-3 tran.2D 3 ini # initial concentration in the central point.

```
> r <- 0.005 # consumption rate
> ini <- 1 # initial value at x=0
> N2 <- ceiling(N/2)
> D.grid <- list()
> # Diffusion on x-interfaces
> D.grid$x.int <- matrix(nrow = N+1, ncol = N, data = runif(N*(N+1)))

> D.grid$y.int <- matrix(nrow = N, ncol = N+1, data = runif(N*(N+1)))
> dx <- 10/N
> dy <- 10/N
> # The model equations
> Diff2Dc <- function (t, y, parms) {
+ CONC <- matrix(nrow = N, ncol = N, data = y)
+ dCONC <- tran.2D(CONC, dx = dx, dy = dy, D.grid = D.grid)$dC + r * CONC
+ return (list(dCONC))
+ }
> # initial condition: 0 everywhere, except in central point
> y <- matrix(nrow = N, ncol = N, data = 0)
> y[N2, N2] <- ini # initial concentration in the central point...
> # solve for 8 time units
> times <- 0:8
> outc <- ode.2D (y = y, func = Diff2Dc, t = times, parms = NULL,
+ dim = c(N, N), lrw = 160000)
> outtimes <- c(1, 3, 5, 7)
> image(outc, ask = FALSE, mfrow = c(2, 2), main = paste("time", outtimes),
> #
> # Outputting Figure A2-4: tran.2D 4
>
+ legend = TRUE, add.contour = TRUE, subset = time %in% outtimes)
> #
> # Outputting Figure A2-5: tran.2D 5
> #
```

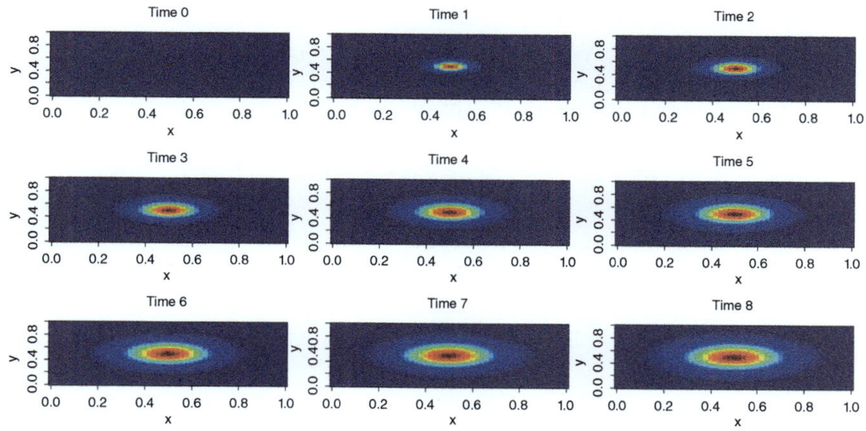

Figure A2-4 tran.2D 4 ini # initial concentration in the central point.

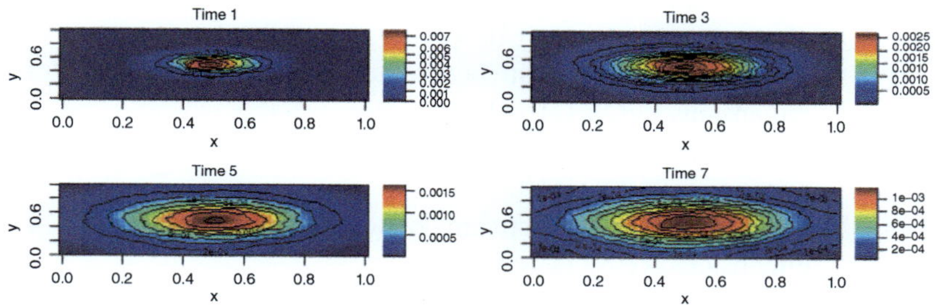

Figure A2-5 tran.2D 5 ini # initial concentration in the central point.

Example 5.4 tran.3D ***General Three-Dimensional Advective–Diffusive Transport***

Description

This program estimates the transport term (i.e. the rate of change of a concentration owing to diffusion and advection) in a three-dimensional rectangular model domain.

Usage

```
tran.3D (C, C.x.up = C[1,,], C.x.down = C[dim(C)[1],,],
    C.y.up = C[ ,1, ], C.y.down=C[ ,dim(C)[2], ],
    C.z.up = C[ , ,1], C.z.down=C[ , ,dim(C)[3]],
    flux.x.up = NULL, flux.x.down = NULL,
    flux.y.up = NULL, flux.y.down = NULL,
    flux.z.up = NULL, flux.z.down = NULL,
    a.bl.x.up = NULL, a.bl.x.down = NULL,
    a.bl.y.up = NULL, a.bl.y.down = NULL,
```

```
a.bl.z.up = NULL, a.bl.z.down = NULL,
D.grid = NULL, D.x = NULL, D.y = D.x, D.z = D.x,
v.grid = NULL, v.x = 0, v.y = 0, v.z = 0,
AFDW.grid = NULL, AFDW.x = 1, AFDW.y = AFDW.x,
AFDW.z = AFDW.x,
VF.grid = NULL, VF.x = 1, VF.y = VF.x, VF.z = VF.x,
A.grid = NULL, A.x = 1, A.y = 1, A.z = 1,
grid = NULL, dx = NULL, dy = NULL, dz = NULL,
full.check = FALSE, full.output = FALSE)
```

Arguments

C	Concentration, expressed per unit volume, defined at the center of\each grid cell; Nx*Ny*Nz array [M/L^3]
C.x.up	Concentration at upstream boundary in *x*-direction; matrix of dimensions Ny*Nz [M/L^3]
C.x.down	Concentration at downstream boundary in *x*-direction; matrix of dimensions Ny*Nz [M/L^3]
C.y.up	Concentration at upstream boundary in *y*-direction; matrix of dimensions Nx*Nz [M/L^3]
C.y.down	Concentration at downstream boundary in *y*-direction; matrix of dimensions Nx*Nz [M/L^3]
C.z.up	Concentration at upstream boundary in *z*-direction; matrix of dimensions Nx*Ny [M/L^3]
C.z.down	Concentration at downstream boundary in *z*-direction; matrix of dimensions Nx*Ny [M/L^3]
flux.x.up	Flux across the upstream boundary in *x*-direction, positive = INTO model domain; matrix of dimensions Ny*Nz [M/L^2/T]
flux.x.down	Flux across the downstream boundary in *x*-direction, positive = OUT of model domain; matrix of dimensions Ny*Nz [M/L^2/T]
flux.y.up	Flux across the upstream boundary in *y*-direction, positive = INTO model domain; matrix of dimensions Nx*Nz [M/L^2/T]
flux.y.down	Flux across the downstream boundary in *y*-direction, positive = OUT of model domain; matrix of dimensions Nx*Nz [M/L^2/T]
flux.z.up	Flux across the upstream boundary in *z*-direction, positive = INTO model domain; matrix of dimensions Nx*Ny [M/L^2/T]
flux.z.down	Flux across the downstream boundary in *z*-direction, positive = OUT of model domain; matrix of dimensions Nx*Ny [M/L^2/T]

`a.bl.x.up` Transfer coefficient across the upstream boundary layer in x-direction `Flux=a.bl.x.up*(C.x.up-C[1,,])`. One value [L/T]

`a.bl.x.down` Transfer coefficient across the downstream boundary layer in x-direction; `Flux=a.bl.x.down*(C[Nx,,]-C.x.down)`. One value [L/T]

`a.bl.y.up` Transfer coefficient across the upstream boundary layer in y-direction; `Flux=a.bl.y.up*(C.y.up-C[,1,])`. One value [L/T]

`a.bl.y.down` Transfer coefficient across the downstream boundary layer in y-direction; `Flux=a.bl.y.down*(C[,Ny,]-C.y.down)`. One value [L/T]

`a.bl.z.up` Transfer coefficient across the upstream boundary layer in y-direction; `Flux=a.bl.y.up*(C.y.up-C[,,1])`. One value [L/T]

`a.bl.z.down` Transfer coefficient across the downstream boundary layer in z-direction; `Flux=a.bl.z.down*(C[,,Nz]-C.z.down)`. One value [L/T]

`D.grid` Diffusion coefficient defined on all grid cell interfaces. Should contain elements `x.int`, `y.int`, `z.int`, arrays with the values on the interfaces in x-, y-, and z-directions, and with dimensions `(Nx+1)*Ny*Nz`, `Nx*(Ny+1)*Nz` and `Nx*Ny*(Nz+1)`, respectively [L^2/T]

`D.x` Diffusion coefficient in x-direction, defined on grid cell interfaces. One value, a vector of length (Nx+1), or a (Nx+1)* Ny *Nz array [L^2/T]

`D.y` Diffusion coefficient in y-direction, defined on grid cell interfaces. One value, a vector of length (Ny+1), or a Nx*(Ny+1)*Nz array [L^2/T]

`D.z` Diffusion coefficient in z-direction, defined on grid cell interfaces. One value, a vector of length (Nz+1), or a Nx*Ny*(Nz+1) array [L^2/T]

`v.grid` Advective velocity defined on all grid cell interfaces. Can be positive (downstream flow) or negative (upstream flow). Should contain elements x.int, y.int, z.int, arrays with the values on the interfaces in x-, y-, and z-directions, and with dimensions (Nx+1)* Ny*Nz, Nx*(Ny+1)*Nz and Nx*Ny*(Nz+1), respectively [L/T]

`v.x` Advective velocity in the x-direction, defined on grid cell interfaces. Can be positive (downstream flow) or negative (upstream flow). One value, a vector of length (Nx+1), or a (Nx+1)*Ny*Nz array [L/T]

`v.y` Advective velocity in the y-direction, defined on grid cell interfaces. Can be positive (downstream flow) or negative (upstream flow). One value, a vector of length (Ny+1), or a Nx*(Ny+1)*Nz array [L/T]

`v.z` Advective velocity in the z-direction, defined on grid cell interfaces. Can be positive (downstream flow) or negative (upstream flow). One value, a vector of length (Nz+1), or a Nx*Ny*(Nz+1) array [L/T]

`AFDW.grid`	Weight used in the finite difference scheme for advection in the x-direction, defined on grid cell interfaces; backward = 1, centered = 0.5, forward = 0; default is backward. Should contain elements x.int, y.int, z.int, arrays with the values on the interfaces in x-, y-, and z-directions, and with dimensions (Nx+1)*Ny*Nz, Nx*(Ny+1)*Nz, and Nx*Ny*(Nz+1), respectively [−]
`AFDW.x`	Weight used in the finite difference scheme for advection in the x-direction, defined on grid cell interfaces; backward = 1, centered = 0.5, forward = 0; default is backward. One value, a vector of length (Nx+1), a prop.1D list created by setup.prop.1D, or a (Nx+1)*Ny*Nz array [−]
`AFDW.y`	Weight used in the finite difference scheme for advection in the y-direction, defined on grid cell interfaces; backward = 1, centered = 0.5, forward = 0; default is backward. One value, a vector of length (Ny+1), a prop.1D list created by setup.prop.1D, or a Nx*(Ny+1)*Nz array [−]
`AFDW.z`	Weight used in the finite difference scheme for advection in the z-direction, defined on grid cell interfaces; backward = 1, centered = 0.5, forward = 0, default is backward. One value, a vector of length (Nz+1), a prop.1D list created by setup.prop.1D, or a Nx*Ny*(Nz+1) array [−]
`VF.grid`	Volume fraction. A list should contain elements `x.int`, `y.int`, `z.int`, arrays with the values on the interfaces in x-, y-, and z-directions, and with dimensions (Nx+1)*Ny*Nz, Nx*(Ny+1)*Nz, and Nx*Ny*(Nz+1), respectively [−]
`VF.x`	Volume fraction at the grid cell interfaces in the x-direction. One value, a vector of length (Nx+1), a `prop.1D` list created by setup.prop.1D, or a (Nx+1)*Ny*Nz array [−]
`VF.y`	Volume fraction at the grid cell interfaces in the y-direction. One value, a vector of length (Ny+1), a `prop.1D` list created by setup.prop.1D, or a Nx*(Ny+1)*Nz array [−]
`VF.z`	Volume fraction at the grid cell interfaces in the z-direction. One value, a vector of length (Nz+1), a `prop.1D` list created by setup.prop.1D, or a Nx*Ny*(Nz+1) array [−]
A.grid	Interface area, a list should contain elements x.int, y.int, z.int, arrays with the values on the interfaces in x-, y-, and z-directions, and with dimensions (Nx+1)*Ny*Nz, Nx*(Ny+1)*Nz, and Nx*Ny*(Nz+1), respectively [L^2]
`A.x`	Interface area defined at the grid cell interfaces in the x-direction. One value, a vector of length (Nx+1), a prop.1D list created by setup.prop.1D, or a (Nx+1)*Ny*Nz array [L^2]
`A.y`	Interface area defined at the grid cell interfaces in the y-direction. One value, a vector of length (Ny+1), a prop.1D list created by setup.prop.1D, or a Nx*(Ny+1)*Nz array [L^2]

A.z	Interface area defined at the grid cell interfaces in the z-direction. One value, a vector of length (Nz+1), a prop.1D list created by setup.prop.1D, or a Nx*Ny*(Nz+1) array [L^2]
dx	Distance between adjacent cell interfaces in the x-direction (thickness of grid cells). One value or vector of length Nx [L]
dy	Distance between adjacent cell interfaces in the y-direction (thickness of grid cells). One value or vector of length Ny [L]
dz	Distance between adjacent cell interfaces in the z-direction (thickness of grid cells). One value or vector of length Nz [L]
grid	Discretization grid, a list containing at least elements dx, dx.aux, dy, dy.aux, dz, dz.aux (see setup.grid.2D) [L]
full.check	Logical flag enabling a full check of the consistency of the arguments (default = FALSE; TRUE slows down execution by 50%)
full.output	Logical flag enabling a full return of the output (default = FALSE; TRUE slows down execution by 20%)

Details

Do not use this with too large grid.

The boundary conditions are either

(1) zero-gradient
(2) fixed concentration
(3) convective boundary layer
(4) fixed flux.

This is also the order of priority. The zero gradient is the default, the fixed flux overrules all others.

Value

A list containing

dC	The rate of change of the concentration C owing to transport, defined in the center of each grid cell, an array with dimension Nx*Ny*Nz [$M/L^3/T$]
C.x.up	Concentration at the upstream interface in x-direction. A matrix of dimension Ny*Nz [M/L^3]. Only when full.output = TRUE
C.x.down	Concentration at the downstream interface in x-direction. A matrix of dimension Ny*Nz [M/L^3]. Only when full.output = TRUE
C.y.up	Concentration at the upstream interface in y-direction. A matrix of dimension Nx*Nz [M/L^3]. Only when full.output = TRUE

`C.y.down`	Concentration at the downstream interface in *y*-direction. A matrix of dimension Nx*Nz [M/L^3]. Only when `full.output = TRUE`
`C.z.up`	Concentration at the upstream interface in *z*-direction. A matrix of dimension Nx*Ny [M/L]. Only when `full.output = TRUE`
`C.z.down`	Concentration at the downstream interface in *z*-direction. A matrix of dimension Nx*Ny [M/L^3]. Only when `full.output = TRUE`
`x.flux`	Flux across the interfaces in *x*-direction of the grid cells. An (Nx+1)*Ny*Nz array [M/L^2/T]. Only when `full.output = TRUE`
`y.flux`	Flux across the interfaces in *y*-direction of the grid cells. An Nx*(Ny+1)*Nz array [M/L^2/T]. Only when `full.output = TRUE`
`z.flux`	Flux across the interfaces in *z*-direction of the grid cells. A Nx*Ny*(Nz+1) array [M/L^2/T]. Only when `full.output= TRUE`
`flux.x.up`	Flux across the upstream boundary in *x*-direction, `positive = INTO` model domain. A matrix of dimension Ny*Nz [M/L^2/T]
`flux.x.down`	Flux across the downstream boundary in *x*-direction, `positive=OUT` model domain. A matrix of dimension Ny*Nz [M/L^2/T]
`flux.y.up`	Flux across the upstream boundary in *y*-direction, `positive = INTO` model domain. A matrix of dimension Nx*Nz [M/L^2/T]
`flux.y.down`	Flux across the downstream boundary in *y*-direction, `positive=OUT` of model domain. A matrix of dimension Nx*Nz [M/L^2/T]
`flux.z.up`	Flux across the upstream boundary in *z*-direction, `positive = INTO` model domain. A matrix of dimension Nx*Ny [M/L^2/T]
`flux.z.down`	Flux across the downstream boundary in *z*-direction, `positive = OUT` of model domain. A matrix of dimension Nx*Ny [M/L^2/T]

Authors

Filip Meysman <filip.meysman@nioz.nl>,
Karline Soetaert <karline.soetaert@nioz.nl>.

See Also

`tran.cylindrical`, `tran.spherical`, for a discretization of 3D transport equations in cylindrical and spherical coordinates.
`tran.1D`, `tran.2D`

Examples

```
# =================================
# Diffusion in 3D; imposed boundary conditions   =
# =================================

diffusion3D <- function(t, Y, par) {
  yy <- array(dim = c(n, n, n), data = Y) # vector to 3D array
  dY <- -r * yy                   # consumption
  BND <- matrix(nrow = n, ncol = n, 1)  # boundary concentration

 dY <- dY + tran.3D(C = yy,
     C.x.up = BND, C.y.up = BND, C.z.up = BND,
     C.x.down = BND, C.y.down = BND, C.z.down = BND,
     D.x = Dx, D.y = Dy, D.z = Dz,
     dx = dx, dy = dy, dz = dz, full.check = TRUE)$dC
  return(list(dY))
}

# parameters
dy <- dx <- dz <- 1  # grid size
Dy <- Dx <- Dz <- 1 # diffusion coeff, X- and Y-direction
r <- 0.025        # consumption rate

n <- 10
y <- array(dim = c(n, n, n), data = 10.)

print(system.time(
  ST3 <- steady.3D(y, func = diffusion3D, parms = NULL,
           pos = TRUE, dimens = c(n, n, n),
           lrw = 2000000, verbose = TRUE)
))

pm <- par(mfrow = c(1,1))
y <- array(dim = c(n, n, n), data = ST3$y)
filled.contour(y[ , ,n/2], color.palette = terrain.colors)

# a selection in the x-direction
image(ST3, mfrow = c(2, 2), add.contour = TRUE, legend = TRUE,
    dimselect = list(x = c(1, 4, 8, 10)))

par(mfrow = pm)
```

In the R domain:

```
>
> # In the R domain:
>
> install.packages("ReacTran")
```

```
Installing package into 'C:/Users/Bert/Documents/R/win-library/3.3'
(as 'lib' is unspecified)
--- Please select a CRAN mirror for use in this session ---
A CRAN mirror is selected.
also installing the dependencies 'rootSolve', 'deSolve'

trying URL 'https://mirrors.tuna.tsinghua.edu.cn/CRAN/bin/windows/contrib/3.3/
rootSolve_1.7.zip'
Content type 'application/zip' length 683839 bytes (667 KB)
downloaded 667 KB

trying URL 'https://mirrors.tuna.tsinghua.edu.cn/CRAN/bin/windows/contrib/3.3/
deSolve_1.20.zip'
Content type 'application/zip' length 2667028 bytes (2.5 MB)
downloaded 2.5 MB

trying URL 'https://mirrors.tuna.tsinghua.edu.cn/CRAN/bin/windows/contrib/3.3/
ReacTran_1.4.3.1.zip'
Content type 'application/zip' length 915996 bytes (894 KB)
downloaded 894 KB

package 'rootSolve' successfully unpacked and MD5 sums checked
package 'deSolve' successfully unpacked and MD5 sums checked
package 'ReacTran' successfully unpacked and MD5 sums checked

The downloaded binary packages are in
C:\Users\Bert\AppData\Local\Temp\RtmpmEkAb2\downloaded_packages
> require(ReacTran)
Loading required package: ReacTran
Loading required package: rootSolve
Loading required package: deSolve
Loading required package: shape
> library(ReacTran)
> ls("package:ReacTran") # Listing all the programs in this package:
 [1] "advection.1D"    "advection.volume.1D"  "fiadeiro"
 [4] "g.cylinder"      "g.sphere"             "g.spheroid"
 [7] "p.exp"           "p.lin"                "p.sig"
[10] "polar2cart"      "setup.compaction.1D"  "setup.grid.1D"
[13] "setup.grid.2D"   "setup.prop.1D"        "setup.prop.2D"
[16] "tran.1D"         "tran.2D"              "tran.3D"
[19] "tran.cylindrical" "tran.polar"          "tran.spherical"
[22] "tran.volume.1D"  "tran.volume.2D"       "tran.volume.3D"
> tran.3D # Outputting the program listing of "tran.3D"
```

```
> function (C, C.up = C[1], C.down = C[length(C)], flux.up = NULL,
  flux.down = NULL, a.bl.up = NULL, a.bl.down = NULL, D = 0,
  v = 0, AFDW = 1, VF = 1, A = 1, dx, full.check = FALSE, full.output = FALSE)
{
  if (is.null(dx))
    stop("error: dx should be inputted ")
  N <- length(C)
  if (N == 0)
    stop("C should be a vector with numeric values")
  if (is.null(C.up))
    C.up <- C[1]
  if (is.null(C.down))
    C.down <- C[N]
  if (!is.list(AFDW))
    AFDW <- list(int = rep_len(AFDW, N + 1))
  if (!is.list(D))
    D <- list(int = rep_len(D, N + 1))
  if (!is.list(v))
    v <- list(int = rep_len(v, N + 1))
  if (!is.list(VF))
    VF <- list(int = rep_len(VF, N + 1), mid = 0.5 * (rep_len(VF,
      N + 1)[1:N] + rep_len(VF, N + 1)[2:(N + 1)]))
  if (!is.list(A))
    A <- list(int = rep_len(A, N + 1), mid = 0.5 * (rep_len(A,
      N + 1)[1:N] + rep_len(A, N + 1)[2:(N + 1)]))
  if (is.list(dx))
    grid <- dx
  if (!is.list(dx))
    grid <- list(dx = rep_len(dx, N), dx.aux = 0.5 * (c(0,
      rep_len(dx, N)) + c(rep_len(dx, N), 0)))
  if (full.check) {
    gn <- names(AFDW)
    if (!"int" %in% gn)
      stop("error: AFDW should be a list that contains 'int', the AFDW values at the
interface of the grid cell ")
    if (is.null(AFDW$int))
      stop("error: AFDW is NULL, should contain (numeric) values")
    if (!is.null(AFDW$int)) {
      if (!((length(AFDW$int) == 1) || (length(AFDW$int) ==
        (N + 1))))
        stop("error: AFDW should be a vector of length 1 or N+1")
    }
    if (any(AFDW$int < 0) || any(AFDW$int > 1))
```

```
    stop("error: the AFDW should always range between 0 and 1")
  gn <- names(D)
  if (!"int" %in% gn)
    stop("error: D should be a list that contains 'int', the D values at the interface
of the grid cell ")
  if (is.null(D$int))
    stop("error: D is NULL, should contain (numeric) values")
  if (!is.null(D$int)) {
    if (!((length(D$int) == 1) || (length(D$int) == (N +
      1))))
     stop("error: D should be a vector of length 1 or N+1")
  }
  if (any(D$int < 0))
    stop("error: the diffusion coefficient should always be positive")
  gn <- names(v)
  if (!"int" %in% gn)
    stop("error: v should be a list that contains 'int', the v values at the interface
of the grid cell ")
  if (is.null(v$int))
   stop("error: the advective velocity v is NULL, should contain (numeric) values")
  if (!is.null(v$int)) {
    if (!((length(v$int) == 1) || (length(v$int) == (N +
      1))))
      stop("error: v should be a vector of length 1 or N+1")
  }
  gn <- names(VF)
  if (!"int" %in% gn)
  stop("error: VF should be a list that contains 'int', the area values at the interface
of the grid cell ")
  if (!"mid" %in% gn)
    stop("error: VF should be a list that contains 'mid', the area at the middle of
the grid cells")
  if (is.null(VF$int) || is.null(VF$mid))
    stop("error: the volume fraction VF should contain (numeric) values")
  if (!is.null(VF$int)) {
    if (!((length(VF$int) == 1) || (length(VF$int) ==
      (N + 1))))
      stop("error: VF$int should be a vector of length 1 or N+1")
  }
  if (!is.null(VF$mid)) {
    if (!((length(VF$mid) == 1) || (length(VF$mid) ==
      (N))))
      stop("error: VF$mid should be a vector of length 1 or N")
  }
  if (any(VF$int < 0) || any(VF$mid < 0) || any(VF$int >
```

```
    1) || any(VF$mid > 1))
     stop("error: the volume fraction should range between 0 and 1")
   gn <- names(A)
   if (!"int" %in% gn)
    stop("error: A should be a list that contains 'int', the area values at the interface
of the grid cell ")
   if (!"mid" %in% gn)
     stop("error: A should be a list that contains 'mid', the area at the middle of
the grid cells")
   if (is.null(A$int) || is.null(A$mid))
     stop("error: the surface area A is NULL, should contain (numeric) values")
   if (!is.null(A$int)) {
     if (!((length(A$int) == 1) || (length(A$int) == (N +
       1))))
       stop("error: A$int should be a vector of length 1 or N+1")
   }
   if (!is.null(A$mid)) {
     if (!((length(A$mid) == 1) || (length(A$mid) == (N))))
       stop("error: A$mid should be a vector of length 1 or N")
   }
   if (any(A$int < 0) || any(A$mid < 0))
     stop("error: the area A should always be positive")
   gn <- names(grid)
   if (!"dx" %in% gn)
     stop("error: grid should be a list that contains 'dx' ")
   if (!"dx.aux" %in% gn)
     stop("error: grid should be a list that contains 'dx.aux' ")
   if (is.null(grid$dx) || is.null(grid$dx.aux))
   stop("error: the grid should be a list with (numeric) values for 'dx' and 'dx.aux' ")
   if (!is.null(grid$dx)) {
     if (!((length(grid$dx) == 1) || (length(grid$dx) ==
       N)))
       stop("error: dx should be a vector of length 1 or N")
   }
   if (!is.null(grid$dx.aux)) {
     if (!((length(grid$dx.aux) == 1) || (length(grid$dx.aux) ==
       (N + 1))))
       stop("error: dx.aux should be a vector of length 1 or N+1")
   }
   if (any(grid$dx <= 0) || any(grid$dx.aux <= 0))
     stop("error: the grid distances dx and dx.aux should always be
         positive")
   if (!is.null(a.bl.up) & !is.null(C.up)) {
     if (a.bl.up < 0)
       stop("error: the boundary layer transfer coefficient should be positive")
   }
```

```
    if (!is.null(a.bl.down) & !is.null(C.down)) {
      if (a.bl.down < 0)
        stop("error: the boundary layer transfer coefficient should be positive")
    }
  }
  if (full.output) {
    if (!is.null(flux.up)) {
      if (v$int[1] >= 0) {
        nom <- flux.up + VF$int[1] * (D$int[1]/grid$dx.aux[1] +
         (1 - AFDW$int[1]) * v$int[1]) * C[1]
        denom <- VF$int[1] * (D$int[1]/grid$dx.aux[1] +
         AFDW$int[1] * v$int[1])
      }
      else {
        nom <- flux.up + VF$int[1] * (D$int[1]/grid$dx.aux[1] +
         AFDW$int[1] * v$int[1]) * C[1]
        denom <- VF$int[1] * (D$int[1]/grid$dx.aux[1] +
         (1 - AFDW$int[1]) * v$int[1])
      }
      C.up <- nom/denom
    }
    if (!is.null(flux.down)) {
      if (v$int[N + 1] >= 0) {
        nom <- flux.down - VF$int[N + 1] * (D$int[N +
         1]/grid$dx.aux[N + 1] + AFDW$int[N + 1] * v$int[N +
         1]) * C[N]
        denom <- -VF$int[N + 1] * (D$int[N + 1]/grid$dx.aux[N +
         1] + (1 - AFDW$int[N + 1]) * v$int[N + 1])
      }
      else {
        nom <- flux.down - VF$int[N + 1] * (D$int[N +
         1]/grid$dx.aux[N + 1] + (1 - AFDW$int[N + 1]) *
         v$int[N + 1]) * C[N]
        denom <- -VF$int[N + 1] * (D$int[N + 1]/grid$dx.aux[N +
        1] + AFDW$int[N + 1] * v$int[N + 1])
      }
      C.down <- nom/denom
    }
    if (!is.null(a.bl.up) & !is.null(C.up)) {
      if (v$int[1] >= 0) {
        nom <- a.bl.up * C.up + VF$int[1] * (D$int[1]/grid$dx.aux[1] +
         (1 - AFDW$int[1]) * v$int[1]) * C[1]
        denom <- a.bl.up + VF$int[1] * (D$int[1]/grid$dx.aux[1] +
         AFDW$int[1] * v$int[1])
      }
      else {
```

```
      nom <- a.bl.up * C.up + VF$int[1] * (D$int[1]/grid$dx.aux[1] +
       AFDW$int[1] * v$int[1]) * C[1]
      denom <- a.bl.up + VF$int[1] * (D$int[1]/grid$dx.aux[1] +
       (1 - AFDW$int[1]) * v$int[1])
    }
    C.up <- nom/denom
  }
  if (!is.null(a.bl.down) & !is.null(C.down)) {
    if (v$int[N + 1] >= 0) {
      nom <- a.bl.down * C.down + VF$int[N + 1] * (D$int[N +
       1]/grid$dx.aux[N + 1] + (1 - AFDW$int[N + 1]) *
       v$int[N + 1]) * C[N]
      denom <- a.bl.down + VF$int[N + 1] * (D$int[N +
       1]/grid$dx.aux[N + 1] + AFDW$int[N + 1] * v$int[N +
       1])
    }
    else {
      nom <- a.bl.down * C.down + VF$int[N + 1] * (D$int[N +
       1]/grid$dx.aux[N + 1] + AFDW$int[N + 1] * v$int[N +
       1]) * C[N]
      denom <- a.bl.down + VF$int[N + 1] * (D$int[N +
       1]/grid$dx.aux[N + 1] + (1 - AFDW$int[N + 1]) *
       v$int[N + 1])
    }
    C.down <- nom/denom
  }
  dif.flux <- as.vector(-VF$int * D$int * diff(c(C.up,
   C, C.down))/grid$dx.aux)
  adv.flux <- rep(0, length.out = length(dif.flux))
  if (any(v$int > 0)) {
    vv <- v$int
    vv[v$int < 0] <- 0
    conc <- AFDW$int * c(C.up, C)
    if (any(AFDW$int < 1))
      conc <- conc + (1 - AFDW$int) * c(C, C.down)
    adv.flux <- adv.flux + as.vector(VF$int * vv * conc)
  }
  if (any(v$int < 0)) {
    vv <- v$int
    vv[v$int > 0] <- 0
    conc <- AFDW$int * c(C, C.down)
    if (any(AFDW$int < 1))
      conc <- conc + (1 - AFDW$int) * c(C.up, C)
    adv.flux <- adv.flux + as.vector(VF$int * vv * conc)
  }
  flux <- dif.flux + adv.flux
```

```
}
else {
 if (!is.null(a.bl.up) & !is.null(C.up)) {
  if (v$int[1] >= 0) {
   nom <- a.bl.up * C.up + VF$int[1] * (D$int[1]/grid$dx.aux[1] +
    (1 - AFDW$int[1]) * v$int[1]) * C[1]
   denom <- a.bl.up + VF$int[1] * (D$int[1]/grid$dx.aux[1] +
    AFDW$int[1] * v$int[1])
  }
  else {
   nom <- a.bl.up * C.up + VF$int[1] * (D$int[1]/grid$dx.aux[1] +
    AFDW$int[1] * v$int[1]) * C[1]
   denom <- a.bl.up + VF$int[1] * (D$int[1]/grid$dx.aux[1] +
    (1 - AFDW$int[1]) * v$int[1])
  }
  C.up <- nom/denom
 }
 if (!is.null(a.bl.down) & !is.null(C.down)) {
  if (v$int[N + 1] >= 0) {
   nom <- a.bl.down * C.down + VF$int[N + 1] * (D$int[N +
    1]/grid$dx.aux[N + 1] + (1 - AFDW$int[N + 1]) *
    v$int[N + 1]) * C[N]
   denom <- a.bl.down + VF$int[N + 1] * (D$int[N +
    1]/grid$dx.aux[N + 1] + AFDW$int[N + 1] * v$int[N +
    1])
  }
  else {
   nom <- a.bl.down * C.down + VF$int[N + 1] * (D$int[N +
    1]/grid$dx.aux[N + 1] + AFDW$int[N + 1] * v$int[N +
    1]) * C[N]
   denom <- a.bl.down + VF$int[N + 1] * (D$int[N +
    1]/grid$dx.aux[N + 1] + (1 - AFDW$int[N + 1]) *
    v$int[N + 1])
  }
  C.down <- nom/denom
 }
 flux <- as.vector(-(VF$int) * D$int * diff(c(C.up, C,
  C.down))/grid$dx.aux)
 if (any(v$int > 0)) {
  vv <- v$int
  vv[v$int < 0] <- 0
  conc <- AFDW$int * c(C.up, C)
  if (any(AFDW$int < 1))
   conc <- conc + (1 - AFDW$int) * c(C, C.down)
  flux <- flux + as.vector(VF$int * vv * conc)
 }
 if (any(v$int < 0)) {
```

```
        vv <- v$int
        vv[v$int > 0] <- 0
        conc <- AFDW$int * c(C, C.down)
        if (any(AFDW$int < 1))
          conc <- conc + (1 - AFDW$int) * c(C.up, C)
        flux <- flux + as.vector(VF$int * vv * conc)
      }
    }
    if (!is.null(flux.up))
      flux[1] <- flux.up
    if (!is.null(flux.down))
      flux[N + 1] <- flux.down
    dC <- -diff(A$int * flux)/A$mid/VF$mid/grid$dx
    if (!full.output) {
      return(list(dC = dC, flux.up = flux[1], flux.down = flux[length(flux)]))
    }
    else {
      return(list(dC = dC, C.up = C.up, C.down = C.down, dif.flux = dif.flux,
        adv.flux = adv.flux, flux = flux, flux.up = flux[1],
        flux.down = flux[length(flux)]))
    }
}
<environment: namespace:ReacTran>
>
```

```
> # Running the program tran.3D
>
> ## =========================================
> ## Diffusion in 3D; imposed boundary conditions
> ## =========================================
> diffusion3D <- function(t, Y, par) {
+ yy <- array(dim = c(n, n, n), data = Y) # vector to 3D array
+ dY <- -r * yy # consumption
+ BND <- matrix(nrow = n, ncol = n, 1) # boundary concentration
+ dY <- dY + tran.3D(C = yy,
+ C.x.up = BND, C.y.up = BND, C.z.up = BND,
+ C.x.down = BND, C.y.down = BND, C.z.down = BND,
+ D.x = Dx, D.y = Dy, D.z = Dz,
+ dx = dx, dy = dy, dz = dz, full.check = TRUE)$dC
+ return(list(dY))
+ }
> # parameters
> dy <- dx <- dz <- 1 # grid size
> Dy <- Dx <- Dz <- 1 # diffusion coeff, X- and Y-direction
> r <- 0.025 # consumption rate
> n <- 10
> y <- array(dim = c(n, n, n), data = 10.)
```

```
> print(system.time(
+ ST3 <- steady.3D(y, func = diffusion3D, parms = NULL,
+ pos = TRUE, dimens = c(n, n, n),
+ lrw = 2000000, verbose = TRUE)
+ ))
> # Outputting:
[1] "Steady-state settings"
 sparseType                message

>
 sparseType                message
1    3D sparse 3D jacobian, calculated internally
[1] "estimated number of nonzero elements: 6910"
[1] "estimated number of function calls: 1001"
[1] "number of species: 1"
[1] "dimensions: 10 10 10"
[1] "cyclic boundaries: 0 0 0"
mean residual derivative 4.73661e-07
[1] "precision at each steady state step"
[1] 1.105000e+01 4.736612e-07
[1] ""
[1] "--------------------"
[1] " Memory requirements"
[1] "--------------------"
  par                         mess  val
1 nnz           the number of nonzero elements: 6400
2 ngp the number of independent groups of state variables:   12
3 nsp    the length of the work array actually required: 82366
  user system elapsed
  0.05  0.03  0.11
>
> pm <- par(mfrow = c(1,1))
> # Outputting:
> #
> # Figure A3-1: tran-3D 1
> #
> y <- array(dim = c(n, n, n), data = ST3$y)
> filled.contour(y[ , ,n/2], color.palette = terrain.colors)
# Outputting:
> #
> # Figure A3-2: tran-3D 2
> #
> # a selection in the x-direction
> image(ST3, mfrow = c(2, 2), add.contour = TRUE, legend = TRUE,
+ dimselect = list(x = c(1, 4, 8, 10)))
> #
```

Figure A3-1 tran-3D 1 Preparation of space for subsequent graphical outputs.

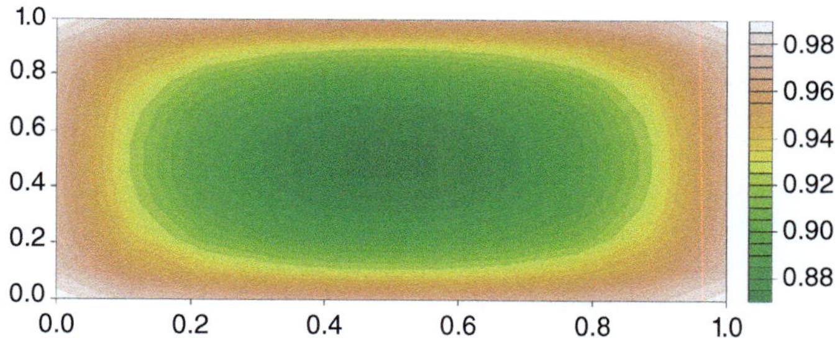

Figure A3-2 trans.3D 2

```
> y <- array(dim = c(n, n, n), data = ST3$y)
                    > filled.contour(y[ , ,n/2], color.palette = terrain.colors)
```

Contours of flow patterns.

 # Outputting:

```
> #
> # Figure A3-3: tran-3D 3
> #
> par(mfrow = pm)
>
```

Example 5.5 tran.cylindrical ***Diffusive Transport in Cylindrical (r, θ, z), and Spherical (r, θ, φ) Coordinates***

Description

This program estimates the transport term (i.e. the rate of change of a concentration owing to diffusion) in a cylindrical (r, θ, z) or spherical (r, θ, φ) coordinate system.

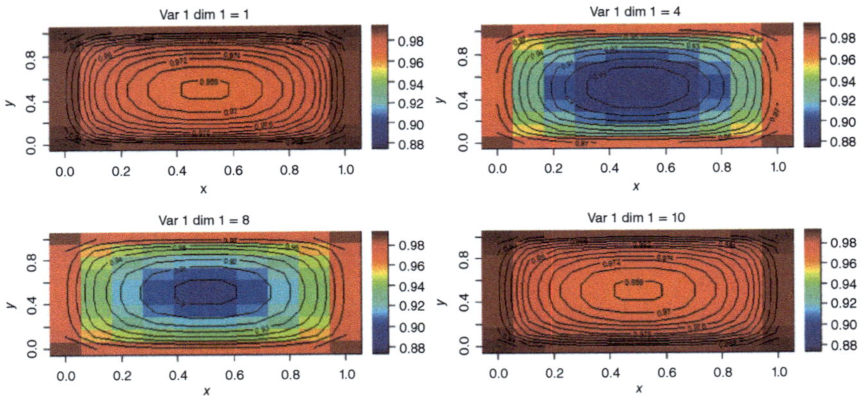

Figure A3-3 trans.3D 3

```
> image(ST3, mfrow = c(2, 2),
                    add.contour = TRUE,
          + legend = TRUE,  dimselect = list(x = c(1, 4, 8, 10)))
```

Contours of flow patterns.

Usage

```
tran.cylindrical (C, C.r.up = NULL, C.r.down = NULL,
                  C.theta.up = NULL, C.theta.down = NULL,
                  C.z.up = NULL, C.z.down = NULL,
                  flux.r.up = NULL, flux.r.down = NULL,
                  flux.theta.up = NULL, flux.theta.down = NULL,
                  flux.z.up = NULL, flux.z.down = NULL,
                  cyclicBnd = NULL,
                  D.r = NULL, D.theta = D.r, D.z = D.r,
                  r = NULL, theta = NULL, z = NULL)

tran.spherical  (C, C.r.up = NULL, C.r.down = NULL,
                  C.theta.up = NULL, C.theta.down = NULL,
                  C.phi.up = NULL, C.phi.down = NULL,
                  flux.r.up = NULL, flux.r.down = NULL,
                  flux.theta.up = NULL, flux.theta.down = NULL,
                  flux.phi.up = NULL, flux.phi.down = NULL,
                  cyclicBnd = NULL,
                  D.r = NULL, D.theta = D.r, D.phi = D.r,
                  r = NULL, theta = NULL, phi = NULL)
```

Arguments

C	Concentration, expressed per unit volume, defined at the center of each grid cell; Nr*Ntheta*Nz (cylindrical) or Nr*Ntheta*Nphi (spherical coordinates) array $[M/L^3]$
C.r.up	Concentration at upstream boundary in $r(x)$-direction; one value $[M/L^3]$
C.r.down	Concentration at downstream boundary in $r(x)$-direction; one value $[M/L^3]$
C.theta.up	Concentration at upstream boundary in theta-direction; one value $[M/L^3]$
C.theta.down	Concentration at downstream boundary in theta-direction; one value $[M/L^3]$
C.z.up	Concentration at upstream boundary in z-direction (cylindrical coordinates); one value $[M/L^3]$
C.z.down	Concentration at downstream boundary in z-direction (cylindrical coordinates); one value $[M/L^3]$
C.phi.up	Concentration at upstream boundary in phi-direction (spherical coordinates); onevalue $[M/L^3]$
C.phi.down	Concentration at downstream boundary in phi-direction (spherical coordinates); one value $[M/L^3]$
flux.r.up	Flux across the upstream boundary in r-direction, positive = INTO model domain; one value $[M/L^2/T]$
flux.r.down	Flux across the downstream boundary in r-direction, positive = OUT of model domain; one value $[M/L^2/T]$
flux.theta.up	Flux across the upstream boundary in theta-direction, positive = INTO model domain; one value $[M/L^2/T]$
flux.theta.down	Flux across the downstream boundary in theta-direction, positive = OUT of model domain; one value $[M/L^2/T]$
flux.z.up	Flux across the upstream boundary in z-direction (cylindrical coordinates); positive = INTO model domain; one value $[M/L^2/T]$
flux.z.down	Flux across the downstream boundary in z-direction, (cylindrical coordinates); positive = OUT of model domain; one value $[M/L^2/T]$
flux.phi.up	Flux across the upstream boundary in phi-direction (spherical coordinates); positive = INTO model domain; one value $[M/L^2/T]$
flux.phi.down	Flux across the downstream boundary in phi-direction, (spherical coordinates); positive = OUT of model domain; one value $[M/L^2/T]$

cyclicBnd	If not NULL, the direction in which a cyclic boundary is defined, i.e. cyclicBnd = 1 for the *r*-direction, cyclicBnd = 2 for the theta-direction and cyclicBnd = c(1,2) for both the *r*- and theta-directions
D.r	Diffusion coefficient in the *r*-direction, defined on grid cell interfaces. One value or a vector of length (Nr+1) [L^2/T]
D.theta	Diffusion coefficient in theta-direction, defined on grid cell interfaces. One value or a vector of length (Ntheta+1), [L^2/T]
D.z	Diffusion coefficient in *z*-direction, defined on grid cell interfaces for cylindrical coordinates. One value or a vector of length (Nz+1) [L^2/T]
D.phi	Diffusion coefficient in phi-direction, defined on grid cell interfaces for cylindrical coordinates. One value or a vector of length (Nphi+1) [L^2/T]
r	Position of adjacent cell interfaces in the *r*-direction. A vector of length Nr+1 [L]
theta	Position of adjacent cell interfaces in the theta-direction. A vector of length Ntheta+1 [L]. Theta should be within [0,2 pi]
z	Position of adjacent cell interfaces in the *z*-direction (cylindrical coordinates). A vector of length Nz+1 [L]
phi	Position of adjacent cell interfaces in the phi-direction (spherical coordinates). A vector of length Nphi+1 [L]. Phi should be within [0,2 pi]

Details

tran.cylindrical performs (diffusive) transport in cylindrical coordinates
tran.spherical performs (diffusive) transport in spherical coordinates
The **boundary conditions** are either

(1) zero gradient
(2) fixed concentration
(3) fixed flux
(4) cyclic boundary.

This is also the order of priority. The cyclic boundary overrules the other. If fixed concentration, fixed flux, and cyclicBnd are NULL then the boundary is zero-gradient.

A cyclic boundary condition has concentration and flux at upstream and downstream boundary the same. It is useful mainly for the theta- and phi-directions.

--- Do not expect too much of this equation: do not try to use it with many boxes ---

Value

A list containing

dC	The rate of change of the concentration *C* due to transport, defined in the center of each grid cell, a Nr*Ntheta*Nz (cylindrical) or Nr*Ntheta*Nphi (spherical coordinates) array [M/L^3/T]
flux.r.up	Flux across the upstream boundary in *r*-direction, positive = INTO model domain. A matrix of dimension Ntheta*Nz (cylindrical) or Ntheta*Nphi (spherical) [M/L^2/T]
flux.r.down	Flux across the downstream boundary in *r*-direction, positive = OUT of model domain. A matrix of dimension Ntheta*Nz (cylindrical) or Ntheta*Nphi (spherical) [M/L^2/T]
flux.theta.up	Flux across the upstream boundary in theta-direction, positive = INTO model domain. A matrix of dimension Nr*Nz (cylindrical) or Nr*Nphi (spherical) [M/L^2/T]
flux.theta.down	Flux across the downstream boundary in theta-direction, positive = OUT of model domain. A matrix of dimension Nr*Nz (cylindrical) or Nr*Nphi (spherical) [M/L^2/T]
flux.z.up	Flux across the upstream boundary in *z*-direction, for cylindrical coordinates; positive = OUT of model domain. A matrix of dimension Nr*Ntheta [M/L^2/T]
flux.z.down	Flux across the downstream boundary in *z*-direction for cylindrical coordinates; positive = OUT of model domain. A matrix of dimension Nr*Ntheta [M/L^2/T]
flux.phi.up	Flux across the upstream boundary in phi-direction, for spherical coordinates; positive = OUT of model domain. A matrix of dimension Nr*Ntheta [M/L^2/T]
flux.phi.down	Flux across the downstream boundary in phi-direction, for spherical coordinates; positive = OUT of model domain. A matrix of dimension Nr*Ntheta [M/L^2/T]

See Also

`tran.polar` for a discretization of 2D transport equations in polar coordinates.
`tran.1D`, `tran.2D`, `tran.3D`

Examples

```
# ================
# Testing the functions =
# ================
#
#Grid definition
r.N      <- 4    # number of cells in r-direction
theta.N <- 6    # number of cells in theta-direction
```

```
z.N     <- 3    # number of cells in z-direction
D       <- 100  # diffusion coefficient

r     <- seq(0, 8, len = r.N+1)          # cell size r-direction [cm]
theta <- seq(0,2*pi, len = theta.N+1) # theta-direction - theta: from 0, 2pi
phi   <- seq(0,2*pi, len = z.N+1)       # phi-direction (0,2pi)
z     <- seq(0,5, len = z.N+1)           # cell size z-direction [cm]

# Initial conditions
C <- array(dim = c(r.N, theta.N, z.N), data = 0)
# Concentration boundary conditions
tran.cylindrical (C = C, D.r = D, D.theta = D,
C.r.up = 1, C.r.down = 1,
C.theta.up = 1, C.theta.down = 1,
C.z.up = 1, C.z.down = 1,
r = r, theta = theta, z = z )

tran.spherical (C = C, D.r = D, D.theta = D,
 C.r.up = 1, C.r.down = 1, C.theta.up = 1, C.theta.down = 1,
 C.phi.up = 1, C.phi.down = 1,
 r = r, theta = theta, phi = phi)

# Flux boundary conditions
tran.cylindrical(C = C, D.r = D, r = r, theta = theta, z = z,
 flux.r.up = 10, flux.r.down = 10,
 flux.theta.up = 10, flux.theta.down = 10,
 flux.z.up = 10, flux.z.down = 10)

tran.spherical(C = C, D.r = D, r = r, theta = theta, phi = phi,
 flux.r.up = 10, flux.r.down = 10,
 flux.theta.up = 10, flux.theta.down = 10,
 flux.phi.up = 10, flux.phi.down = 10)

# cyclic boundary conditions
tran.cylindrical(C = C, D.r = D, r = r, theta = theta, z = z,
  cyclicBnd = 1:3)
tran.spherical(C = C, D.r = D, r = r, theta = theta, phi = phi,
 cyclicBnd = 1:3)

# zero-gradient boundary conditions
tran.cylindrical(C = C, D.r = D, r = r, theta = theta, z = z)
tran.spherical(C = C, D.r = D, r = r, theta = theta, phi = phi)

# ==========================================
# A model with diffusion and first-order consumption =
# ==========================================
```

```
N <- 10          # number of grid cells
rr <- 0.005      # consumption rate
D <- 400

r      <- seq (2, 4, len = N+1)
theta  <- seq (0, 2*pi, len = N+1)
z      <- seq (0, 3, len = N+1)
phi    <- seq (0, 2*pi, len = N+1)

# The model equations
Diffcylin <- function (t, y, parms) {
  CONC <- array(dim = c(N, N, N), data = y)
  tran  <- tran.cylindrical(CONC,
         D.r = D, D.theta = D, D.z = D,
         r = r, theta = theta, z = z,
         C.r.up = 0, C.r.down = 1,
         cyclicBnd = 2)
 dCONC <- tran$dC - rr * CONC
 return (list(dCONC))
}

Diffspher <- function (t, y, parms) {
   CONC <- array(dim = c(N, N, N), data = y)
     tran <- tran.spherical (CONC,
          D.r = D, D.theta = D, D.phi = D,
          r = r, theta = theta, phi = phi,
          C.r.up = 0, C.r.down = 1,
          cyclicBnd = 2:3)
 dCONC <- tran$dC - rr * CONC
 return (list(dCONC))
}

# initial condition: 0 everywhere, except in central point
y   <- array(dim = c(N, N, N), data = 0)
N2  <- ceiling(N/2)

y[N2, N2, N2] <- 100  # initial concentration in the central point...

# solve to steady-state; cyclicBnd = 2,
outcyl <- steady.3D (y = y, func = Diffcylin, parms = NULL,
dim = c(N, N, N), lrw = 1e6, cyclicBnd = 2)

STDcyl <- array(dim = c(N, N, N), data = outcyl$y)
image(STDcyl[,,1])

# For spherical coordinates, cyclic Bnd = 2, 3
outspher <- steady.3D (y = y, func = Diffspher, parms = NULL, pos=TRUE,
          dim = c(N, N, N), lrw = 1e6, cyclicBnd = 2:3)
```

```
#STDspher <- array(dim = c(N, N, N), data = outspher$y)
#image(STDspher[,,1])

## Not run:
  image(outspher)

## End(Not run)
```

In the R domain:

```
>
> # In the R domain:
>
> install.packages("ReacTran")
Installing package into 'C:/Users/Bert/Documents/R/win-library/3.3'
(as 'lib' is unspecified)
--- Please select a CRAN mirror for use in this session ---
A CRAN mirror is selected.
also installing the dependencies 'rootSolve', 'deSolve'

trying URL 'https://mirrors.tuna.tsinghua.edu.cn/CRAN/bin/windows/contrib/3.3/
rootSolve_1.7.zip'
Content type 'application/zip' length 683839 bytes (667 KB)
downloaded 667 KB

trying URL 'https://mirrors.tuna.tsinghua.edu.cn/CRAN/bin/windows/contrib/3.3/
deSolve_1.20.zip'
Content type 'application/zip' length 2667028 bytes (2.5 MB)
downloaded 2.5 MB

trying URL 'https://mirrors.tuna.tsinghua.edu.cn/CRAN/bin/windows/contrib/3.3/
ReacTran_1.4.3.1.zip'
Content type 'application/zip' length 915996 bytes (894 KB)
downloaded 894 KB

package 'rootSolve' successfully unpacked and MD5 sums checked
package 'deSolve' successfully unpacked and MD5 sums checked
package 'ReacTran' successfully unpacked and MD5 sums checked

The downloaded binary packages are in
C:\Users\Bert\AppData\Local\Temp\RtmpmEkAb2\downloaded_packages
 > require(ReacTran)
Loading required package: ReacTran
Loading required package: rootSolve
Loading required package: deSolve
Loading required package: shape
```

```
> library(ReacTran)
> ls("package:ReacTran") # Listing all the programs in this package:
 [1] "advection.1D"      "advection.volume.1D"  "fiadeiro"
 [4] "g.cylinder"        "g.sphere"             "g.spheroid"
 [7] "p.exp"             "p.lin"                "p.sig"
[10] "polar2cart"        "setup.compaction.1D"  "setup.grid.1D"
[13] "setup.grid.2D"     "setup.prop.1D"        "setup.prop.2D"
[16] "tran.1D"           "tran.2D"              "tran.3D"
[19] "tran.cylindrical"  "tran.polar"           "tran.spherical"
[22] "tran.volume.1D"    "tran.volume.2D"       "tran.volume.3D"

> tran.cylindrical # Outputting the program listing of "tran.cylindrical"
>
> ## ==================================================
> ## Testing the functions
> ## ==================================================

> # Grid definition

> r.N <- 4 # number of cells in r-direction
> theta.N <- 6 # number of cells in theta-direction
> z.N <- 3 # number of cells in z-direction
> D <- 100 # diffusion coefficient
> r <- seq(0, 8, len = r.N+1) # cell size r-direction [cm]
> theta <- seq(0,2*pi, len = theta.N+1) # theta-direction - theta: from 0, 2pi
> phi <- seq(0,2*pi, len = z.N+1) # phi-direction (0,2pi)
> z <- seq(0,5, len = z.N+1) # cell size z-direction [cm]
> # Initial conditions
> C <- array(dim = c(r.N, theta.N, z.N), data = 0)
> # Concentration boundary conditions
> tran.cylindrical (C = C, D.r = D, D.theta = D,
+ C.r.up = 1, C.r.down = 1,
+ C.theta.up = 1, C.theta.down = 1,
+ C.z.up = 1, C.z.down = 1,
+ r = r, theta = theta, z = z )
$dC
, , 1

           [,1]     [,2]     [,3]     [,4]     [,5]       [,6]
[1,] 426107.503  72.0000  72.0000  72.0000  72.0000 426107.503
[2,]  47409.278  72.0000  72.0000  72.0000  72.0000  47409.278
[3,]  17113.420  72.0000  72.0000  72.0000  72.0000  17113.420
[4,]   8823.745 129.1429 129.1429 129.1429 129.1429   8823.745

, , 2
```

```
            [,1]     [,2]     [,3]     [,4]     [,5]       [,6]
[1,] 426035.503  0.00000  0.00000  0.00000  0.00000 426035.503
[2,]  47337.278  0.00000  0.00000  0.00000  0.00000  47337.278
[3,]  17041.420  0.00000  0.00000  0.00000  0.00000  17041.420
[4,]   8751.745 57.14286 57.14286 57.14286 57.14286   8751.745

, , 3

            [,1]     [,2]     [,3]     [,4]     [,5]       [,6]
[1,] 426107.503  72.0000  72.0000  72.0000  72.0000 426107.503
[2,]  47409.278  72.0000  72.0000  72.0000  72.0000  47409.278
[3,]  17113.420  72.0000  72.0000  72.0000  72.0000  17113.420
[4,]   8823.745 129.1429 129.1429 129.1429 129.1429   8823.745

$flux.r.up

     [,1] [,2] [,3]
[1,]  100  100  100
[2,]  100  100  100
[3,]  100  100  100
[4,]  100  100  100
[5,]  100  100  100
[6,]  100  100  100

$flux.r.down
     [,1] [,2] [,3]
[1,] -100 -100 -100
[2,] -100 -100 -100
[3,] -100 -100 -100
[4,] -100 -100 -100
[5,] -100 -100 -100
[6,] -100 -100 -100

$flux.theta.up
         [,1]     [,2]     [,3]
[1,] 9230.769 9230.769 9230.769
[2,] 3076.923 3076.923 3076.923
[3,] 1846.154 1846.154 1846.154
[4,] 1318.681 1318.681 1318.681

$flux.theta.down
          [,1]      [,2]      [,3]
[1,] -9230.769 -9230.769 -9230.769
[2,] -3076.923 -3076.923 -3076.923
[3,] -1846.154 -1846.154 -1846.154
[4,] -1318.681 -1318.681 -1318.681
```

```
$flux.z.up
     [,1] [,2] [,3] [,4] [,5] [,6]
[1,]  120  120  120  120  120  120
[2,]  120  120  120  120  120  120
[3,]  120  120  120  120  120  120
[4,]  120  120  120  120  120  120

$flux.z.down
     [,1] [,2] [,3] [,4] [,5] [,6]
[1,] -120 -120 -120 -120 -120 -120
[2,] -120 -120 -120 -120 -120 -120
[3,] -120 -120 -120 -120 -120 -120
[4,] -120 -120 -120 -120 -120 -120

> tran.spherical (C = C, D.r = D, D.theta = D,
+ C.r.up = 1, C.r.down = 1, C.theta.up = 1, C.theta.down = 1,
+ C.phi.up = 1, C.phi.down = 1,
+ r = r, theta = theta, phi = phi)
$dC
, , 1

         [,1]      [,2]       [,3]       [,4]       [,5]      [,6]
[1,]  0.00000 151293676 45412970.6 21643848.5 12640411.2 8750598.0
[2,]  0.00000  16810408  5045885.6  2404872.1  1404490.1  972288.7
[3,]  0.00000   6051747  1816518.8   865753.9   505616.4  350023.9
[4,] 65.30612   3087691   926860.6   441776.5   258032.9  178648.9

, , 2

         [,1]     [,2]     [,3]     [,4]     [,5]       [,6]
[1,]  0.00000  0.00000  0.00000  0.00000  0.00000 464556.742
[2,]  0.00000  0.00000  0.00000  0.00000  0.00000  51617.416
[3,]  0.00000  0.00000  0.00000  0.00000  0.00000  18582.270
[4,] 65.30612 65.30612 65.30612 65.30612 65.30612   9546.056

, , 3

         [,1]      [,2]       [,3]       [,4]       [,5]      [,6]
[1,]  0.00000 151293676 45412970.6 21643848.5 12640411.2 8750598.0
[2,]  0.00000  16810408  5045885.6  2404872.1  1404490.1  972288.7
[3,]  0.00000   6051747  1816518.8   865753.9   505616.4  350023.9
[4,] 65.30612   3087691   926860.6   441776.5   258032.9  178648.9

$flux.r.up
     [,1] [,2] [,3]
[1,]  100  100  100
```

```
[2,] 100  100  100
[3,] 100  100  100
[4,] 100  100  100
[5,] 100  100  100
[6,] 100  100  100

$flux.r.down
      [,1] [,2] [,3]
[1,] -100 -100 -100
[2,] -100 -100 -100
[3,] -100 -100 -100
[4,] -100 -100 -100
[5,] -100 -100 -100
[6,] -100 -100 -100

$flux.theta.up
         [,1]     [,2]     [,3]
[1,] 9230.769 9230.769 9230.769
[2,] 3076.923 3076.923 3076.923
[3,] 1846.154 1846.154 1846.154
[4,] 1318.681 1318.681 1318.681

$flux.theta.down
          [,1]      [,2]      [,3]
[1,] -9230.769 -9230.769 -9230.769
[2,] -3076.923 -3076.923 -3076.923
[3,] -1846.154 -1846.154 -1846.154
[4,] -1318.681 -1318.681 -1318.681

$flux.phi.up
          [,1]      [,2]     [,3]      [,4]      [,5] [,6]
[1,] 0 213034.42 106542.22 71055.94 53321.163 42686.998
[2,] 0 71011.47   35514.07 23685.31 17773.721 14228.999
[3,] 0 42606.88   21308.44 14211.19 10664.233  8537.400
[4,] 0 30433.49   15220.32 10150.85  7617.309  6098.143

$flux.phi.down
   [,1]       [,2]       [,3]      [,4]       [,5]       [,6]
[1,] 0 -213034.42 -106542.22 -71055.94 -53321.163 -42686.998
[2,] 0  -71011.47  -35514.07 -23685.31 -17773.721 -14228.999
[3,] 0  -42606.88  -21308.44 -14211.19 -10664.233  -8537.400
[4,] 0  -30433.49  -15220.32 -10150.85  -7617.309  -6098.143

> # Flux boundary conditions
> tran.cylindrical(C = C, D.r = D, r = r, theta = theta, z = z,
+ flux.r.up = 10, flux.r.down = 10,
```

```
+ flux.theta.up = 10, flux.theta.down = 10,
+ flux.z.up = 10, flux.z.down = 10)

$dC
, , 1
            [,1]      [,2]      [,3]      [,4]      [,5]       [,6]
[1,] 467.53846 6.0000000 6.0000000 6.0000000 6.0000000 -455.53846
[2,] 159.84615 6.0000000 6.0000000 6.0000000 6.0000000 -147.84615
[3,]  98.30769 6.0000000 6.0000000 6.0000000 6.0000000  -86.30769
[4,]  66.21978 0.2857143 0.2857143 0.2857143 0.2857143  -65.64835

, , 2
            [,1]      [,2]      [,3]      [,4]      [,5]       [,6]
[1,] 461.53846  0.000000  0.000000  0.000000  0.000000 -461.53846
[2,] 153.84615  0.000000  0.000000  0.000000  0.000000 -153.84615
[3,]  92.30769  0.000000  0.000000  0.000000  0.000000  -92.30769
[4,]  60.21978 -5.714286 -5.714286 -5.714286 -5.714286  -71.64835

, , 3
            [,1]      [,2]      [,3]      [,4]      [,5]       [,6]
[1,] 455.53846  -6.00000  -6.00000  -6.00000  -6.00000 -467.53846
[2,] 147.84615  -6.00000  -6.00000  -6.00000  -6.00000 -159.84615
[3,]  86.30769  -6.00000  -6.00000  -6.00000  -6.00000  -98.30769
[4,]  54.21978 -11.71429 -11.71429 -11.71429 -11.71429  -77.64835

$flux.r.up
     [,1] [,2] [,3]
[1,]   10   10   10
[2,]   10   10   10
[3,]   10   10   10
[4,]   10   10   10
[5,]   10   10   10
[6,]   10   10   10

$flux.r.down
     [,1] [,2] [,3]
[1,]   10   10   10
[2,]   10   10   10
[3,]   10   10   10
[4,]   10   10   10
[5,]   10   10   10
[6,]   10   10   10

$flux.theta.up
     [,1] [,2] [,3]
[1,]   10   10   10
```

```
[2,] 10  10  10
[3,] 10  10  10
[4,] 10  10  10

$flux.theta.down
     [,1] [,2] [,3]
[1,] 10  10  10
[2,] 10  10  10
[3,] 10  10  10
[4,] 10  10  10

$flux.z.up
     [,1] [,2] [,3] [,4] [,5] [,6]
[1,]  10  10  10  10  10  10
[2,]  10  10  10  10  10  10
[3,]  10  10  10  10  10  10
[4,]  10  10  10  10  10  10

$flux.z.down
     [,1] [,2] [,3] [,4] [,5] [,6]
[1,]  10  10  10  10  10  10
[2,]  10  10  10  10  10  10
[3,]  10  10  10  10  10  10
[4,]  10  10  10  10  10  10

> tran.spherical(C = C, D.r = D, r = r, theta = theta, phi = phi,
+ flux.r.up = 10, flux.r.down = 10,
+ flux.theta.up = 10, flux.theta.down = 10,
+ flux.phi.up = 10, flux.phi.down = 10)
$dC
, , 1

          [,1]     [,2]      [,3]      [,4]      [,5]      [,6]
[1,] 21302.192 7101.842 4262.4391 3046.0294 2370.6181 1437.8461
[2,]  7100.731 2367.281 1420.8130 1015.3431  790.2060  479.2820
[3,]  4260.438 1420.368  852.4878  609.2059  474.1236  287.5692
[4,]  3036.640 1008.018  602.3893  428.6164  332.1291  198.8760

, , 2

          [,1]      [,2]      [,3]      [,4]      [,5]      [,6]
[1,]  0.000000  0.000000  0.000000  0.000000  0.000000 -503.2698
[2,]  0.000000  0.000000  0.000000  0.000000  0.000000 -167.7566
[3,]  0.000000  0.000000  0.000000  0.000000  0.000000 -100.6540
[4,] -6.530612 -6.530612 -6.530612 -6.530612 -6.530612  -78.4263
```

```
, , 3

             [,1]      [,2]       [,3]       [,4]       [,5]       [,6]
[1,] -21302.192 -7101.842 -4262.4391 -3046.0294 -2370.6181 -2444.3858
[2,]  -7100.731 -2367.281 -1420.8130 -1015.3431  -790.2060  -814.7953
[3,]  -4260.438 -1420.368  -852.4878  -609.2059  -474.1236  -488.8772
[4,]  -3049.701 -1021.079  -615.4505  -441.6777  -345.1903  -355.7286

$flux.r.up
     [,1] [,2] [,3]
[1,]   10   10   10
[2,]   10   10   10
[3,]   10   10   10
[4,]   10   10   10
[5,]   10   10   10
[6,]   10   10   10

$flux.r.down
     [,1] [,2] [,3]
[1,]   10   10   10
[2,]   10   10   10
[3,]   10   10   10
[4,]   10   10   10
[5,]   10   10   10
[6,]   10   10   10

$flux.theta.up
     [,1] [,2] [,3]
[1,]   10   10   10
[2,]   10   10   10
[3,]   10   10   10
[4,]   10   10   10

$flux.theta.down
     [,1] [,2] [,3]
[1,]   10   10   10
[2,]   10   10   10
[3,]   10   10   10
[4,]   10   10   10

$flux.phi.up
     [,1] [,2] [,3] [,4] [,5] [,6]
[1,]   10   10   10   10   10   10
[2,]   10   10   10   10   10   10
[3,]   10   10   10   10   10   10
[4,]   10   10   10   10   10   10
```

```
$flux.phi.down
     [,1] [,2] [,3] [,4] [,5] [,6]
[1,]   10   10   10   10   10   10
[2,]   10   10   10   10   10   10
[3,]   10   10   10   10   10   10
[4,]   10   10   10   10   10   10

> # cyclic boundary conditions
> tran.cylindrical(C = C, D.r = D, r = r, theta = theta, z = z,
+ cyclicBnd = 1:3)
$dC
, , 1

     [,1] [,2] [,3] [,4] [,5] [,6]
[1,]    0    0    0    0    0    0
[2,]    0    0    0    0    0    0
[3,]    0    0    0    0    0    0
[4,]    0    0    0    0    0    0

, , 2

     [,1] [,2] [,3] [,4] [,5] [,6]
[1,]    0    0    0    0    0    0
[2,]    0    0    0    0    0    0
[3,]    0    0    0    0    0    0
[4,]    0    0    0    0    0    0

, , 3

     [,1] [,2] [,3] [,4] [,5] [,6]
[1,]    0    0    0    0    0    0
[2,]    0    0    0    0    0    0
[3,]    0    0    0    0    0    0
[4,]    0    0    0    0    0    0

$flux.r.up
     [,1] [,2] [,3]
[1,]    0    0    0
[2,]    0    0    0
[3,]    0    0    0
[4,]    0    0    0
[5,]    0    0    0
[6,]    0    0    0

$flux.r.down
     [,1] [,2] [,3]
```

```
[1,]   0    0   0
[2,]   0    0   0
[3,]   0    0   0
[4,]   0    0   0
[5,]   0    0   0
[6,]   0    0   0

$flux.theta.up
     [,1] [,2] [,3]
[1,]   0    0    0
[2,]   0    0    0
[3,]   0    0    0
[4,]   0    0    0

$flux.theta.down
     [,1] [,2] [,3]
[1,]   0    0    0
[2,]   0    0    0
[3,]   0    0    0
[4,]   0    0    0

$flux.z.up
     [,1] [,2] [,3] [,4] [,5] [,6]
[1,]   0    0    0    0    0    0
[2,]   0    0    0    0    0    0
[3,]   0    0    0    0    0    0
[4,]   0    0    0    0    0    0

$flux.z.down
     [,1] [,2] [,3] [,4] [,5] [,6]
[1,]   0    0    0    0    0    0
[2,]   0    0    0    0    0    0
[3,]   0    0    0    0    0    0
[4,]   0    0    0    0    0    0

> tran.spherical(C = C, D.r = D, r = r, theta = theta, phi = phi,
+ cyclicBnd = 1:3)
$dC

, , 1

     [,1] [,2] [,3] [,4] [,5] [,6]
[1,]   0    0    0    0    0    0
[2,]   0    0    0    0    0    0
[3,]   0    0    0    0    0    0
[4,]   0    0    0    0    0    0
```

```
, , 2

     [,1] [,2] [,3] [,4] [,5] [,6]
[1,] 0    0    0    0    0    0
[2,] 0    0    0    0    0    0
[3,] 0    0    0    0    0    0
[4,] 0    0    0    0    0    0

, , 3

     [,1] [,2] [,3] [,4] [,5] [,6]
[1,] 0    0    0    0    0    0
[2,] 0    0    0    0    0    0
[3,] 0    0    0    0    0    0
[4,] 0    0    0    0    0    0

$flux.r.up
     [,1] [,2] [,3]
[1,]    0    0    0
[2,]    0    0    0
[3,]    0    0    0
[4,]    0    0    0
[5,]    0    0    0
[6,]    0    0    0

$flux.r.down
     [,1] [,2] [,3]
[1,]    0    0    0
[2,]    0    0    0
[3,]    0    0    0
[4,]    0    0    0
[5,]    0    0    0
[6,]    0    0    0

$flux.theta.up

     [,1] [,2] [,3]
[1,]    0    0    0
[2,]    0    0    0
[3,]    0    0    0
[4,]    0    0    0

$flux.theta.down
     [,1] [,2] [,3]
[1,]    0    0    0
[2,]    0    0    0
```

```
[3,]    0    0    0
[4,]    0    0    0

$flux.phi.up
     [,1] [,2] [,3] [,4] [,5] [,6]
[1,]    0    0    0    0    0    0
[2,]    0    0    0    0    0    0
[3,]    0    0    0    0    0    0
[4,]    0    0    0    0    0    0

$flux.phi.down
     [,1] [,2] [,3] [,4] [,5] [,6]
[1,]    0    0    0    0    0    0
[2,]    0    0    0    0    0    0
[3,]    0    0    0    0    0    0
[4,]    0    0    0    0    0    0

> # zero-gradient boundary conditions
> tran.cylindrical(C = C, D.r = D, r = r, theta = theta, z = z)
$dC
, , 1

     [,1] [,2] [,3] [,4] [,5] [,6]
[1,]    0    0    0    0    0    0
[2,]    0    0    0    0    0    0
[3,]    0    0    0    0    0    0
[4,]    0    0    0    0    0    0

, , 2

     [,1] [,2] [,3] [,4] [,5] [,6]
[1,]    0    0    0    0    0    0
[2,]    0    0    0    0    0    0
[3,]    0    0    0    0    0    0
[4,]    0    0    0    0    0    0

, , 3

     [,1] [,2] [,3] [,4] [,5] [,6]
[1,]    0    0    0    0    0    0
[2,]    0    0    0    0    0    0
[3,]    0    0    0    0    0    0
[4,]    0    0    0    0    0    0

$flux.r.up
     [,1] [,2] [,3]
[1,]    0    0    0
```

```
[2,]  0  0  0
[3,]  0  0  0
[4,]  0  0  0
[5,]  0  0  0
[6,]  0  0  0

$flux.r.down
     [,1] [,2] [,3]
[1,]  0    0    0
[2,]  0    0    0
[3,]  0    0    0
[4,]  0    0    0
[5,]  0    0    0
[6,]  0    0    0

$flux.theta.up
     [,1] [,2] [,3]
[1,]  0    0    0
[2,]  0    0    0
[3,]  0    0    0
[4,]  0    0    0

$flux.theta.down
     [,1] [,2] [,3]
[1,]  0    0    0
[2,]  0    0    0
[3,]  0    0    0
[4,]  0    0    0

$flux.z.up
     [,1] [,2] [,3] [,4] [,5] [,6]
[1,] 0    0    0    0    0    0
[2,] 0    0    0    0    0    0
[3,] 0    0    0    0    0    0
[4,] 0    0    0    0    0    0

$flux.z.down
     [,1] [,2] [,3] [,4] [,5] [,6]
[1,] 0    0    0    0    0    0
[2,] 0    0    0    0    0    0
[3,] 0    0    0    0    0    0
[4,] 0    0    0    0    0    0

> tran.spherical(C = C, D.r = D, r = r, theta = theta, phi = phi)
$dC
, , 1
```

```
     [,1] [,2] [,3] [,4] [,5] [,6]
[1,]    0    0    0    0    0    0
[2,]    0    0    0    0    0    0
[3,]    0    0    0    0    0    0
[4,]    0    0    0    0    0    0

, , 2

     [,1] [,2] [,3] [,4] [,5] [,6]
[1,]    0    0    0    0    0    0
[2,]    0    0    0    0    0    0
[3,]    0    0    0    0    0    0
[4,]    0    0    0    0    0    0

, , 3

     [,1] [,2] [,3] [,4] [,5] [,6]
[1,]    0    0    0    0    0    0
[2,]    0    0    0    0    0    0
[3,]    0    0    0    0    0    0
[4,]    0    0    0    0    0    0

$flux.r.up
     [,1] [,2] [,3]
[1,]    0    0    0
[2,]    0    0    0
[3,]    0    0    0
[4,]    0    0    0
[5,]    0    0    0
[6,]    0    0    0

$flux.r.down
     [,1] [,2] [,3]
[1,]    0    0    0
[2,]    0    0    0
[3,]    0    0    0
[4,]    0    0    0
[5,]    0    0    0
[6,]    0    0    0

$flux.theta.up
     [,1] [,2] [,3]
[1,]    0    0    0
[2,]    0    0    0
[3,]    0    0    0
[4,]    0    0    0
```

```
$flux.theta.down
     [,1] [,2] [,3]
[1,]    0    0    0
[2,]    0    0    0
[3,]    0    0    0
[4,]    0    0    0

$flux.phi.up
     [,1] [,2] [,3] [,4] [,5] [,6]
[1,] 0    0    0    0    0    0
[2,] 0    0    0    0    0    0
[3,] 0    0    0    0    0    0
[4,] 0    0    0    0    0    0

$flux.phi.down
     [,1] [,2] [,3] [,4] [,5] [,6]
[1,] 0    0    0    0    0    0
[2,] 0    0    0    0    0    0
[3,] 0    0    0    0    0    0
[4,] 0    0    0    0    0    0

> # ====================================
> # A model with diffusion and first-order consumption =
> # ====================================
> N <- 10 # number of grid cells
> rr <- 0.005 # consumption rate
> D <- 400
> r <- seq (2, 4, len = N+1)
> theta <- seq (0, 2*pi, len = N+1)
> z <- seq (0, 3, len = N+1)
> phi <- seq (0, 2*pi, len = N+1)
> # The model equations
> Diffcylin <- function (t, y, parms) {
+ CONC <- array(dim = c(N, N, N), data = y)
+ tran <- tran.cylindrical(CONC,
+ D.r = D, D.theta = D, D.z = D,
+ r = r, theta = theta, z = z,
+ C.r.up = 0, C.r.down = 1,
+ cyclicBnd = 2)
+ dCONC <- tran$dC - rr * CONC
+ return (list(dCONC))
+ }
> Diffspher <- function (t, y, parms) {
+ CONC <- array(dim = c(N, N, N), data = y)
```

```
+ tran <- tran.spherical (CONC,
+ D.r = D, D.theta = D, D.phi = D,
+ r = r, theta = theta, phi = phi,
+ C.r.up = 0, C.r.down = 1,
+ cyclicBnd = 2:3)
+ dCONC <- tran$dC - rr * CONC
+ return (list(dCONC))
+ }
> # initial condition: 0 everywhere, except in central point
> y <- array(dim = c(N, N, N), data = 0)
> N2 <- ceiling(N/2)
> y[N2, N2, N2] <- 100 # initial concentration in the central point...

> # solve to steady-state; cyclicBnd = 2,
> outcyl <- steady.3D (y = y, func = Diffcylin, parms = NULL,
+ dim = c(N, N, N), lrw = 1e6, cyclicBnd = 2)
> STDcyl <- array(dim = c(N, N, N), data = outcyl$y)
> image(STDcyl[,,1])
> #
> # Outputting: Figure A4-1: Cylindrical
> #
> # For spherical coordinates, cyclic Bnd = 2, 3
> outspher <- steady.3D (y = y, func = Diffspher, parms = NULL, pos=TRUE,
+ dim = c(N, N, N), lrw = 1e6, cyclicBnd = 2:3)
> #STDspher <- array(dim = c(N, N, N), data = outspher$y)
> #image(STDspher[,,1])
> ## Not run:
> image(outspher)
> #
> # Outputting: Figure A4-2: Spherical

>
```

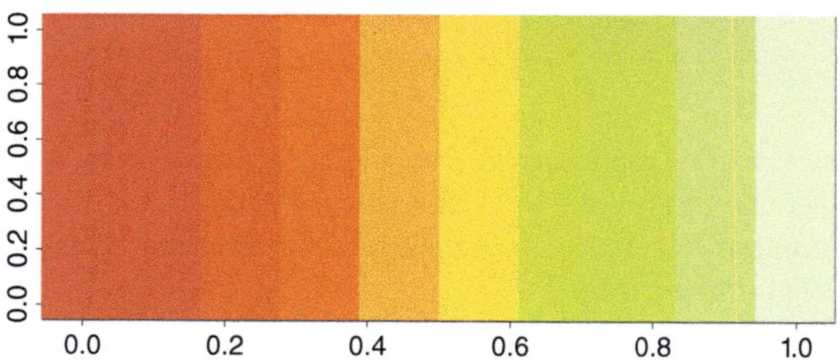

Figure A4-1 Cylindrical A model with diffusion and first-order consumption diffusive transport in cylindrical (r, θ, z), and spherical (r, θ, φ) coordinates.

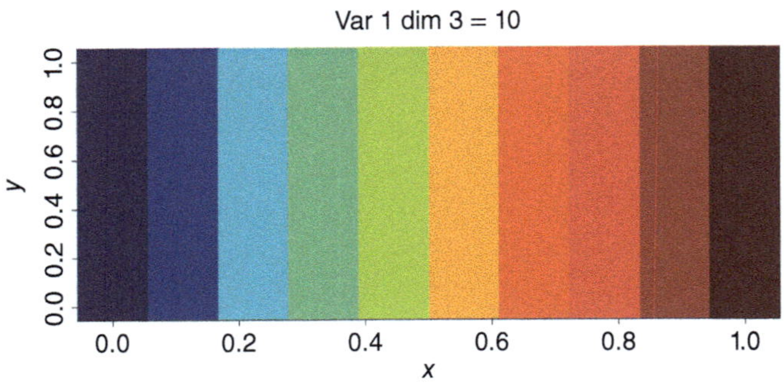

Figure A4-2 Spherical A model with diffusion and first-order consumption diffusive transport in cylindrical (r, θ, z), and spherical (r, θ, φ) coordinates.

Example 5.6 tran.polar ***Diffusive Transport in Polar (r, θ) Coordinates***

Description

This program, in R, estimates the transport term (i.e. the rate of change of a concentration owing to diffusion) in a polar (r, θ) coordinate system.

Usage

```
tran.polar (C, C.r.up = NULL, C.r.down = NULL,
            C.theta.up = NULL, C.theta.down = NULL,
            flux.r.up = NULL, flux.r.down = NULL,
            flux.theta.up = NULL, flux.theta.down = NULL,
            cyclicBnd = NULL, D.r = 1, D.theta = D.r,
            r = NULL, theta = NULL, full.output = FALSE)

polar2cart (out, r, theta, x = NULL, y = NULL)
```

Arguments

C	Concentration, expressed per unit volume, defined at the center of each grid cell; Nr*Ntheta matrix $[M/L^3]$
C.r.up	Concentration at upstream boundary in $r(x)$-direction; vector of length Ntheta $[M/L^3]$
C.r.down	Concentration at downstream boundary in $r(x)$-direction; vector of length Ntheta $[M/L^3]$
C.theta.up	Concentration at upstream boundary in theta-direction; vector of length Nr $[M/L^3]$
C.theta.down	Concentration at downstream boundary in theta-direction; vector of length Nr $[M/L^3]$

flux.r.up	Flux across the upstream boundary in r-direction, positive = INTO model domain; vector of length Ntheta [M/L²/T]
flux.r.down	Flux across the downstream boundary in r-direction, positive = OUT of model domain; vector of length Ntheta [M/L²/T]
flux.theta.up	Flux across the upstream boundary in theta-direction, positive = INTO model domain; vector of length Nr [M/L²/T]
flux.theta.down	Flux across the downstream boundary in theta-direction, positive = OUT of model domain; vector of length Nr [M/L²/T]
cyclicBnd	If not NULL, the direction in which a cyclic boundary is defined, i.e. cyclicBnd = 1 for the r-direction, cyclicBnd = 2 for the theta-direction and cyclicBnd = c(1,2) for both the r- and theta-directions
D.r	Diffusion coefficient in r-direction, defined on grid cell interfaces. One value, a vector of length (Nr+1), a prop.1D list created by setup.prop.1D, or a (Nr+1)*Ntheta matrix [L²/T]
D.theta	Diffusion coefficient in theta-direction, defined on grid cell interfaces. One value, a vector of length (Ntheta+1), a prop.1D list created by setup.prop.1D, or a Nr*(Ntheta+1) matrix [L²/T]
r	Position of adjacent cell interfaces in the r-direction. A vector of length Nr+1 [L]
theta	Position of adjacent cell interfaces in the theta-direction. A vector of length Ntheta+1 [L]. Theta should be within [0, 2 pi]
full.output	Logical flag enabling a full return of the output (default = FALSE; TRUE slows down execution by 20%)
out	Output as returned by tran.polar, and which is to be mapped from polar to cartesian coordinates
x	The Cartesian x-coordinates to whicht the polar coordinates are to be mapped
y	The Cartesian y-coordinates to which the polar coordinates are to be mapped

Details

tran.polar performs (simplified) transport in polar coordinates

The boundary conditions are either

(1) zero gradient
(2) fixed concentration
(3) fixed flux
(4) cyclic boundary.

This is also the order of priority. The cyclic boundary overrules the other. If fixed concentration, fixed flux, and cyclicBnd are `NULL`, then the boundary is zero-gradient.

A cyclic boundary condition has concentration and flux at upstream and downstream boundary the same.

`polar2cart` maps the polar coordinates to Cartesian coordinates.

If `x` and `y` are not provided, then it will create an (x,y) grid based on `r`: `seq(-maxr, maxr, length.out=Nr)`, where `maxr` is the maximum value of `r` and `Nr` is the number of elements in `r`.

Value

A list containing

`dC`	The rate of change of the concentration C due to transport, defined in the center of each grid cell, a `Nr*Ntheta` matrix $[M/L^3/T]$
`C.r.up`	Concentration at the upstream interface in r-direction. A vector of length Ntheta $[M/L^3]$. Only when `full.output = TRUE`
`C.r.down`	Concentration at the downstream interface in r-direction. A vector of length Ntheta $[M/L^3]$. Only when `full.output = TRUE`
`C.theta.up`	Concentration at the upstream interface in theta-direction. A vector of length Nr $[M/L^3]$. Only when `full.output = TRUE`
`C.theta.down`	Concentration at the downstream interface in theta-direction. A vector of length Nr $[M/L^3]$. Only when `full.output = TRUE`
`r.flux`	Flux across the interfaces in x-direction of the grid cells. A (Nr+1)*Ntheta matrix $[M/L^2/T]$. Only when `full.output = TRUE`
`theta.flux`	Flux across the interfaces in y-direction of the grid cells. A Nr*(Ntheta+1) matrix $[M/L^2/T]$. Only when `full.output = TRUE`
`flux.r.up`	Flux across the upstream boundary in r-direction, positive = `INTO` model domain. A vector of length Ntheta $[M/L^2/T]$
`flux.r.down`	Flux across the downstream boundary in r-direction, positive = `OUT` of model domain. A vector of length Ntheta $[M/L^2/T]$
`flux.theta.up`	Flux across the upstream boundary in theta-direction, positive = `INTO` model domain. A vector of length Nr $[M/L^2/T]$

flux.theta.down	Flux across the downstream boundary in theta-direction, positive = OUT of model domain. A vector of length Nr $[M/L^2/T]$

See Also

tran.cylindrical, tran.spherical for a discretization of 3D transport equations in cylindrical and spherical coordinates.
tran.1D, tran.2D, tran.3D

Examples

```
# ===============
# Testing the functions =
# ===============
# Parameters
F      <- 100         # input flux [micromol cm-2 yr-1]
D      <- 400         # mixing coefficient [cm2 yr-1]

# Grid definition
r.N <- 4 # number of cells in r-direction
theta.N <- 6 # number of cells in theta-direction
r.L <- 8 # domain size r-direction [cm]
r <- seq(0, r.L,len = r.N+1) # cell size r-direction [cm]
theta <- seq(0, 2*pi,len = theta.N+1) # theta-direction - theta: from 0, 2pi

# Initial conditions
C <- matrix(nrow = r.N, ncol = theta.N, data = 0)

# Boundary conditions: fixed concentration
C.r.up          <- rep(1, times = theta.N)
C.r.down        <- rep(0, times = theta.N)
C.theta.up      <- rep(1, times = r.N)
C.theta.down    <- rep(0, times = r.N)

# Concentration boundary conditions
tran.polar(C = C, D.r = D, D.theta = D,
  r = r, theta = theta,
  C.r.up = C.r.up, C.r.down = C.r.down,
  C.theta.up = C.theta.up, C.theta.down = C.theta.down)

# Flux boundary conditions
flux.r.up <- rep(200, times = theta.N)
flux.r.down <- rep(-200, times = theta.N)
```

```
flux.theta.up <- rep(200, times = r.N)
flux.theta.down <- rep(-200, times = r.N)

tran.polar(C = C, D.r = D, r = r, theta = theta,
  flux.r.up = flux.r.up, flux.r.down = flux.r.down,
  flux.theta.up = flux.theta.up, flux.theta.down = flux.theta.down,
  full.output = TRUE)
## ====================================
## A model with diffusion and first-order consumption  =
## ====================================
N    <- 50       # number of grid cells
XX   <- 4        # total size
rr   <- 0.005    # consumption rate
ini  <- 1        # initial value at x=0
D    <- 400

r        <- seq (2, 4, len = N+1)
theta    <- seq(0, 2*pi, len = N+1)
theta.m <- 0.5*(theta[-1]+theta[-(N+1)])

# The model equations

Diffpolar <- function (t, y, parms) {
  CONC   <- matrix(nrow = N, ncol = N, data = y)
  tran     <- tran.polar(CONC, D.r = D, D.theta = D, r = r, theta = theta,
             C.r.up = 0, C.r.down = 1*sin(5*theta.m),
             cyclicBnd = 2, full.output=TRUE )
  dCONC <- tran$dC - rr * CONC
  return (list(dCONC))
}

# solve to steady-state; cyclicBnd = 2, because of C.theta.up, C.theta.down
out <- steady.2D (y = rep(0, N*N), func = Diffpolar, parms = NULL,
             dim = c(N, N), lrw = 1e6, cyclicBnd = 2)

image(out)

cart <- polar2cart(out, r = r, theta = theta,
                              x = seq(-4, 4, len = 100),
                              y = seq(-4, 4, len = 100))
image(cart)
```

In the R domain:

```
> # In the R domain:
> install.packages("ReacTran")
```

```
Installing package into 'C:/Users/Bert/Documents/R/win-library/3.3'
(as 'lib' is unspecified)
--- Please select a CRAN mirror for use in this session ---
A CRAN mirror is selected.
also installing the dependencies 'rootSolve', 'deSolve'
trying URL 'https://mirrors.tuna.tsinghua.edu.cn/CRAN/bin/windows/contrib/3.3/
rootSolve_1.7.zip'
Content type 'application/zip' length 683839 bytes (667 KB)
downloaded 667 KB

trying URL 'https://mirrors.tuna.tsinghua.edu.cn/CRAN/bin/windows/contrib/3.3/
deSolve_1.20.zip'
Content type 'application/zip' length 2667028 bytes (2.5 MB)
downloaded 2.5 MB

trying URL 'https://mirrors.tuna.tsinghua.edu.cn/CRAN/bin/windows/contrib/3.3/
ReacTran_1.4.3.1.zip'
Content type 'application/zip' length 915996 bytes (894 KB)
downloaded 894 KB

package 'rootSolve' successfully unpacked and MD5 sums checked
package 'deSolve' successfully unpacked and MD5 sums checked
package 'ReacTran' successfully unpacked and MD5 sums checked
The downloaded binary packages are in
C:\Users\Bert\AppData\Local\Temp\RtmpmEkAb2\downloaded_packages
> require(ReacTran)
Loading required package: ReacTran
Loading required package: rootSolve
Loading required package: deSolve
Loading required package: shape
> library(ReacTran)
> ls("package:ReacTran") # Listing all the programs in this package:
 [1] "advection.1D"      "advection.volume.1D"  "fiadeiro"
 [4] "g.cylinder"        "g.sphere"             "g.spheroid"
 [7] "p.exp"             "p.lin"                "p.sig"
[10] "polar2cart"        "setup.compaction.1D"  "setup.grid.1D"
[13] "setup.grid.2D"     "setup.prop.1D"        "setup.prop.2D"
[16] "tran.1D"           "tran.2D"              "tran.3D"
[19] "tran.cylindrical"  "tran.polar"           "tran.spherical"
[22] "tran.volume.1D"    "tran.volume.2D"       "tran.volume.3D"
```

In the R domain:

```
>
> install.packages("ReacTran")
Installing package into 'C:/Users/Bert/Documents/R/win-library/3.3'
```

```
(as 'lib' is unspecified)
--- Please select a CRAN mirror for use in this session ---
```
A CRAN mirror is selected.

```
also installing the dependencies 'rootSolve', 'deSolve'

trying URL 'https://mirrors.tuna.tsinghua.edu.cn/CRAN/bin/windows/contrib/3.3/
rootSolve_1.7.zip'
Content type 'application/zip' length 683839 bytes (667 KB)
downloaded 667 KB

trying URL 'https://mirrors.tuna.tsinghua.edu.cn/CRAN/bin/windows/contrib/3.3/
deSolve_1.20.zip'
Content type 'application/zip' length 2667028 bytes (2.5 MB)
downloaded 2.5 MB

trying URL 'https://mirrors.tuna.tsinghua.edu.cn/CRAN/bin/windows/contrib/3.3/
ReacTran_1.4.3.1.zip'
Content type 'application/zip' length 915996 bytes (894 KB)
downloaded 894 KB

package 'rootSolve' successfully unpacked and MD5 sums checked
package 'deSolve' successfully unpacked and MD5 sums checked
package 'ReacTran' successfully unpacked and MD5 sums checked

The downloaded binary packages are in
C:\Users\Bert\AppData\Local\Temp\RtmpmEkAb2\downloaded_packages
> require(ReacTran)
Loading required package: ReacTran
Loading required package: rootSolve
Loading required package: deSolve
Loading required package: shape
> library(ReacTran)
> ls("package:ReacTran") # Listing all the programs in this package:
 [1] "advection.1D"    "advection.volume.1D"  "fiadeiro"
 [4] "g.cylinder"      "g.sphere"             "g.spheroid"
 [7] "p.exp"           "p.lin"                "p.sig"
[10] "polar2cart"      "setup.compaction.1D"  "setup.grid.1D"
[13] "setup.grid.2D"   "setup.prop.1D"        "setup.prop.2D"
[16] "tran.1D"         "tran.2D"              "tran.3D"
[19] "tran.cylindrical" "tran.polar"          "tran.spherical"
[22] "tran.volume.1D"  "tran.volume.2D"       "tran.volume.3D"

> tran.polar # Outputting the program listing of "tran.polar"
>
function (C, C.r.up = NULL, C.r.down = NULL, C.theta.up = NULL,
```

```
  C.theta.down = NULL, flux.r.up = NULL, flux.r.down = NULL,
  flux.theta.up = NULL, flux.theta.down = NULL, cyclicBnd = NULL,
  D.r = 1, D.theta = D.r, r = NULL, theta = NULL, full.output = FALSE)
{
  checkDim <- function(dr, dtheta, nx = "dr", ntheta = "dtheta") {
    if (length(dr) != 1)
      stop(nx, ", should have length 1")
    if (length(dtheta) != 1)
      stop(ntheta, ", should have length 1")
  }
checkDim2 <- function(dr, dtheta, nx = "dr", ntheta = "dtheta") {
    if (length(dr) != dimC[1] + 1)
      stop(nx, ", should have length equal to dim(C)[1]+1 = ",
        dimC[1] + 1, " it is ", length(dr))
    if (length(dtheta) != dimC[2] + 1)
      stop(ntheta, ", should have length equal to dim(C)[2]+1 = ",
        dimC[2] + 1, " it is ", length(dtheta))
  }
  dimC <- dim(C)
  Type <- 2
  if (max(theta) > 2 * pi)
    stop("theta should be < 2pi")
  if (min(theta) < 0)
    stop("theta should be > 0 ")
 checkDim2(r, theta, "r, the grid in x-direction", "theta, the grid in theta-direction")
  Nr <- dimC[1]
  r.c <- 0.5 * (r[-1] + r[-(Nr + 1)])
  dr <- diff(r)
  drint <- c(r.c[1] - r[1], diff(r.c), r[Nr + 1] - r.c[Nr])
  divr <- 1/r
  divr[is.na(divr)] <- 0
  Ntheta <- dimC[2]
  theta.c <- 0.5 * (theta[-1] + theta[-(Ntheta + 1)])
  dtheta <- diff(theta)
  dthetaint <- c(theta.c[1] - theta[1], diff(theta.c), theta[Ntheta +
    1] - theta.c[Ntheta])
  checkDim(D.r, D.theta, "D.r, the diffusion in x-direction ",
    "D.theta, the diffusion in theta-direction ")
  if (!is.null(cyclicBnd)) {
  }
  if (!is.null(C.r.up)) {
    if (length(C.r.up) != 1 && length(C.r.up) != Ntheta)
      stop("'C.r.up' should have length 1 or equal to ",
        Ntheta)
  }
  else if (!is.null(flux.r.up)) {
```

```
  if (length(flux.r.up) != 1 && length(flux.r.up) != Ntheta)
    stop("'flux.r.up' should have length 1 or equal to ",
      Ntheta)
}
else if (!1 %in% cyclicBnd)
  C.r.up = C[1, ]
if (!is.null(C.r.down)) {
  if (length(C.r.down) != 1 && length(C.r.down) != Ntheta)
    stop("'C.r.down' should have length 1 or equal to ",
      Ntheta)
}
else if (!is.null(flux.r.down)) {
  if (length(flux.r.down) != 1 && length(flux.r.down) !=
    Ntheta)
    stop("'flux.r.down' should have length 1 or equal to ",
      Ntheta)
}
else if (!1 %in% cyclicBnd)
  C.r.down = C[Nr, ]
if (!is.null(C.theta.up)) {
  if (length(C.theta.up) != 1 && length(C.theta.up) !=
    Nr)
    stop("'C.theta.up' should have length 1 or equal to ",
      Nr)
}
else if (!is.null(flux.theta.up)) {
  if (length(flux.theta.up) != 1 && length(flux.theta.up) !=
    Nr)
    stop("'flux.theta.up' should have length 1 or equal to ",
      Nr)
}
else if (!2 %in% cyclicBnd)
  stop("'flux.theta.up' OR 'C.theta.up' should be specified")
if (!is.null(C.theta.down)) {
  if (length(C.theta.down) != 1 && length(C.theta.down) !=
    Nr)
    stop("'C.theta.down' should have length 1 or equal to ",
      Nr)
}
else if (!is.null(flux.theta.down)) {
  if (length(flux.theta.down) != 1 && length(flux.theta.down) !=
    Nr)
    stop("'flux.theta.down' should have length 1 or equal to ",
      Nr)
}
```

```
  else if (!2 %in% cyclicBnd)
  stop("'flux.theta.down' OR 'C.theta.down' should be specified")
  Flux.r <- 0
  Flux.theta <- 0
  if (1 %in% cyclicBnd) {
    C.r.up <- (C[1, ] * drint[Nr + 1] + C[Nr, ] * drint[1])/(drint[1] +
      drint[Nr + 1])
    C.r.down <- C.r.up
  }
  if (is.null(C.r.up))
    C.r.up <- flux.r.up/D.r * drint[1] + C[1, ]
  if (is.null(C.r.down))
    C.r.down <- -flux.r.down/D.r * drint[Nr + 1] + C[Nr,
      ]
  Flux.r <- D.r * (rbind(C.r.up, C, deparse.level = 0) - rbind(C,
    C.r.down, deparse.level = 0))/drint
  if (2 %in% cyclicBnd) {
    C.theta.up <- (C[, 1] * dthetaint[Ntheta + 1] + C[, Ntheta] *
      dthetaint[1])/(dthetaint[1] + dthetaint[Ntheta +
      1])
    C.theta.down <- C.theta.up
  }
  if (is.null(C.theta.up))
    C.theta.up <- flux.theta.up/D.theta * dthetaint[1] *
      r.c + C[, 1]
  if (is.null(C.theta.down))
    C.theta.down <- -flux.theta.down/D.theta * dthetaint[Ntheta +
      1] * r.c + C[, Ntheta]
  Flux.theta <- D.theta * (cbind(C.theta.up, C, deparse.level = 0) -
    cbind(C, C.theta.down, deparse.level = 0))/matrix(data = dthetaint,
    nrow = Nr, ncol = (Ntheta + 1), byrow = TRUE)/r.c
  dFdtheta <- 0
  dFdr <- -diff(Flux.r * r)/dr/r.c
  dFdtheta <- -t(diff(t(Flux.theta))/dtheta)/r.c
  if (full.output)
    return(list(dC = dFdr + dFdtheta, C.r.up = C.r.up, C.r.down = C.r.down,
      C.theta.up = C.theta.up, C.theta.down = C.theta.down,
      r.flux = Flux.r, theta.flux = Flux.theta, flux.r.up = Flux.r[1,
        ], flux.r.down = Flux.r[Nr + 1, ], flux.theta.up = Flux.theta[,
        1], flux.theta.down = Flux.theta[, Ntheta + 1]))
  else return(list(dC = dFdr + dFdtheta, flux.r.up = Flux.r[1,
    ], flux.r.down = Flux.r[Nr + 1, ], flux.theta.up = Flux.theta[,
    1], flux.theta.down = Flux.theta[, Ntheta + 1]))
}
<environment: namespace:ReacTran>
```

```
>
> # Running the program "tran.polar":
>
> #===============
> # Testing the functions =
> #===============
> # Parameters
> F <- 100 # input flux [micromol cm-2 yr-1]
> D <- 400 # mixing coefficient [cm2 yr-1]
> # Grid definition
> r.N <- 4 # number of cells in r-direction
> theta.N <- 6 # number of cells in theta-direction
> r.L <- 8 # domain size r-direction [cm]
> r <- seq(0, r.L,len = r.N+1) # cell size r-direction [cm]
> theta <- seq(0, 2*pi,len = theta.N+1) # theta-direction - theta: from 0,
+ # to 2pi
> # Initial conditions
> C <- matrix(nrow = r.N, ncol = theta.N, data = 0)
> # Boundary conditions: fixed concentration
> C.r.up <- rep(1, times = theta.N)
> C.r.down <- rep(0, times = theta.N)
> C.theta.up <- rep(1, times = r.N)
> C.theta.down <- rep(0, times = r.N)
> # Concentration boundary conditions
> tran.polar(C = C, D.r = D, D.theta = D,
+ r = r, theta = theta,
+ C.r.up = C.r.up, C.r.down = C.r.down,
+ C.theta.up = C.theta.up, C.theta.down = C.theta.down)
> # Outputting:

$dC
           [,1]  [,2] [,3] [,4] [,5] [,6]
[1,] 1704142.01    0    0    0    0    0
[2,]  189349.11    0    0    0    0    0
[3,]   68165.68    0    0    0    0    0
[4,]   34778.41    0    0    0    0    0

$flux.r.up
[1] 400 400 400 400 400 400

$flux.r.down
[1] 0 0 0 0 0 0

$flux.theta.up
[1] 36923.077 12307.692 7384.615 5274.725
```

```
$flux.theta.down
[1] 0 0 0 0

> # Flux boundary conditions
> flux.r.up <- rep(200, times = theta.N)
> flux.r.down <- rep(-200, times = theta.N)
> flux.theta.up <- rep(200, times = r.N)
> flux.theta.down <- rep(-200, times = r.N)
> tran.polar(C = C, D.r = D, r = r, theta = theta,
+ flux.r.up = flux.r.up, flux.r.down = flux.r.down,
+ flux.theta.up = flux.theta.up, flux.theta.down = flux.theta.down,
+ full.output = TRUE)

> # Outputting:

$dC
         [,1]     [,2]     [,3]     [,4]     [,5]     [,6]
[1,] 9230.769   0.0000   0.0000   0.0000   0.0000 9230.769
[2,] 3076.923   0.0000   0.0000   0.0000   0.0000 3076.923
[3,] 1846.154   0.0000   0.0000   0.0000   0.0000 1846.154
[4,] 1432.967 114.2857 114.2857 114.2857 114.2857 1432.967

$C.r.up
[1] 0.5 0.5 0.5 0.5 0.5 0.5

$C.r.down
[1] 0.5 0.5 0.5 0.5 0.5 0.5

$C.theta.up
[1] 0.005416667 0.016250000 0.027083333 0.037916667

$C.theta.down
[1] 0.005416667 0.016250000 0.027083333 0.037916667

$r.flux
     [,1] [,2] [,3] [,4] [,5] [,6]
[1,]  200  200  200  200  200  200
[2,]    0    0    0    0    0    0
[3,]    0    0    0    0    0    0
[4,]    0    0    0    0    0    0
[5,] -200 -200 -200 -200 -200 - 200

$theta.flux
     [,1] [,2] [,3] [,4] [,5] [,6] [,7]
[1,]  200    0    0    0    0    0 -200
[2,]  200    0    0    0    0    0 -200
```

```
[3,] 200    0    0   0    0   0 -200
[4,] 200    0    0   0    0   0 -200

$flux.r.up
[1] 200 200 200 200 200 200

$flux.r.down
[1] -200 -200 -200 -200 -200 -200

$flux.theta.up
[1] 200 200 200 200

$flux.theta.down
[1] -200 -200 -200 -200
```

```
> # Outputting:
> #
> # Figure A5-1: tran.polar 1
> #
>
> ## ==================================================
> ## A model with diffusion and first-order consumption
> ## ==================================================
> N <- 50 # number of grid cells
> XX <- 4 # total size
> rr <- 0.005 # consumption rate
> ini <- 1 # initial value at x=0
> D <- 400
> r <- seq (2, 4, len = N+1)
> theta <- seq(0, 2*pi, len = N+1)
```

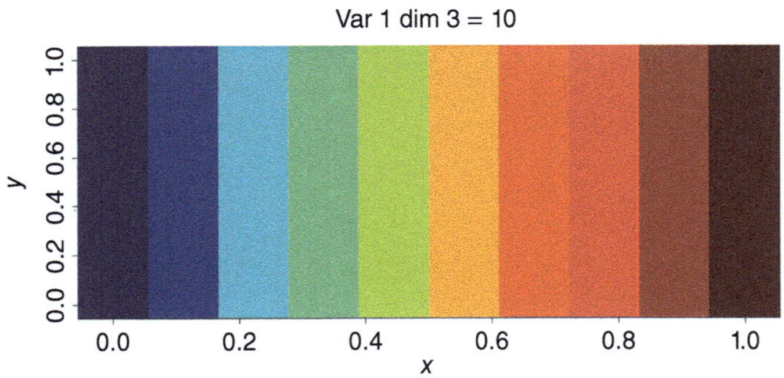

Figure A5-1 tran.polar 1 Diffusive transport in polar (r, θ) coordinates.

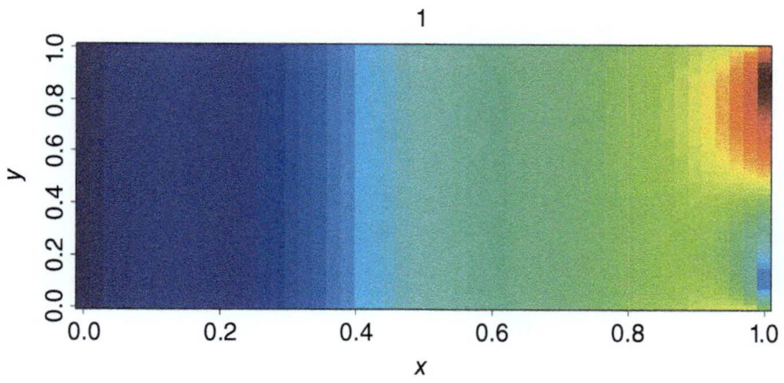

Figure A5-2 tran.polar 2 Diffusive transport in polar (r, θ) coordinates.

```
> theta.m <- 0.5*(theta[-1]+theta[-(N+1)])
> # The model equations
> Diffpolar <- function (t, y, parms) {
+ CONC <- matrix(nrow = N, ncol = N, data = y)
+ tran <- tran.polar(CONC, D.r = D, D.theta = D, r = r, theta = theta,
+ C.r.up = 0, C.r.down = 1*sin(5*theta.m),
+ cyclicBnd = 2, full.output=TRUE )
+ dCONC <- tran$dC - rr * CONC
+ return (list(dCONC))
+ }
>
> # solve to steady-state; cyclicBnd = 2, because of C.theta.up,
+ # C.theta.down
> out <- steady.2D (y = rep(0, N*N), func = Diffpolar, parms = NULL,
+ dim = c(N, N), lrw = 1e6, cyclicBnd = 2)
> image(out)
>
> # Figure A5-2: tran.polar 2
>
> cart <- polar2cart(out, r = r, theta = theta,
+ x = seq(-4, 4, len = 100),
+ y = seq(-4, 4, len = 100))
>
> image(cart)
>
> # Figure A5-3: tran.polar 3
>
>
>
```

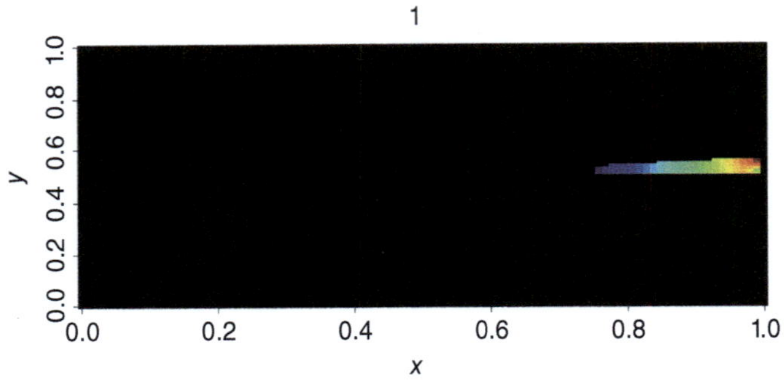

Figure A5-3 tran.polar 3 Diffusive transport in polar (r, θ) coordinates.

Example 5.7 tran.volume.1D, tran.volume.2D, tran.volume.3D, *1D, 2D, and 3D Volumetric Advective–Diffusive Transport in an Aquatic System*

Description

This program estimates the volumetric transport term (i.e. the rate of change of the concentration owing to advection and diffusion) in a 1D, 2D, or 3D model of an aquatic system (e.g. river, estuary, and open sea).

Volumetric transport implies the use of flows (mass per unit of time) rather than fluxes (mass per unit of area per unit of time) as is done in tran.1D, tran.2D, or tran.3D.

The tran.volume.xD programs are suited for modeling channels (like rivers and estuaries) where the cross-sectional area changes, but where this area change needs not to be explicitly modeled as such.

Another difference with tran.1D is that the tran.volume.1D routine also allows lateral water or lateral mass input (such as input from side rivers or diffusive lateral ground water inflow).

The tran.volume.2D program can check for water balance and assume an influx or efflux in case the net flows in and out of a box are not = 0.

Usage

```
tran.volume.1D(C, C.up = C[1], C.down = C[length(C)],
               C.lat = C, F.up = NULL, F.down = NULL, F.lat = NULL,
               Disp,flow = 0, flow.lat = NULL, AFDW = 1,
               V = NULL, full.check = FALSE, full.output = FALSE)

tran.volume.2D(C, C.x.up = C[1, ], C.x.down = C[nrow(C), ],
               C.y.up = C[, 1], C.y.down = C[, ncol(C)],
               C.z = C, masscons = TRUE,
               F.x.up = NULL, F.x.down = NULL,
               F.y.up = NULL, F.y.down = NULL,
```

```
                Disp.grid = NULL, Disp.x = NULL, Disp.y = Disp.x,
                flow.grid = NULL, flow.x = NULL, flow.y = NULL,
                AFDW.grid = NULL, AFDW.x = 1, AFDW.y = AFDW.x,
                V = NULL, full.check = FALSE, full.output = FALSE)

tran.volume.3D(C, C.x.up = C[1, , ], C.x.down = C[dim(C)[1], , ],
                C.y.up = C[, 1, ], C.y.down = C[, dim(C)[2], ],
                C.z.up = C[, , 1], C.z.down = C[, , dim(C)[3]],
                F.x.up = NULL, F.x.down = NULL,
                F.y.up = NULL, F.y.down = NULL,
                F.z.up = NULL, F.z.down = NULL,
                Disp.grid = NULL,
                Disp.x = NULL, Disp.y = Disp.x, Disp.z = Disp.x,
                flow.grid = NULL, flow.x = 0, flow.y = 0, flow.z = 0,
                AFDW.grid = NULL, AFDW.x = 1, AFDW.y = AFDW.x,
                AFDW.z = AFDW.x,
                V = NULL, full.check = FALSE, full.output = FALSE)
```

Arguments

C	Tracer concentration, defined at the center of the grid cells. A vector of length N [M/L^3] (tran.volume.1D), a matrix of dimension Nr*Nc (tran.volume.2D) or an Nx*Ny*Nz array (tran.volume.3D) [M/L^3]
C.up	Tracer concentration at the upstream interface. One value [M/L^3]
C.down	Tracer concentration at downstream interface. One value [M/L^3]
C.lat	Tracer concentration in the lateral input, defined at grid cell centers. One value, a vector of length N, or a list as defined by setup.prop.1D [M/L^3]. The default is C.lat = C (a zero-gradient condition). Setting C.lat=0, together with a positive F.lat will lead to dilution of the tracer concentration in the grid cells
C.x.up	Concentration at upstream boundary in x-direction; vector of length Ny (2D) or matrix of dimensions Ny*Nz (3D) [M/L^3]
C.x.down	Concentration at downstream boundary in x-direction; vector of length Ny (2D) or matrix of dimensions Ny*Nz (3D) [M/L^3]
C.y.up	Concentration at upstream boundary in y-direction; vector of length Nx (2D) or matrix of dimensions Nx*Nz (3D) [M/L^3]
C.y.down	Concentration at downstream boundary in y-direction; vector of length Nx (2D) or matrix of dimensions Nx*Nz (3D) [M/L^3]
C.z.up	Concentration at upstream boundary in z-direction; matrix of dimensions Nx*Ny [M/L^3]

C.z.down	Concentration at downstream boundary in z-direction; matrix of dimensions Nx*Ny [M/L^3]
C.z	Concentration at boundary in z-direction for 2D models where masscons =TRUE. Matrix of dimensions Nx*Ny [M/L^3]
masscons	When TRUE, will check flow balance in 2D model. The flow in the third direction will then be estimated
F.up	Total tracer input at the upstream interface. One value [M/T]
F.down	Total tracer input at downstream interface. One value [M/T]
F.lat	Total lateral tracer input, defined at grid cell centers. One value, a vector of length N, or a 1D list property as defined by setup.prop.1D [M/T]
F.x.up	Total tracer input at the upstream interface in x-direction. Positive = INTO model domain. A vector of length Ny (2D) or a matrix of dimensions Ny*Nz (3D) [M/T]
F.x.down	Total tracer input at the downstream interface in x-direction. Positive = INTO model domain. A vector of length Ny (2D) or a matrix of dimensions Ny*Nz (3D) [M/T]
F.y.up	Total tracer input at the upstream interface in y-direction. Positive = INTO model domain. A vector of length Nx (2D) or a matrix of dimensions Nx*Nz (3D) [M/T]
F.y.down	Total tracer input at the downstream interface in y-direction. Positive = INTO model domain. A vector of length Nx (2D) or a matrix of dimensions Nx*Nz (3D) [M/T]
F.z.up	Total tracer input at the upstream interface in z-direction. Positive = INTO model domain. A matrix of dimensions Nx*Ny [M/T]
F.z.down	Total tracer input at downstream interface in z-direction. Positive = INTO model domain. A matrix of dimensions Nx*Ny [M/T]
Disp.grid	BULK dispersion coefficients defined on all grid cell interfaces. For tran.volume.2D, should contain two matrices, x.int (dimension (Nx+1)*Ny) and y.int (dimension Nx*(Ny+1)). For tran.volume.3D should contain three arrays x.int (dim = (Nx+1)*Ny*Nz), y.int (dim = Nx*(Ny+1)*Nz), and z.int (dim = Nx*Ny*(Nz+1))
Disp	BULK dispersion coefficient, defined on grid cell interfaces. One value, a vector of length N+1, or a 1D list property as defined by setup.prop.1D [L^3/T]
Disp.x	BULK dispersion coefficient in x-direction, defined on grid cell interfaces. One value, a vector of length (Nx+1), a prop.1D list created by setup.prop.1D, a (Nx+1)*Ny matrix (2D) or a Nx*(Ny+1)*Nz array (3D) [L^3/T]
Disp.y	BULK dispersion coefficient in y-direction, defined on grid cell interfaces. One value, a vector of length (Ny+1), a prop.1D list created by setup.prop.1D, or a Nx*(Ny+1) matrix (2D) or a Nx*(Ny+1)*Nz array (3D) [L^3/T]

Disp.z	BULK dispersion coefficient in z-direction, defined on grid cell interfaces. One value, a vector of length (Nz+1), or a Nx*Ny*(Nz+1) array [L^3/T]
flow	Water flow rate, defined on grid cell interfaces. One value, a vector of length N+1, or a list as defined by `setup.prop.1D` [L^3/T]. If `flow`, `lat` is not `NULL` the flow should be one value containing the flow rate at the upstream boundary. If `flow.lat` is `NULL` then flow can be either one value, a vector or a list
flow.lat	Lateral water flow rate [L^3/T] into each volume box, defined at grid cell centers. One value, a vector of length N, or a list as defined by `setup.prop.1D`. If `flow.lat` has a value, then flow should be the flow rate at the upstream interface (one value). For each grid cell, the `flow` at the downstream side of a grid cell is then estimated by water balance (adding `flow.lat` in the cell to flow rate at the upstream side of the grid cell). If `flow.lat` is `NULL`, then it is determined by water balance from `flow`
flow.grid	Flow rates defined on all grid cell interfaces. Can be positive (downstream flow) or negative (upstream flow). Should contain elements `x.int`, `y.int`, `z.int` (3D), arrays with the values on the interfaces in x-, y- and z-directions [L^3/T]
flow.x	Flow rates in the x-direction, defined on grid cell interfaces. Can be positive (downstream flow) or negative (upstream flow). One value, a vector of length `(Nx+1)`, a `prop.1D` list created by `setup.prop.1D` (2D), a `(Nx+1)*Ny` matrix (2D) or a `(Nx+1)*Ny*Nz` array (3D) [L^3/T]
flow.y	Flow rates in the y-direction, defined on grid cell interfaces. Can be positive (downstream flow) or negative (upstream flow). One value, a vector of length (Ny+1), a `prop.1D` list created by `setup.prop.1D` (2D), a `Nx*(Ny+1)` matrix (2D) or a `Nx*(Ny+1)*Nz` array [L^3/T]
flow.z	Flow rates in the z-direction, defined on grid cell interfaces. Can be positive (downstream flow) or negative (upstream flow). One value, a vector of length `(Nz+1)`, or a `Nx*Ny*(Nz+1)` array [L^3/T]
AFDW	Weight used in the finite difference scheme for advection, defined on grid cell interfaces; backward = 1, centered = 0.5, forward = 0; default is backward. One value, a vector of length N+1, or a list as defined by `setup.prop.1D [-]`
AFDW.grid	Weight used in the finite difference scheme for advection in the x-direction, defined on grid cell interfaces; backward = 1, centered = 0.5, forward = 0; default is backward. For `tran.volume.3D` should contain elements `x.int`, `y.int`, `z.int` (3D), for `tran.volume.2D` should contain elements `x.int` and `y.int [-]`

AFDW.x	Weight used in the finite difference scheme for advection in the x-direction, defined on grid cell interfaces; backward = 1, centered = 0.5, forward = 0; default is backward. One value, a vector of length (Nx+1), a prop.1D list created by setup.prop.1D, a (Nx+1)*Ny Matrix (2D) or a (Nx+1)*Ny*Nz array (3D) [-]
AFDW.y	Weight used in the finite difference scheme for advection in the y-direction, defined on grid cell interfaces; backward = 1, centered = 0.5, forward = 0; default is backward. One value, a vector of length (Ny+1), a prop.1D list created by setup.prop.1D, a Nx*(Ny+1) matrix (2D) or a Nx*(Ny+1)*Nz array [-]
AFDW.z	Weight used in the finite difference scheme for advection in the z-direction, defined on grid cell interfaces; backward = 1, centered = 0.5, forward = 0; default is backward. One value, a vector of length (Nz+1), a prop.1D list created by setup.prop.1D, or a Nx*Ny*(Nz+1) array [-]
V	Grid cell volume, defined at grid cell centers [L^3]. One value, a vector of length N, or a list as defined by setup.prop.1D
full.check	Logical flag enabling a full check of the consistency of the arguments (default = FALSE; TRUE slows down execution by 50%)
full.output	Logical flag enabling a full return of the output (default = FALSE; TRUE slows down execution by 20%)

Details

The boundary conditions are of type

(1) zero gradient (default)
(2) fixed concentration
(3) fixed input

The *bulk dispersion coefficient* (Disp) and the *flow rate* (flow) can be either one value or a vector of length N+1, defined at all grid cell interfaces, including upstream and downstream boundary.

The spatial discretization is given by the volume of each box (V), which can be one value or a vector of length N+1, defined at the center of each grid cell.

The water flow is mass conservative. Over each volume box, the routine calculates internally the downstream outflow of water in terms of the upstream inflow and the lateral inflow.

Value

dC	The rate of change of the concentration C owing to transport, defined in the center of each grid cell [$M/L^3/T$]

F.up	Mass flow across the upstream boundary, positive = INTO model domain. One value [M/T]
F.down	Mass flow across the downstream boundary, positive = OUT of model domain. One value [M/T]
F.lat	Lateral mass input per volume box, positive = INTO model domain. A vector of length *N* [M/T]
flow	Water flow across the interface of each grid cell. A vector of length N+1 [L^3/T]. Only provided when (full.output = TRUE
flow.up	Water flow across the upstream (external) boundary, positive = INTO model domain. One value [L^3/T]. Only provided when (full.output = TRUE)
flow.down	Water flow across the downstream (external) boundary, positive = OUT of model domain. One value [L^3/T]. Only provided when (full.output = TRUE)
flow.lat	Lateral water input on each volume box, positive = INTO model domain. A vector of length *N* [L^3/T]. Only provided when (full.output = TRUE)
F	Mass flow across at the interface of each grid cell. A vector of length N+1 [M/T]. Only provided when (full.output = TRUE)

Authors

Filip Meysman <filip.meysman@nioz.nl>,
Karline Soetaert <karline.soetaert@nioz.nl>.

See Also

tran.1D
advection.volume.1D, for more sophisticated advection schemes.

Examples

```
## ===================================================
## EXAMPLE : organic carbon (OC) decay in a widening estuary
## ===================================================

# Two scenarios are simulated: the baseline includes only input
# of organic matter upstream. The second scenario simulates the
# input of an important side river half way the estuary.
#================#
# Model formulation    #
#================#
river.model <- function (t = 0, OC, pars = NULL) {
tran <- tran.volume.1D(C = OC, F.up = F.OC, F.lat = F.lat,
```

```
        Disp = Disp, flow = flow.up, flow.lat = flow.lat,
         V = Volume, full.output = TRUE)
reac <- - k*OC
return(list(dCdt = tran$dC + reac, Flow = tran$flow))
}

#===============#
# Parameter definition  #
#===============#

# Initializing morphology estuary:

nbox <- 500 # number of grid cells
lengthEstuary <- 100000 # length of estuary [m]
BoxLength <- lengthEstuary/nbox # [m]
Distance <- seq(BoxLength/2, by = BoxLength, len =nbox) # [m]
Int.Distance <- seq(0, by = BoxLength, len = (nbox+1)) # [m]


# Cross-sectional area: sigmoid function of estuarine distance [m2]
CrossArea <- 4000 + 72000 * Distance^5 /(Distance^5+50000^5)

 # Volume of boxes (m3)
Volume <- CrossArea*BoxLength

# Transport coefficients
Disp <- 1000 # m3/s, bulk dispersion coefficient
flow.up <- 180 # m3/s, main river upstream inflow
flow.lat.0 <- 180 # m3/s, side river inflow

F.OC <- 180        # input organic carbon [mol s-1]
F.lat.0 <- 180     # lateral input organic carbon [mol s-1]

k <- 10/(365*24*3600) # decay constant organic carbon [s-1]

#===========#
# Model solution    #
#===========#
#scenario 1: without lateral input
F.lat <- rep(0, length.out = nbox)
flow.lat <- rep(0, length.out = nbox)

Conc1 <- steady.1D(runif(nbox), fun = river.model, nspec = 1, name = "OC")

#scenario 2: with lateral input
F.lat <- F.lat.0 * dnorm(x =Distance/lengthEstuary,
```

```
                mean = Distance[nbox/2]/lengthEstuary,
                sd = 1/20, log = FALSE)/nbox

flow.lat <- flow.lat.0 * dnorm(x = Distance/lengthEstuary,
                               mean = Distance[nbox/2]/lengthEstuary,
                               sd = 1/20, log = FALSE)/nbox
Conc2 <- steady.1D(runif(nbox), fun = river.model, nspec = 1, name = "OC")

#============#
# Plotting output  #
#============#

# use S3 plot method

plot(Conc1, Conc2, grid = Distance/1000, which = "OC",
   mfrow = c(2, 1), lwd = 2, xlab = "distance [km]",
  main = "Organic carbon decay in the estuary",
  ylab = "OC Concentration [mM]")

plot(Conc1, Conc2, grid = Int.Distance/1000, which = "Flow",
   mfrow = NULL, lwd = 2, xlab = "distance [km]",
   main = "Longitudinal change in the water flow rate",
   ylab = "Flow rate [m3 s-1]")

legend ("topright", lty = 1:2, col = 1:2, lwd = 2,
     c("baseline", "+ side river input"))
```

In the R domain:

```
> # In the R domain:
> install.packages("ReacTran")
Installing package into 'C:/Users/Bert/Documents/R/win-library/3.3'
(as 'lib' is unspecified)
--- Please select a CRAN mirror for use in this session ---
A CRAN mirror is selected.
also installing the dependencies 'rootSolve', 'deSolve'
trying URL 'https://mirrors.tuna.tsinghua.edu.cn/CRAN/bin/windows/contrib/3.3/
rootSolve_1.7.zip'

Content type 'application/zip' length 683839 bytes (667 KB)
downloaded 667 KB
trying URL 'https://mirrors.tuna.tsinghua.edu.cn/CRAN/bin/windows/contrib/3.3/
deSolve_1.20.zip'
Content type 'application/zip' length 2667028 bytes (2.5 MB)
downloaded 2.5 MB
trying URL 'https://mirrors.tuna.tsinghua.edu.cn/CRAN/bin/windows/contrib/3.3/
ReacTran_1.4.3.1.zip'
```

```
Content type 'application/zip' length 915996 bytes (894 KB)
downloaded 894 KB

package 'rootSolve' successfully unpacked and MD5 sums checked
package 'deSolve' successfully unpacked and MD5 sums checked
package 'ReacTran' successfully unpacked and MD5 sums checked

The downloaded binary packages are in
C:\Users\Bert\AppData\Local\Temp\RtmpmEkAb2\downloaded_packages
> require(ReacTran)
Loading required package: ReacTran
Loading required package: rootSolve
Loading required package: deSolve
Loading required package: shape
> library(ReacTran)

> ls("package:ReacTran") # Listing all the programs in this package:
 [1] "advection.1D"      "advection.volume.1D"   "fiadeiro"
 [4] "g.cylinder"        "g.sphere"              "g.spheroid"
 [7] "p.exp"             "p.lin"                 "p.sig"
[10] "polar2cart"        "setup.compaction.1D"   "setup.grid.1D"
[13] "setup.grid.2D"     "setup.prop.1D"         "setup.prop.2D"
[16] "tran.1D"           "tran.2D"               "tran.3D"
[19] "tran.cylindrical"  "tran.polar"            "tran.spherical"
[22] "tran.volume.1D'    "tran.volume.2D"        "tran.volume.3D"


> # In the R domain:
>
> install.packages("ReacTran")
Installing package into 'C:/Users/Bert/Documents/R/win-library/3.3'
(as 'lib' is unspecified)
--- Please select a CRAN mirror for use in this session ---
A CRAN mirror is selected.
also installing the dependencies 'rootSolve', 'deSolve'
trying URL 'https://mirrors.tuna.tsinghua.edu.cn/CRAN/bin/windows/contrib/3.3/
rootSolve_1.7.zip'
Content type 'application/zip' length 683839 bytes (667 KB)
downloaded 667 KB
trying URL 'https://mirrors.tuna.tsinghua.edu.cn/CRAN/bin/windows/contrib/3.3/
deSolve_1.20.zip'
Content type 'application/zip' length 2667028 bytes (2.5 MB)
downloaded 2.5 MB
trying URL 'https://mirrors.tuna.tsinghua.edu.cn/CRAN/bin/windows/contrib/3.3/
ReacTran_1.4.3.1.zip'
Content type 'application/zip' length 915996 bytes (894 KB)
downloaded 894 KB
```

```
package 'rootSolve' successfully unpacked and MD5 sums checked
package 'deSolve' successfully unpacked and MD5 sums checked
package 'ReacTran' successfully unpacked and MD5 sums checked

The downloaded binary packages are in
C:\Users\Bert\AppData\Local\Temp\RtmpmEkAb2\downloaded_packages
> require(ReacTran)
Loading required package: ReacTran
Loading required package: rootSolve
Loading required package: deSolve
Loading required package: shape
> library(ReacTran)
> ls("package:ReacTran") # Listing all the programs in this package:
 [1] "advection.1D"    "advection.volume.1D"  "fiadeiro"
 [4] "g.cylinder"      "g.sphere"             "g.spheroid"
 [7] "p.exp"           "p.lin"                "p.sig"
[10] "polar2cart"      "setup.compaction.1D"  "setup.grid.1D"
[13] "setup.grid.2D"   "setup.prop.1D"        "setup.prop.2D"
[16] "tran.1D"         "tran.2D"              "tran.3D"
[19] "tran.cylindrical" "tran.polar"          "tran.spherical"
[22] "tran.volume.1D"  "tran.volume.2D"       "tran.volume.3D"

> tran.volume.1D # Outputting the program listing of "tran.volume.1D"

function (C, C.up = C[1], C.down = C[length(C)], C.lat = C, F.up = NULL,
  F.down = NULL, F.lat = NULL, Disp, flow = 0, flow.lat = NULL,
  AFDW = 1, V = NULL, full.check = FALSE, full.output = FALSE)
{
  Rep <- function(x, length.out) if (!is.null(x))
    return(rep(x, length.out = length.out))
  else return(NULL)
  N <- length(C)
  if (N == 0)
    stop("Error: C should be a vector with numeric values")
  if (!is.list(AFDW))
    AFDW <- list(int = AFDW)
  if (!is.list(F.lat))
    F.lat <- list(mid = Rep(F.lat, length.out = N))
  if (!is.list(Disp))
    Disp <- list(int = rep(Disp, length.out = N + 1))
  if (!is.list(flow))
    flow <- list(int = flow)
  if (!is.list(flow.lat))
    flow.lat <- list(mid = flow.lat)
  if (!is.list(V))
    V <- list(mid = rep(V, length.out = N))
```

```
  if (!is.list(C.lat))
    C.lat <- list(mid = rep(C.lat, length.out = length(C)))
  if (full.check) {
    gn <- names(AFDW)
    if (!"int" %in% gn)
      stop("error: AFDW should be a list that contains 'int', the AFDW values at the
interface of the grid cells")
    if (is.null(AFDW$int) || length(AFDW$int) == 0)
      stop("error: AFDW should contain (numeric) values")
    if (any(AFDW$int < 0) || any(AFDW$int > 1))
      stop("error: AFDW values should always range between 0 and 1")
    gn <- names(V)
    if (!"mid" %in% gn)
      stop("error: V should be a list that contains 'mid'")
    if (is.null(V$mid))
      stop("error: the argument V should contain (numeric) values")
    if (any(V$mid < 0))
      stop("error: the volume V should always be larger than zero")
    gn <- names(Disp)
    if (!"int" %in% gn)
      stop("error: Disp should be a list that contains 'int', the dispersion coefficient
at the grid cell interfaces")
    if (is.null(Disp$int))
      stop("error: the argument Disp should be contain (numeric) values")
    if (any(Disp$int < 0))
      stop("error: the bulk dispersion coefficient Disp should always be positive")
    gn <- names(flow)
    if (!"int" %in% gn)
      stop("error: flow should be a list that contains 'int'")
    if (is.null(flow$int))
      stop("error: the argument flow should contain (numeric) values")
    gn <- names(flow.lat)
    if (!"mid" %in% gn)
      stop("error: flow.lat should be a list that contains 'mid'")
    gn <- names(C.lat)
    if (!"mid" %in% gn)
      stop("error: the argument C.lat should be a list that contains 'mid'")
    if (any(C.lat$int < 0))
      stop("error: the concentration C.lat should always be positive")
    gn <- names(F.lat)
    if (!"mid" %in% gn)
      stop("error: the argument F.lat should be a list that contains 'mid'")
  }
  if (is.null(flow.lat$mid)) {
    flow$int <- rep(flow$int, length.out = N + 1)
```

```
  flow.lat$mid <- diff(flow$int)
 }
 else {
  if (!is.vector(flow.lat$mid) || length(flow.lat$mid) ==
    1)
    flow.lat$mid <- rep(flow.lat$mid, length.out = N)
  if (length(flow.lat$mid) != N)
    stop("flow.lat should be one number or a vector of length = N")
  if (!(length(flow$int) = 1))
    stop("flow should be of length = 1 if flow.lat is specified")

  f1 <- flow$int[1]
  flow$int <- c(f1, f1 + cumsum(flow.lat$mid))
 }
 if (is.null(F.lat$mid)) {
  if (is.null(C.lat$mid))
    stop("C.lat and F.lat cannot be both NULL")
  v1 <- flow.lat$mid
  F.lat$mid <- 0
  if (any(flow.lat$mid > 0)) {
    v1[v1 < 0] <- 0
    F.lat$mid <- C.lat$mid * v1
  }
  if (any(flow.lat$mid < 0)) {
    ii <- which(flow.lat$mid < 0)
    F.lat$mid[ii] <- C[ii] * flow.lat$mid[ii]
  }
 }
 else {
  C.lat$mid <- F.lat$mid/flow.lat$mid
 }
 Dif.F <- as.vector(-Disp$int * diff(c(C.up, C, C.down)))
 v1 <- flow$int
 Adv.F <- 0
 if (any(v1 > 0)) {
  v1[v1 < 0] <- 0
  conc <- AFDW$int * c(C.up, C)
  if (any(AFDW$int < 1))
    conc <- conc + (1 - AFDW$int) * c(C, C.down)
  Adv.F <- Adv.F + as.vector(v1 * conc)
 }
 else Adv.F <- 0
 if (any(flow$int < 0)) {
  v1 <- flow$int
```

```
    v1[v1 > 0] <- 0
    conc <- AFDW$int * c(C, C.down)
    if (any(AFDW$int < 1))
      conc <- conc + (1 - AFDW$int) * c(C.up, C)
    Adv.F <- Adv.F + as.vector(v1 * conc)
  }
  F <- as.vector(Dif.F + Adv.F)
  if (!is.null(F.up))
    F[1] <- F.up
  if (!is.null(F.down))
    F[length(F)] <- F.down
  dC <- -diff(F)/V$mid + F.lat$mid/V$mid
  if (!full.output) {
    return(list(dC = dC, F.up = F[1], F.down = F[length(F)]))
  }
  else {
    return(list(dC = dC, flow = flow$int, flow.up = flow$int[1],
      flow.down = flow$int[N + 1], flow.lat = flow.lat$mid,
      F = F, F.up = F[1], F.down = F[length(F)], F.lat = F.lat$mid))
  }
}
<environment: namespace:ReacTran>
>
> # Running the program "tran.volume.1D":
>
>   <environment: namespace:ReacTran>
> q()
>
> ## ==========================================#
> ## EXAMPLE : organic carbon (OC) decay in a widening estuary    #
> ## ==========================================#
>
> # Two scenarios are simulated: the baseline includes only input
> # of organic matter upstream. The second scenario simulates the
> # input of an important side river half way the estuary.
>
> #==============#
> # Model formulation      #
> #==============#
>
> river.model <- function (t = 0, OC, pars = NULL) {
+ tran <- tran.volume.1D(C = OC, F.up = F.OC, F.lat = F.lat,
+ Disp = Disp, flow = flow.up, flow.lat = flow.lat,
+ V = Volume, full.output = TRUE)
+ reac <- - k*OC
+ return(list(dCdt = tran$dC + reac, Flow = tran$flow))
+ }
```

```
> #===============#
> # Parameter definition    #
> #===============#
>
> # Initializing morphology estuary:
> nbox <- 500                  # number of grid cells
> lengthEstuary <- 100000       # length of estuary [m]
> BoxLength <- lengthEstuary/nbox # [m]
> Distance <- seq(BoxLength/2, by = BoxLength, len =nbox) # [m]
> Int.Distance <- seq(0, by = BoxLength, len = (nbox+1)) # [m]
> # Cross-sectional area: sigmoid function of estuarine distance [m2]
> CrossArea <- 4000 + 72000 * Distance^5 /(Distance^5+50000^5)
> # Volume of boxes (m3)
> Volume <- CrossArea*BoxLength
> # Transport coefficients
> Disp <- 1000                 # m3/s, bulk dispersion coefficient
> flow.up <- 180               # m3/s, main river upstream inflow
> flow.lat.0 <- 180            # m3/s, side river inflow
> F.OC <- 180                  # input organic carbon [mol s-1]
> F.lat.0 <- 180               # lateral input organic carbon [mol s-1]
> k <- 10/(365*24*3600) # decay constant organic carbon [s-1]

> #===========#
> # Model solution    #
> #===========#
>
> #scenario 1: without lateral input
> F.lat <- rep(0, length.out = nbox)
> flow.lat <- rep(0, length.out = nbox)
> Conc1 <- steady.1D(runif(nbox), fun = river.model, nspec = 1,
# name = "OC")
> #scenario 2: with lateral input
> F.lat <- F.lat.0 * dnorm(x =Distance/lengthEstuary,
+ mean = Distance[nbox/2]/lengthEstuary,
+ sd = 1/20,
+ log = FALSE)/nbox
> flow.lat <- flow.lat.0 * dnorm(x = Distance/lengthEstuary,
+ mean = Distance[nbox/2]/lengthEstuary,
+ sd = 1/20,
+ log = FALSE)/nbox
> Conc2 <- steady.1D(runif(nbox), fun = river.model, nspec = 1,
+ name = "OC")
>

> #============#
> # Plotting output   #
> #============#
```

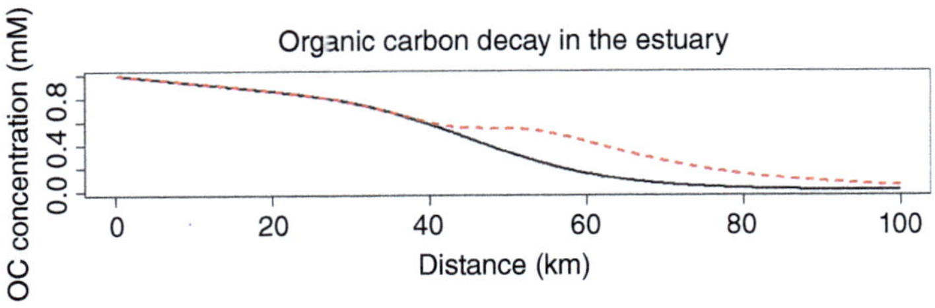

Figure A6-1 tran.volume.1D-1 1D diffusive transport in an aquatic system rate of organic carbon decay in the estuary away from the source.

```
>
> # use S3 plot method
> plot(Conc1, Conc2, grid = Distance/1000, which = "OC",
+ mfrow = c(2, 1), lwd = 2, xlab = "distance [km]",
+ main = "Organic carbon decay in the estuary",
+ ylab = "OC Concentration [mM]")
> #
> # Outputting: Figure A6-1: tran.volume.1D-1
> #
> plot(Conc1, Conc2, grid = Int.Distance/1000, which = "Flow",
+ mfrow = NULL, lwd = 2, xlab = "distance [km]",
+ main = "Longitudinal change in the water flow rate",
+ ylab = "Flow rate [m3 s-1]")
> #
> # Outputting: Figure A6-2: tran.volume.1D-2
> #
```

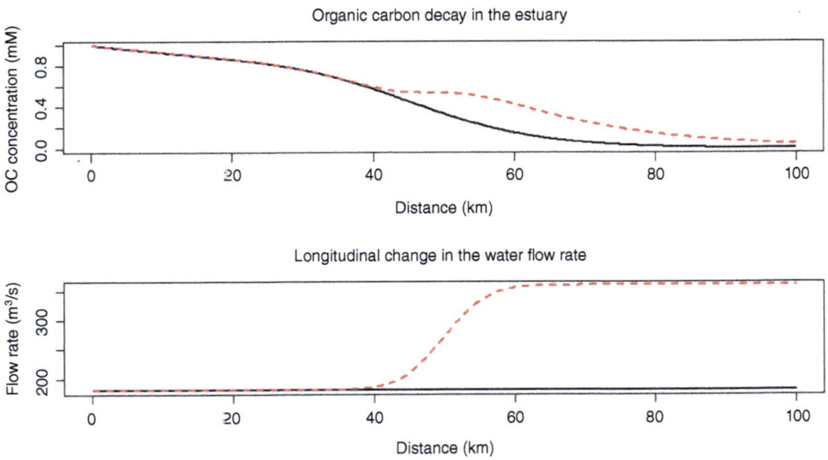

Figure A6-2 tran.volume.1D-2 1D diffusive transport in an aquatic system rate of organic carbon decay in the estuary compared with longitudinal change in the water flow rate as a function of distance from the source.

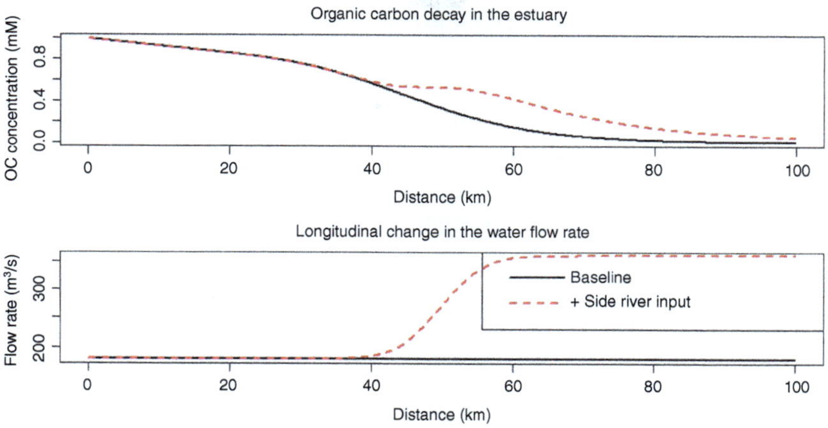

Figure A6-3 tran.volume.1D-3 1D diffusive transport in an aquatic system rate of organic carbon decay in the estuary compared with longitudinal change in the water flow rate as a function of distance from the source, with added legend.

```
> legend ("topright", lty = 1:2, col = 1:2, lwd = 2,
+ c("baseline", "+ side river input"))
> #
> # Outputting: Figure A6-3 tran.volume.1D-3
> #
>
```

Example 5.8 *Modeling Framework for Cellular Communities in Their Environments*

Version	1.6
Package	'BacArena' **May 23, 2017**
Title	Modeling Framework for Cellular Communities in their Environments
Version	1.6
Authors	Eugen Bauer Johannes Zimmermann
Maintainer	Johannes Zimmermann <j.zimmermann@iem.uni-kiel.de>
Description	Can be used for simulation of organisms living in communities. Each organism is represented individually and genome scale metabolic models determine the uptake and release of compounds. Biological processes such as movement, diffusion, chemotaxis, and kinetics are available along with data analysis techniques
URL	https://github.com/euba/BacArena

Bug Reports	https://github.com/euba/BacArena/issues
Depends	R (>= 3.0.0), sybil (>= 1.3.0), ReacTran (>= 1.4.2), deSolve (>= 1.12), Matrix (>= 1.2)
Imports	igraph, methods, utils, stats, graphics, ggplot2, reshape2, glpkAPI, plyr, Rcpp
Suggests	sybilSBML, parallel, knitr, rmarkdown
LinkingTo	Rcpp, RcppArmadillo, RcppEigen
License	GPL-3
VignetteBuilder	knitr
RoxygenNote	6.0.1
LazyData	True
NeedsCompilation	Yes
Repository	CRAN
Date/Publication	2017-05-23 17:28:04 UTC

R **topics demonstrated:**

addDefaultMed	4
addEssentialMed	4
addEval	5
addOrg	6
addSubs	7
Arena-class	8
Arena-constructor	9
Bac-class	10
Bac-Constructor	10
BacArena	11
cellgrowth	11
changeDiff	12
changeFobj	13
changeOrg	14
changeSub	15
checkCorr	16
checkPhen	17
checkPhen_par	18
chemotaxis	18
colpal1	19
colpal2	20
colpal3	20
colpal4	21
colpal5	21
colpal6	22
constrain	22
consume	23
createGradient	24

dat2mat	25
diffuse	26
diffusePDE	26
diffuseR	27
diffuse_par	28
emptyHood	28
Eval-class	29
Eval-constructor	30
evalArena	30
extractMed	31
findFeeding	32
findFeeding2	33
findFeeding3	34
findFeeding3rep	34
findInArena	35
findRxnFlux	35
flushSubs	36
fluxVarSim	37
getArena	38
getCorrM	39
getPhenoMat	40
getPhenotype	41
getSubHist	42
getSublb	42
getVarSubs	43
growExp	44
growLin	44
growth	45
growth_par	46
Human-class	47
Human-constructor	47
lsd	48
lysis	48
minePheno	49
move	50
NemptyHood	51
openArena	52
optimizeLP	52
Organism-class	53
Organism-constructor	54
plotAbundance	55
plotCurves	56
plotCurves2	57
plotFluxVar	58
plotFVA	59

plotGrowthCurve	59
plotInterNum	60
plotPhenCurve	60
plotPhenNum	61
plotReaActivity	61
plotShadowCost	62
plotSpecActivity	62
plotSubCurve	63
plotSubDist	64
plotSubDist2	64
plotSubUsage	65
plotSubVar	65
plotTotFlux	66
redEval	66
reset_screen	67
rmSubs	68
selPheno	68
setKinetics	69
sihumi_test	70
simBac	70
simBac_par	71
simEnv	72
simEnv_par	73
simHum	75
statPheno	76
stirEnv	77
Substance-class	78
Substance-constructor	79
unit_conversion	80
usd	80

Example 5.9 `plotCurves` ***Function for Plotting the Overall Change as Curves***

Description

The generic function `plotCurves()` plots the growth curves and concentration changes of substances from simulation steps in an `Eval` object.

Usage

```
plotCurves(object, medplot = object@mediac, retdata = F, remove = F,
           legend = F, graph = T)

## S4 method for signature 'Eval'
```

```
plotCurves(object, medplot = object@mediac, retdata = F, remove = F,
          legend = F, graph = T)
```

Arguments

object	An object of class Eval
medplot	A character vector giving the name of substances which should be plotted
retdata	A boolean variable indicating if the data used to generate the plots should be returned
remove	A boolean variable indicating if substances, which don't change in their concentration should be removed from the plot
legend	Boolean variable indicating if legend should be plotted
graph	True if graphic should be plotted

Details

The parameter retdata can be used to access the data used for the returned plots to create own custom plots.

Value

Returns two graphs in one plot: the growth curves and the curves of concentration changes. Optional the data to generate the original plots can be returned.

See Also

Eval-class and Arena-class

Examples

```
get Escherichia coli core metabolic model

data(Ec_core, envir = environment())
bac <- Bac(Ec_core,deathrate=0.05,
          minweight=0.05,growtype="exponential") #initialize a bacterium
arena <- Arena(n=20,m=20) #initialize the environment
arena <- addOrg(arena,bac,amount=10) #add 10 organisms
arena <- addSubs(arena,40) #add all possible substances
eval  <- simEnv(arena,5)
plotCurves(eval)
```

In the R domain:
R version 3.3.3 (2017-03-06) – "Another Canoe"

Copyright (C) 2017 The R Foundation for Statistical Computing
Platform: x86_64-w64-mingw32/x64 (64-bit)

R is a free software and comes with ABSOLUTELY NO WARRANTY.
You are welcome to redistribute it under certain conditions.
Type 'license()' or 'licence()' for distribution details.

Natural language support but running in an English locale.

R is a collaborative project with many contributors.
Type 'contributors()' for more information and
'citation()' on how to cite R or R
packages in publications.

Type 'demo()' for some demos, 'help()' for online help, or
'help.start()' for an HTML browser interface to help.
Type 'q()' to quit R.

```
>
> install.packages("BacArena")
Installing package into 'C:/Users/Bert/Documents/R/win-library/3.3'
(as 'lib' is unspecified)
--- Please select a CRAN mirror for use in this session ---
```

A CRAN mirror is selected

```
trying URL 'https://mirrors.tuna.tsinghua.edu.cn/CRAN/bin/windows/contrib/3.3/
BacArena_1.6.zip'
Content type 'application/zip' length 5091073 bytes (4.9 MB)
downloaded 4.9 MB
```

package 'BacArena' successfully unpacked and MD5 sums checked

The downloaded binary packages are in
C:\Users\Bert\AppData\Local\Temp\Rtmp8sWwS0\downloaded_packages
> library(BacArena)
Loading required package: sybil
Loading required package: Matrix
Loading required package: lattice
Loading required package: ReacTran
Loading required package: rootSolve
Loading required package: deSolve
Loading required package: shape
BacArena paper: https://doi.org/10.1371/journal.pcbi.1005544
Tutorial: https://CRAN.R-project.org/package=BacArena/vignettes/BacArena-Introduction.pdf
Development and help: https://github.com/euba/bacarena

```
> ls("package:BacArena")
 [1] "addDefaultMed"     "addEssentialMed" "addEval"          "addOrg"
 [5] "addSubs"           "Arena"           "Bac"              "cellgrowth"
 [9] "changeDiff"        "changeFobj"      "changeOrg"        "changeSub"
[13] "checkCorr"         "checkPhen"       "checkPhen_par"    "chemotaxis"
[17] "colpal1"           "colpal2"         "colpal3"          "colpal4"
[21] "colpal5"           "colpal6"         "constrain"        "consume"
[25] "createGradient"    "dat2mat"         "diffuse"          "diffuse_par"
[29] "diffusePDE"        "diffuseR"        "emptyHood"        "Eval"
[33] "evalArena"         "extractMed"      "findFeeding"      "findFeeding2"
[37] "findFeeding3"      "findFeeding3rep" "findInArena"      "findRxnFlux"
[41] "flushSubs"         "fluxVarSim"      "getArena"         "getCorrM"
[45] "getPhenoMat"       "getPhenotype"    "getSubHist"       "getSublb"
[49] "getVarSubs"        "growExp"         "growLin"          "growth"
[53] "growth_par"        "Human"           "lsd"              "lysis"
[57] "minePheno"         "move"            "NemptyHood"       "openArena"
[61] "optimizeLP"        "Organism"        "plotAbundance"    "plotCurves"
[65] "plotCurves2"       "plotFluxVar"     "plotFVA"          "plotGrowthCurve"
[69] "plotInterNum"      "plotPhenCurve"   "plotPhenNum"      "plotReaActivity"
[73] "plotShadowCost"    "plotSpecActivity" "plotSubCurve"    "plotSubDist"
[77] "plotSubDist2"      "plotSubUsage"    "plotSubVar"       "plotTotFlux"
[81] "redEval"           "reset_screen"    "rmSubs"           "selPheno"
[85] "setKinetics"       "sihumi_test"     "simBac"           "simBac_par"
[89] "simEnv"            "simEnv_par"      "simHum"           "statPheno"
[93] "stirEnv"           "Substance"       "unit_conversion"  "usd"
> plotCurves # Outputting the listing of the R program plotCurves
nonstandardGenericFunction for "plotCurves" defined from package
"BacArena"
function (object, medplot = object@mediac, retdata = F, remove = F,
          legend = F, graph = T)
{
  standardGeneric("plotCurves")
}
<environment: 0x0000000017361628>
Methods may be defined for arguments: object, medplot, retdata, remove, legend,
graph
Use showMethods("plotCurves") for currently available ones.
> data(Ec_core, envir = environment()) # get Escherichia coli core
+ # metabolic model
> bac <- Bac(Ec_core,deathrate=0.05,
+ minweight=0.05,growtype="exponential")
# initialize a bacterium
Loading required package: glpkAPI
using GLPK version 4.47
> arena <- Arena(n=20,m=20) #initialize the environment
> arena <- addOrg(arena,bac,amount=10) #add 10 organisms
```

```
> arena <- addSubs(arena,40) #add all possible substances
> eval  <- simEnv(arena,5)

iteration: 1  organisms: 10  biomass: 5.25312 pg
       count biomass %
ecoli_core_model  10 5.25312 0
   time total: 0.49    diffusion: 0.22 ( 44.9 %)

iteration: 2   organisms: 15  biomass: 11.59317 pg
       count biomass    %
ecoli_core_model  15 11.59317 120.6912
   time total: 0.5    diffusion: 0.21 ( 42 %)

iteration: 3   organisms: 30  biomass: 22.54044 pg
       count biomass    %
ecoli_core_model  30 22.54044 94.42863
   time total: 0.63    diffusion: 0.2 ( 31.7 %)

iteration: 4   organisms: 60  biomass: 44.95924 pg
       count biomass    %
ecoli_core_model  60 44.95924 99.46032
   time total: 0.93    diffusion: 0.19 ( 20.4 %)

iteration: 5   organisms: 120     biomass: 90.11333 pg
       count biomass    %
ecoli_core_model  120 90.11333 100.4334
   time total: 1.46    diffusion: 0.17 ( 11.6 %)

> plotCurves(eval)
>
> # Outputting: Figure A7-1 plotCurves-1
>
```

Example 5.10 plotTotFlux ***Function for Plotting the Overall Change in Reaction Activity***

Description

The generic function plotTotFlux() plots the time course of reactions with high variation in activity for an Eval object.

Usage

```
plotTotFlux(object, legendpos = "topright", num = 20)
## S4 method for signature 'Eval'
plotTotFlux(object, legendpos = "topright", num = 20)
```

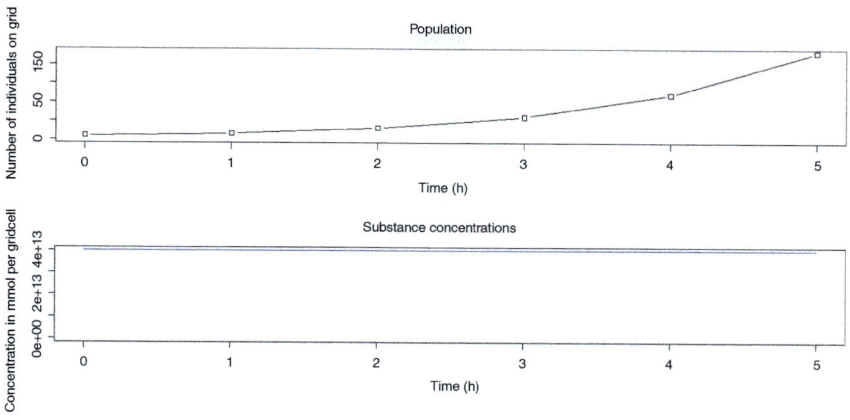

Figure A7-1 plotCurves-1 Diffusion and reaction.

Arguments

object	An object of class Eval.
legendpos	A character variable declaring the position of the legend
num	An integer defining the number of substrates to be plot

Examples

```
data(Ec_core, envir = environment())
#get Escherichia coli core metabolic model
bac <- Bac(Ec_core,deathrate=0.05,
minweight=0.05,growtype="exponential")
#initialize a bacterium
arena <- Arena(n=20,m=20) #initialize the environment
arena <- addOrg(arena,bac,amount=10) #add 10 organisms
arena <- addSubs(arena,40) #add all possible substances
eval  <- simEnv(arena,5)
plotTotFlux(eval)
```

In the R domain:

```
>
> data(Ec_core, envir = environment())
# get Escherichia coli core metabolic model
> bac <- Bac(Ec_core,deathrate=0.05,
+ minweight=0.05,growtype="exponential")
+ #initialize a bacterium
> arena <- Arena(n=20,m=20) #initialize the environment
> arena <- addOrg(arena,bac,amount=10) #add 10 organisms
```

```
> arena <- addSubs(arena,40) #add all possible substances
> eval   <- simEnv(arena,5)

iteration: 1   organisms: 10  biomass: 4.797873 pg
        count biomass %
        ecoli_core_model  10 4.797873 0
        time total: 0.48    diffusion: 0.22 (45.8 %)

iteration: 2   organisms: 15  biomass: 10.76978 pg
        count biomass    %
        ecoli_core_model  15 10.76978 124.4699
        time total: 0.52    diffusion: 0.22 (42.3 %)

iteration: 3   organisms: 30  biomass: 21.56735 pg
        count biomass    %
        ecoli_core_model  30 21.56735 100.258
        time total: 2.34    diffusion: 1.91 (81.6 %)

iteration: 4   organisms: 60  biomass: 43.73724 pg
        count biomass    %
        ecoli_core_model  60 43.73724 102.7938
        time total: 0.96    diffusion: 0.21 (21.9 %)

iteration: 5   organisms: 120    biomass: 88.57083 pg
        count biomass    %
        ecoli_core_model  120 88.57083 102.5067
        time total: 1.49    diffusion: 0.21 (14.1 %)

> plotTotFlux(eval)
 [1] "ecoli_core_model.EX_glu_L(e)" "ecoli_core_model.GLUt2r"
"ecoli_core_model.EX_fum(e)"
 [4] "ecoli_core_model.FUMt2_2"     "ecoli_core_model.D_LACt2"
"ecoli_core_model.EX_lac_D(e)"
 [7] "ecoli_core_model.LDH_D"       "ecoli_core_model.CO2t"
"ecoli_core_model.EX_co2(e)"
[10] "ecoli_core_model.ACONTa"      "ecoli_core_model.ACONTb"
"ecoli_core_model.CS"    eaction
[13] "ecoli_core_model.ICDHyr"      "ecoli_core_model.GLUDy"
"ecoli_core_model.ACALDt"
[16] "ecoli_core_model.EX_acald(e)" "ecoli_core_model.EX_pyr(e)"
"ecoli_core_model.PYRt2r"
[19] "ecoli_core_model.EX_nh4(e)"   "ecoli_core_model.NH4t"
>
> # Outputting: Figure A8-1 plotTotFlux Diffusion and Reaction
```

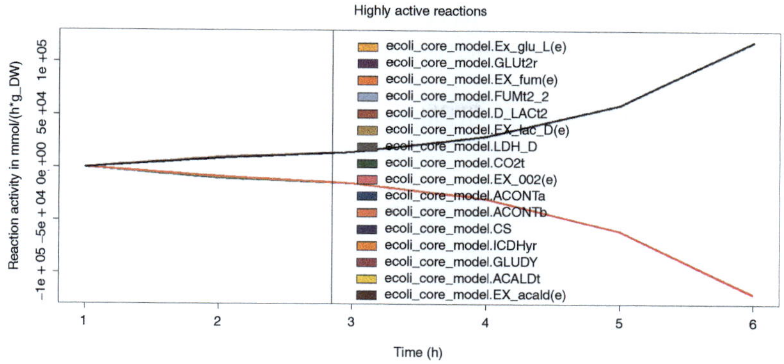

Figure A8-1 plotTotFlux Diffusion and reaction.

5.4 Some Final Remarks on Solving Partial Differential Equations Using the R Package ReacTran

Abstract

The R-package ReacTran (Soetaert and Meysman 2009) contains functions to solve reactive transport equations, as used, for example, in the environmental sciences.

Essentially, it

(1) Provides functions that subdivide the spatial extent into a number of discrete grid cells.
(2) Approximates the advective–diffusive transport term by finite differences or finite volumes.

The main package vignette (Soetaert and Meysman 2010) explains how ReacTran may be used to model reaction-transport phenomena.

However, the functions from ReacTran may be used to solve more general types of partial differential equations (≤order 2).

This vignette shows how the package may be used to solve partial differential equations of the parabolic, hyperbolic, and elliptic type, providing an example each.

Keywords: Partial Differential Equations, hyperbolic, parabolic, elliptic, R.

5.4.1 Partial Differential Equations

In partial differential equations (PDE), the function has several independent variables (e.g. time, depth, etc.) and may contain their partial derivatives.

A first step to solve PDE is to discretize one or more of the independent variables.

Usually, the independent variable "time" is not discretized, while other variables (e.g. spatial axes) are discretized, so that a set of ODE is obtained, which may be solved with appropriate initial value solvers from the package deSolve (Soetaert, Petzoldt, and Setzer 2010). This technique, the method-of-lines, applies to hyperbolic and parabolic PDEs.

For time-invariant problems, usually all independent variables are discretized and the resulting algebraic equations solved with root solving functions from package rootSolve (Soetaert 2009).

Functions tran.1D, tran.2D, and tran.3D from the R package ReacTran implement finite difference approximations of the general diffusive–advective transport equation, which for 1D is:

$$-(1/A_x\xi_x)\cdot[(\partial/\partial x)A_x\cdot\{-D(\partial\xi_x C/\partial x)\}-(\partial/\partial x)(A_x\cdot v\cdot\xi_x C)]$$

Here, D is the "diffusion coefficient", v is the "advection rate", and A_x and ξ are the surface area and volume fraction, respectively.

Assuming that A, ξ, D, and v are constant along x, one may rewrite this in a more general form: $D\partial^2C/\partial x^2 - u\partial C/\partial x$ in which the first term is a second-order and the second term a first-order derivative.

R-function tran.1D is defined as:

```
tran.1D <- function (C, C.up = C[1], C.down = C[length(C)],
                     flux.up = NULL, flux.down = NULL, a.bl.up = NULL,
                     a.bl.down = NULL, D = 0,
                     v = 0, AFDW = 1, VF = 1, A = 1, dx, ...)
```

where C.up and C.down are the upstream and downstreams boundary values, respectively, flux.up and flux.down are the upstream and downstream fluxes, respectively, v and D are the advection and diffusion coefficient, respectively, A is the surface area, x contains the grid, and VF is the volume fraction (ξ).

For the other arguments, see the help file of tran.1D.

A suitable *grid* may be generated with functions setup.grid.1D and setup.grid.2D (there is no corresponding 3D function), while a *property* may be added to this grid using functions setup.prop.1D and setup.prop.2D. These two latter functions are useful to have the variable surface areas, volume fractions, advection, and diffusion rates being defined at the correct places of the grid. These functions are defined as:

```
setup.grid.1D  <- function(x.up = 0, x.down = NULL, L = NULL, N = NULL,
                           dx.1 = NULL, ...)
setup.grid.2D  <- function(x.grid = NULL, y.grid = NULL)
setup.prop.1D <- function (func = NULL, value = NULL, xy = NULL,
                           interpolate = "spline", grid, ...)
setup.prop.2D <- function (func = NULL, value = NULL, grid, y.func = func,
                           y.value = value, ...)
```

5.4.2 A Parabolic PDE

As an example of the parabolic type, consider the 1D diffusion–reaction model, in spherical, cylindrical, and Cartesian coordinates, defined for r in [0, 10]:

$$\partial C/\partial t = (1/r^2)(\partial/\partial r)\{r^2 \cdot D \cdot (\partial C/\partial r)\} - Q$$

$$\partial C/\partial t = (1/r)(\partial/\partial r)\{r \cdot D \cdot (\partial C/\partial r)\} - Q$$

$$\partial C/\partial t = (\partial/\partial r)\{D \cdot (\partial C/\partial r)\} - Q$$

with t the time, r the (radial) distance from the origin, Q the consumption rate, and with boundary conditions (values at the edges of the model):

$$(\partial C/\partial r)_{r=0} = 0$$

$$C_{r=10} = C_{\text{ext}}$$

To solve this model in R, first the 1D model Grid is defined; it divides 10 cm (L) into 1000 equally sized boxes (N).

```
Grid <- setup.grid.1D(N = 1000, L = 10)
```

Next, the properties r and r^2 are defined on this grid:

```
r <- setup.prop.1D(grid = Grid, func = function(r) r)
r2 <- setup.prop.1D(grid = Grid, func = function(r) r^2)
```

The model equation includes a transport term, approximated by ReacTran function tran.1D() and a consumption term (Q); only the downstream boundary condition, prescribed as a concentration (C.down) which has to be specified, as the zero-gradient at the upstream boundary is the default:

```
library(ReacTran)
pde1D <- function(t, C, parms, A = 1) {
   tran <- tran.1D(C = C, A = A, D = D, C.down = Cext,
     dx = Grid)$dC
   list(tran - Q)
}
```

The model parameters are defined:

```
D <- 1
Q <- 1
Cext <- 20
```

5.4.2.1 Steady-State Solution

In the first application, the model is solved to *steady-state*, which retrieves the condition where the concentrations are invariant, e.g. for the cylindrical coordinate case:

$$0 = (1/r^2) \cdot (\partial/\partial r)\{r^2 \cdot D \cdot (\partial C/\partial r)\} - Q$$

The R package ReacTran (**Soetaert and Meysman** 2012) contains functions for … R-**package** ReacTran: Reactive Transport Modelling in R … and Herman, **2009**. In R, steady-state conditions may be estimated using functions from the package rootSolve which implement among others a Newton–Raphson algorithm (Press, Teukolsky, Vetterling, and Flannery 1992). For one-dimensional models, steady.1D should be used. The initial estimate of the steady-state solution (y) is unimportant; here, one may take simply N random numbers. Argument nspec = 1 informs the solver that only one component is described.

Although a system of 1000 equations needs to be solved, this takes only a fraction of a second:

```
library(rootSolve)
Cartesian <- steady.1D(y = runif(Grid$N),
  func = pde1D, parms = NULL, nspec = 1, A = 1)
Cylindrical <- steady.1D(y = runif(Grid$N),
  func = pde1D, parms = NULL, nspec = 1, A = r)

print(system.time(
  Spherical <- steady.1D(y = runif(Grid$N),
    func = pde1D, parms = NULL, nspec = 1, A = r2)
))

user  system  elapsed
0.02    0.00     0.00
```

The values of the state variables (y) are plotted against the radial distance, in the middle of the grid cells (Grid$x.mid).

```
plot (Grid$x.mid, Cartesian$y, type = "l", main = "steady-state PDE",
  lwd = 3, xlab = "x", ylab = "C", col = "darkgreen", lty = 1)
lines(Grid$x.mid, Cylindrical$y, lwd = 3, col="blue", lty = 2)
lines(Grid$x.mid, Spherical$y, lwd = 3, col = "red", lty = 3)
legend("bottomright", c("cartesian", "cylindrical", "spherical"),
  col = c("darkgreen", "blue", "red"), lwd = 3, lty = 1:3)
```

The analytical solutions compare well with the numerical approximation for all cases studied.

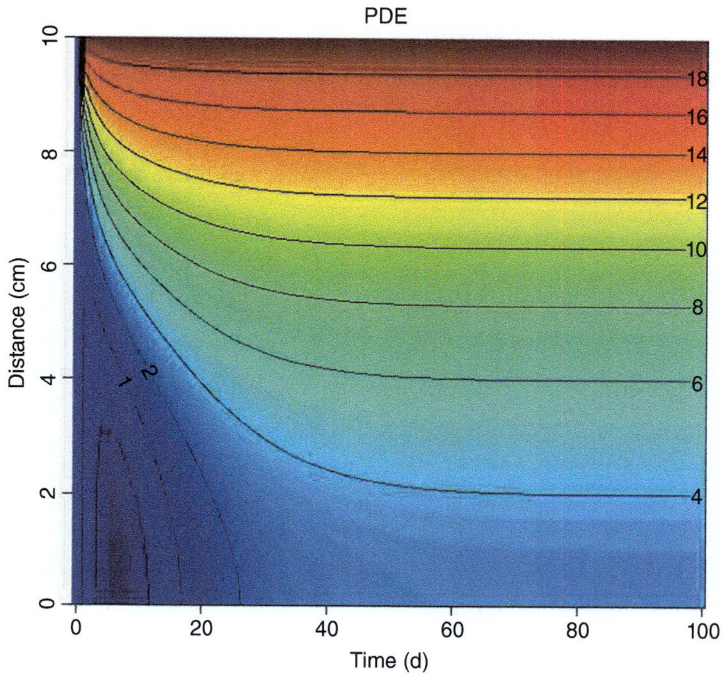

Figure 5.2 Dynamic solution of the 1D diffusion–reaction model.

5.4.2.2 The Method of Lines

Next, the model (for spherical coordinates) is run dynamically for 100 time units using the deSolve() function ode.1D, and starting with an initially uniform distribution (y = rep(1, Grid$N)):

```
require(deSolve)
times <- seq(0, 100, by = 1)
system.time(
  out <- ode.1D(y = rep(1, Grid$N), times = times, func = pde1D,
   parms = NULL, nspec = 1, A = r2)
)

user  system  elapsed
0.54   0.04    0.56
```

Here, out is a matrix, whose first column contains the output times, and the next columns the values of the state variables in the different boxes:

```
tail(out[, 1:4], n = 3)
```

```
          time        1        2        3
[99,]      98 3.332278 3.332303 3.332366
[100,]     99 3.332383 3.332408 3.332471
[101,]    100 3.332478 3.332503 3.332566
```

Plotting the result using a blue-yellow-red color scheme, and using deSolve's S3 method `image`:

```
image(out, grid = Grid$x.mid, xlab = "time, days",
      ylab = "Distance, cm", main = "PDE", add.contour = TRUE)
```

Further Reading

https://en.wikipedia.org/wiki/Advection

Soetaert, K. and Herman, P.M.J. (2009). *A Practical Guide to Ecological Modelling – Using R as a Simulation Platform*. Springer.

Soetaert, K. and Meysman, F., Centre for Estuarine and Marine Ecology, Netherlands Institute of Ecology, The Netherlands. R-package ReacTran: reactive transport modelling in R Project. https://cran.r-project.org/web/packages/ReacTran/vignettes/ReacTran.pdf.

6

Solving Partial Differential Equations, Generally Applicable to Modeling Simultaneous Mass Transfer and Chemical Reactions, Using the R Package ReacTran

6.1 Partial Differential Equations (PDE)

In partial differential equations (PDE), the function has several independent variables (e.g. time and distance) and may also contain their partial derivatives. A first step to solve PDE is to discretize one or more of the independent variables. Usually, the independent variable "time" is not discretized, while other variables (e.g. spatial axes) are discretized, so that a set of ordinary differential equations (ODE) is obtained, which then may be solved with appropriate initial values solvers from the package deSolve (Soetaert, Petzoldt, and Setzer 2010). This technique, the ***method-of-lines***, applies to hyperbolic and parabolic PDEs.

For time-invariant problems, usually all independent variables are discretized, and the resulting algebraic equations solved with root-solving functions from package rootSolve (Soetaert 2009). Functions tran.1D, tran.2D, and tran.3D from the R package ReacTran implement finite difference approximations of the general diffusive–advective transport equation, which, for 1D is

$$-(1/A_x\xi_x)\cdot[(\partial/\partial x)A_x\cdot\{-D(\partial\xi_x C/\partial x)\}-(\partial/\partial x)(A_x\cdot v\cdot\xi_x C)]$$

Here D is the "diffusion coefficient," v is the "advection rate," and A_x and ξ are the surface area and volume fraction, respectively.

Assuming that A, ξ, D, and v are constant along x, one may rewrite this in a more general form:

$$D\partial^2C/\partial x^2-u\partial C/\partial x \tag{6.1}$$

in which the first term is a second-order, the second term a first-order derivative.

R-function tran.1D is defined as

```
tran.1D <- function (C, C.up = C[1], C.down = C[length(C)],
                     flux.up = NULL, flux.down = NULL, a.bl.up = NULL,
                     a.bl.down = NULL, D = 0,
                     v = 0, AFDW = 1, VF = 1, A = 1, dx, ...)
```

Simultaneous Mass Transfer and Chemical Reactions in Engineering Science, First Edition. Bertram K. C. Chan.

where

- C.up and C.down are the upstream and downstream boundary values,
- flux.up and flux.down are the upstream and downstream fluxes,
- v and D are the advection and diffusion coefficients, respectively,
- A is the surface area,
- x contains the grid, and
- VF is the volume fraction (ξ).

For the other arguments, see the help file of tran.1D. A suitable *grid* may be generated with functions setup.grid.1D and setup.grid.2D (there is no corresponding 3D function), while a *property* may be added to this grid using functions setup.prop.1D, and setup.prop.2D. These latter two functions are useful to have the variable surface areas, volume fractions, advection, and diffusion rates being defined at the correct places of the grid.

These functions are defined as

```
setup.grid.1D <- function(x.up = 0, x.down = NULL, L = NULL, N = NULL,
                          dx.1 = NULL, ...)
setup.grid.2D <- function(x.grid = NULL, y.grid = NULL)
setup.prop.1D <- function (func = NULL, value = NULL, xy = NULL,
                           interpolate = "spline", grid, ...)
setup.prop.2D <- function (func = NULL, value = NULL, grid, y.func = func,
                           y.value = value, ...)
```

6.2 A Parabolic PDE

As an example of the parabolic type, consider the 1D diffusion-reaction model, in spherical, cylindrical, and Cartesian coordinates, defined, respectively, for r in $[0, 10]$:

$$\partial C/\partial t = (1/r^2)(\partial/\partial r)\{r^2 \cdot D \cdot (\partial C/\partial r)\} - Q \tag{6.2}$$

$$\partial C/\partial t = (1/r)(\partial/\partial r)\{r \cdot D \cdot (\partial C/\partial r)\} - Q \tag{6.3}$$

$$\partial C/\partial t = (\partial/\partial r)\{D \cdot (\partial C/\partial r)\} - Q \tag{6.4}$$

with t the time, r the (radial) distance from the origin, Q the consumption rate, and with boundary conditions (values at the edges of the model):

$$(\partial C/\partial r)_{r=0} = 0 \tag{6.5}$$

$$C_{r=10} = C_{\text{ext}} \tag{6.6}$$

To solve this model in R, first the 1D model Grid is defined; it divides 10 cm (L) into 1000 equally sized boxes (N).

```
Grid <- setup.grid.1D(N = 1000, L = 10)
```

Next the properties r and r^2 are defined on this grid:

```
r <- setup.prop.1D(grid = Grid, func = function(r) r)
r2 <- setup.prop.1D(grid = Grid, func = function(r) r^2)
```

The model equation includes a transport term, approximated by ReacTran function tran.1D() and a consumption term (Q); only the downstream boundary condition, prescribed as a concentration (C.down) which has to be specified, as the zero gradient at the upstream boundary is the default:

```
library(ReacTran)
pde1D <- function(t, C, parms, A = 1) {
     tran <- tran.1D(C = C, A = A, D = D, C.down = Cext,
          dx = Grid)$dC
     list(tran - Q)
}
```

The model parameters are defined:

```
D <- 1
Q <- 1
Cext <- 20
```

6.3 Steady-State Solution

In a first application, the model is solved to *steady state*, which retrieves the condition where the concentrations are invariant, e.g. for the cylindrical coordinate case:

$$0 = (1/r^2) \cdot (\partial/\partial r)\{r^2 \cdot D \cdot (\partial C/\partial r)\} - Q \tag{6.7}$$

The R package ReacTran (Soetaert and Meysman 2012) contains functions for ... R-**package** ReacTran: Reactive Transport Modelling in R ... and Herman, **2009**. In R, steady-state conditions may be estimated using functions from the package rootSolve, which implement among others a Newton–Raphson algorithm (Press, Teukolsky, Vetterling, and Flannery 1992). For one-dimensional models, steady.1D should be used. The initial estimate of the steady-state solution (y) is unimportant; here one may take simply N random numbers.

Argument nspec = 1 informs the solver that only one component is described. Although a system of 1000 equations needs to be solved, this takes only a fraction of a second:

```
library(rootSolve)
Cartesian <- steady.1D(y = runif(Grid$N),
   func = pde1D, parms = NULL, nspec = 1, A = 1)
```

```
Cylindrical <- steady.1D(y = runif(Grid$N),
   func = pde1D, parms = NULL, nspec = 1, A = r)

print(system.time(
  Spherical <- steady.1D(y = runif(Grid$N),
      func = pde1D, parms = NULL, nspec = 1, A = r2)
))

user   system   elapsed
0.02     0.00      0.00
```

The values of the state variables (y) are plotted against the radial distance in the middle of the grid cells (Grid$x.mid).

```
plot (Grid$x.mid, Cartesian$y, type = "l", main = "steady-state PDE",
  lwd = 3, xlab = "x", ylab = "C", col = "darkgreen", lty = 1)
lines(Grid$x.mid, Cylindrical$y, lwd = 3, col="blue", lty = 2)
lines(Grid$x.mid, Spherical$y, lwd = 3, col = "red", lty = 3)
legend("bottomright", c("cartesian", "cylindrical", "spherical"),
  col = c("darkgreen", "blue", "red"), lwd = 3, lty = 1:3)
> #
> # Outputting: Figure 6.1: tran.1D: Steady-state Solution of the 1-D
> #                              Diffusion-Reaction Model
> #
```

The analytical solutions compare well with the numerical approximation for all three cases (Figure 6.1):

```
max(abs(Q/6/D*(r2$mid - 10^2) + Cext - Spherical$y))
[1] 5.82081e-05
max(abs(Q/4/D*(r2$mid - 10^2) + Cext - Cylindrical$y))
```

***Remark*:** Note that there is no automatic error checking/control here, so to reduce this error, the number of boxes may be increased.

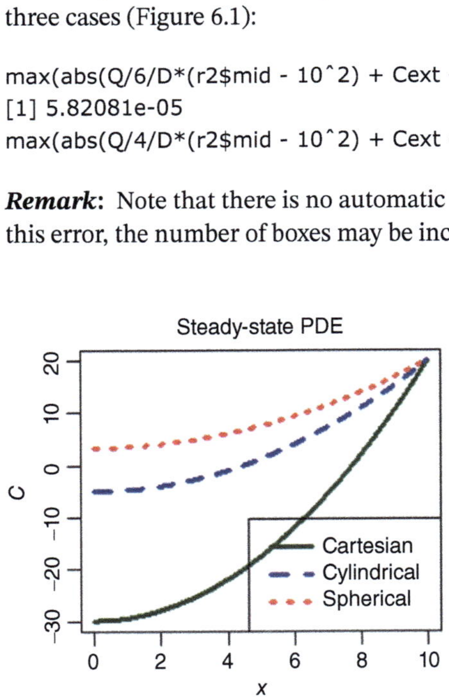

Figure 6.1 tran.1D: Steady-state solution of the 1D diffusion-reaction model.

6.3.1 The Method of Lines

Next, the model (for spherical coordinates) is run dynamically for 100-time units using the deSolve() function ode.1D and starting with an initially uniform distribution (y = rep(1, Grid$N)) (Figure 6.2):

```
require(deSolve)
times <- seq(0, 100, by = 1)
system.time(
  out <- ode.1D(y = rep(1, Grid$N), times = times, func = pde1D,
    parms = NULL, nspec = 1, A = r2)
)
```

```
user  system  elapsed
0.54    0.04     0.56
```

Here, out is a matrix, whose first column contains the output times and the next columns the values of the state variables in the different boxes:

```
tail(out[, 1:4], n = 3)
```

```
       time        1        2        3
[99,]    98 3.332278 3.332303 3.332366
[100,]   99 3.332383 3.332408 3.332471
[101,]  100 3.332478 3.332503 3.332566
```

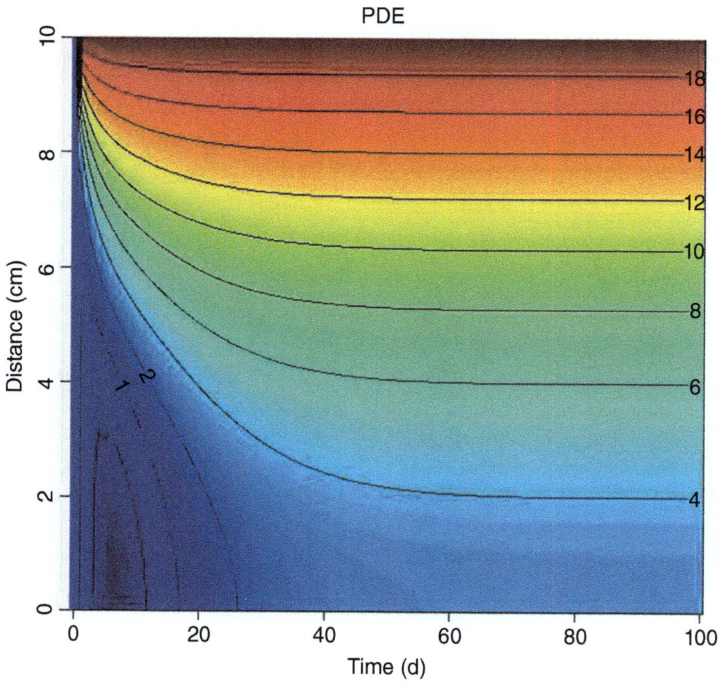

Figure 6.2 Dynamic solution of the 1D diffusion-reaction model.

Plotting the result using a blue-yellow-red color scheme and using deSolve's S3 method `image`:

```
image(out, grid = Grid$x.mid, xlab = "time, days",
      ylab = "Distance, cm", main = "PDE", add.contour = TRUE)
```

6.3.2 A Hyperbolic PDE

The equation for a wave traveling in one direction (x) is given by

$$\partial^2 u/\partial t^2 = c^2 \partial^2 u/\partial x^2 \tag{6.8}$$

where c is the propagation speed of the wave, and u is the variable that changes as the wave passes. This equation is second order in both t and x. The wave equation is the prototype of a "hyperbolic" PDE.

For it to be solved in R, the equation is rewritten as two simultaneous/coupled equations, first order in time:

$$\partial u_1/\partial t = u_2 \tag{6.9}$$

$$\partial u_2/\partial t = c^2 \partial^2 u_1/\partial x^2 \tag{6.10}$$

One may solve the equation with the following initial and boundary conditions:

$$u_1(0, x) = \exp^{-0.05x^2} \tag{6.11}$$

$$u_2(0, x) = 0 \tag{6.12}$$

$$u_{t,-\infty} = 0 \tag{6.13}$$

$$u_{t,\infty} = 0 \tag{6.14}$$

where the first condition represents a Gaussian pulse.

The implementation in R starts with defining the box size `dx` and the grid, `xgrid`. To comply with the boundary conditions (which are defined at ∞), the grid needs to be taken large enough such that u remains effectively 0 at the boundaries, for all run times.

Here, the grid extends from −100 to 100:

```
dx    <- 0.2
xgrid <- setup.grid.1D(-100, 100, dx.1 = dx)
x     <- xgrid$x.mid
N     <- xgrid$N
```

The initial condition, `yini` and output `times` are defined next:

```
uini <- exp(-0.05 * x^2)
vini <- rep(0, N)
yini <- c(uini, vini)
times <- seq (from = 0, to = 50, by = 1)
```

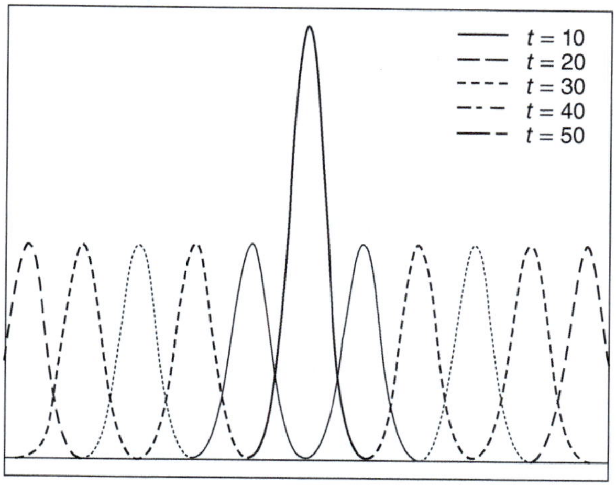

Figure 6.3 The 1D wave equation; black = initial condition; grey = several time lines.

The wave equation derivative function first extracts, from state variable vector y the two properties u1, u2, both of length N, after which the ReacTran function tran.1D() performs the transport of u1. The squared velocity (c^2) is taken as 1 (D=1).

The wave equation may be solved efficiently with a nonstiff solver such as the Runge–Kutta method ode45.

Now plot the results (Figure 6.3); initial condition in black, the values for selected time points in dark gray; a legend with times is written.

S3-method imagemay also be used to generate persp-like plots, for example, see Figure 6.4:

```
par(mar=c(0,0,0,0))
image(out, which = "u", method = "persp", main = "",
```

Figure 6.4 The 1D wave equation as a persp plot.

```
border = NA, col = "lightblue", box = FALSE, shade = 0.5, theta = 0,
phi = 60)
```

Outputting: **Figure 6.4**

6.4 The General 3D Advective–Diffusive Transport PDE

One may also like to try the following “movie”:

```
plot.1D(out, grid = x, which = "u", type = "l",
lwd = 2, ylim = c(0,1), ask = TRUE)
```

6.4.1 An Elliptic PDE

The final example describes again a diffusion-reaction system with production p, consumption rC, and diffusive transport (diffusion coefficients Dx, Dy) of a substance C in two dimensions (x, y); the boundaries are prescribed as zero gradient (the default).

$$\partial C/\partial t = (\partial/\partial x)[D_x \cdot \partial C/\partial x] + \partial/\partial y[D_y \cdot (\partial C/\partial y)] - rC + p_{xy} \tag{6.15}$$

The parameter p_{xy} is the production rate, which is zero everywhere except for 50 randomly positioned spots where it is 1.

Transport is performed by ReacTran function tran.2D(); the state variable vector (y) is recast in matrix form (CONC) before it is transported. The first-order consumption rate -r * CONC is added to the rate of change owing to transport (Tran$dC). Production, p is added to 50 cells indexed by ii and are randomly selected from the grid. The function returns a list, containing the derivatives, as a vector:

```
require(ReacTran)
pde2D <- function (t, y, parms) {
CONC <- matrix(nr = n, nc = n, y)
Tran <- tran.2D(CONC, D.x = Dx, D.y = Dy, dx = dx, dy = dy)
dCONC <- Tran$dC - r * CONC
dCONC[ii]<- dCONC[ii] + p
return(list(as.vector(dCONC)))
}
```

Before running the model, the grid sizes (dx, dx), diffusion coefficients (Dx, Dy), first-order consumption rate (r) are defined. There are 100 boxes in x- and y-direction (n). Furthermore, one assumes that the substance is produced in 50 randomly chosen cells (ii) at a constant rate (p):

```
n <- 100
dy <- dx <- 1
```

```
Dy <- Dx <- 2
r <- 0.001
p <- runif(50)
ii <- trunc(cbind(runif(50) * n, runif(50) * n) + 1)
```

The steady state is found using rootSolve's function steady.2D. It takes as arguments among other the dimensionality of the problem (dimens) and the length of the work array needed by the solver (lrw = 600000). It takes less than 0.5 seconds to solve this 10 000 state variable model (Figure 6.5).

```
require(rootSolve)
Conc0 <- matrix(nr = n, nc = n, 10)
print(system.time(ST <- steady.2D(y = Conc0, func = pde2D,
    parms = NULL, dimens = c(n, n), lrw = 6e+05)))

user    system    elapsed
0.16      0.00       0.15

image(ST, main = "steady-state 2D PDE")
```

****

In the R domain:

```
R version 3.3.3 (2017-03-06) -- "Another Canoe"
Copyright (C) 2017 The R Foundation for Statistical Computing
Platform: x86_64-w64-mingw32/x64 (64-bit)
```

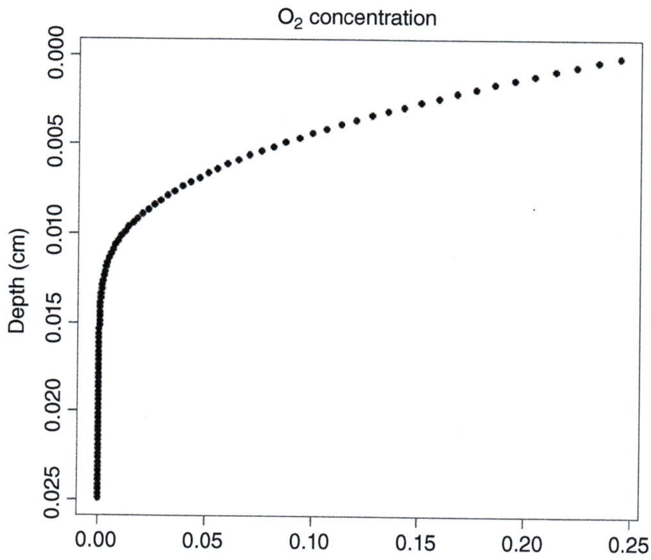

Figure 6.5 rootSolve's function steady.2D.

```
R is free software and comes with ABSOLUTELY NO WARRANTY.
You are welcome to redistribute it under certain conditions.
Type 'license()' or 'licence()' for distribution details.

  Natural language support but running in an English locale.

R is a collaborative project with many contributors.
Type 'contributors()' for more information and
'citation()' on how to cite R or R packages in publications.

Type 'demo()' for some demos, 'help()' for on-line help, or
'help.start()' for an HTML browser interface to help.
Type 'q()' to quit R.

[Previously saved workspace restored]

> install.packages("ReacTran")
Installing package into 'C:/Users/Bert/Documents/R/win-library/3.3'
(as 'lib' is unspecified)

--- Please select a CRAN mirror for use in this session ---
trying URL 'https://mirror-
hk.koddos.net/CRAN/bin/windows/contrib/3.3/ReacTran_1.4.3.1.zip'
Content type 'application/zip' length 915996 bytes (894 KB)
downloaded 894 KB

package 'ReacTran' successfully unpacked and MD5 sums checked

The downloaded binary packages are in
C:\Users\Bert\AppData\Local\Temp\Rtmps5qDpw\downloaded_packages
> require(ReacTran)
Loading required package: ReacTran
Loading required package: rootSolve
Loading required package: deSolve
Loading required package: shape
> library(ReacTran)
> ls("package:ReacTran")
 [1] "advection.1D"        "advection.volume.1D" "fiadeiro"
 [4] "g.cylinder"          "g.sphere"            "g.spheroid"
 [7] "p.exp"               "p.lin"               "p.sig"
[10] "polar2cart"          "setup.compaction.1D" "setup.grid.1D"
[13] "setup.grid.2D"       "setup.prop.1D"       "setup.prop.2D"
[16] "tran.1D"             "tran.2D"             "tran.3D"
[19] "tran.cylindrical"    "tran.polar"          "tran.spherical"
[22] "tran.volume.1D"      "tran.volume.2D"      "tran.volume.3D"
>
```

```
> library(rootSolve)
>
> ls("package:rootSolve")
 [1] "gradient"     "hessian"      "jacobian.band" "jacobian.full"
 [5] "multiroot"    "multiroot.1D" "runsteady"     "steady"
 [9] "steady.1D"    "steady.2D"    "steady.3D"     "steady.band"
[13] "stode"        "stodes"       "uniroot.all"
>
> Cartesian <- steady.1D(y = runif(Grid$N),
+   func = pde1D, parms = NULL, nspec = 1, A = 1)
> Cylindrical <- steady.1D(y = runif(Grid$N),
+   func = pde1D, parms = NULL, nspec = 1, A = r)
>
> print(system.time(
+   Spherical <- steady.1D(y = runif(Grid$N),
+     func = pde1D, parms = NULL, nspec = 1, A = r2)
+ ))
>
> plot (Grid$x.mid, Cartesian$y, type = "l", main = "steady-state PDE",
+
+   lwd = 3, xlab = "x", ylab = "C", col = "darkgreen", lty = 1)
> lines(Grid$x.mid, Cylindrical$y, lwd = 3, col="blue", lty = 2)
>
> lines(Grid$x.mid, Spherical$y, lwd = 3, col = "red", lty = 3)
>
> legend("bottomright", c("cartesian", "cylindrical", "spherical"),
+
+   col = c("darkgreen", "blue", "red"), lwd = 3, lty = 1:3)
>
> max(abs(Q/6/D*(r2$mid - 10^2) + Cext - Spherical$y))
>
>
> Grid <- setup.grid.1D(N = 1000, L = 10)
>
> r <- setup.prop.1D(grid = Grid, func = function(r) r)
>
> r2 <- setup.prop.1D(grid = Grid, func = function(r) r^2)
> library(ReacTran)
> pde1D <- function(t, C, parms, A = 1) {
+ tran <- tran.1D(C = C, A = A, D = D, C.down = Cext,
+             dx = Grid)$dC
+       list(tran - Q)
+ }
> D <- 1
> Q <- 1
> Cext <- 20
> library(rootSolve)
```

```
> Cartesian <- steady.1D(y = runif(Grid$N),
+   func = pde1D, parms = NULL, nspec = 1, A = 1)
> Cylindrical <- steady.1D(y = runif(Grid$N),
+   func = pde1D, parms = NULL, nspec = 1, A = r)
> print(system.time(
+   Spherical <- steady.1D(y = runif(Grid$N),
+      func = pde1D, parms = NULL, nspec = 1, A = r2)
+ ))

   user  system elapsed
      0       0       0
>
> plot (Grid$x.mid, Cartesian$y, type = "l", main = "steady-state PDE",
+
+   lwd = 3, xlab = "x", ylab = "C", col = "darkgreen", lty = 1)
>
> lines(Grid$x.mid, Cylindrical$y, lwd = 3, col="blue", lty = 2)
>
> lines(Grid$x.mid, Spherical$y, lwd = 3, col = "red", lty = 3)
>
> legend("bottomright", c("cartesian", "cylindrical", "spherical"),
+
+    col = c("darkgreen", "blue", "red"), lwd = 3, lty = 1:3)
>
```

> ***Outputting:*** **Figure 6.5**

Example 6.1 tran.3D ***General Three-Dimensional Advective–Diffusive Transport***

Description

This program estimates the transport term (i.e. the rate of change of a concentration owing to diffusion and advection) in a three-dimensional rectangular model domain.

Usage

```
tran.3D (C, C.x.up = C[1,,], C.x.down = C[dim(C)[1],,],
         C.y.up = C[ ,1, ], C.y.down=C[ ,dim(C)[2], ],
         C.z.up = C[ , ,1], C.z.down=C[ , ,dim(C)[3]],
         flux.x.up = NULL, flux.x.down = NULL,
         flux.y.up = NULL, flux.y.down = NULL,
         flux.z.up = NULL, flux.z.down = NULL,
         a.bl.x.up = NULL, a.bl.x.down = NULL,
```

```
a.bl.y.up = NULL, a.bl.y.down = NULL,
a.bl.z.up = NULL, a.bl.z.down = NULL,
D.grid = NULL, D.x = NULL, D.y = D.x, D.z = D.x,
v.grid = NULL, v.x = 0, v.y = 0, v.z = 0,
AFDW.grid = NULL, AFDW.x = 1, AFDW.y = AFDW.x,
AFDW.z = AFDW.x,
VF.grid = NULL, VF.x = 1, VF.y = VF.x, VF.z = VF.x,
A.grid = NULL, A.x = 1, A.y = 1, A.z = 1,
grid = NULL, dx = NULL, dy = NULL, dz = NULL,
full.check = FALSE, full.output = FALSE)
```

Arguments

C	Concentration, expressed per unit volume, defined at the center of each grid cell; Nx*Ny*Nz array [M/L^3]
C.x.up	Concentration at upstream boundary in x-direction; matrix of dimensions Ny*Nz [M/L^3]
C.x.down	Concentration at downstream boundary in x-direction; matrix of dimensions Ny*Nz [M/L^3]
C.y.up	Concentration at upstream boundary in y-direction; matrix of dimensions Nx*Nz [M/L^3]
C.y.down	Concentration at downstream boundary in y-direction; matrix of dimensions Nx*Nz [M/L^3]
C.z.up	Concentration at upstream boundary in z-direction; matrix of dimensions Nx*Ny [M/L^3]
C.z.down	Concentration at downstream boundary in z-direction; matrix of dimensions Nx*Ny [M/L^3]
flux.x.up	Flux across the upstream boundary in x-direction, positive = INTO model domain; matrix of dimensions Ny*Nz [M/L^2/T]
flux.x.down	Flux across the downstream boundary in x-direction, positive = OUT of model domain; matrix of dimensions Ny*Nz [M/L^2/T]
flux.y.up	Flux across the upstream boundary in y-direction, positive = INTO model domain; matrix of dimensions Nx*Nz [M/L^2/T]
flux.y.down	Flux across the downstream boundary in y-direction, positive = OUT of model domain; matrix of dimensions Nx*Nz [M/L^2/T]
flux.z.up	Flux across the upstream boundary in z-direction, positive = INTO model domain; matrix of dimensions Nx*Ny [M/L^2/T]
flux.z.down	Flux across the downstream boundary in z-direction, positive = OUT of model domain; matrix of dimensions Nx*Ny [M/L^2/T]
a.bl.x.up	Transfer coefficient across the upstream boundary layer in x-direction; `Flux=a.bl.x.up*(C.x.up-C[1,,])`. One value [L/T]

`a.bl.x.down`	Transfer coefficient across the downstream boundary layer in *x*-direction; `Flux=a.bl.x.down*(C[Nx,,]-C.x.down)`. One value [L/T]
`a.bl.y.up`	Transfer coefficient across the upstream boundary layer in *y*-direction; `Flux=a.bl.y.up*(C.y.up-C[,1,])`. One value [L/T]
`a.bl.y.down`	Transfer coefficient across the downstream boundary layer in *y*-direction; `Flux=a.bl.y.down*(C[,Ny,]-C.y.down)`. One value [L/T]
`a.bl.z.up`	Transfer coefficient across the upstream boundary layer in *y*-direction; `Flux=a.bl.y.up*(C.y.up-C[,,1])`. One value [L/T]
`a.bl.z.down`	Transfer coefficient across the downstream boundary layer in *z*-direction; `Flux=a.bl.z.down*(C[,,Nz]-C.z.down)`. One value [L/T]
`D.grid`	Diffusion coefficient defined on all grid cell interfaces. Should contain elements `x.int`, `y.int`, `z.int`, arrays with the values on the interfaces in `x`-, `y`-, and `z`-directions, and with dimensions `(Nx+1)*Ny*Nz`, `Nx*(Ny+1)*Nz`, and `Nx*Ny*(Nz+1)`, respectively [L^2/T]
`D.x`	Diffusion coefficient in *x*-direction, defined on grid cell interfaces. One value, a vector of length (Nx+1), or a (Nx+1)*Ny*Nz array [L^2/T]
`D.y`	Diffusion coefficient in *y*-direction, defined on grid cell interfaces. One value, a vector of length (Ny+1), or a Nx*(Ny+1)*Nz array [L^2/T]
`D.z`	Diffusion coefficient in *z*-direction, defined on grid cell interfaces. One value, a vector of length (Nz+1), or a Nx*Ny*(Nz+1) array [L^2/T]
`v.grid`	Advective velocity defined on all grid cell interfaces. Can be positive (downstream flow) or negative (upstream flow). Should contain elements `x.int`, `y.int`, `z.int`, arrays with the values on the interfaces in *x*-, *y*-, and *z*-directions, and with dimensions (Nx+1)*Ny*Nz, Nx*(Ny+1)*Nz, and Nx*Ny*(Nz+1), respectively [L/T]
`v.x`	Advective velocity in the *x*-direction, defined on grid cell interfaces. Can be positive (downstream flow) or negative (upstream flow). One value, a vector of length (Nx+1), or a (Nx+1)*Ny*Nz array [L/T]
`v.y`	Advective velocity in the *y*-direction, defined on grid cell interfaces. Can be positive (downstream flow) or negative (upstream flow). One value, a vector of length (Ny+1), or a Nx*(Ny+1)*Nz array [L/T]
`v.z`	Advective velocity in the *z*-direction, defined on grid cell interfaces. Can be positive (downstream flow) or negative

	(upstream flow). One value, a vector of length (Nz+1), or a Nx*Ny*(Nz+1) array [L/T]
AFDW.grid	Weight used in the finite difference scheme for advection in the *x*-direction, defined on grid cell interfaces; backward = 1, centered = 0.5, forward = 0; default is backward. Should contain elements x.int, y.int, z.int, arrays with the values on the interfaces in *x*-, *y*-, and *z*-directions, and with dimensions (Nx+1)*Ny*Nz, Nx*(Ny+1)*Nz, and Nx*Ny*(Nz+1), respectively [-]
AFDW.x	Weight used in the finite difference scheme for advection in the *x*-direction, defined on grid cell interfaces; backward = 1, centered = 0.5, forward = 0; default is backward. One value, a vector of length (Nx+1), a prop.1D list created by setup.prop.1D, or a (Nx+1)*Ny*Nz array [-]
AFDW.y	Weight used in the finite difference scheme for advection in the *y*-direction, defined on grid cell interfaces; backward = 1, centered = 0.5, forward = 0; default is backward. One value, a vector of length (Ny+1), a prop.1D list created by setup.prop.1D, or a Nx*(Ny+1)*Nz array [-]
AFDW.z	Weight used in the finite difference scheme for advection in the *z*-direction, defined on grid cell interfaces; backward = 1, centered = 0.5, forward = 0, default is backward. One value, a vector of length (Nz+1), a prop.1D list created by setup.prop.1D, or a Nx*Ny*(Nz+1) array [-]
VF.grid	Volume fraction. A list should contain elements x.int, y.int, z.int, arrays with the values on the interfaces in *x*-, *y*-, and *z*-directions, and with dimensions (Nx+1)*Ny*Nz, Nx*(Ny+1)*Nz, and Nx*Ny*(Nz+1), respectively [-]
VF.x	Volume fraction at the grid cell interfaces in the *x*-direction. One value, a vector of length (Nx+1), a prop.1D list created by setup.prop.1D, or a (Nx+1)*Ny*Nz array [-]
VF.y	Volume fraction at the grid cell interfaces in the *y*-direction. One value, a vector of length (Ny+1), a prop.1D list created by setup.prop.1D, or a Nx*(Ny+1)*Nz array [-]
VF.z	Volume fraction at the grid cell interfaces in the *z*-direction. One value, a vector of length (Nz+1), a prop.1D list created by setup.prop.1D, or a Nx*Ny*(Nz+1) array [-]
A.grid	Interface area, a list should contain elements x.int, y.int, z.int, arrays with the values on the interfaces in *x*-, *y*-, and *z*-directions, and with dimensions (Nx+1)*Ny*Nz, Nx*(Ny+1)*Nz, and Nx*Ny*(Nz+1), respectively [L^2]
A.x	Interface area defined at the grid cell interfaces in the *x*-direction. One value, a vector of length (Nx+1), a prop.1D list created by setup.prop.1D, or a (Nx+1)*Ny*Nz array [L^2]

A.y	Interface area defined at the grid cell interfaces in the *y*-direction. One value, a vector of length (Ny+1), a prop.1D list created by setup.prop.1D, or a Nx*(Ny+1)*Nz array [L^2]
A.z	Interface area defined at the grid cell interfaces in the *z*-direction. One value, a vector of length (Nz+1), a prop.1D list created by setup.prop.1D, or a Nx*Ny*(Nz+1) array [L^2]
dx	Distance between adjacent cell interfaces in the *x*-direction (thickness of grid cells). One value or vector of length Nx [L]
dy	Distance between adjacent cell interfaces in the *y*-direction (thickness of grid cells). One value or vector of length Ny [L]
dz	Distance between adjacent cell interfaces in the *z*-direction (thickness of grid cells). One value or vector of length Nz [L]
grid	Discretization grid, a list containing at least elements dx, dx.aux, dy, dy.aux, dz, dz.aux (see setup.grid.2D) [L]
full.check	Logical flag enabling a full check of the consistency of the arguments (default = FALSE; TRUE slows down execution by 50%)
full.output	Logical flag enabling a full return of the output (default = FALSE; TRUE slows down execution by 20%)

Details

Do not use this with too large grid.

The boundary conditions are either

(1) zero gradient
(2) fixed concentration
(3) convective boundary layer
(4) fixed flux.

This is also the order of priority. The zero gradient is the default, the fixed flux overrules all others.

Value

A list containing

dC	The rate of change of the concentration *C* owing to transport, defined in the center of each grid cell, an array with dimension Nx*Ny*Nz [$M/L^3/T$]
C.x.up	Concentration at the upstream interface in *x*-direction. A matrix of dimension Ny*Nz [M/L^3]. Only when full.output = TRUE
C.x.down	Concentration at the downstream interface in *x*-direction. A matrix of dimension Ny*Nz [M/L^3]. Only when full.output = TRUE
C.y.up	Concentration at the upstream interface in *y*-direction. A matrix of dimension Nx*Nz [M/L^3]. Only when full.output = TRUE

`C.y.down`	Concentration at the downstream interface in y-direction. A matrix of dimension Nx*Nz [M/L^3]. Only when `full.output = TRUE`
`C.z.up`	Concentration at the upstream interface in z-direction. A matrix of dimension Nx*Ny [M/L]. Only when `full.output = TRUE`
`C.z.down`	Concentration at the downstream interface in z-direction. A matrix of dimension Nx*Ny [M/L^3]. Only when `full.output = TRUE`
`x.flux`	Flux across the interfaces in x-direction of the grid cells. An (Nx+1)*Ny*Nz array [M/L^2/T]. Only when `full.output = TRUE`
`y.flux`	Flux across the interfaces in y-direction of the grid cells. An Nx*(Ny+1)*Nz array [M/L^2/T]. Only when `full.output = TRUE`
`z.flux`	Flux across the interfaces in z-direction of the grid cells. A Nx*Ny*(Nz+1) array [M/L^2/T]. Only when `full.output=TRUE`
`flux.x.up`	Flux across the upstream boundary in x-direction, `positive = INTO` model domain. A matrix of dimension Ny*Nz [M/L^2/T]
`flux.x.down`	Flux across the downstream boundary in x-direction, `positive= OUT` model domain. A matrix of dimension Ny*Nz [M/L^2/T]
`flux.y.up`	Flux across the upstream boundary in y-direction, `positive = INTO` model domain. A matrix of dimension Nx*Nz [M/L^2/T]
`flux.y.down`	Flux across the downstream boundary in y-direction, `positive= OUT` of model domain. A matrix of dimension Nx*Nz [M/L^2/T]
flux.z.up	Flux across the upstream boundary in z-direction, `positive = INTO` model domain. A matrix of dimension Nx*Ny [M/L^2/T]
flux.z.down	Flux across the downstream boundary in z-direction, `positive = OUT` of model domain. A matrix of dimension Nx*Ny [M/L^2/T]

Authors

Filip Meysman <filip.meysman@nioz.nl>,
Karline Soetaert <karline.soetaert@nioz.nl>.

6.5 The General 3D Advective–Diffusive Transport PDE

6.5.1 The Advection Equation

In Cartesian coordinates the advection *operator* is

$$\mathbf{u} \cdot \nabla = u_x \frac{\partial}{\partial x} + u_y \frac{\partial}{\partial y} + u_z \frac{\partial}{\partial z} \tag{6.16}$$

where

$\mathbf{u}$ = (u_x, u_y, u_z) is the *velocity field*
∇ = the del operator

Note that Cartesian coordinates are used here.

The advection equation for a conserved quantity described by a scalar field ψ is expressed mathematically by a *continuity equation*:

$$\frac{\partial \psi}{\partial t} + \nabla \cdot (\psi \mathbf{u}) = 0 \tag{6.17}$$

where $\nabla\cdot$ is the divergence operator, and $\mathbf{u}$ is the velocity vector field.

Frequently, it is assumed that the flow is incompressible, that is, the velocity field satisfies the equation

$$\nabla \cdot \mathbf{u} = 0 \tag{6.18}$$

and $\mathbf{u}$ is said to be ***solenoidal***. Under such conditions, the above equation may be written as

$$\frac{\partial \psi}{\partial t} + \mathbf{u} \cdot \nabla \psi = 0 \tag{6.19}$$

In particular, if the flow is steady, then $\partial/\partial t = 0$, and

$$\mathbf{u} \cdot \nabla \psi = 0 \tag{6.20}$$

which shows that ψ is constant along a *streamline*. Hence,

$$\partial \psi / \partial t = 0 \tag{6.21}$$

so ψ does ***not*** vary in time.

If a vector quantity $\mathbf{a}$ (such as a magnetic field) is being advected by the solenoidal velocity field $\mathbf{u}$, the advection equation above becomes (Figure 6.6):

$$\frac{\partial \mathbf{a}}{\partial \boldsymbol{t}} + (\mathbf{u} \cdot \nabla)\mathbf{a} = 0 \tag{6.22}$$

6.5.2 Solving the Advection Equation

The advection equation is not simple to solve numerically: the system is a hyperbolic PDE and interest typically centers on discontinuous "shock" solutions, which

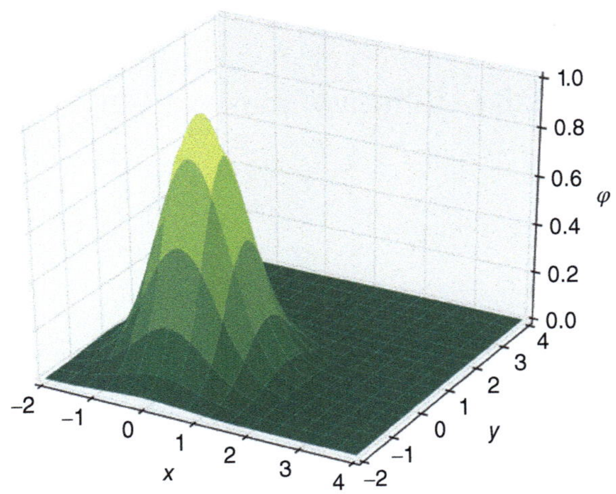

Figure 6.6 A simulation of the advection equation where $\mathbf{u} = (\sin t, \cos t)$ is solenoidal. Source: Shiyu Ji/Wikimedia Commons/CC BY-SA 4.0.

are difficult for numerical schemes to solve. Even with one space dimension and a constant velocity field, the system remains difficult to simulate. The 1D equation takes the form:

$$\frac{\partial \psi}{\partial t} + u_x \frac{\partial \psi}{\partial x} = 0 \tag{6.23}$$

where $\psi = \psi(x, t)$ is the scalar field being advected, and u_x is the x component of the vector $\mathbf{u} = (u_x, 0, 0)$.

6.5.3 The Advection Operator in the Incompressible Navier–Stokes Equations

Numerical simulation can be aided by considering the *skew symmetric* form for the advection operator:

$$\frac{1}{2}\mathbf{u} \cdot \nabla \mathbf{u} + \frac{1}{2}\nabla(\mathbf{uu}) \tag{6.24}$$

where

$$\nabla(\mathbf{uu}) = [\nabla(\mathbf{u}u_x), \nabla(\mathbf{u}u_y), \nabla(\mathbf{u}u_z)] \tag{6.25}$$

and **u** is the same as above.

Since skew symmetry implies only *imaginary eigenvalues*, this form reduces the "blow up" and "spectral blocking" often experienced in numerical solutions with sharp discontinuities.

Using vector calculus identities, these operators can also be expressed in other ways, available in more software packages for more coordinate systems.

$$\mathbf{u} \cdot \nabla \mathbf{u} = \nabla \left(\frac{\|\mathbf{u}\|^2}{2} \right) + (\nabla \times \mathbf{u}) \times \mathbf{u} \tag{6.26}$$

$$\frac{1}{2}\mathbf{u} \cdot \nabla \mathbf{u} + \frac{1}{2}\nabla(\mathbf{uu}) = \nabla \left(\frac{\|\mathbf{u}\|^2}{2} \right) + (\nabla \times \mathbf{u}) \times \mathbf{u} + \frac{1}{2}\mathbf{u}(\nabla \cdot \mathbf{u}) \tag{6.27}$$

This form also makes visible that the ***skew symmetric*** operator introduces error when the velocity field diverges. Solving the advection equation by numerical methods is challenging and there is a large scientific literature on this subject!

Example 6.2 `tran.3D` ***General Three-Dimensional Advective–Diffusive Transport***

Description

This program estimates the transport term (i.e. the rate of change of a concentration owing to diffusion and advection) in a three-dimensional rectangular model domain.

Usage

```
tran.3D (C, C.x.up = C[1,,], C.x.down = C[dim(C)[1],,],
         C.y.up = C[ ,1, ], C.y.down=C[ ,dim(C)[2], ],
         C.z.up = C[ , ,1], C.z.down=C[ , ,dim(C)[3]],
```

```
flux.x.up = NULL, flux.x.down = NULL,
flux.y.up = NULL, flux.y.down = NULL,
flux.z.up = NULL, flux.z.down = NULL,
a.bl.x.up = NULL, a.bl.x.down = NULL,
a.bl.y.up = NULL, a.bl.y.down = NULL,
a.bl.z.up = NULL, a.bl.z.down = NULL,
D.grid = NULL, D.x = NULL, D.y = D.x, D.z = D.x,
v.grid = NULL, v.x = 0, v.y = 0, v.z = 0,
AFDW.grid = NULL, AFDW.x = 1, AFDW.y = AFDW.x,
AFDW.z = AFDW.x,
VF.grid = NULL, VF.x = 1, VF.y = VF.x, VF.z = VF.x,
A.grid = NULL, A.x = 1, A.y = 1, A.z = 1,
grid = NULL, dx = NULL, dy = NULL, dz = NULL,
full.check = FALSE, full.output = FALSE)
```

Arguments

C	Concentration, expressed per unit volume, defined at the center of each grid cell; Nx*Ny*Nz array [M/L^3]
C.x.up	Concentration at upstream boundary in *x*-direction; matrix of dimensions Ny*Nz [M/L^3]
C.x.down	Concentration at downstream boundary in *x*-direction; matrix of dimensions Ny*Nz [M/L^3]
C.y.up	Concentration at upstream boundary in *y*-direction; matrix of dimensions Nx*Nz [M/L^3]
C.y.down	Concentration at downstream boundary in *y*-direction; matrix of dimensions Nx*Nz [M/L^3]
C.z.up	Concentration at upstream boundary in *z*-direction; matrix of dimensions Nx*Ny [M/L^3]
C.z.down	Concentration at downstream boundary in *z*-direction; matrix of dimensions Nx*Ny [M/L^3]
flux.x.up	Flux across the upstream boundary in *x*-direction, positive = INTO model domain; matrix of dimensions Ny*Nz [M/L^2/T]
flux.x.down	Flux across the downstream boundary in *x*-direction, positive = OUT of model domain; matrix of dimensions Ny*Nz [M/L^2/T]
flux.y.up	Flux across the upstream boundary in *y*-direction, positive = INTO model domain; matrix of dimensions Nx*Nz [M/L^2/T]
flux.y.down	Flux across the downstream boundary in *y*-direction, positive = OUT of model domain; matrix of dimensions Nx*Nz [M/L^2/T]
flux.z.up	Flux across the upstream boundary in *z*-direction, positive = INTO model domain; matrix of dimensions Nx*Ny [M/L^2/T]

flux.z.down	Flux across the downstream boundary in *z*-direction, positive = OUT of model domain; matrix of dimensions Nx*Ny [M/L^2/T]
a.bl.x.up	Transfer coefficient across the upstream boundary layer in *x*-direction Flux=a.bl.x.up*(C.x.up-C[1,,]). One value [L/T]
a.bl.x.down	Transfer coefficient across the downstream boundary layer in *x*-direction; Flux=a.bl.x.down*(C[Nx,,]-C.x.down). One value [L/T]
a.bl.y.up	Transfer coefficient across the upstream boundary layer in *y*-direction; Flux=a.bl.y.up*(C.y.up-C[,1,]). One value [L/T]
a.bl.y.down	Transfer coefficient across the downstream boundary layer in *y*-direction; Flux=a.bl.y.down*(C[,Ny,]-C.y.down). One value [L/T]
a.bl.z.up	Transfer coefficient across the upstream boundary layer in *y*-direction; Flux=a.bl.y.up*(C.y.up-C[,,1]). One value [L/T]
a.bl.z.down	Transfer coefficient across the downstream boundary layer in *z*-direction; Flux=a.bl.z.down*(C[,,Nz]-C.z.down). One value [L/T]
D.grid	Diffusion coefficient defined on all grid cell interfaces. Should contain elements x.int, y.int, z.int, arrays with the values on the interfaces in x-, y-, and z-directions, and with dimensions (Nx+1)*Ny*Nz, Nx*(Ny+1)*Nz, and Nx*Ny*(Nz+1), respectively [L2/T]
D.x	Diffusion coefficient in *x*-direction, defined on grid cell interfaces. One value, a vector of length (Nx+1), or a (Nx+1)*Ny*Nz array [L^2/T]
D.y	Diffusion coefficient in *y*-direction, defined on grid cell interfaces. One value, a vector of length (Ny+1), or a Nx*(Ny+1)*Nz array [L^2/T]
D.z	Diffusion coefficient in *z*-direction, defined on grid cell interfaces. One value, a vector of length (Nz+1), or a Nx*Ny*(Nz+1) array [L^2/T]
v.grid	Advective velocity defined on all grid cell interfaces. Can be positive (downstream flow) or negative (upstream flow). Should contain elements x.int, y.int, z.int, arrays with the values on the interfaces in *x*-, *y*-, and *z*-directions, and with dimensions (Nx+1)*Ny*Nz, Nx*(Ny+1)*Nz, and Nx*Ny*(Nz+1), respectively [L/T]
v.x	Advective velocity in the *x*-direction, defined on grid cell interfaces. Can be positive (downstream flow) or negative (upstream flow). One value, a vector of length (Nx+1), or a (Nx+1)*Ny*Nz array [L/T]
v.y	Advective velocity in the *y*-direction, defined on grid cell interfaces. Can be positive (downstream flow) or negative (upstream flow). One value, a vector of length (Ny+1), or a Nx*(Ny+1)*Nz array [L/T]

`v.z`	Advective velocity in the *z*-direction, defined on grid cell interfaces. Can be positive (downstream flow) or negative (upstream flow). One value, a vector of length (Nz+1), or a Nx*Ny*(Nz+1) array [L/T]
`AFDW.grid`	Weight used in the finite difference scheme for advection in the *x*-direction, defined on grid cell interfaces; backward = 1, centered = 0.5, forward = 0; default is backward. Should contain elements x.int, y.int, z.int, arrays with the values on the interfaces in *x*-, *y*-, and *z*-directions, and with dimensions (Nx+1)*Ny*Nz, Nx*(Ny+1)*Nz, and Nx*Ny*(Nz+1), respectively [–]
`AFDW.x`	Weight used in the finite difference scheme for advection in the *x*-direction, defined on grid cell interfaces; backward = 1, centered = 0.5, forward = 0; default is backward. One value, a vector of length (Nx+1), a prop.1D list created by `setup.prop.1D`, or a (Nx+1)*Ny*Nz array [–]
`AFDW.y`	Weight used in the finite difference scheme for advection in the *y*-direction, defined on grid cell interfaces; backward = 1, centered = 0.5, forward = 0; default is backward. One value, a vector of length (Ny+1), a prop.1D list created by `setup.prop.1D`, or a Nx*(Ny+1)*Nz array [–]
`AFDW.z`	Weight used in the finite difference scheme for advection in the *z*-direction, defined on grid cell interfaces; backward = 1, centered = 0.5, forward = 0, default is backward. One value, a vector of length `(Nz+1)`, a prop.1D list created by `setup.prop.1D, or a Nx*Ny*(Nz+1)` array [–]
`VF.grid`	Volume fraction. A list should contain elements `x.int`, `y.int`, `z.int`, arrays with the values on the interfaces in *x*-, *y*-, and *z*-directions, and with dimensions (Nx+1)*Ny*Nz, Nx*(Ny+1)*Nz, and Nx*Ny*(Nz+1), respectively [–]
`VF.x`	Volume fraction at the grid cell interfaces in the *x*-direction. One value, a vector of length (Nx+1), a `prop.1D` list created by `setup.prop.1D`, or a (Nx+1)*Ny*Nz array [–]
`VF.y`	Volume fraction at the grid cell interfaces in the *y*-direction. One value, a vector of length (Ny+1), a `prop.1D` list created by `setup.prop.1D`, or a Nx*(Ny+1)*Nz array [–]
`VF.z`	Volume fraction at the grid cell interfaces in the *z*-direction. One value, a vector of length (Nz+1), a `prop.1D` list created by `setup.prop.1D`, or a Nx*Ny*(Nz+1) `array [–]`
`A.grid`	Interface area, a list should contain elements x.int, y.int, z.int, arrays with the values on the interfaces in *x*-, *y*-, and *z*-directions, and with dimensions (Nx+1)*Ny*Nz, Nx*(Ny+1)*Nz, and Nx*Ny*(Nz+1), respectively [L^2]

A.x	Interface area defined at the grid cell interfaces in the x-direction. One value, a vector of length (Nx+1), a prop.1D list created by `setup.prop.1D`, or a (Nx+1)*Ny*Nz array [L^2]
A.y	Interface area defined at the grid cell interfaces in the y-direction. One value, a vector of length (Ny+1), a prop.1D list created by `setup.prop.1D`, or a Nx*(Ny+1)*Nz array [L^2]
A.z	Interface area defined at the grid cell interfaces in the z-direction. One value, a vector of length (Nz+1), a `prop.1D` list created by `setup.prop.1D`, or a Nx*Ny*(Nz+1) array [L^2]
dx	Distance between adjacent cell interfaces in the x-direction (thickness of grid cells). One value or vector of length Nx [L]
dy	Distance between adjacent cell interfaces in the y-direction (thickness of grid cells). One value or vector of length Ny [L]
dz	Distance between adjacent cell interfaces in the z-direction (thickness of grid cells). One value or vector of length Nz [L]
grid	Discretization grid, a list containing at least elements dx, dx.aux, dy, dy.aux, dz, dz.aux (see `setup.grid.2D`) [L]
full.check	Logical flag enabling a full check of the consistency of the arguments (default = `FALSE`; `TRUE` slows down execution by 50%)
full.output	Logical flag enabling a full return of the output (`default` = `FALSE`; `TRUE` slows down execution by 20%)

Details

Do not use this with grids which are too large.

The boundary conditions are either

(1) zero gradient
(2) fixed concentration
(3) convective boundary layer
(4) fixed flux.

This is also the order of priority. The zero gradient is the default, the fixed flux overrules all others.

Value

A list containing

dC	The rate of change of the concentration C owing to transport, defined in the center of each grid cell, an array with dimension Nx*Ny*Nz [$M/L^3/T$]
C.x.up	Concentration at the upstream interface in x-direction. A matrix of dimension Ny*Nz [M/L^3]. Only when `full.output` = `TRUE`

`C.x.down`	Concentration at the downstream interface in *x*-direction. A matrix of dimension Ny*Nz [M/L^3]. Only when `full.output = TRUE`
`C.y.up`	Concentration at the upstream interface in *y*-direction. A matrix of dimension Nx*Nz [M/L^3]. Only when `full.output = TRUE`
`C.y.down`	Concentration at the downstream interface in *y*-direction. A matrix of dimension Nx*Nz [M/L^3]. Only when `full.output = TRUE`
`C.z.up`	Concentration at the upstream interface in *z*-direction. A matrix of dimension Nx*Ny [M/L]. Only when `full.output = TRUE`
`C.z.down`	Concentration at the downstream interface in *z*-direction. A matrix of dimension Nx*Ny [M/L^3]. Only when `full.output = TRUE`
`x.flux`	Flux across the interfaces in *x*-direction of the grid cells. An (Nx+1)*Ny*Nz array [M/L^2/T]. Only when `full.output = TRUE`
`y.flux`	Flux across the interfaces in *y*-direction of the grid cells. An Nx*(Ny+1)*Nz array [M/L^2/T]. Only when `full.output = TRUE`
`z.flux`	Flux across the interfaces in *z*-direction of the grid cells. A Nx*Ny*(Nz+1) array [M/L^2/T]. Only when `full.output= TRUE`
`flux.x.up`	Flux across the upstream boundary in *x*-direction, `positive = INTO` model domain. A matrix of dimension Ny*Nz [M/L^2/T]
`flux.x.down`	Flux across the downstream boundary in *x*-direction, `positive=OUT` model domain. A matrix of dimension Ny*Nz [M/L^2/T]
`flux.y.up`	Flux across the upstream boundary in *y*-direction, `positive = INTO` model domain. A matrix of dimension Nx*Nz [M/L^2/T]
`flux.y.down`	Flux across the downstream boundary in *y*-direction, `positive=OUT` of model domain. A matrix of dimension Nx*Nz [M/L^2/T]
`flux.z.up`	Flux across the upstream boundary in *z*-direction, `positive = INTO` model domain. A matrix of dimension Nx*Ny [M/L^2/T]
`flux.z.down`	Flux across the downstream boundary in *z*-direction, `positive = OUT` of model domain. A matrix of dimension Nx*Ny [M/L^2/T]

Authors

Filip Meysman <filip.meysman@nioz.nl>,
Karline Soetaert <karline.soetaert@nioz.nl>.

See Also

tran.cylindrical, tran.spherical, for a discretization of 3D transport equations in cylindrical and spherical coordinates.

tran.1D, tran.2D

Examples

```
# ===============================
#  Diffusion in 3D; imposed boundary conditions  =
# ===============================
diffusion3D <- function(t, Y, par) {
   yy <- array(dim = c(n, n, n), data = Y) # vector to 3D array
   dY <- -r * yy                           # consumption
   BND <- matrix(nrow = n, ncol = n, 1)  # boundary concentration

   dY <- dY + tran.3D(C = yy,
           C.x.up = BND, C.y.up = BND, C.z.up = BND,
           C.x.down = BND, C.y.down = BND, C.z.down = BND,
           D.x = Dx, D.y = Dy, D.z = Dz,
           dx = dx, dy = dy, dz = dz, full.check = TRUE)$dC
   return(list(dY))
}

# parameters
dy <- dx <- dz <- 1  # grid size
Dy <- Dx <- Dz <- 1  # diffusion coeff, X- and Y-direction
r <- 0.025           # consumption rate

n <- 10
y <- array(dim = c(n, n, n), data = 10.)

print(system.time(
  ST3 <- steady.3D(y, func = diffusion3D, parms = NULL,
                   pos = TRUE, dimens = c(n, n, n),
                   lrw = 2000000, verbose = TRUE)
))

pm <- par(mfrow = c(1,1))
y <- array(dim = c(n, n, n), data = ST3$y)
filled.contour(y[ , ,n/2], color.palette = terrain.colors)
```

```
# a selection in the x-direction
image(ST3, mfrow = c(2, 2), add.contour = TRUE, legend = TRUE,
       dimselect = list(x = c(1, 4, 8, 10)))

par(mfrow = pm)
```

In the R domain (Figures 6.7, 6.8):

```
>
> # In the R domain:
>
> install.packages("ReacTran")
Installing package into 'C:/Users/Bert/Documents/R/win-library/3.3'
(as 'lib' is unspecified)
--- Please select a CRAN mirror for use in this session ---
A CRAN mirror is selected.
also installing the dependencies 'rootSolve', 'deSolve'

trying URL
'https://mirrors.tuna.tsinghua.edu.cn/CRAN/bin/windows/contrib/3.3/
rootSolve_1.7.zip'
Content type 'application/zip' length 683839 bytes (667 KB)
downloaded 667 KB

trying URL
'https://mirrors.tuna.tsinghua.edu.cn/CRAN/bin/windows/contrib/3.3/
deSolve_1.20.zip'
Content type 'application/zip' length 2667028 bytes (2.5 MB)
downloaded 2.5 MB

trying URL
'https://mirrors.tuna.tsinghua.edu.cn/CRAN/bin/windows/contrib/3.3/
ReacTran_1.4.3.1.zip'
Content type 'application/zip' length 915996 bytes (894 KB)
downloaded 894 KB
package 'rootSolve' successfully unpacked and MD5 sums checked
package 'deSolve' successfully unpacked and MD5 sums checked
package 'ReacTran' successfully unpacked and MD5 sums checked

The downloaded binary packages are in
C:\Users\Bert\AppData\Local\Temp\RtmpmEkAb2\downloaded_packages
> require(ReacTran)
Loading required package: ReacTran
Loading required package: rootSolve
Loading required package: deSolve
Loading required package: shape
> library(ReacTran)
```

```
> ls("package:ReacTran") # Listing all the programs in this package:
 [1] "advection.1D"      "advection.volume.1D" "fiadeiro"
 [4] "g.cylinder"        "g.sphere"            "g.spheroid"
 [7] "p.exp"             "p.lin"               "p.sig"
[10] "polar2cart"        "setup.compaction.1D" "setup.grid.1D"
[13] "setup.grid.2D"     "setup.prop.1D"       "setup.prop.2D"
[16] "tran.1D"           "tran.2D"             "tran.3D"
[19] "tran.cylindrical"  "tran.polar"          "tran.spherical"
[22] "tran.volume.1D"    "tran.volume.2D"      "tran.volume.3D"
> tran.3D # Outputting the program listing of "tran.3D"
> function (C, C.up = C[1], C.down = C[length(C)], flux.up = NULL,
  flux.down = NULL, a.bl.up = NULL, a.bl.down = NULL, D = 0,
  v = 0, AFDW = 1, VF = 1, A = 1, dx, full.check = FALSE, full.output = FALSE)

{  if (is.null(dx))
      stop("error: dx should be inputted ")
   N <- length(C)
   if (N == 0)
      stop("C should be a vector with numeric values")
   if (is.null(C.up))
      C.up <- C[1]
   if (is.null(C.down))
      C.down <- C[N]
   if (!is.list(AFDW))
      AFDW <- list(int = rep_len(AFDW, N + 1))
   if (!is.list(D))
      D <- list(int = rep_len(D, N + 1))
   if (!is.list(v))
      v <- list(int = rep_len(v, N + 1))
   if (!is.list(VF))
      VF <- list(int = rep_len(VF, N + 1), mid = 0.5 * (rep_len(VF,
         N + 1)[1:N] + rep_len(VF, N + 1)[2:(N + 1)]))
   if (!is.list(A))
      A <- list(int = rep_len(A, N + 1), mid = 0.5 * (rep_len(A,
         N + 1)[1:N] + rep_len(A, N + 1)[2:(N + 1)]))
   if (is.list(dx))
      grid <- dx
   if (!is.list(dx))
      grid <- list(dx = rep_len(dx, N), dx.aux = 0.5 * (c(0,
         rep_len(dx, N)) + c(rep_len(dx, N), 0)))
   if (full.check) {
      gn <- names(AFDW)
      if (!"int" %in% gn)
         stop("error: AFDW should be a list that contains 'int', the AFDW
values at the interface of the grid cell ")
      if (is.null(AFDW$int))
         stop("error: AFDW is NULL, should contain (numeric) values")
```

```
    if (!is.null(AFDW$int)) {
      if (!((length(AFDW$int) == 1) || (length(AFDW$int) ==
        (N + 1))))
        stop("error: AFDW should be a vector of length 1 or N+1")
    }
    if (any(AFDW$int < 0) || any(AFDW$int > 1))
      stop("error: the AFDW should always range between 0 and 1")
    gn <- names(D)
    if (!"int" %in% gn)

 stop("error: D should be a list that contains 'int', the D values at the
interface of the grid cell ")
    if (is.null(D$int))
      stop("error: D is NULL, should contain (numeric) values")
    if (!is.null(D$int)) {
      if (!((length(D$int) == 1) || (length(D$int) == (N +
        1))))
        stop("error: D should be a vector of length 1 or N+1")
    }
    if (any(D$int < 0))
      stop("error: the diffusion coefficient should always be positive")
    gn <- names(v)
    if (!"int" %in% gn)
      stop("error: v should be a list that contains 'int', the v values at the
interface of the grid cell ")
    if (is.null(v$int))
      stop("error: the advective velocity v is NULL, should contain (numeric)
values")
    if (!is.null(v$int)) {
      if (!((length(v$int) == 1) || (length(v$int) == (N +
        1))))
        stop("error: v should be a vector of length 1 or N+1")
    }
    gn <- names(VF)
    if (!"int" %in% gn)
      stop("error: VF should be a list that contains 'int', the area values at the
interface of the grid cell ")
    if (!"mid" %in% gn)
      stop("error: VF should be a list that contains 'mid', the area at the
middle of the grid cells")
    if (is.null(VF$int) || is.null(VF$mid))
      stop("error: the volume fraction VF should contain (numeric) values")
    if (!is.null(VF$int)) {
      if (!((length(VF$int) == 1) || (length(VF$int) ==
        (N + 1))))
        stop("error: VF$int should be a vector of length 1 or N+1")
    }
```

```
    if (!is.null(VF$mid)) {
      if (!((length(VF$mid) == 1) || (length(VF$mid) ==
        (N))))
        stop("error: VF$mid should be a vector of length 1 or N")
    }
    if (any(VF$int < 0) || any(VF$mid < 0) || any(VF$int >
      1) || any(VF$mid > 1))
      stop("error: the volume fraction should range between 0 and 1")
    gn <- names(A)
    if (!"int" %in% gn)
      stop("error: A should be a list that contains 'int', the area values at the
interface of the grid cell ")
    if (!"mid" %in% gn)
      stop("error: A should be a list that contains 'mid', the area at the
middle of the grid cells")
    if (is.null(A$int) || is.null(A$mid))
      stop("error: the surface area A is NULL, should contain (numeric) values")
    if (!is.null(A$int)) {
      if (!((length(A$int) == 1) || (length(A$int) == (N +
        1))))
        stop("error: A$int should be a vector of length 1 or N+1")
    }
    if (!is.null(A$mid)) {
      if (!((length(A$mid) == 1) || (length(A$mid) == (N))))
        stop("error: A$mid should be a vector of length 1 or N")
    }
    if (any(A$int < 0) || any(A$mid < 0))
      stop("error: the area A should always be positive")
    gn <- names(grid)
    if (!"dx" %in% gn)
      stop("error: grid should be a list that contains 'dx' ")
    if (!"dx.aux" %in% gn)
      stop("error: grid should be a list that contains 'dx.aux' ")
    if (is.null(grid$dx) || is.null(grid$dx.aux))
      stop("error: the grid should be a list with (numeric)
values for 'dx' and 'dx.aux' ")
    if (!is.null(grid$dx)) {
      if (!((length(grid$dx) == 1) || (length(grid$dx) ==
        N)))
        stop("error: dx should be a vector of length 1 or N")
    }
    if (!is.null(grid$dx.aux)) {
      if (!((length(grid$dx.aux) == 1) || (length(grid$dx.aux) ==
        (N + 1))))
        stop("error: dx.aux should be a vector of length 1 or N+1")
    }
```

```
if (any(grid$dx <= 0) || any(grid$dx.aux <= 0))
        stop("error: the grid distances dx and dx.aux should always be positive")
    if (!is.null(a.bl.up) & !is.null(C.up)) {
      if (a.bl.up < 0)
        stop("error: the boundary layer transfer coefficient should be positive")
    }
    if (!is.null(a.bl.down) & !is.null(C.down)) {
      if (a.bl.down < 0)
        stop("error: the boundary layer transfer coefficient should be positive")
    }
  }
  if (full.output) {
    if (!is.null(flux.up)) {
      if (v$int[1] >= 0) {
        nom <- flux.up + VF$int[1] * (D$int[1]/grid$dx.aux[1] +
          (1 - AFDW$int[1]) * v$int[1]) * C[1]
        denom <- VF$int[1] * (D$int[1]/grid$dx.aux[1] +
          AFDW$int[1] * v$int[1])
      }
      else {
        nom <- flux.up + VF$int[1] * (D$int[1]/grid$dx.aux[1] +
          AFDW$int[1] * v$int[1]) * C[1]
        denom <- VF$int[1] * (D$int[1]/grid$dx.aux[1] +
          (1 - AFDW$int[1]) * v$int[1])
      }
      C.up <- nom/denom
    }
    if (!is.null(flux.down)) {
      if (v$int[N + 1] >= 0) {
        nom <- flux.down - VF$int[N + 1] * (D$int[N +
          1]/grid$dx.aux[N + 1] + AFDW$int[N + 1] * v$int[N +
          1]) * C[N]
        denom <- -VF$int[N + 1] * (D$int[N + 1]/grid$dx.aux[N +
          1] + (1 - AFDW$int[N + 1]) * v$int[N + 1])
      }
      else {
        nom <- flux.down - VF$int[N + 1] * (D$int[N +
          1]/grid$dx.aux[N + 1] + (1 - AFDW$int[N + 1]) *
          v$int[N + 1]) * C[N]

        denom <- -VF$int[N + 1] * (D$int[N + 1]/grid$dx.aux[N +
          1] + AFDW$int[N + 1] * v$int[N + 1])
      }
      C.down <- nom/denom
    }
```

```
  if (!is.null(a.bl.up) & !is.null(C.up)) {
    if (v$int[1] >= 0) {
      nom <- a.bl.up * C.up + VF$int[1] * (D$int[1]/grid$dx.aux[1] +
        (1 - AFDW$int[1]) * v$int[1]) * C[1]
      denom <- a.bl.up + VF$int[1] * (D$int[1]/grid$dx.aux[1] +
        AFDW$int[1] * v$int[1])
    }
    else {
      nom <- a.bl.up * C.up + VF$int[1] * (D$int[1]/grid$dx.aux[1] +
        AFDW$int[1] * v$int[1]) * C[1]
      denom <- a.bl.up + VF$int[1] * (D$int[1]/grid$dx.aux[1] +
        (1 - AFDW$int[1]) * v$int[1])
    }
    C.up <- nom/denom
  }
  if (!is.null(a.bl.down) & !is.null(C.down)) {
    if (v$int[N + 1] >= 0) {
      nom <- a.bl.down * C.down + VF$int[N + 1] * (D$int[N +
        1]/grid$dx.aux[N + 1] + (1 - AFDW$int[N + 1]) *
        v$int[N + 1]) * C[N]
      denom <- a.bl.down + VF$int[N + 1] * (D$int[N +
        1]/grid$dx.aux[N + 1] + AFDW$int[N + 1] * v$int[N +
        1])
    }
    else {
      nom <- a.bl.down * C.down + VF$int[N + 1] * (D$int[N +
        1]/grid$dx.aux[N + 1] + AFDW$int[N + 1] * v$int[N +
        1]) * C[N]
      denom <- a.bl.down + VF$int[N + 1] * (D$int[N +
        1]/grid$dx.aux[N + 1] + (1 - AFDW$int[N + 1]) *
        v$int[N + 1])
    }
    C.down <- nom/denom
  }
  dif.flux <- as.vector(-VF$int * D$int * diff(c(C.up,
    C, C.down))/grid$dx.aux)
  adv.flux <- rep(0, length.out = length(dif.flux))
  if (any(v$int > 0)) {
vv <- v$int
    vv[v$int < 0] <- 0
    conc <- AFDW$int * c(C.up, C)
    if (any(AFDW$int < 1))
      conc <- conc + (1 - AFDW$int) * c(C, C.down)
    adv.flux <- adv.flux + as.vector(VF$int * vv * conc)
  }
```

```
    if (any(v$int < 0)) {
      vv <- v$int
      vv[v$int > 0] <- 0
      conc <- AFDW$int * c(C, C.down)
      if (any(AFDW$int < 1))
        conc <- conc + (1 - AFDW$int) * c(C.up, C)
      adv.flux <- adv.flux + as.vector(VF$int * vv * conc)
    }
    flux <- dif.flux + adv.flux
  }
  else {
    if (!is.null(a.bl.up) & !is.null(C.up)) {
      if (v$int[1] >= 0) {
        nom <- a.bl.up * C.up + VF$int[1] * (D$int[1]/grid$dx.aux[1] +
          (1 - AFDW$int[1]) * v$int[1]) * C[1]
        denom <- a.bl.up + VF$int[1] * (D$int[1]/grid$dx.aux[1] +
          AFDW$int[1] * v$int[1])
      }
      else {
        nom <- a.bl.up * C.up + VF$int[1] * (D$int[1]/grid$dx.aux[1] +
          AFDW$int[1] * v$int[1]) * C[1]
        denom <- a.bl.up + VF$int[1] * (D$int[1]/grid$dx.aux[1] +
          (1 - AFDW$int[1]) * v$int[1])
      }
      C.up <- nom/denom
    }
    if (!is.null(a.bl.down) & !is.null(C.down)) {
      if (v$int[N + 1] >= 0) {
        nom <- a.bl.down * C.down + VF$int[N + 1] * (D$int[N +
          1]/grid$dx.aux[N + 1] + (1 - AFDW$int[N + 1]) *
          v$int[N + 1]) * C[N]
        denom <- a.bl.down + VF$int[N + 1] * (D$int[N +
          1]/grid$dx.aux[N + 1] + AFDW$int[N + 1] * v$int[N +
          1])
      }
      else {
          nom <- a.bl.down * C.down + VF$int[N + 1] * (D$int[N +
        1]/grid$dx.aux[N + 1] + AFDW$int[N + 1] * v$int[N +
        1]) * C[N]
        denom <- a.bl.down + VF$int[N + 1] * (D$int[N +
          1]/grid$dx.aux[N + 1] + (1 - AFDW$int[N + 1]) *
          v$int[N + 1])
      }
      C.down <- nom/denom
    }
    flux <- as.vector(-(VF$int) * D$int * diff(c(C.up, C,
      C.down))/grid$dx.aux)
```

```
        if (any(v$int > 0)) {
            vv <- v$int
            vv[v$int < 0] <- 0
            conc <- AFDW$int * c(C.up, C)
            if (any(AFDW$int < 1))
                conc <- conc + (1 - AFDW$int) * c(C, C.down)
            flux <- flux + as.vector(VF$int * vv * conc)
        }
        if (any(v$int < 0)) {
            vv <- v$int
            vv[v$int > 0] <- 0
            conc <- AFDW$int * c(C, C.down)
            if (any(AFDW$int < 1))
                conc <- conc + (1 - AFDW$int) * c(C.up, C)
            flux <- flux + as.vector(VF$int * vv * conc)
        }
    }
    if (!is.null(flux.up))
        flux[1] <- flux.up
    if (!is.null(flux.down))
        flux[N + 1] <- flux.down
    dC <- -diff(A$int * flux)/A$mid/VF$mid/grid$dx
    if (!full.output) {
        return(list(dC = dC, flux.up = flux[1], flux.down = flux[length(flux)]))
    }
    else {
        return(list(dC = dC, C.up = C.up, C.down = C.down, dif.flux = dif.flux,
            adv.flux = adv.flux, flux = flux, flux.up = flux[1],
            flux.down = flux[length(flux)]))
    }
}

<environment: namespace:ReacTran>
>

> # Running the program tran.3D
>
> ## ===============================================
> ## Diffusion in 3D; imposed boundary conditions
> ## ===============================================
> diffusion3D <- function(t, Y, par) {
+ yy <- array(dim = c(n, n, n), data = Y) # vector to 3D array
+ dY <- -r * yy # consumption
+ BND <- matrix(nrow = n, ncol = n, 1) # boundary concentration
+ dY <- dY + tran.3D(C = yy,
+ C.x.up = BND, C.y.up = BND, C.z.up = BND,
+ C.x.down = BND, C.y.down = BND, C.z.down = BND,
```

```
+ D.x = Dx, D.y = Dy, D.z = Dz,
+ dx = dx, dy = dy, dz = dz, full.check = TRUE)$dC
+ return(list(dY))
+ }
> # parameters
> dy <- dx <- dz <- 1     # grid size
> Dy <- Dx <- Dz <- 1     # diffusion coeff, X- and Y-direction
> r <- 0.025              # consumption rate
> n <- 10
> y <- array(dim = c(n, n, n), data = 10.)
> print(system.time(
+ ST3 <- steady.3D(y, func = diffusion3D, parms = NULL,
+ pos = TRUE, dimens = c(n, n, n),
+ lrw = 2000000, verbose = TRUE)
+ ))
> # Outputting:
[1] "Steady-state settings"
  sparseType                          message

>
  sparseType                          message
1        3D sparse 3D jacobian, calculated internally
[1] "estimated number of nonzero elements:  6910"
[1] "estimated number of function calls:  1001"
[1] "number of species:  1"
[1] "dimensions:  10 10 10"
[1] "cyclic boundaries:  0 0 0"
mean residual derivative 4.73661e-07
[1] "precision at each steady state step"
[1] 1.105000e+01 4.736612e-07
[1] ""
[1] "--------------------"
[1] " Memory requirements"
[1] "--------------------"
   par                                        mess   val
1  nnz                  the number of nonzero elements:  6400
2  ngp  the number of independent groups of state variables:     12
3  nsp       the length of the work array actually required: 82366

   user  system elapsed
   0.05    0.03    0.11
>
> pm <- par(mfrow = c(1,1))
>
> #
> y <- array(dim = c(n, n, n), data = ST3$y)
> filled.contour(y[ , ,n/2], color.palette = terrain.colors)
>
```

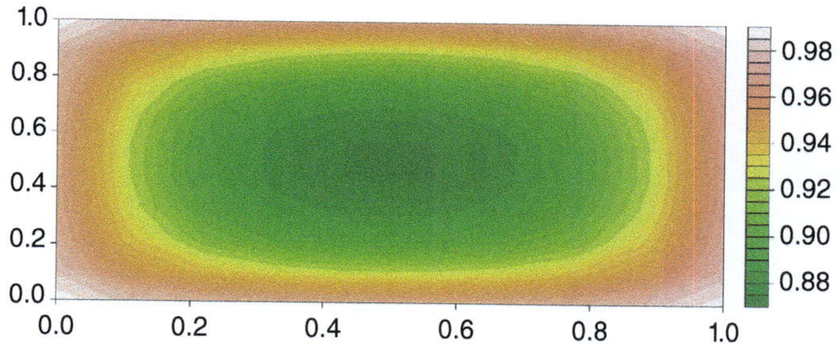

Figure 6.7 Contours of flow patterns.

```
trans.3D 2 > y <- array(dim = c(n, n, n), data = ST3$y)
           > filled.contour(y[ , ,n/2], color.palette = terrain.colors)
```

```
> # Outputting:
> #
> # Figure 6.7: tran-3D 2
> #
> # a selection in the x-direction
> image(ST3, mfrow = c(2, 2), add.contour = TRUE, legend = TRUE,
+ dimselect = list(x = c(1, 4, 8, 10)))
>
> # Outputting:
> #
> # Figure 6.8: tran-3D 3
> #
> par(mfrow = pm)
>
```

See Also

tran.cylindrical, tran.spherical, for a discretization of 3D transport equations in cylindrical and spherical coordinates.

tran.1D, tran.2D

Examples

```
# ==================================
# Diffusion in 3D; imposed boundary conditions   =
# ==================================

diffusion3D <- function(t, Y, par) {
   yy <- array(dim = c(n, n, n), data = Y) # vector to 3D array
   dY <- -r * yy                           # consumption
   BND <- matrix(nrow = n, ncol = n, 1)    # boundary concentration
```

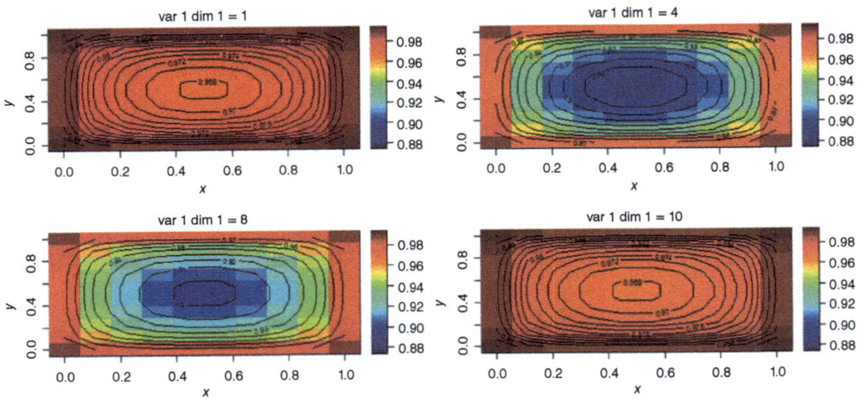

Figure 6.8 Contours of flow patterns.

```
trans.3D 3 > image(ST3, mfrow = c(2, 2),
                                      add.contour = TRUE,
                         + legend = TRUE,  dimselect = list(x = c(1, 4, 8, 10)))
```

```
    dY <- dY + tran.3D(C = yy,
           C.x.up = BND, C.y.up = BND, C.z.up = BND,
           C.x.down = BND, C.y.down = BND, C.z.down = BND,
           D.x = Dx, D.y = Dy, D.z = Dz,
           dx = dx, dy = dy, dz = dz, full.check = TRUE)$dC
    return(list(dY))
}

# parameters
dy <- dx <- dz <- 1   # grid size
Dy <- Dx <- Dz <- 1  # diffusion coeff, X- and Y-direction
r <- 0.025            # consumption rate

n <- 10
y <- array(dim = c(n, n, n), data = 10.)

print(system.time(
   ST3 <- steady.3D(y, func = diffusion3D, parms = NULL,
                    pos = TRUE, dimens = c(n, n, n),
                    lrw = 2000000, verbose = TRUE)
))

pm <- par(mfrow = c(1,1))
y <- array(dim = c(n, n, n), data = ST3$y)
filled.contour(y[ , ,n/2], color.palette = terrain.colors)

# a selection in the x-direction
```

```
image(ST3, mfrow = c(2, 2), add.contour = TRUE, legend = TRUE,
      dimselect = list(x = c(1, 4, 8, 10)))

par(mfrow = pm)
```

In the R domain (Figures 6.10–6.12):

```
>
> # In the R domain:
>
> install.packages("ReacTran")
Installing package into 'C:/Users/Bert/Documents/R/win-library/3.3'
(as 'lib' is unspecified)
--- Please select a CRAN mirror for use in this session ---
A CRAN mirror is selected.
also installing the dependencies 'rootSolve', 'deSolve'

trying URL
'https://mirrors.tuna.tsinghua.edu.cn/CRAN/bin/windows/contrib/3.3/
rootSolve_1.7.zip'
Content type 'application/zip' length 683839 bytes (667 KB)
downloaded 667 KB

trying URL
'https://mirrors.tuna.tsinghua.edu.cn/CRAN/bin/windows/contrib/3.3/
deSolve_1.20.zip'

Content type 'application/zip' length 2667028 bytes (2.5 MB)
downloaded 2.5 MB

trying URL
'https://mirrors.tuna.tsinghua.edu.cn/CRAN/bin/windows/contrib/3.3/
ReacTran_1.4.3.1.zip'
Content type 'application/zip' length 915996 bytes (894 KB)
downloaded 894 KB

package 'rootSolve' successfully unpacked and MD5 sums checked
package 'deSolve' successfully unpacked and MD5 sums checked
package 'ReacTran' successfully unpacked and MD5 sums checked

The downloaded binary packages are in
C:\Users\Bert\AppData\Local\Temp\RtmpmEkAb2\downloaded_packages
> require(ReacTran)
Loading required package: ReacTran
Loading required package: rootSolve
Loading required package: deSolve
```

```
Loading required package: shape
> library(ReacTran)

> ls("package:ReacTran") # Listing all the programs in this package:
 [1] "advection.1D"       "advection.volume.1D"    "fiadeiro"
 [4] "g.cylinder"         "g.sphere"               "g.spheroid"
 [7] "p.exp"              "p.lin"                  "p.sig"
[10] "polar2cart"         "setup.compaction.1D"    "setup.grid.1D"
[13] "setup.grid.2D"      "setup.prop.1D"          "setup.prop.2D"
[16] "tran.1D"            "tran.2D" "tran.3D"
[19] "tran.cylindrical"   "tran.polar"             "tran.spherical"
[22] "tran.volume.1D"     "tran.volume.2D"         "tran.volume.3D"
> tran.3D # Outputting the program listing of "tran.3D"
> function (C, C.up = C[1], C.down = C[length(C)], flux.up = NULL,
   flux.down = NULL, a.bl.up = NULL, a.bl.down = NULL, D = 0,
   v = 0, AFDW = 1, VF = 1, A = 1, dx, full.check = FALSE, full.output = FALSE)
{
   if (is.null(dx))
      stop("error: dx should be inputted ")
   N <- length(C)
   if (N == 0)
      stop("C should be a vector with numeric values")
   if (is.null(C.up))
      C.up <- C[1]
   if (is.null(C.down))
      C.down <- C[N]
   if (!is.list(AFDW))
      AFDW <- list(int = rep_len(AFDW, N + 1))
   if (!is.list(D))
      D <- list(int = rep_len(D, N + 1))
   if (!is.list(v))
      v <- list(int = rep_len(v, N + 1))
   if (!is.list(VF))
      VF <- list(int = rep_len(VF, N + 1), mid = 0.5 * (rep_len(VF,
         N + 1)[1:N] + rep_len(VF, N + 1)[2:(N + 1)]))
   if (!is.list(A))
      A <- list(int = rep_len(A, N + 1), mid = 0.5 * (rep_len(A,
         N + 1)[1:N] + rep_len(A, N + 1)[2:(N + 1)]))
if (is.list(dx))
      grid <- dx
   if (!is.list(dx))
      grid <- list(dx = rep_len(dx, N), dx.aux = 0.5 * (c(0,
         rep_len(dx, N)) + c(rep_len(dx, N), 0)))
   if (full.check) {
      gn <- names(AFDW)
      if (!"int" %in% gn)
```

```
        stop("error: AFDW should be a list that contains 'int', the AFDW
values at the interface of the grid cell ")
    if (is.null(AFDW$int))
        stop("error: AFDW is NULL, should contain (numeric) values")
    if (!is.null(AFDW$int)) {
        if (!((length(AFDW$int) == 1) || (length(AFDW$int) ==
            (N + 1))))
            stop("error: AFDW should be a vector of length 1 or N+1")
    }
    if (any(AFDW$int < 0) || any(AFDW$int > 1))
        stop("error: the AFDW should always range between 0 and 1")
    gn <- names(D)
    if (!"int" %in% gn)
        stop("error: D should be a list that contains 'int', the D values at the
interface of the grid cell ")
    if (is.null(D$int))
        stop("error: D is NULL, should contain (numeric) values")
    if (!is.null(D$int)) {
        if (!((length(D$int) == 1) || (length(D$int) == (N +
            1))))
            stop("error: D should be a vector of length 1 or N+1")
    }
    if (any(D$int < 0))
        stop("error: the diffusion coefficient should always be positive")
    gn <- names(v)
    if (!"int" %in% gn)
        stop("error: v should be a list that contains 'int', the v values at the
interface of the grid cell ")
    if (is.null(v$int))
        stop("error: the advective velocity v is NULL, should contain (numeric)
values")
    if (!is.null(v$int)) {
        if (!((length(v$int) == 1) || (length(v$int) == (N +
            1))))
        stop("error: v should be a vector of length 1 or N+1")
    }
    gn <- names(VF)
    if (!"int" %in% gn)
        stop("error: VF should be a list that contains 'int', the area values at the
interface of the grid cell ")
    if (!"mid" %in% gn)
        stop("error: VF should be a list that contains 'mid', the area at the
middle of the grid cells")
    if (is.null(VF$int) || is.null(VF$mid))
        stop("error: the volume fraction VF should contain (numeric) values")
    if (!is.null(VF$int)) {
```

```
      if (!((length(VF$int) == 1) || (length(VF$int) ==
         (N + 1))))
         stop("error: VF$int should be a vector of length 1 or N+1")
   }
   if (!is.null(VF$mid)) {
      if (!((length(VF$mid) == 1) || (length(VF$mid) ==
         (N))))
         stop("error: VF$mid should be a vector of length 1 or N")
   }
   if (any(VF$int < 0) || any(VF$mid < 0) || any(VF$int >
      1) || any(VF$mid > 1))
      stop("error: the volume fraction should range between 0 and 1")
   gn <- names(A)
   if (!"int" %in% gn)
      stop("error: A should be a list that contains 'int', the area values at the
interface of the grid cell ")
   if (!"mid" %in% gn)
      stop("error: A should be a list that contains 'mid', the area at the
middle of the grid cells")
   if (is.null(A$int) || is.null(A$mid))
      stop("error: the surface area A is NULL, should contain (numeric) values")
   if (!is.null(A$int)) {
      if (!((length(A$int) == 1) || (length(A$int) == (N +
         1))))
         stop("error: A$int should be a vector of length 1 or N+1")
   }
   if (!is.null(A$mid)) {
      if (!((length(A$mid) == 1) || (length(A$mid) == (N))))
         stop("error: A$mid should be a vector of length 1 or N")
   }
   if (any(A$int < 0) || any(A$mid < 0))
      stop("error: the area A should always be positive")
   gn <- names(grid)
   if (!"dx" %in% gn)
      stop("error: grid should be a list that contains 'dx' ")
   if (!"dx.aux" %in% gn)
      stop("error: grid should be a list that contains 'dx.aux' ")
   if (is.null(grid$dx) || is.null(grid$dx.aux))
      stop("error: the grid should be a list with (numeric)
values for 'dx' and 'dx.aux' ")
   if (!is.null(grid$dx)) {
      if (!((length(grid$dx) == 1) || (length(grid$dx) ==
         N)))
         stop("error: dx should be a vector of length 1 or N")
   }
   if (!is.null(grid$dx.aux)) {
```

```
      if (!((length(grid$dx.aux) == 1) || (length(grid$dx.aux) ==
        (N + 1))))
        stop("error: dx.aux should be a vector of length 1 or N+1")
    }
    if (any(grid$dx <= 0) || any(grid$dx.aux <= 0))
      stop("error: the grid distances dx and dx.aux should always be positive")
    if (!is.null(a.bl.up) & !is.null(C.up)) {
      if (a.bl.up < 0)
        stop("error: the boundary layer transfer coefficient should be positive")
    }
    if (!is.null(a.bl.down) & !is.null(C.down)) {
      if (a.bl.down < 0)
        stop("error: the boundary layer transfer coefficient should be positive")
    }
  }
  if (full.output) {
    if (!is.null(flux.up)) {
      if (v$int[1] >= 0) {
        nom <- flux.up + VF$int[1] * (D$int[1]/grid$dx.aux[1] +
          (1 - AFDW$int[1]) * v$int[1]) * C[1]
        denom <- VF$int[1] * (D$int[1]/grid$dx.aux[1] +
          AFDW$int[1] * v$int[1])
    }
      else {
        nom <- flux.up + VF$int[1] * (D$int[1]/grid$dx.aux[1] +
          AFDW$int[1] * v$int[1]) * C[1]
        denom <- VF$int[1] * (D$int[1]/grid$dx.aux[1] +
          (1 - AFDW$int[1]) * v$int[1])
      }
      C.up <- nom/denom
    }
    if (!is.null(flux.down)) {
      if (v$int[N + 1] >= 0) {
        nom <- flux.down - VF$int[N + 1] * (D$int[N +
          1]/grid$dx.aux[N + 1] + AFDW$int[N + 1] * v$int[N +
          1]) * C[N]
        denom <- -VF$int[N + 1] * (D$int[N + 1]/grid$dx.aux[N +
          1] + (1 - AFDW$int[N + 1]) * v$int[N + 1])
      }
      else {
        nom <- flux.down - VF$int[N + 1] * (D$int[N +
          1]/grid$dx.aux[N + 1] + (1 - AFDW$int[N + 1]) *
          v$int[N + 1]) * C[N]
        denom <- -VF$int[N + 1] * (D$int[N + 1]/grid$dx.aux[N +
          1] + AFDW$int[N + 1] * v$int[N + 1])
      }
```

```
      C.down <- nom/denom
    }
    if (!is.null(a.bl.up) & !is.null(C.up)) {
      if (v$int[1] >= 0) {
        nom <- a.bl.up * C.up + VF$int[1] * (D$int[1]/grid$dx.aux[1] +
          (1 - AFDW$int[1]) * v$int[1]) * C[1]
        denom <- a.bl.up + VF$int[1] * (D$int[1]/grid$dx.aux[1] +
          AFDW$int[1] * v$int[1])
      }
      else {
        nom <- a.bl.up * C.up + VF$int[1] * (D$int[1]/grid$dx.aux[1] +
          AFDW$int[1] * v$int[1]) * C[1]
        denom <- a.bl.up + VF$int[1] * (D$int[1]/grid$dx.aux[1] +
          (1 - AFDW$int[1]) * v$int[1])
      }
      C.up <- nom/denom
    }
    if (!is.null(a.bl.down) & !is.null(C.down)) {

     if (v$int[N + 1] >= 0) {
        nom <- a.bl.down * C.down + VF$int[N + 1] * (D$int[N +
          1]/grid$dx.aux[N + 1] + (1 - AFDW$int[N + 1]) *
          v$int[N + 1]) * C[N]
        denom <- a.bl.down + VF$int[N + 1] * (D$int[N +
          1]/grid$dx.aux[N + 1] + AFDW$int[N + 1] * v$int[N +
          1])
      }
      else {
        nom <- a.bl.down * C.down + VF$int[N + 1] * (D$int[N +
          1]/grid$dx.aux[N + 1] + AFDW$int[N + 1] * v$int[N +
          1]) * C[N]
        denom <- a.bl.down + VF$int[N + 1] * (D$int[N +
          1]/grid$dx.aux[N + 1] + (1 - AFDW$int[N + 1]) *
          v$int[N + 1])
      }
      C.down <- nom/denom
    }
    dif.flux <- as.vector(-VF$int * D$int * diff(c(C.up,
      C, C.down))/grid$dx.aux)
    adv.flux <- rep(0, length.out = length(dif.flux))
    if (any(v$int > 0)) {
      vv <- v$int
      vv[v$int < 0] <- 0
      conc <- AFDW$int * c(C.up, C)
      if (any(AFDW$int < 1))
        conc <- conc + (1 - AFDW$int) * c(C, C.down)
      adv.flux <- adv.flux + as.vector(VF$int * vv * conc)
```

```
    }
    if (any(v$int < 0)) {
      vv <- v$int
      vv[v$int > 0] <- 0
      conc <- AFDW$int * c(C, C.down)
      if (any(AFDW$int < 1))
        conc <- conc + (1 - AFDW$int) * c(C.up, C)
      adv.flux <- adv.flux + as.vector(VF$int * vv * conc)
    }
    flux <- dif.flux + adv.flux
  }
  else {
    if (!is.null(a.bl.up) & !is.null(C.up)) {
      if (v$int[1] >= 0) {
        nom <- a.bl.up * C.up + VF$int[1] * (D$int[1]/grid$dx.aux[1] +
          (1 - AFDW$int[1]) * v$int[1]) * C[1]
        denom <- a.bl.up + VF$int[1] * (D$int[1]/grid$dx.aux[1] +
          AFDW$int[1] * v$int[1])
      }
      else {
        nom <- a.bl.up * C.up + VF$int[1] * (D$int[1]/grid$dx.aux[1] +
          AFDW$int[1] * v$int[1]) * C[1]
        denom <- a.bl.up + VF$int[1] * (D$int[1]/grid$dx.aux[1] +
          (1 - AFDW$int[1]) * v$int[1])
      }
      C.up <- nom/denom
    }
    if (!is.null(a.bl.down) & !is.null(C.down)) {
      if (v$int[N + 1] >= 0) {
        nom <- a.bl.down * C.down + VF$int[N + 1] * (D$int[N +
          1]/grid$dx.aux[N + 1] + (1 - AFDW$int[N + 1]) *
          v$int[N + 1]) * C[N]
        denom <- a.bl.down + VF$int[N + 1] * (D$int[N +
          1]/grid$dx.aux[N + 1] + AFDW$int[N + 1] * v$int[N +
          1])
      }
      else {
        nom <- a.bl.down * C.down + VF$int[N + 1] * (D$int[N +
          1]/grid$dx.aux[N + 1] + AFDW$int[N + 1] * v$int[N +
          1]) * C[N]
        denom <- a.bl.down + VF$int[N + 1] * (D$int[N +
          1]/grid$dx.aux[N + 1] + (1 - AFDW$int[N + 1]) *
          v$int[N + 1])
      }
      C.down <- nom/denom
    }
    flux <- as.vector(-(VF$int) * D$int * diff(c(C.up, C,
```

```
            C.down))/grid$dx.aux)
        if (any(v$int > 0)) {
            vv <- v$int
            vv[v$int < 0] <- 0
            conc <- AFDW$int * c(C.up, C)
            if (any(AFDW$int < 1))
                conc <- conc + (1 - AFDW$int) * c(C, C.down)
            flux <- flux + as.vector(VF$int * vv * conc)
        }
    if (any(v$int < 0)) {
            vv <- v$int
            vv[v$int > 0] <- 0
            conc <- AFDW$int * c(C, C.down)
            if (any(AFDW$int < 1))
                conc <- conc + (1 - AFDW$int) * c(C.up, C)
            flux <- flux + as.vector(VF$int * vv * conc)
        }
    }
    if (!is.null(flux.up))
        flux[1] <- flux.up
    if (!is.null(flux.down))
        flux[N + 1] <- flux.down
    dC <- -diff(A$int * flux)/A$mid/VF$mid/grid$dx
    if (!full.output) {
        return(list(dC = dC, flux.up = flux[1], flux.down = flux[length(flux)]))
    }
    else {
        return(list(dC = dC, C.up = C.up, C.down = C.down, dif.flux = dif.flux,
            adv.flux = adv.flux, flux = flux, flux.up = flux[1],
            flux.down = flux[length(flux)]))
    }
}
<environment: namespace:ReacTran>
>
```

```
> # Running the program tran.3D
>
> ## =========================================
> ## Diffusion in 3D; imposed boundary conditions
> ## =========================================
> diffusion3D <- function(t, Y, par) {
+ yy <- array(dim = c(n, n, n), data = Y) # vector to 3D array
+ dY <- -r * yy # consumption
+ BND <- matrix(nrow = n, ncol = n, 1) # boundary concentration
+ dY <- dY + tran.3D(C = yy,
+ C.x.up = BND, C.y.up = BND, C.z.up = BND,
+ C.x.down = BND, C.y.down = BND, C.z.down = BND,
+ D.x = Dx, D.y = Dy, D.z = Dz,
+ dx = dx, dy = dy, dz = dz, full.check = TRUE)$dC
```

```
+ return(list(dY))
+ }

> # parameters
> dy <- dx <- dz <- 1 # grid size
> Dy <- Dx <- Dz <- 1 # diffusion coeff, X- and Y-direction
> r <- 0.025 # consumption rate
> n <- 10
> y <- array(dim = c(n, n, n), data = 10.)
> print(system.time(
+ ST3 <- steady.3D(y, func = diffusion3D, parms = NULL,
+ pos = TRUE, dimens = c(n, n, n),
+ lrw = 2000000, verbose = TRUE)
+ ))
> # Outputting:
[1] "Steady-state settings"
  sparseType                          message

>
  sparseType                          message
1        3D sparse 3D jacobian, calculated internally
[1] "estimated number of nonzero elements:  6910"
[1] "estimated number of function calls:  1001"
[1] "number of species:  1"
[1] "dimensions:  10 10 10"
[1] "cyclic boundaries:  0 0 0"
mean residual derivative 4.73661e-07
[1] "precision at each steady state step"
[1] 1.105000e+01 4.736612e-07
[1] ""
[1] "--------------------"
[1] " Memory requirements"
[1] "--------------------"
  par                                 mess   val
1  nnz                the number of nonzero elements:  6400
2  ngp  the number of independent groups of state variables:     12
3  nsp      the length of the work array actually required: 82366
  user  system elapsed
  0.05    0.03    0.11
>
> pm <- par(mfrow = c(1,1))

> y <- array(dim = c(n, n, n), data = ST3$y)
> filled.contour(y[ , ,n/2], color.palette = terrain.colors)
> #
> # a selection in the x-direction
> image(ST3, mfrow = c(2, 2), add.contour = TRUE, legend = TRUE,
+ dimselect = list(x = c(1, 4, 8, 10)))
> #
```

Example 6.3 tran.cylindrical ***Diffusive Transport in Cylindrical (r, θ, z), and Spherical (r, θ, φ) Coordinates***

Description

This program estimates the transport term (i.e. the rate of change of a concentration owing to diffusion) in a cylindrical (r, θ, z) or spherical (r, θ, φ) coordinate system.

Usage

```
tran.cylindrical (C, C.r.up = NULL, C.r.down = NULL,
                  C.theta.up = NULL, C.theta.down = NULL,
                  C.z.up = NULL, C.z.down = NULL,
                  flux.r.up = NULL, flux.r.down = NULL,
                  flux.theta.up = NULL, flux.theta.down = NULL,
                  flux.z.up = NULL, flux.z.down = NULL,
                  cyclicBnd = NULL,
                  D.r = NULL, D.theta = D.r, D.z = D.r,
                  r = NULL, theta = NULL, z = NULL)

tran.spherical  (C, C.r.up = NULL, C.r.down = NULL,
                  C.theta.up = NULL, C.theta.down = NULL,
                  C.phi.up = NULL, C.phi.down = NULL,
                  flux.r.up = NULL, flux.r.down = NULL,
                  flux.theta.up = NULL, flux.theta.down = NULL,
                  flux.phi.up = NULL, flux.phi.down = NULL,
                  cyclicBnd = NULL,
                  D.r = NULL, D.theta = D.r, D.phi = D.r,
                  r = NULL, theta = NULL, phi = NULL)
```

Arguments

C	Concentration, expressed per unit volume, defined at the center of each grid cell; Nr*Ntheta*Nz (cylindrical) or Nr*Ntheta*Nphi (spherical coordinates) array $[M/L^3]$
C.r.up	Concentration at upstream boundary in $r(x)$-direction; one value $[M/L^3]$
C.r.down	Concentration at downstream boundary in $r(x)$-direction; one value $[M/L^3]$
C.theta.up	Concentration at upstream boundary in theta-direction; one value $[M/L^3]$
C.theta.down	Concentration at downstream boundary in theta-direction; one value $[M/L^3]$
C.z.up	Concentration at upstream boundary in z-direction (cylindrical coordinates); one value $[M/L^3]$
C.z.down	Concentration at downstream boundary in z-direction (cylindrical coordinates); one value $[M/L^3]$
C.phi.up	Concentration at upstream boundary in phi-direction (spherical coordinates); one value $[M/L^3]$

C.phi.down	Concentration at downstream boundary in phi-direction (spherical coordinates); one value [M/L^3]
flux.r.up	Flux across the upstream boundary in r-direction, positive = INTO model domain; one value [M/L^2/T]
flux.r.down	Flux across the downstream boundary in r-direction, positive = OUT of model domain; one value [M/L^2/T]
flux.theta.up	Flux across the upstream boundary in theta-direction, positive = INTO model domain; one value [M/L^2/T]
flux.theta.down	Flux across the downstream boundary in theta-direction, positive = OUT of model domain; one value [M/L^2/T]
flux.z.up	Flux across the upstream boundary in z-direction (cylindrical coordinates); positive = INTO model domain; one value [M/L^2/T]
flux.z.down	Flux across the downstream boundary in z-direction, (cylindrical coordinates); positive = OUT of model domain; one value [M/L^2/T]
flux.phi.up	Flux across the upstream boundary in phi-direction (spherical coordinates); positive = INTO model domain; one value [M/L^2/T]
flux.phi.down	Flux across the downstream boundary in phi-direction (spherical coordinates); positive = OUT of model domain; one value [M/L^2/T]
cyclicBnd	If not NULL, the direction in which a cyclic boundary is defined, i.e. cyclicBnd = 1 for the r-direction, cyclicBnd = 2 for the theta-direction and cyclicBnd = c(1,2) for both the r- and theta-directions
D.r	Diffusion coefficient in the r-direction, defined on grid cell interfaces. One value or a vector of length (Nr+1) [L^2/T]
D.theta	Diffusion coefficient in theta-direction, defined on grid cell interfaces. One value or a vector of length (Ntheta+1) [L^2/T]
D.z	Diffusion coefficient in z-direction, defined on grid cell interfaces for cylindrical coordinates. One value or a vector of length (Nz+1) [L^2/T]
D.phi	Diffusion coefficient in phi-direction, defined on grid cell interfaces for cylindrical coordinates. One value or a vector of length (Nphi+1) [L^2/T]
r	Position of adjacent cell interfaces in the r-direction. A vector of length Nr+1 [L]
theta	Position of adjacent cell interfaces in the theta-direction. A vector of length Ntheta+1 [L]. Theta should be within [0,2 pi]
z	Position of adjacent cell interfaces in the z-direction (cylindrical coordinates). A vector of length Nz+1 [L]
phi	Position of adjacent cell interfaces in the phi-direction (spherical coordinates). A vector of length Nphi+1 [L]. Phi should be within [0,2 pi]

Details

tran.cylindrical performs (diffusive) transport in cylindrical coordinates
tran.spherical performs (diffusive) transport in spherical coordinates

The **boundary conditions** are either

(1) zero gradient
(2) fixed concentration
(3) fixed flux
(4) cyclic boundary.

This is also the order of priority. The cyclic boundary overrules the other. If fixed concentration, fixed flux, and cyclicBnd are NULL then the boundary is zero gradient.

A cyclic boundary condition has concentration and flux at upstream and downstream boundary the same. It is useful mainly for the theta- and phi-directions.

--- Do not expect too much of this equation: do not try to use it with many boxes ---

Value

A list containing

dC	The rate of change of the concentration C due to transport, defined in the center of each grid cell, a Nr*Ntheta*Nz (cylindrical) or Nr*Ntheta*Nphi (spherical coordinates) array [$M/L^3/T$]
flux.r.up	Flux across the upstream boundary in r-direction, positive = INTO model domain. A matrix of dimension Ntheta*Nz (cylindrical) or Ntheta*Nphi (spherical) [M/L2/T]
flux.r.down	Flux across the downstream boundary in r-direction, positive = OUT of model domain. A matrix of dimension Ntheta*Nz (cylindrical) or Ntheta*Nphi (spherical) [$M/L^2/T$]
flux.theta.up	Flux across the upstream boundary in theta-direction, positive = INTO model domain. A matrix of dimension Nr*Nz (cylindrical) or Nr*Nphi (spherical) [$M/L^2/T$]
flux.theta.down	Flux across the downstream boundary in theta-direction, positive = OUT of model domain. A matrix of dimension Nr*Nz (cylindrical) or Nr*Nphi (spherical) [$M/L^2/T$]
flux.z.up	Flux across the upstream boundary in z-direction, for cylindrical coordinates; positive = OUT of model domain. A matrix of dimension Nr*Ntheta [$M/L^2/T$]
flux.z.down	Flux across the downstream boundary in z-direction for cylindrical coordinates; positive = OUT of model domain. A matrix of dimension Nr*Ntheta [$M/L^2/T$]
flux.phi.up	Flux across the upstream boundary in phi-direction, for spherical coordinates; positive = OUT of model domain. A matrix of dimension Nr*Ntheta [$M/L^2/T$]

flux.phi.down — Flux across the downstream boundary in phi-direction, for spherical coordinates; positive = OUT of model domain. A matrix of dimension Nr*Ntheta [$M/L^2/T$]

See Also

tran.polar for a discretization of 2D transport equations in polar coordinates.
tran.1D, tran.2D, tran.3D

Examples

```
# ================
# Testing the functions =
# ================
#
#Grid definition
r.N      <- 4        # number of cells in r-direction
theta.N <- 6        # number of cells in theta-direction
z.N      <- 3        # number of cells in z-direction

D        <- 100  # diffusion coefficient


r        <- seq(0, 8, len = r.N+1)           # cell size r-direction [cm]
theta <- seq(0,2*pi, len = theta.N+1) # theta-direction - theta: from 0, 2pi
phi    <- seq(0,2*pi, len = z.N+1)       # phi-direction (0,2pi)
z        <- seq(0,5, len = z.N+1)             # cell size z-direction [cm]

# Initial conditions
C <- array(dim = c(r.N, theta.N, z.N), data = 0)
# Concentration boundary conditions
tran.cylindrical (C = C, D.r = D, D.theta = D,
C.r.up = 1, C.r.down = 1,
C.theta.up = 1, C.theta.down = 1,
C.z.up = 1, C.z.down = 1,
r = r, theta = theta, z = z )

tran.spherical (C = C, D.r = D, D.theta = D,
  C.r.up = 1, C.r.down = 1, C.theta.up = 1, C.theta.down = 1,
  C.phi.up = 1, C.phi.down = 1,
  r = r, theta = theta, phi = phi)

# Flux boundary conditions
tran.cylindrical(C = C, D.r = D, r = r, theta = theta, z = z,
  flux.r.up = 10, flux.r.down = 10,
  flux.theta.up = 10, flux.theta.down = 10,
  flux.z.up = 10, flux.z.down = 10)
```

```
tran.spherical(C = C, D.r = D, r = r, theta = theta, phi = phi,
  flux.r.up = 10, flux.r.down = 10,
  flux.theta.up = 10, flux.theta.down = 10,
  flux.phi.up = 10, flux.phi.down = 10)

# cyclic boundary conditions
tran.cylindrical(C = C, D.r = D, r = r, theta = theta, z = z,
  cyclicBnd = 1:3)
tran.spherical(C = C, D.r = D, r = r, theta = theta, phi = phi,
  cyclicBnd = 1:3)

# zero-gradient boundary conditions
tran.cylindrical(C = C, D.r = D, r = r, theta = theta, z = z)
tran.spherical(C = C, D.r = D, r = r, theta = theta, phi = phi)

# ==========================================
# A model with diffusion and first-order consumption  =
# ==========================================

N <- 10                    # number of grid cells
rr <- 0.005                # consumption rate
D <- 400

r          <- seq (2, 4, len = N+1)
theta      <- seq (0, 2*pi, len = N+1)
z          <- seq (0, 3, len = N+1)
phi        <- seq (0, 2*pi, len = N+1)

# The model equations
Diffcylin <- function (t, y, parms) {
   CONC <- array(dim = c(N, N, N), data = y)
   tran    <- tran.cylindrical(CONC,
           D.r = D, D.theta = D, D.z = D,
           r = r, theta = theta, z = z,
           C.r.up = 0, C.r.down = 1,
           cyclicBnd = 2)
  dCONC <- tran$dC - rr * CONC
  return (list(dCONC))
}

Diffspher <- function (t, y, parms) {
   CONC <- array(dim = c(N, N, N), data = y)
       tran <- tran.spherical (CONC,
            D.r = D, D.theta = D, D.phi = D,
            r = r, theta = theta, phi = phi,
            C.r.up = 0, C.r.down = 1,
```

```
          cyclicBnd = 2:3)
  dCONC <- tran$dC - rr * CONC
  return (list(dCONC))
}

# initial condition: 0 everywhere, except in central point
y    <- array(dim = c(N, N, N), data = 0)
N2   <- ceiling(N/2)

y[N2, N2, N2] <- 100    # initial concentration in the central point...

# solve to steady-state; cyclicBnd = 2,
outcyl <- steady.3D (y = y, func = Diffcylin, parms = NULL,
dim = c(N, N, N), lrw = 1e6, cyclicBnd = 2)

STDcyl <- array(dim = c(N, N, N), data = outcyl$y)
image(STDcyl[,,1])

# For spherical coordinates, cyclic Bnd = 2, 3
outspher <- steady.3D (y = y, func = Diffspher, parms = NULL, pos=TRUE,
                       dim = c(N, N, N), lrw = 1e6, cyclicBnd = 2:3)

#STDspher <- array(dim = c(N, N, N), data = outspher$y)
#image(STDspher[,,1])

## Not run:
    image(outspher)

## End(Not run)
```

In the R domain (Figures 6.9 and 6.10):

```
>
> # In the R domain:
>
> install.packages("ReacTran")
Installing package into 'C:/Users/Bert/Documents/R/win-library/3.3'
(as 'lib' is unspecified)
--- Please select a CRAN mirror for use in this session ---
A CRAN mirror is selected.
also installing the dependencies 'rootSolve', 'deSolve'
trying URL 'https://mirrors.tuna.tsinghua.edu.cn/CRAN/bin/windows/contrib/
3.3/rootSolve_1.7.zip'
Content type 'application/zip' length 683839 bytes (667 KB)
downloaded 667 KB
```

```
trying URL 'https://mirrors.tuna.tsinghua.edu.cn/CRAN/bin/windows/contrib/
3.3/deSolve_1.20.zip'
Content type 'application/zip' length 2667028 bytes (2.5 MB)
downloaded 2.5 MB

trying URL 'https://mirrors.tuna.tsinghua.edu.cn/CRAN/bin/windows/contrib/
3.3/ReacTran_1.4.3.1.zip'
Content type 'application/zip' length 915996 bytes (894 KB)
downloaded 894 KB

package 'rootSolve' successfully unpacked and MD5 sums checked
package 'deSolve' successfully unpacked and MD5 sums checked
package 'ReacTran' successfully unpacked and MD5 sums checked

The downloaded binary packages are in

C:\Users\Bert\AppData\Local\Temp\RtmpmEkAb2\downloaded_packages
> require(ReacTran)
Loading required package: ReacTran
Loading required package: rootSolve
Loading required package: deSolve
Loading required package: shape
> library(ReacTran)
> ls("package:ReacTran") # Listing all the programs in this package:
 [1] "advection.1D"      "advection.volume.1D"  "fiadeiro"
 [4] "g.cylinder"        "g.sphere"             "g.spheroid"
 [7] "p.exp"             "p.lin"                "p.sig"
[10] "polar2cart"        "setup.compaction.1D"  "setup.grid.1D"
[13] "setup.grid.2D"     "setup.prop.1D"        "setup.prop.2D"
[16] "tran.1D"           "tran.2D"              "tran.3D"
[19] "tran.cylindrical"  "tran.polar"           "tran.spherical"
[22] "tran.volume.1D"    "tran.volume.2D"       "tran.volume.3D"

> tran.cylindrical # Outputting the program listing of "tran.cylindrical "
>
> ## ================================================
> ## Testing the functions
> ## ================================================

> # Grid definition

> r.N <- 4 # number of cells in r-direction
> theta.N <- 6 # number of cells in theta-direction
> z.N <- 3 # number of cells in z-direction
> D <- 100 # diffusion coefficient
```

```
> r <- seq(0, 8, len = r.N+1) # cell size r-direction [cm]
> theta <- seq(0,2*pi, len = theta.N+1) # theta-direction - theta: from 0, 2pi
> phi <- seq(0,2*pi, len = z.N+1) # phi-direction (0,2pi)
> z <- seq(0,5, len = z.N+1) # cell size z-direction [cm]
> # Initial conditions
> C <- array(dim = c(r.N, theta.N, z.N), data = 0)
> # Concentration boundary conditions
> tran.cylindrical (C = C, D.r = D, D.theta = D,
+ C.r.up = 1, C.r.down = 1,
+ C.theta.up = 1, C.theta.down = 1,
+ C.z.up = 1, C.z.down = 1,
+ r = r, theta = theta, z = z )
$dC
, , 1

           [,1]     [,2]     [,3]     [,4]     [,5]       [,6]
[1,] 426107.503  72.0000  72.0000  72.0000  72.0000 426107.503
[2,]  47409.278  72.0000  72.0000  72.0000  72.0000  47409.278
[3,]  17113.420  72.0000  72.0000  72.0000  72.0000  17113.420
[4,]   8823.745 129.1429 129.1429 129.1429 129.1429   8823.745

, , 2

           [,1]     [,2]     [,3]     [,4]     [,5]       [,6]
[1,] 426035.503  0.00000  0.00000  0.00000  0.00000 426035.503
[2,]  47337.278  0.00000  0.00000  0.00000  0.00000  47337.278
[3,]  17041.420  0.00000  0.00000  0.00000  0.00000  17041.420
[4,]   8751.745 57.14286 57.14286 57.14286 57.14286   8751.745

, , 3

           [,1]     [,2]     [,3]     [,4]     [,5]       [,6]
[1,] 426107.503  72.0000  72.0000  72.0000  72.0000 426107.503
[2,]  47409.278  72.0000  72.0000  72.0000  72.0000  47409.278
[3,]  17113.420  72.0000  72.0000  72.0000  72.0000  17113.420
[4,]   8823.745 129.1429 129.1429 129.1429 129.1429   8823.745


$flux.r.up
     [,1] [,2] [,3]
[1,]  100  100  100
[2,]  100  100  100
[3,]  100  100  100
[4,]  100  100  100
[5,]  100  100  100
[6,]  100  100  100
```

```
$flux.r.down
     [,1] [,2] [,3]
[1,] -100 -100 -100
[2,] -100 -100 -100
[3,] -100 -100 -100
[4,] -100 -100 -100
[5,] -100 -100 -100
[6,] -100 -100 -100

$flux.theta.up
         [,1]     [,2]     [,3]
[1,] 9230.769 9230.769 9230.769
[2,] 3076.923 3076.923 3076.923
[3,] 1846.154 1846.154 1846.154
[4,] 1318.681 1318.681 1318.681

$flux.theta.down
          [,1]      [,2]      [,3]
[1,] -9230.769 -9230.769 -9230.769
[2,] -3076.923 -3076.923 -3076.923
[3,] -1846.154 -1846.154 -1846.154
[4,] -1318.681 -1318.681 -1318.681

$flux.z.up
     [,1] [,2] [,3] [,4] [,5] [,6]
[1,]  120  120  120  120  120  120
[2,]  120  120  120  120  120  120
[3,]  120  120  120  120  120  120
[4,]  120  120  120  120  120  120

$flux.z.down
     [,1] [,2] [,3] [,4] [,5] [,6]
[1,] -120 -120 -120 -120 -120 -120
[2,] -120 -120 -120 -120 -120 -120
[3,] -120 -120 -120 -120 -120 -120
[4,] -120 -120 -120 -120 -120 -120

> tran.spherical (C = C, D.r = D, D.theta = D,
+ C.r.up = 1, C.r.down = 1, C.theta.up = 1, C.theta.down = 1,
+ C.phi.up = 1, C.phi.down = 1,
+ r = r, theta = theta, phi = phi)
$dC
, , 1
```

```
          [,1]       [,2]       [,3]       [,4]       [,5]      [,6]
[1,]  0.00000 151293676 45412970.6 21643848.5 12640411.2 8750598.0
[2,]  0.00000  16810408  5045885.6  2404872.1  1404490.1  972288.7
[3,]  0.00000   6051747  1816518.8   865753.9   505616.4  350023.9
[4,] 65.30612   3087691   926860.6   441776.5   258032.9  178648.9

, , 2

         [,1]     [,2]     [,3]     [,4]     [, 5]      [,6]
[1,]  0.00000  0.00000  0.00000  0.00000  0.00000 464556.742
[2,]  0.00000  0.00000  0.00000  0.00000  0.00000  51617.416
[3,]  0.00000  0.00000  0.00000  0.00000  0.00000  18582.270
[4,] 65.30612 65.30612 65.30612 65.30612 65.30612   9546.056

, , 3

          [,1]       [,2]       [,3]        ,4]       [,5]      [,6]
[1,]  0.00000 151293676 45412970.6 21643848.5 12640411.2 8750598.0
[2,]  0.00000  16810408 5 045885.6  2404872.1  1404490.1  972288.7
[3,]  0.00000   6051747  1816518.8   865753.9   505616.4  350023.9
[4,] 65.30612   3087691   926860.6   441776.5   258032.9  178648.9

$flux.r.up
     [,1] [,2] [,3]
[1,]  100  100  100
[2,]  100  100  100
[3,]  100  100  100
[4,]  100  100  100
[5,]  100  100  100
[6,]  100  100  100

$flux.r.down
     [,1] [,2] [,3]
[1,] -100 -100 -100
[2,] -100 -100 -100
[3,] -100 -100 -100
[4,] -100 -100 -100
[5,] -100 -100 -100
[6,] -100 -100 -100

$flux.theta.up
         [,1]     [,2]     [,3]
[1,] 9230.769 9230.769 9230.769
[2,] 3076.923 3076.923 3076.923
```

```
[3,] 1846.154 1846.154 1846.154
[4,] 1318.681 1318.681 1318.681

$flux.theta.down
          [,1]      [,2]      [,3]
[1,] -9230.769 -9230.769 -9230.769
[2,] -3076.923 -3076.923 -3076.923
[3,] -1846.154 -1846.154 -1846.154
[4,] -1318.681 -1318.681 -1318.681

$flux.phi.up
     [,1]      [,2]      [,3]     [,4]      [,5]      [,6]
[1,]    0 213034.42 106542.22 71055.94 53321.163 42686.998
[2,]    0  71011.47  35514.07 23685.31 17773.721 14228.999
[3,]    0  42606.88  21308.44 14211.19 10664.233  8537.400
[4,]    0  30433.49  15220.32 10150.85  7617.309  6098.143

$flux.phi.down
     [,1]       [,2]       [,3]      [,4]       [,5]       [,6]
[1,]    0 -213034.42 -106542.22 -71055.94 -53321.163 -42686.998
[2,]    0  -71011.47  -35514.07 -23685.31 -17773.721 -14228.999
[3,]    0  -42606.88  -21308.44 -14211.19 -10664.233  -8537.400
[4,]    0  -30433.49  -15220.32 -10150.85  -7617.309  -6098.143

> # Flux boundary conditions
> tran.cylindrical(C = C, D.r = D, r = r, theta = theta, z = z,
+ flux.r.up = 10, flux.r.down = 10,
+ flux.theta.up = 10, flux.theta.down = 10,
+ flux.z.up = 10, flux.z.down = 10)
$dC
, , 1

          [,1]      [,2]      [,3]      [,4]      [,5]       [,6]
[1,] 467.53846 6.0000000 6.0000000 6.0000000 6.0000000 -455.53846
[2,] 159.84615 6.0000000 6.0000000 6.0000000 6.0000000 -147.84615
[3,]  98.30769 6.0000000 6.0000000 6.0000000 6.0000000  -86.30769
[4,]  66.21978 0.2857143 0.2857143 0.2857143 0.2857143  -65.64835
, , 2

          [,1]      [,2]      [,3]      [,4]      [,5]       [,6]
[1,] 461.53846  0.000000  0.000000  0.000000  0.000000 -461.53846
[2,] 153.84615  0.000000  0.000000  0.000000  0.000000 -153.84615
[3,]  92.30769  0.000000  0.000000  0.000000  0.000000  -92.30769
[4,]  60.21978 -5.714286 -5.714286 -5.714286 -5.714286  -71.64835
, , 3
```

```
          [,1]      [,2]      [,3]      [,4]      [,5]       [,6]
[1,] 455.53846  -6.00000  -6.00000  -6.00000  -6.00000 -467.53846
[2,] 147.84615  -6.00000  -6.00000  -6.00000  -6.00000 -159.84615
[3,]  86.30769  -6.00000  -6.00000  -6.00000  -6.00000  -98.30769
[4,]  54.21978 -11.71429 -11.71429 -11.71429 -11.71429   77.64835


$flux.r.up
     [,1] [,2] [,3]
[1,]   10   10   10
[2,]   10   10   10
[3,]   10   10   10
[4,]   10   10   10
[5,]   10   10   10
[6,]   10   10   10

$flux.r.down
     [,1] [,2] [,3]
[1,]   10   10   10
[2,]   10   10   10
[3,]   10   10   10
[4,]   10   10   10
[5,]   10   10   10
[6,]   10   10   10


$flux.theta.up
     [,1] [,2] [,3]
[1,]   10   10   10
[2,]   10   10   10
[3,]   10   10   10
[4,]   10   10   10

$flux.theta.down
     [,1] [,2] [,3]
[1,]   10   10   10
[2,]   10   10   10
[3,]   10   10   10
[4,]   10   10   10

$flux.z.up
     [,1] [,2] [,3] [,4] [,5] [,6]
[1,]   10   10   10   10   10   10
[2,]   10   10   10   10   10   10
[3,]   10   10   10   10   10   10
[4,]   10   10   10   10   10   10
```

```
$flux.z.down
     [,1] [,2] [,3] [,4] [,5] [,6]
[1,]   10   10   10   10   10   10
[2,]   10   10   10   10   10   10
[3,]   10   10   10   10   10   10
[4,]   10   10   10   10   10   10

> tran.spherical(C = C, D.r = D, r = r, theta = theta, phi = phi,
+ flux.r.up = 10, flux.r.down = 10,
+ flux.theta.up = 10, flux.theta.down = 10,
+ flux.phi.up = 10, flux.phi.down = 10)
$dC

, , 1

          [,1]     [,2]      [,3]      [,4]      [,5]      [,6]
[1,] 21302.192 7101.842 4262.4391 3046.0294 2370.6181 1437.8461
[2,]  7100.731 2367.281 1420.8130 1015.3431  790.2060  479.2820
[3,]  4260.438 1420.368  852.4878  609.2059  474.1236  287.5692
[4,]  3036.640 1008.018  602.3893  428.6164  332.1291  198.8760

, , 2

          [,1]      [,2]      [,3]      [,4]      [,5]      [,6]
[1,]  0.000000  0.000000  0.000000  0.000000  0.000000 -503.2698
[2,]  0.000000  0.000000  0.000000  0.000000  0.000000 -167.7566
[3,]  0.000000  0.000000  0.000000  0.000000  0.000000 -100.6540
[4,] -6.530612 -6.530612 -6.530612 -6.530612 -6.530612  -78.4263

, , 3

           [,1]      [,2]       [,3]       [,4]       [,5]       [,6]
[1,] -21302.192 -7101.842 -4262.4391 -3046.0294 -2370.6181 -2444.3858
[2,]  -7100.731 -2367.281 -1420.8130 -1015.3431  -790.2060  -814.7953
[3,]  -4260.438 -1420.368  -852.4878  -609.2059  -474.1236  -488.8772
[4,]  -3049.701 -1021.079  -615.4505  -441.6777  -345.1903  -355.7286


$flux.r.up
     [,1] [,2] [,3]
[1,]   10   10   10
[2,]   10   10   10
[3,]   10   10   10
[4,]   10   10   10
```

```
[5,]  10  10  10
[6,]  10  10  10

$flux.r.down
      [,1] [,2] [,3]
[1,]  10  10  10
[2,]  10  10  10
[3,]  10  10  10
[4,]  10  10  10
[5,]  10  10  10
[6,]  10  10  10

$flux.theta.up
      [,1] [,2] [,3]
[1,]  10  10  10
[2,]  10  10  10
[3,]  10  10  10
[4,]  10  10  10

$flux.theta.down
      [,1] [,2] [,3]
[1,]  10  10  10
[2,]  10  10  10
[3,]  10  10  10
[4,]  10  10  10

$flux.phi.up
     [,1] [,2]  [,3] [,4] [,5] [,6]
[1,]  10  10  10  10  10  10
[2,]  10  10  10  10  10  10
[3,]  10  10  10  10  10  10
[4,]  10  10  10  10  10  10

$flux.phi.down
     [,1] [,2]  [,3] [,4] [,5] [,6]
[1,]  10  10  10  10  10  10
[2,]  10  10  10  10  10  10
[3,]  10  10  10  10  10  10
[4,]  10  10  10  10  10  10

> # cyclic boundary conditions
> tran.cylindrical(C = C, D.r = D, r = r, theta = theta, z = z,
+ cyclicBnd = 1:3)
$dC
```

```
, , 1

     [,1] [,2] [,3] [,4] [,5] [,6]
[1,]    0    0    0    0    0    0
[2,]    0    0    0    0    0    0
[3,]    0    0    0    0    0    0
[4,]    0    0    0    0    0    0

, , 2

     [,1] [,2] [,3] [,4] [,5] [,6]
[1,]    0    0    0    0    0    0
[2,]    0    0    0    0    0    0
[3,]    0    0    0    0    0    0
[4,]    0    0    0    0    0    0

, , 3

     [,1] [,2] [,3] [,4] [,5] [,6]
[1,]    0    0    0    0    0    0
[2,]    0    0    0    0    0    0
[3,]    0    0    0    0    0    0
[4,]    0    0    0    0    0    0

$flux.r.up
     [,1] [,2] [,3]
[1,]    0    0    0
[2,]    0    0    0
[3,]    0    0    0
[4,]    0    0    0
[5,]    0    0    0
[6,]    0    0    0

$flux.r.down
     [,1] [,2] [,3]
[1,]    0    0    0
[2,]    0    0    0
[3,]    0    0    0
[4,]    0    0    0
[5,]    0    0    0
[6,]    0    0    0

$flux.theta.up
     [,1] [,2] [,3]
[1,]    0    0    0
[2,]    0    0    0
```

```
[3,]  0  0  0
[4,]  0  0  0

$flux.theta.down
     [,1] [,2] [,3]
[1,]  0  0  0
[2,]  0  0  0
[3,]  0  0  0
[4,]  0  0  0

$flux.z.up
     [,1] [,2] [,3] [,4] [,5] [,6]
[1,]  0  0  0  0  0  0
[2,]  0  0  0  0  0  0
[3,]  0  0  0  0  0  0
[4,]  0  0  0  0  0  0

$flux.z.down
     [,1] [,2] [,3] [,4] [,5] [,6]
[1,]  0  0  0  0  0  0
[2,]  0  0  0  0  0  0
[3,]  0  0  0  0  0  0
[4,]  0  0  0  0  0  0

> tran.spherical(C = C, D.r = D, r = r, theta = theta, phi = phi,
+ cyclicBnd = 1:3)
$dC
, , 1

     [,1] [,2] [,3] [,4] [,5] [,6]
[1,]  0  0  0  0  0  0
[2,]  0  0  0  0  0  0
[3,]  0  0  0  0  0  0
[4,]  0  0  0  0  0  0

, , 2

     [,1] [,2] [,3] [,4] [,5] [,6]
[1,]  0  0  0  0  0  0
[2,]  0  0  0  0  0  0
[3,]  0  0  0  0  0  0
[4,]  0  0  0  0  0  0

, , 3

     [,1] [,2] [,3] [,4] [,5] [,6]
[1,]  0  0  0  0  0  0
```

```
[2,]    0    0    0    0    0    0
[3,]    0    0    0    0    0    0
[4,]    0    0    0    0    0    0

$flux.r.up
     [,1] [,2] [,3]
[1,]    0    0    0
[2,]    0    0    0
[3,]    0    0    0
[4,]    0    0    0
[5,]    0    0    0
[6,]    0    0    0

$flux.r.down
     [,1] [,2] [,3]
[1,]    0    0    0
[2,]    0    0    0
[3,]    0    0    0
[4,]    0    0    0
[5,]    0    0    0
[6,]    0    0    0

$flux.theta.up
     [,1] [,2] [,3]
[1,]    0    0    0
[2,]    0    0    0
[3,]    0    0    0
[4,]    0    0    0

$flux.theta.down
     [,1] [,2] [,3]
[1,]    0    0    0
[2,]    0    0    0
[3,]    0    0    0
[4,]    0    0    0

$flux.phi.up
     [,1] [,2] [,3] [,4] [,5] [,6]
[1,]    0    0    0    0    0    0
[2,]    0    0    0    0    0    0
[3,]    0    0    0    0    0    0
[4,]    0    0    0    0    0    0

$flux.phi.down
     [,1] [,2] [,3] [,4] [,5] [,6]
[1,]    0    0    0    0    0    0
[2,]    0    0    0    0    0    0
```

```
[3,]  0  0  0  0  0  0
[4,]  0  0  0  0  0  0

> # zero-gradient boundary conditions
> tran.cylindrical(C = C, D.r = D, r = r, theta = theta, z = z)
$dC
, , 1

     [,1] [,2] [,3] [,4] [,5] [,6]
[1,]    0    0    0    0    0    0
[2,]    0    0    0    0    0    0
[3,]    0    0    0    0    0    0
[4,]    0    0    0    0    0    0

, , 2

     [,1] [,2] [,3] [,4] [,5] [,6]
[1,]    0    0    0    0    0    0
[2,]    0    0    0    0    0    0
[3,]    0    0    0    0    0    0
[4,]    0    0    0    0    0    0

, , 3

     [,1] [,2] [,3] [,4] [,5] [,6]
[1,]    0    0    0    0    0    0
[2,]    0    0    0    0    0    0
[3,]    0    0    0    0    0    0
[4,]    0    0    0    0    0    0

$flux.r.up
     [,1] [,2] [,3]
[1,]    0    0    0
[2,]    0    0    0
[3,]    0    0    0
[4,]    0    0    0
[5,]    0    0    0
[6,]    0    0    0

$flux.r.down
     [,1] [,2] [,3]
[1,]    0    0    0
[2,]    0    0    0
[3,]    0    0    0
[4,]    0    0    0
[5,]    0    0    0
[6,]    0    0    0
```

```
$flux.theta.up
     [,1] [,2] [,3]
[1,]    0    0    0
[2,]    0    0    0
[3,]    0    0    0
[4,]    0    0    0

$flux.theta.down
     [,1] [,2] [,3]
[1,]    0    0    0
[2,]    0    0    0
[3,]    0    0    0
[4,]    0    0    0

$flux.z.up
     [,1] [,2] [,3] [,4] [,5] [,6]
[1,]    0    0    0    0    0    0
[2,]    0    0    0    0    0    0
[3,]    0    0    0    0    0    0
[4,]    0    0    0    0    0    0

$flux.z.down
     [,1] [,2] [,3] [,4] [,5] [,6]
[1,]    0    0    0    0    0    0
[2,]    0    0    0    0    0    0
[3,]    0    0    0    0    0    0
[4,]    0    0    0    0    0    0

> tran.spherical(C = C, D.r = D, r = r, theta = theta, phi = phi)
$dC
, , 1

     [,1] [,2] [,3] [,4] [,5] [,6]
[1,]    0    0    0    0    0    0
[2,]    0    0    0    0    0    0
[3,]    0    0    0    0    0    0
[4,]    0    0    0    0    0    0

, , 2

     [,1] [,2] [,3] [,4] [,5] [,6]
[1,]    0    0    0    0    0    0
[2,]    0    0    0    0    0    0
[3,]    0    0    0    0    0    0
[4,]    0    0    0    0    0    0
```

```
, , 3

     [,1] [,2] [,3] [,4] [,5] [,6]
[1,]    0    0    0    0    0    0
[2,]    0    0    0    0    0    0
[3,]    0    0    0    0    0    0
[4,]    0    0    0    0    0    0


$flux.r.up
     [,1] [,2] [,3]
[1,]    0    0    0
[2,]    0    0    0
[3,]    0    0    0
[4,]    0    0    0
[5,]    0    0    0
[6,]    0    0    0

$flux.r.down
     [,1] [,2] [,3]
[1,]    0    0    0
[2,]    0    0    0
[3,]    0    0    0
[4,]    0    0    0
[5,]    0    0    0
[6,]    0    0    0

$flux.theta.up
     [,1] [,2] [,3]
[1,]    0    0    0
[2,]    0    0    0
[3,]    0    0    0
[4,]    0    0    0

$flux.theta.down
     [,1] [,2] [,3]
[1,]    0    0    0
[2,]    0    0    0
[3,]    0    0    0
[4,]    0    0    0

$flux.phi.up
     [,1] [,2] [,3] [,4] [,5] [,6]
[1,]    0    0    0    0    0    0
[2,]    0    0    0    0    0    0
```

```
[3,]  0  0  0  0  0  0
[4,]  0  0  0  0  0  0

$flux.phi.down
     [,1] [,2] [,3] [,4] [,5] [,6]
[1,]  0    0    0    0    0    0
[2,]  0    0    0    0    0    0
[3,]  0    0    0    0    0    0
[4,]  0    0    0    0    0    0

> # ==================================
> # A model with diffusion and first-order consumption  =
> # ==================================
> N <- 10 # number of grid cells
> rr <- 0.005 # consumption rate
> D <- 400
> r <- seq (2, 4, len = N+1)

> theta <- seq (0, 2*pi, len = N+1)
> z <- seq (0, 3, len = N+1)
> phi <- seq (0, 2*pi, len = N+1)
> # The model equations
> Diffcylin <- function (t, y, parms) {
+ CONC <- array(dim = c(N, N, N), data = y)
+ tran <- tran.cylindrical(CONC,
+ D.r = D, D.theta = D, D.z = D,
+ r = r, theta = theta, z = z,
+ C.r.up = 0, C.r.down = 1,
+ cyclicBnd = 2)
+ dCONC <- tran$dC - rr * CONC
+ return (list(dCONC))
+ }
> Diffspher <- function (t, y, parms) {
+ CONC <- array(dim = c(N, N, N), data = y)
+ tran <- tran.spherical (CONC,
+ D.r = D, D.theta = D, D.phi = D,
+ r = r, theta = theta, phi = phi,
+ C.r.up = 0, C.r.down = 1,
+ cyclicBnd = 2:3)
+ dCONC <- tran$dC - rr * CONC
+ return (list(dCONC))
+ }
> # initial condition: 0 everywhere, except in central point
> y <- array(dim = c(N, N, N), data = 0)
> N2 <- ceiling(N/2)
> y[N2, N2, N2] <- 100 # initial concentration in the central point...
> # solve to steady-state; cyclicBnd = 2,
```

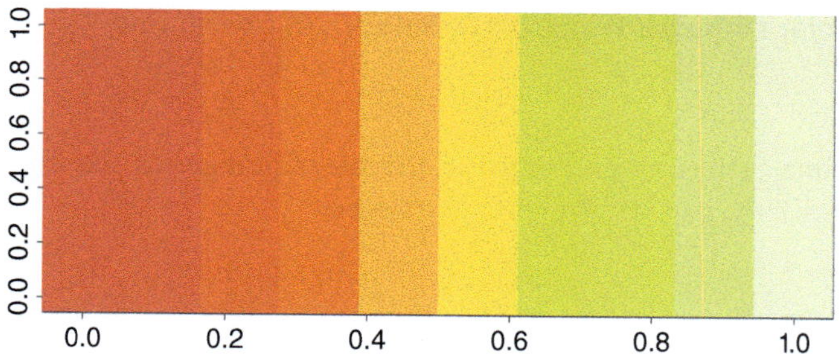

Figure 6.9 Cylindrical A model with diffusion and first-order consumption diffusive transport in cylindrical (r, θ, z), and spherical (r, θ, φ) coordinates.

```
> outcyl <- steady.3D (y = y, func = Diffcylin, parms = NULL,
+ dim = c(N, N, N), lrw = 1e6, cyclicBnd = 2)
> STDcyl <- array(dim = c(N, N, N), data = outcyl$y)
> image(STDcyl[,,1])
> #
> # Outputting: Figure 6-9: Cylindrical

> #
> # For spherical coordinates, cyclic Bnd = 2, 3
> outspher <- steady.3D (y = y, func = Diffspher, parms = NULL, pos=TRUE,
+ dim = c(N, N, N), lrw = 1e6, cyclicBnd = 2:3)
> #STDspher <- array(dim = c(N, N, N), data = outspher$y)
> #image(STDspher[,,1])
> ## Not run:
> image(outspher)
> #
> # Outputting: Figure 6-10: Spherical

>
```

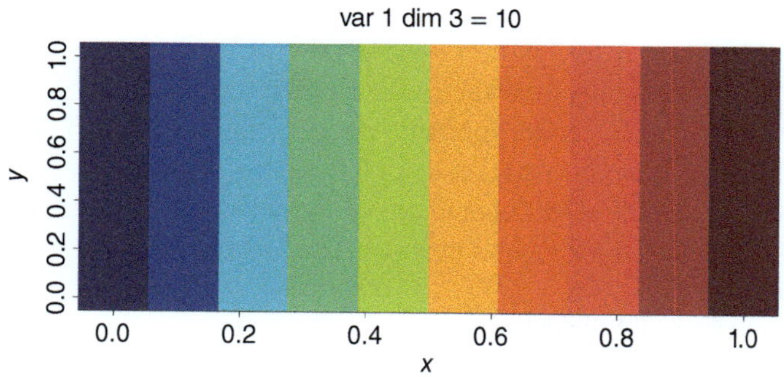

Figure 6.10 Spherical A model with diffusion and first-order consumption diffusive transport in cylindrical (r, θ, z), and spherical (r, θ, φ) coordinates.

Example 6.4 tran.polar ***Diffusive Transport in Polar (r, θ) Coordinates***

Description

This program, in R, estimates the transport term (i.e. the rate of change of a concentration owing to diffusion) in a polar (r, θ) coordinate system.

Usage

```
tran.polar (C, C.r.up = NULL, C.r.down = NULL,
            C.theta.up = NULL, C.theta.down = NULL,
            flux.r.up = NULL, flux.r.down = NULL,
            flux.theta.up = NULL, flux.theta.down = NULL,
            cyclicBnd = NULL, D.r = 1, D.theta = D.r,
            r = NULL, theta = NULL, full.output = FALSE)

polar2cart (out, r, theta, x = NULL, y = NULL)
```

Arguments

C	Concentration, expressed per unit volume, defined at the center of each grid cell; Nr*Ntheta matrix [M/L^3]
C.r.up	Concentration at upstream boundary in $r(x)$-direction; vector of length Ntheta [M/L^3]
C.r.down	Concentration at downstream boundary in $r(x)$-direction; vector of length Ntheta [M/L^3]
C.theta.up	Concentration at upstream boundary in theta-direction; vector of length Nr [M/L^3]
C.theta.down	Concentration at downstream boundary in theta-direction; vector of length Nr [M/L^3]
flux.r.up	Flux across the upstream boundary in r-direction, positive = INTO model domain; vector of length Ntheta [$M/L^2/T$]
flux.r.down	Flux across the downstream boundary in r-direction, positive = OUT of model domain; vector of length Ntheta [$M/L^2/T$]
flux.theta.up	Flux across the upstream boundary in theta-direction, positive = INTO model domain; vector of length Nr [$M/L^2/T$]
flux.theta.down	Flux across the downstream boundary in theta-direction, positive = OUT of model domain; vector of length Nr [$M/L^2/T$]
cyclicBnd	If not NULL, the direction in which a cyclic boundary is defined, i.e. cyclicBnd = 1 for the r-direction, cyclicBnd = 2 for the theta-direction, and cyclicBnd = c(1,2) for both the r- and theta-directions

D.r	Diffusion coefficient in *r*-direction, defined on grid cell interfaces. One value, a vector of length (Nr+1), a prop.1D list created by setup.prop.1D, or a (Nr+1)* Ntheta matrix [L^2/T]
D.theta	Diffusion coefficient in theta-direction, defined on grid cell interfaces. One value, a vector of length (Ntheta+1), a prop.1D list created by setup.prop.1D, or a Nr*(Ntheta+1) matrix [L^2/T]
r	Position of adjacent cell interfaces in the *r*-direction. A vector of length Nr+1 [L]
theta	Position of adjacent cell interfaces in the theta-direction. A vector of length Ntheta+1 [L]. Theta should be within [0, 2 pi]
full.output	Logical flag enabling a full return of the output (default = FALSE; TRUE slows down execution by 20%)
out	Output as returned by tran.polar, and which is to be mapped from polar to Cartesian coordinates
x	The Cartesian *x*-coordinates to which the polar coordinates are to be mapped
y	The Cartesian *y*-coordinates to which the polar coordinates are to be mapped

Details

tran.polar performs (simplified) transport in polar coordinates.

The boundary conditions are either

(1) zero gradient
(2) fixed concentration
(3) fixed flux
(4) cyclic boundary.

This is also the order of priority. The cyclic boundary overrules the other. If fixed concentration, fixed flux, and cyclicBnd are NULL then the boundary is zero gradient.

A cyclic boundary condition has concentration and flux at upstream and downstream boundary the same.

polar2cart maps the polar coordinates to Cartesian coordinates.

If x and y are not provided, then it will create an (*x*,*y*) grid based on r: seq(-maxr, maxr, length.out=Nr), where maxr is the maximum value of r, and Nr is the number of elements in r.

Value

A list containing

dC	The rate of change of the concentration C due to transport, defined in the center of each grid cell, a Nr*Ntheta matrix [M/L^3/T]
C.r.up	Concentration at the upstream interface in r-direction. A vector of length Ntheta [M/L^3]. Only when full.output = TRUE
C.r.down	Concentration at the downstream interface in r-direction. A vector of length Ntheta [M/L^3]. Only when full.output = TRUE
C.theta.up	Concentration at the upstream interface in theta-direction. A vector of length Nr [M/L^3]. Only when full.output = TRUE
C.theta.down	Concentration at the downstream interface in theta-direction. A vector of length Nr [M/L^3]. Only when full.output = TRUE
r.flux	Flux across the interfaces in x-direction of the grid cells. A (Nr+1)*Ntheta matrix [M/L^2/T]. Only when full.output = TRUE
theta.flux	Flux across the interfaces in y-direction of the grid cells. A Nr*(Ntheta+1) matrix [M/L^2/T]. Only when full.output = TRUE
flux.r.up	Flux across the upstream boundary in r-direction, positive = INTO model domain. A vector of length Ntheta [M/L^2/T]
flux.r.down	Flux across the downstream boundary in r-direction, positive = OUT of model domain. A vector of length Ntheta [M/L^2/T]
flux.theta.up	Flux across the upstream boundary in theta-direction, positive = INTO model domain. A vector of length Nr [M/L^2/T]
flux.theta.down	Flux across the downstream boundary in theta-direction, positive = OUT of model domain. A vector of length Nr [M/L2/T]

See Also

tran.cylindrical, tran.spherical for a discretization of 3D transport equations in cylindrical and spherical coordinates.
tran.1D, tran.2D, tran.3D

Examples

```
# ===============
# Testing the functions  =
# ===============
# Parameters
F          <- 100            # input flux [micromol cm-2 yr-1]
D          <- 400            # mixing coefficient [cm2 yr-1]
```

```
# Grid definition
r.N <- 4 # number of cells in r-direction
theta.N <- 6 # number of cells in theta-direction
r.L <- 8 # domain size r-direction [cm]
r <- seq(0, r.L,len = r.N+1) # cell size r-direction [cm]
theta <- seq(0, 2*pi,len = theta.N+1) # theta-direction - theta: from 0, 2pi

# Initial conditions
C <- matrix(nrow = r.N, ncol = theta.N, data = 0)

# Boundary conditions: fixed concentration
C.r.up          <- rep(1, times = theta.N)
C.r.down        <- rep(0, times = theta.N)
C.theta.up      <- rep(1, times = r.N)
C.theta.down    <- rep(0, times = r.N)

# Concentration boundary conditions
tran.polar(C = C, D.r = D, D.theta = D,
  r = r, theta = theta,
  C.r.up = C.r.up, C.r.down = C.r.down,
  C.theta.up = C.theta.up, C.theta.down = C.theta.down)

# Flux boundary conditions
flux.r.up <- rep(200, times = theta.N)
flux.r.down <- rep(-200, times = theta.N)
flux.theta.up <- rep(200, times = r.N)
flux.theta.down <- rep(-200, times = r.N)

tran.polar(C = C, D.r = D, r = r, theta = theta,
  flux.r.up = flux.r.up, flux.r.down = flux.r.down,
  flux.theta.up = flux.theta.up, flux.theta.down = flux.theta.down,
  full.output = TRUE)

## ==================================
## A model with diffusion and first-order consumption  =
## ==================================
N     <- 50          # number of grid cells
XX    <- 4           # total size
rr    <- 0.005       # consumption rate
ini   <- 1           # initial value at x=0
D     <- 400

r        <- seq (2, 4, len = N+1)
theta    <- seq(0, 2*pi, len = N+1)
theta.m <- 0.5*(theta[-1]+theta[-(N+1)])
```

```
# The model  equations

 Diffpolar <- function (t, y, parms) {
   CONC <- matrix(nrow = N, ncol = N, data = y)
   tran   <- tran.polar(CONC, D.r = D, D.theta = D, r = r, theta = theta,
            C.r.up = 0, C.r.down = 1*sin(5*theta.m),
            cyclicBnd = 2, full.output=TRUE )
  dCONC <- tran$dC - rr * CONC
  return (list(dCONC))
}

# solve to steady-state; cyclicBnd = 2, because of C.theta.up, C.theta.down
out <- steady.2D (y = rep(0, N*N), func = Diffpolar, parms = NULL,
                     dim = c(N, N), lrw = 1e6, cyclicBnd = 2)

image(out)

cart <- polar2cart(out, r = r, theta = theta,
                          x = seq(-4, 4, len = 100),
                          y = seq(-4, 4, len = 100))
image(cart)
```

In the R domain:

```
> # In the R domain:
> install.packages("ReacTran")
Installing package into 'C:/Users/Bert/Documents/R/win-library/3.3'
(as 'lib' is unspecified)
--- Please select a CRAN mirror for use in this session ---
A CRAN mirror is selected.
also installing the dependencies 'rootSolve', 'deSolve'
trying URL 'https://mirrors.tuna.tsinghua.edu.cn/CRAN/bin/windows/contrib/
3.3/rootSolve_1.7.zip'
Content type 'application/zip' length 683839 bytes (667 KB)
downloaded 667 KB

trying URL 'https://mirrors.tuna.tsinghua.edu.cn/CRAN/bin/windows/contrib/3.3/
deSolve_1.20.zip'
Content type 'application/zip' length 2667028 bytes (2.5 MB)
downloaded 2.5 MB

trying URL 'https://mirrors.tuna.tsinghua.edu.cn/CRAN/bin/windows/contrib/3.3/
ReacTran_1.4.3.1.zip'
Content type 'application/zip' length 915996 bytes (894 KB)
downloaded 894 KB package 'rootSolve' successfully unpacked and MD5 sums
checked
```

```
package 'deSolve' successfully unpacked and MD5 sums checked
package 'ReacTran' successfully unpacked and MD5 sums checked
The downloaded binary packages are in
C:\Users\Bert\AppData\Local\Temp\RtmpmEkAb2\downloaded_packages
> require(ReacTran)
Loading required package: ReacTran
Loading required package: rootSolve
Loading required package: deSolve
Loading required package: shape
> library(ReacTran)
> ls("package:ReacTran") # Listing all the programs in this package:
 [1] "advection.1D"       "advection.volume.1D"  "fiadeiro"
 [4] "g.cylinder"         "g.sphere"             "g.spheroid"
 [7] "p.exp"              "p.lin"                "p.sig"
[10] "polar2cart"         "setup.compaction.1D"  "setup.grid.1D"
[13] "setup.grid.2D"      "setup.prop.1D"        "setup.prop.2D"
[16] "tran.1D"            "tran.2D"              "tran.3D"
[19] "tran.cylindrical"   "tran.polar"           "tran.spherical"
[22] "tran.volume.1D"     "tran.volume.2D"       "tran.volume.3D"
```

In the R domain (Figures 6.11–6.13):

```
>
> install.packages("ReacTran")
Installing package into 'C:/Users/Bert/Documents/R/win-library/3.3'
(as 'lib' is unspecified)
--- Please select a CRAN mirror for use in this session ---
A CRAN mirror is selected.
also installing the dependencies 'rootSolve', 'deSolve'

trying URL 'https://mirrors.tuna.tsinghua.edu.cn/CRAN/bin/windows/contrib/3.3/
rootSolve_1.7.zip'
Content type 'application/zip' length 683839 bytes (667 KB)
downloaded 667 KB

trying URL 'https://mirrors.tuna.tsinghua.edu.cn/CRAN/bin/windows/contrib/3.3/
deSolve_1.20.zip'
Content type 'application/zip' length 2667028 bytes (2.5 MB)
downloaded 2.5 MB

trying URL 'https://mirrors.tuna.tsinghua.edu.cn/CRAN/bin/windows/contrib/3.3/
ReacTran_1.4.3.1.zip'
Content type 'application/zip' length 915996 bytes (894 KB)
downloaded 894 KB
package 'rootSolve' successfully unpacked and MD5 sums checked
package 'deSolve' successfully unpacked and MD5 sums checked
```

```
package 'ReacTran' successfully unpacked and MD5 sums checked
The downloaded binary packages are in
C:\Users\Bert\AppData\Local\Temp\RtmpmEkAb2\downloaded_packages

> require(ReacTran)
Loading required package: ReacTran
Loading required package: rootSolve
Loading required package: deSolve
Loading required package: shape
> library(ReacTran)
> ls("package:ReacTran") # Listing all the programs in this package:
 [1] "advection.1D"        "advection.volume.1D"  "fiadeiro"
 [4] "g.cylinder"          "g.sphere"             "g.spheroid"
 [7] "p.exp"               "p.lin"                "p.sig"
[10] "polar2cart"          "setup.compaction.1D"  "setup.grid.1D"
[13] "setup.grid.2D"       "setup.prop.1D"        "setup.prop.2D"
[16] "tran.1D"             "tran.2D"              "tran.3D"
[19] "tran.cylindrical"    "tran.polar" "tran.spherical"
[22] "tran.volume.1D"      "tran.volume.2D"       "tran.volume.3D"

> tran.polar # Outputting the program listing of "tran.polar"
>
function (C, C.r.up = NULL, C.r.down = NULL, C.theta.up = NULL,
   C.theta.down = NULL, flux.r.up = NULL, flux.r.down = NULL,
   flux.theta.up = NULL, flux.theta.down = NULL, cyclicBnd = NULL,
   D.r = 1, D.theta = D.r, r = NULL, theta = NULL, full.output = FALSE)
{
   checkDim <- function(dr, dtheta, nx = "dr", ntheta = "dtheta") {
      if (length(dr) != 1)
         stop(nx, ", should have length 1")
      if (length(dtheta) != 1)
         stop(ntheta, ", should have length 1")
   }
   checkDim2 <- function(dr, dtheta, nx = "dr", ntheta = "dtheta") {
      if (length(dr) != dimC[1] + 1)
         stop(nx, ", should have length equal to dim(C)[1]+1 = ",
            dimC[1] + 1, " it is ", length(dr))
      if (length(dtheta) != dimC[2] + 1)
         stop(ntheta, ", should have length equal to dim(C)[2]+1 = ",
            dimC[2] + 1, " it is ", length(dtheta))
   }
   dimC <- dim(C)
   Type <- 2
   if (max(theta) > 2 * pi)
      stop("theta should be < 2pi")
   if (min(theta) < 0)
```

```
    stop("theta should be > 0 ")
checkDim2(r, theta, "r, the grid in x-direction", "theta, the grid in theta-direction")
 Nr <- dimC[1]
 r.c <- 0.5 * (r[-1] + r[-(Nr + 1)])
 dr <- diff(r)
 drint <- c(r.c[1] - r[1], diff(r.c), r[Nr + 1] - r.c[Nr])
 divr <- 1/r
 divr[is.na(divr)] <- 0
 Ntheta <- dimC[2]
 theta.c <- 0.5 * (theta[-1] + theta[-(Ntheta + 1)])
 dtheta <- diff(theta)
 dthetaint <- c(theta.c[1] - theta[1], diff(theta.c), theta[Ntheta +
    1] - theta.c[Ntheta])
 checkDim(D.r, D.theta, "D.r, the diffusion in x-direction ",
    "D.theta, the diffusion in theta-direction ")
 if (!is.null(cyclicBnd)) {
 }
 if (!is.null(C.r.up)) {
    if (length(C.r.up) != 1 && length(C.r.up) != Ntheta)
       stop("'C.r.up' should have length 1 or equal to ",
          Ntheta)
 }
 else if (!is.null(flux.r.up)) {
    if (length(flux.r.up) != 1 && length(flux.r.up) != Ntheta)
       stop("'flux.r.up' should have length 1 or equal to ",
          Ntheta)
 }
 else if (!1 %in% cyclicBnd)
    C.r.up = C[1, ]
 if (!is.null(C.r.down)) {
    if (length(C.r.down) != 1 && length(C.r.down) != Ntheta)
       stop("'C.r.down' should have length 1 or equal to ",
        Ntheta)
 }
 else if (!is.null(flux.r.down)) {
    if (length(flux.r.down) != 1 && length(flux.r.down) !=
       Ntheta)
       stop("'flux.r.down' should have length 1 or equal to ",
          Ntheta)
 }
 else if (!1 %in% cyclicBnd)
    C.r.down = C[Nr, ]
 if (!is.null(C.theta.up)) {
    if (length(C.theta.up) != 1 && length(C.theta.up) !=
       Nr)
       stop("'C.theta.up' should have length 1 or equal to ",
```

```
          Nr)
  }
  else if (!is.null(flux.theta.up)) {
     if (length(flux.theta.up) != 1 && length(flux.theta.up) !=
        Nr)
        stop("'flux.theta.up' should have length 1 or equal to ",
           Nr)
  }
  else if (!2 %in% cyclicBnd)
     stop("'flux.theta.up' OR 'C.theta.up' should be specified")
  if (!is.null(C.theta.down)) {
     if (length(C.theta.down) != 1 && length(C.theta.down) !=
        Nr)
        stop("'C.theta.down' should have length 1 or equal to ",
           Nr)
  }
  else if (!is.null(flux.theta.down)) {
     if (length(flux.theta.down) != 1 && length(flux.theta.down) !=
        Nr)
        stop("'flux.theta.down' should have length 1 or equal to ",
           Nr)
  }
  else if (!2 %in% cyclicBnd)
     stop("'flux.theta.down' OR 'C.theta.down' should be specified")
  Flux.r <- 0
  Flux.theta <- 0

if (1 %in% cyclicBnd) {
     C.r.up <- (C[1, ] * drint[Nr + 1] + C[Nr, ] * drint[1])/(drint[1] +
        drint[Nr + 1])
  C.r.down <- C.r.up
  }
  if (is.null(C.r.up))
     C.r.up <- flux.r.up/D.r * drint[1] + C[1, ]
  if (is.null(C.r.down))
     C.r.down <- -flux.r.down/D.r * drint[Nr + 1] + C[Nr,
        ]
  Flux.r <- D.r * (rbind(C.r.up, C, deparse.level = 0) - rbind(C,
     C.r.down, deparse.level = 0))/drint
  if (2 %in% cyclicBnd) {
     C.theta.up <- (C[, 1] * dthetaint[Ntheta + 1] + C[, Ntheta] *
        dthetaint[1])/(dthetaint[1] + dthetaint[Ntheta +
        1])
     C.theta.down <- C.theta.up
  }
  if (is.null(C.theta.up))
     C.theta.up <- flux.theta.up/D.theta * dthetaint[1] *
```

```
            r.c + C[, 1]
    if (is.null(C.theta.down))
        C.theta.down <- -flux.theta.down/D.theta * dthetaint[Ntheta +
            1] * r.c + C[, Ntheta]
    Flux.theta <- D.theta * (cbind(C.theta.up, C, deparse.level = 0) -
        cbind(C, C.theta.down, deparse.level = 0))/matrix(data = dthetaint,
        nrow = Nr, ncol = (Ntheta + 1), byrow = TRUE)/r.c
    dFdtheta <- 0
    dFdr <- -diff(Flux.r * r)/dr/r.c
    dFdtheta <- -t(diff(t(Flux.theta))/dtheta)/r.c
    if (full.output)
        return(list(dC = dFdr + dFdtheta, C.r.up = C.r.up, C.r.down = C.r.down,
            C.theta.up = C.theta.up, C.theta.down = C.theta.down,
            r.flux = Flux.r, theta.flux = Flux.theta, flux.r.up = Flux.r[1,
                ], flux.r.down = Flux.r[Nr + 1, ], flux.theta.up = Flux.theta[,
                1], flux.theta.down = Flux.theta[, Ntheta + 1]))
    else return(list(dC = dFdr + dFdtheta, flux.r.up = Flux.r[1,
        ], flux.r.down = Flux.r[Nr + 1, ], flux.theta.up = Flux.theta[,
        1], flux.theta.down = Flux.theta[, Ntheta + 1]))
}

<environment: namespace:ReacTran>
>
> # Running the program "tran.polar":
>
> #===============
> # Testing the functions =
> #===============
> # Parameters
> F <- 100 # input flux [micromol cm-2 yr-1]
> D <- 400 # mixing coefficient [cm2 yr-1]
> # Grid definition
> r.N <- 4 # number of cells in r-direction
> theta.N <- 6 # number of cells in theta-direction
> r.L <- 8 # domain size r-direction [cm]
> r <- seq(0, r.L,len = r.N+1) # cell size r-direction [cm]
> theta <- seq(0, 2*pi,len = theta.N+1) # theta-direction - theta: from 0,
+ # to 2pi
> # Initial conditions
> C <- matrix(nrow = r.N, ncol = theta.N, data = 0)
> # Boundary conditions: fixed concentration
> C.r.up <- rep(1, times = theta.N)
> C.r.down <- rep(0, times = theta.N)
> C.theta.up <- rep(1, times = r.N)
> C.theta.down <- rep(0, times = r.N)
> # Concentration boundary conditions
> tran.polar(C = C, D.r = D, D.theta = D,
```

```
+ r = r, theta = theta,
+ C.r.up = C.r.up, C.r.down = C.r.down,
+ C.theta.up = C.theta.up, C.theta.down = C.theta.down)
> # Outputting:

$dC
           [,1]  [,2] [,3] [,4] [,5] [,6]
[1,] 1704142.01    0    0    0    0    0
[2,]  189349.11    0    0    0    0    0
[3,]   68165.68    0    0    0    0    0
[4,]   34778.41    0    0    0    0    0

$flux.r.up
[1] 400 400 400 400 400 400

$flux.r.down
[1] 0 0 0 0 0 0
$flux.theta.up
[1] 36923.077 12307.692  7384.615  5274.725

$flux.theta.down
[1] 0 0 0 0

> # Flux boundary conditions
> flux.r.up <- rep(200, times = theta.N)
> flux.r.down <- rep(-200, times = theta.N)
> flux.theta.up <- rep(200, times = r.N)
> flux.theta.down <- rep(-200, times = r.N)
> tran.polar(C = C, D.r = D, r = r, theta = theta,
+ flux.r.up = flux.r.up, flux.r.down = flux.r.down,
+ flux.theta.up = flux.theta.up, flux.theta.down = flux.theta.down,
+ full.output = TRUE)

> # Outputting:

$dC
         [,1]     [,2]     [,3]     [,4]     [,5]     [,6]
[1,] 9230.769   0.0000   0.0000   0.0000   0.0000 9230.769
[2,] 3076.923   0.0000   0.0000   0.0000   0.0000 3076.923
[3,] 1846.154   0.0000   0.0000   0.0000   0.0000 1846.154
[4,] 1432.967 114.2857 114.2857 114.2857 114.2857 1432.967

$C.r.up
[1] 0.5 0.5 0.5 0.5 0.5 0.5

$C.r.down
[1] 0.5 0.5 0.5 0.5 0.5 0.5
```

```
$C.theta.up
[1] 0.005416667 0.016250000 0.027083333 0.037916667

$C.theta.down
[1] 0.005416667 0.016250000 0.027083333 0.037916667

$r.flux
     [,1] [,2] [,3] [,4] [,5] [,6]
[1,]  200  200  200  200  200  200
[2,]    0    0    0    0    0    0
[3,]    0    0    0    0    0    0
[4,]    0    0    0    0    0    0
[5,] -200 -200 -200 -200 -200 -200

$theta.flux
     [,1] [,2] [,3] [,4] [,5] [,6] [,7]
[1,]  200    0    0    0    0    0 -200
[2,]  200    0    0    0    0    0 -200
[3,]  200    0    0    0    0    0 -200
[4,]  200    0    0    0    0    0 -200

$flux.r.up
[1] 200 200 200 200 200 200

$flux.r.down
[1] -200 -200 -200 -200 -200 -200

$flux.theta.up
[1] 200 200 200 200

$flux.theta.down
[1] -200 -200 -200 -200

> # Outputting:
> #
> # Figure 6-11: tran.polar 1
> #
>


> ## ========================================
> ## A model with diffusion and first-order consumption
> ## ========================================
> N <- 50 # number of grid cells
> XX <- 4 # total size
> rr <- 0.005 # consumption rate
> ini <- 1 # initial value at x=0
> D <- 400
```

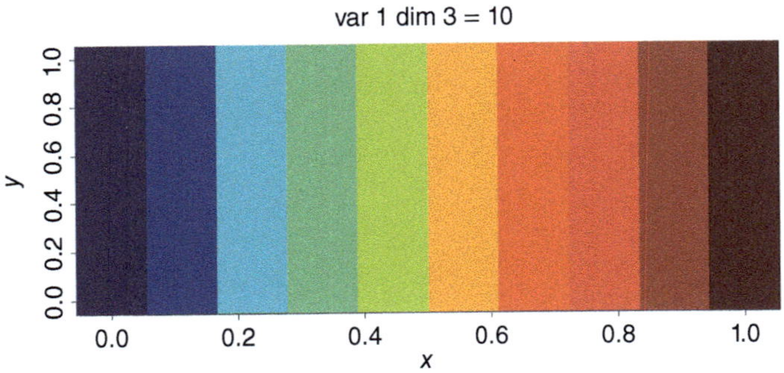

Figure 6.11 tran.polar 1 Diffusive transport in polar (r, θ) coordinates.

```
> r <- seq (2, 4, len = N+1)
> theta <- seq(0, 2*pi, len = N+1)
> theta.m <- 0.5*(theta[-1]+theta[-(N+1)])
> # The model equations
> Diffpolar <- function (t, y, parms) {
+ CONC <- matrix(nrow = N, ncol = N, data = y)
+ tran <- tran.polar(CONC, D.r = D, D.theta = D, r = r, theta = theta,
+ C.r.up = 0, C.r.down = 1*sin(5*theta.m),
+ cyclicBnd = 2, full.output=TRUE )
+ dCONC <- tran$dC - rr * CONC

+ return (list(dCONC))
+ }
>
> # solve to steady-state; cyclicBnd = 2, because of C.theta.up,
+ # C.theta.down
> out <- steady.2D (y = rep(0, N*N), func = Diffpolar, parms = NULL,
+ dim = c(N, N), lrw = 1e6, cyclicBnd = 2)
> image(out)
>
> # Figure 6-12: tran.polar 2
```

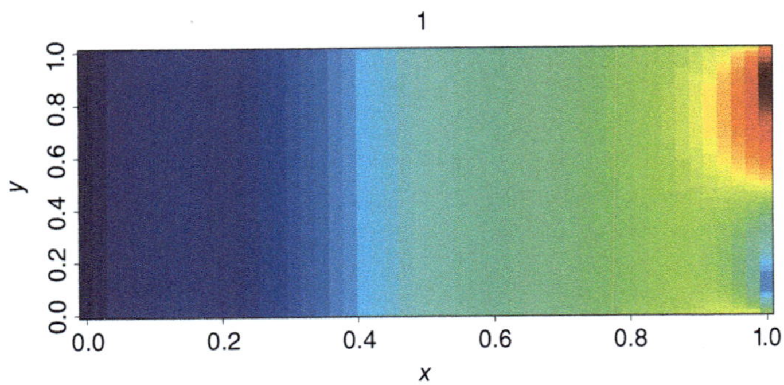

Figure 6.12 tran.polar 2 Diffusive transport in polar (r, θ) coordinates.

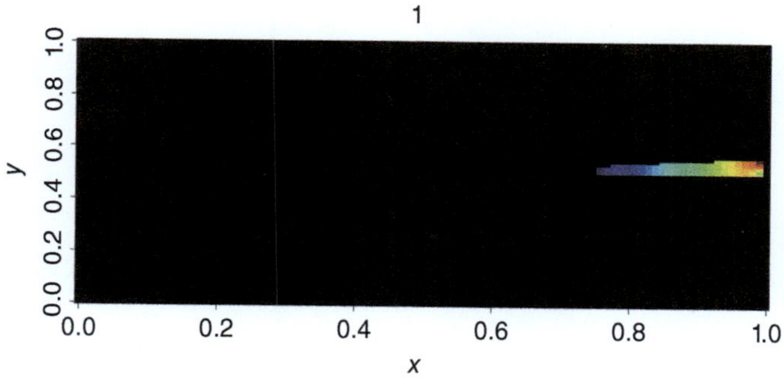

Figure 6.13 tran.polar 3 Diffusive transport in polar (r, θ) coordinates.

```
>
> cart <- polar2cart(out, r = r, theta = theta,
+          x = seq(-4, 4, len = 100),
+       y = seq(-4, 4, len = 100))
>
> image(cart)
>
> # Figure 6-13: tran.polar 3
>
>
>
```

Further Reading

Soetaert, K. and Herman, P.M.J. (2009). *A Practical Guide to Ecological Modelling – Using* R *as a Simulation Platform*. Springer.

Soetaert, K. and Meysman, K., Centre for Estuarine and Marine Ecology, Netherlands Institute of Ecology, The Netherlands (2009). R-package ReacTran:reactive transport modelling in R – R Project. https://cran.r-project.org/web/packages/ReacTran/vignettes/ReacTran.pdf (accessed 16 February 2022).

References

1 Chan, B.K.C. (2016). Supplemental chapter: 'Research-level applications of R'. In: *Biostatistics for Epidemiology and Public Health Using* R. New York: Springer Publishing Company www.springerpub.com/chan-biostatistics.

2 (a) Danckwerts, P.V. (1970). *Gas Liquid Reactions*. New York: McGraw-Hill; (b) van Elk, E.P. (2001). Gas–liquid reactions – influence of liquid bulk and mass transfer on process performance. http://www.vanelk.nl/edwin/cv/thesis.pdf (accessed 21 February 2022).

3 van Elk, E.P., Borman, P.C., Kuipers, J.A.M., and Versteeg, G.F. Received 14 April 1999; received in revised form 8 November 1999; accepted 29 November 1999. See also: *Chemical Engineering Science* 54 (21): 4679–5354 (1999).

4 Danckwerts, P.V. (1950). Absorption by simultaneous diffusion and chemical reaction. *Trans. Faraday Soc.* 46: 300.

5 Danckwerts, P.V. (1950). Unsteady-state diffusion or heat-conduction with moving boundary. *Trans. Faraday Soc.* 46: 712.

6 Danckwerts, P.V. (1967). Gas-absorption accompanied by first-order reaction: Concentration of product, temperature-rise and depletion of reactant. *Chem. Eng. Sci.* 22: 472.

7 Roper, G.H., Hatch, T.F. Jr., and Pigford, R.L. (1962). Theory of absorption and reaction of two gases in a liquid. *Ind. Eng. Chem. Fundam.* 1 (2): 144–152.

8 Hatta, S. (1932). On the absorption velocity of gases by liquids. *Technol. Rep. Tohoku Imp. Univ.* 10: 119.

9 Sherwood, T.K., and Pigford, R.L. (1952). Simultaneous absorption and chemical reaction. In: *Absorption and Extraction* (Chapter IX), McGraw-Hill Series in Chemical Engineering, Kirkpatrick, S.D., Consulting Editor, 317–390. New York: McGraw-Hill.

10 Sherwood, T.K., and Pigford, R.L. (1952). "*Absorption and Extraction*", 334. New York: McGraw-Hill Book Company.

11 Choy, B., and Reible, D.D. (1999). *Diffusion Models of Environmental Transport*. New York: CRC Press.

12 Statistics Canada. http://www.statcan.gc.ca/edu/power-pouvoir/ch3/5214785-eng .htm#a1.

13 Venables, W.N., and Smith, D.M., and the R Development Core Team (2004). *An Introduction to* R. Bristol, UK: Network Theory, Ltd.

Simultaneous Mass Transfer and Chemical Reactions in Engineering Science, First Edition. Bertram K. C. Chan.

14 Aragon, T.J. (2011). *Applied Epidemiology Using* R *(epir)*. Berkeley, CA: UC Berkeley School of Public Health, and San Francisco Department of Public Health.

15 Virasakdi, C. (2008). *Analysis of Epidemiological Data Using* R *and Epicalc*. Thailand: Epidemiology Unit, Prince of Songkla University: cvirasak@medicine.psu.ac.th.

16 Dalgaard, P. (2002). *Introductory Statistics with* R, Springer Statistics and Computing Series. New York: Springer Science+Business Media.

17 Everitt, B.S., and Hothorn, T. (2006). *A Handbook of Statistical Analysis Using* R. Boca Raton, FL: Chapman & Hall/CRC.

18 BMI Notes. (2012). Body mass index. http://en.wikipedia.org/wiki/Body_mass_index (accessed 02 March 2022).

19 Centers for Disease Control and Prevention. (2005). Antiretroviral postexposure Prophylaxis after sexual, injection-drug use, or other nonoccupational exposure to HIV in the United States: recommendations from the U.S. Department of Health and Human Services. MMWR Recomm. Rep. 2005 Jan;54(RR-2):1–20. Available from: http://www.cdc.gov/mmwr/preview/mmwrhtml/rr5402a1.htm.

20 Teetor, P. (2011). R *Cookbook*. Sebastopol, CA: O'Reilly Media.

21 Daniel, W.W. (2005). *Biostatistics – A Foundation for Analysis in the Health Sciences*. New York: Wiley.

22 CRAN. The Comprehensive R Archive Network. http://cran.r-project.org (accessed 19 December 2022).

23 ANOVA. http://en.wikipedia.org/wiki/Analysis_of_variance.

24 Hatch, T.F. Jr., and Pigford, R.L. (1962). Simultaneous absorption carbon dioxide and ammonia in water. *Ind. Eng. Chem. Fundam.* 1 (3): 209–214.

25 Danckwerts, P.V. (1970). *Gas–Liquid Reactions*, McGraw-Hill Chemical Engineering Series. New York: McGraw-Hill Book Company.

26 Berg, H.C. (1977). *Random Walks in Biology*. Princeton.

27 Bird, R.B., Steward, W.E., and Lightfoot, E.N. (2011). *Transport Phenomena*, 2e. Wiley.

28 Schlichting, H., and Gersten, K. (1999). *Boundary Layer Theory*, 8e. Springer.

29 Sherwood, T.K. and Pigford, R.L. (1952). *Absorption and Extraction*, 121. McGraw-Hill.

30 *ftp://ftp.feq.ufu.br/Claudio/Perry/DOCS/Chap14.pdf.

31 Bryan, K. The existence and uniqueness theorem for ODEs. https://www.rose-hulman.edu/~bryan/lottamath/uniqde.pdf (accessed 27 February 2022).

32 For example: Coddington, E.A., and Levinson, N. (1955). *Theory of Ordinary Differential Equations*. New York: McGraw-Hill Book Company.

33 Schultz, M.H. (1973). *Spline Analysis*. Englewood Cliffs, NJ: Prentice-Hall.

34 For example: Hardy, G.H. (2009). *A Course of Pure Mathematics*, 10e. Cambridge, UK: Cambridge University Press (First Edition 1908).

35 Sherwood, T.K., and Pigford, R.L. (1952). *Absorption and Extraction*, McGraw-Hill Series in Chemical Engineering, Kirkpatrick, S.D., Consulting Editor, 320ff. New York: McGraw-Hill.

36 Sherwood, T.K., and Pigford, R.L. (1952). *Absorption and Extraction*, McGraw-Hill Series in Chemical Engineering, Kirkpatrick, S.D., Consulting Editor, 324ff. New York: McGraw-Hill.

37 Hatch, T.F. Jr., and Pigford, R.L. (1962). Simultaneous absorption and of carbon dioxide and ammonia in water. *Ind. Eng. Chem. Fundam.* 1 (2): 214–209.

Further Reading

Backhurst, J.R., Harker, J.H., Coulson, J.M., and Richardson, J.F. (1994). *Chemical Engineering*, vol. 4. Solutions to the Problems in Volume 1. London, UK: Butterworth Heinemann.

Berg, H.C. (1977). *Random Walks in Biology*. Princeton University Press.

ceeserver.cee.cornell.edu

Chan, B.K.C. (2020). *Simultaneous Mass Transfer and Chemical Reactions in Engineering Science – Solution Methods and Chemical Engineering Applications*. Cambridge, MA/Amsterdam, The Netherlands: Elsevier.

Coulson, J.M., and Richardson, J.F. (1996). *Chemical Engineering: Volume 1: Fluid Flow, Heat Transfer and Mass Transfer*, 5e. London, UK: Butterworth Heinemann.

Crocker, B.B., S.M., P.E., Consulting Chemical Engineer; Fellow, American Institute of Chemical Engineers; Member, Air Pollution Control Association. (Phase Separation).

Danckwerts, P.V., and Kennedy, B.E. (1954). Kinetics of liquid-film process in gas absorption. Part I: Models of the absorption process. *Trans. Inst. Chem. Eng.* 32: S49–S52.

Fair, J.R., Ph.D., P.E., Professor of Chemical Engineering, University of Texas; Fellow, American Institute of Chemical Engineers; Member, American Chemical Society, American Society for Engineering Education, National Society of Professional Engineers. (Section Editor, Absorption, Gas-Liquid Contacting).

Steinmeyer, D.E., M.A., M.S., P.E., Distinguished Fellow, Monsanto Company; Fellow, American Institute of Chemical Engineers; Member, American Chemical Society. (Liquid-in Gas Dispersions).

Foust, A.S., Wenzel, L.A., Clump, C.W., et al. (1980). *Principles of Unit Operations*, 2e. Melbourne, FL: Krieger Publishing Company.

Franks, R.G.E. (1967). *Mathematical Modeling in Chemical Engineering*, 4–6. New York: Wiley.

Hatch, T.F. Jr., and Pigford, R.L. (1962). Simultaneous absorption carbon dioxide and ammonia in water. *Ind. Eng. Chem. Fundam.* 1 (3): 209–214.

Higbie, R. (1935). The rate of absorption of a pure gas into a still liquid during short period of exposure. *Trans. Am. Inst. Chem. Eng.* 31: 365–388.

Levinspiel, O. (1972). *Chemical Reaction Engineering*, 2e. New York: Wiley. pp. 210–213, 320–326.

Simultaneous Mass Transfer and Chemical Reactions in Engineering Science, First Edition. Bertram K. C. Chan.

Luyben, W.L. (1989). *Process Modeling, Simulation, and Control for Chemical Engineers*, 2e, 22–24. Columbus, OH: McGraw-Hill Education.

Lynn, S., Straatemeier, J.R., and Kramers, H. (1955). Absorption studies in the light of the penetration theory I. Long wetted-wall columns. *Chem. Eng. Sci.* 4: 49–57.

Majumder, S.K., and Das, C., Dept. of Chemical Engineering, Indian Institute of Technology, Guwahati, Guwahati – 781039, Assam, India "National Programme on Technology Enchanced Learning (NPTEL) – Phase II, Course Name: Mass Transfer Operation-I" http://nptel.ac.in/courses/103103035/1.

McCabe, W.L., and Smith, J. (195X). *Unit Operations of Chemical Engineering*, Series in Chemical Engineering, 1e. New York: The McGraw-Hill Companies, Inc.

McCabe, W.L., Smith, J.C., and Harriott, P. (2005). *Unit Operations of Chemical Engineering*, Series in Chemical Engineering, 7e. New York: The McGraw-Hill Companies, Inc.

Olutoye, M.A. (2007). Modelling of a gas-absorption packed column for carbon dioxide-sodium hydroxide system. https://www.researchgate.net/publication/254071709_Modelling_of_a_Gas-Absorption_Packed_Column_for_Carbon_Dioxide-Sodium_Hydroxide_System (accessed 23 April 2018).

Olutoye, M.A., and Mohammed, A. (2006). Modelling of a gas-absorption packed column for carbon dioxide-sodium hydroxide system. *AU J.T.* 10 (2): 132–140. https://www.researchgate.net/publication/254071709.

Penney, W.R., Ph.D., P.E., Professor of Chemical Engineering, University of Arkansas; Member, American Institute of Chemical Engineers. (Gas-in-Liquid Dispersions).

Perry. (2005). *Chemical Engineering Handbook*, Section 14 Gas Absorption and Gas–Liquid System Design*, 7e. New York: McGraw-Hill Book Company. ftp://ftp.feq.ufu.br/Claudio/Perry/DOCS/Chap14.pdf.

Roper, G.H., Hatch, T.F. Jr., and Pigford, R.L. (1962). Theory of absorption and reaction of two gases in a liquid. *Ind. Eng. Chem. Fundam.* 1 (2): 144–152.

Sherwood, T.K., and Pigford, R.L. (1952). *Absorption and Extraction*, McGraw-Hill Series in Chemical Engineering. New York: Kirkpatrick, S.D., Consulting Editor, McGraw-Hill Book Company.

van Elk, E.P., Borman, P.C., Kuipers, J.A.M., and Versteeg, G.F. Received 14 April 1999; received in revised form 8 November 1999; accepted 29 November 1999.

Index

a

absorption
  acceleration factors 140–141
  accompanied by chemical reaction 13, 199
  into agitated liquid 17–19
  into a quiescent liquid 11–13
  simultaneous absorption of two reacting gases, carbon dioxide and H_2S 18
advection 221
  vs. convection 222–224
  equation 577–579
  operator 579
advection.1D 401
Analysis of Variance (ANOVA) 95, 96, 109
automated coding systems 34

b

biomolecular reaction 2, 3, 114–115
bio-transporter 2, 114
bivariate 87, 90, 92
bivariate normal distribution 92
Body Mass Index (BMI) 59, 61
boundary conditions 4, 5, 10, 12, 13, 26, 27, 29, 127, 139, 145–147, 153, 154, 156, 158, 162–163, 188–190, 201–205, 208, 210–212, 217, 229, 261, 312, 327, 348, 365, 387, 399, 431, 481, 496, 517, 534, 557, 562, 563, 576, 583, 608, 629
boundary layer theory model 22, 131–132
bubble column 126

c

calculation of packed height requirement 141–144
carbamate 112
carbon dioxide 111–112, 115, 134, 135
carotenoid transport 3, 115
chemical absorption 21–29
chemical reactions, accompanying diffusion, basic equation 13
chemisorption 136
chlorination 3, 115
chlorine absorption 141
column, wetted wall 8, 10, 21, 127, 136
complementary error function 122, 129

d

Danckwerts model, Danckwerts surface renewal theory model 129–130
data and data processing 32–38
data capture 34–36
data coding 33–34
data editing 34, 35
data entry
  and analysis using the spreadsheet interface in R 79, 80
  at the R command prompt 73–83
  using several available R functions 77–79

Simultaneous Mass Transfer and Chemical Reactions in Engineering Science, First Edition. Bertram K. C. Chan.

data-frame for R computation 73–75
data quality 36
differential material balance 137–138
diffusion 117
coefficient 5, 6, 12, 18, 19, 113, 117–122, 198, 200, 203, 229, 386, 399, 556, 561, 562, 568
equation 12, 18, 118, 194
length 122
diffusive models of environmental transport 30
diffusivity 9, 11, 12, 14–16, 21, 25, 113, 116–118, 122, 129, 132, 195, 198

e
entering data at the R command prompt 73–83
environmental transport 30, 116
enzymes 2, 114
equilibrium 5, 6, 10, 21, 26–28, 123–127, 143, 195, 215–218
equimolar counter-diffusion 132
error function 12, 27, 122, 128, 129, 211, 212, 217
error sum of squares (SSE) 94
existence theorem of the governing simultaneous semi-linear parabolic partial differential equations 183–188

f
fast reaction 18, 22, 112, 200
Fick's first law of diffusion 117–121
Fick's laws of diffusion 117–122
Fick's second law of diffusion 119–122
film model 4–6, 20, 123
film-renewal model 126
film theory of mass transfer 7–9
first-order irreversible reaction 14, 19
first-order reactions 4, 5, 13–14, 18, 19, 29, 196, 200, 213
first session using R 40–52
flat plate experiment 131–132
fluid-fluid interface 131
Function list() 84–86
Function Source() 81–82

g
gas absorption column 135–137
gas absorption packed column 136, 141
gas transfer rates 123

h
height of tower equation 138–139
heme 2, 114
Henry's law 123, 124, 195, 215
heteroskedasticity 50
Higbie penetration theory model 11, 22, 127–129
Hinterland Model 4

i
imputations 35–36
industrial chemical and chemical engineering processes 3
instantaneous mass transfer coefficient 129
instantaneous reaction 6, 14–16, 193, 194
intestinal absorption 3, 115
irreversible reactions 13–17

j
jet, laminar 111, 199

k
K_g 125
K_l 125

l
laminar jet 111, 199
Lewis–Whitman two-film model 123–126
ligand 2, 114
lipid transporter 3, 115
liquid-film coefficient 195

m
macro model 3, 6
mass transfer coefficient 9, 11, 122–124, 129, 132, 134, 140, 143, 157, 158, 202, 204, 212–213

mass transfer past solids under turbulent flow 132
mass transfer under laminar flow conditions 132
mathematical models of the theory of simultaneous mass transfer and chemical reactions 22–23
mathematical operations using R 54–56
mathematical theory of simultaneous mass transfer and chemical reactions 20–30
mathematics of advection 222, 392–393
metabolic reactions 2, 114,
metal complexes 2, 114
meteorology 222, 392
micro model 3–5
models of mass transfer 123–144
multivariate data analysis 92–109
multivariates 57

n

NA "Not Available" for missing values in datasets 67–68
Nernst One-Film Theory Model 123–126
Number of Transfer Units (NTU) 141
numerical solution 4, 29–30, 111–218, 224, 394, 579

o

oxygen transport 2, 114

p

packed tower/column design 141–142
penetration model 4–6, 11, 127
physical absorption 4, 5, 8–10, 15, 20–22, 28, 112–113, 140, 218
plotCurves 379, 381
plotTotFlux 383, 385
Prandtl's experimental mass transfer from a flat plate 131–132

q

qualifications, for statistician 32
quality assurance 36, 38
quality control 36
quality management in statistical agencies 36–37
quiescent liquid 11–17

r

R programming language 31, 65
  as a calculator 54–72
  command prompt 73–83
  dim() in 65
  matrix() in R 66
  and statistics 38–40
  use of factors in 57–59
reaction, gas-liquid 1–20, 136
reactive transport equation 224, 226, 385, 395, 555
ReacTran 3, 115, 226, 386–388, 391, 395–396, 555–641
regression analysis 90

s

Sherwood number (N_{Sh}) 4, 5, 132
simultaneous absorption of two reacting gases, carbon dioxide and H_2S 18
simultaneous biomolecular reactions and mass transfer 2–6, 114–115
single film control 126
sodium hydroxide 115, 134
solenoidal 223, 224, 393, 394, 578
stagnant film model 4–6
Stanton number 113
sum of squares due to regression (SSR) 94
surface renewal theory of mass transfer 9–11, 129–130

t

theoretical models of mass transfer 123–144
theory of simultaneous gas absorption and chemical reactions in mass transfer operations 22, 23
time-averaged mass transfer coefficient 129
total sum of squares (SST) 93, 94

tran.cylindrical 226, 237, 313, 327, 350, 397, 483, 496, 519, 585, 595, 606, 630
tran.2D 226, 231, 257, 313, 328, 361, 386, 396, 401, 483, 497, 519, 530, 556, 561, 568, 585, 595, 609, 630
tran.3D 226, 231, 262, 308, 328, 361, 386, 396, 401, 428, 432, 477, 497, 519, 530, 556, 561, 570, 579, 609, 630
tran.polar 226, 262, 328, 346, 348, 397, 432, 497, 516, 517, 609, 628, 629
trans.1D 227, 397
tran.spherical 226, 313, 324, 327, 350, 397, 483, 496, 519, 585, 595, 608, 630
tran.volume.1D 360
tran.volume.2D 360
tran.volume.3D 360
two-film model 124–126
two reacting gases, simultaneous absorption 18, 22–23

u

uniqueness theorem of the governing simultaneous semi-linear parabolic partial differential equations 188–192
univariate, bivariate, and multivariate data analysis 87–99
univariate data analysis 87–89
unsteady-state absorption in a stagnant liquid 25–28
unsteady-state mass transfer 10, 127, 130
uptake by intestinal mucosal cells 3, 115

v

values in R, and computations using vectors and matrices 56–57
variables 1, 15, 52, 57, 58, 62–63, 76, 82, 87, 90, 92–94, 146, 147, 171, 176, 177, 183, 188, 201, 206, 207, 386, 388, 389, 556, 558, 559, 561, 562, 564, 565
vector y 567

w

wetted-wall column 8, 10, 21, 127, 136
Whitman Stagnant Film Model 3, 4

x

x vector 56, 62–63, 64